DESIGNING ENGINEERS

AN INTRODUCTORY TEXT

SUSAN McCAHAN
University of Toronto

PHILIP ANDERSON
University of Toronto

MARK KORTSCHOT
University of Toronto

PETER E. WEISS
University of Toronto

KIMBERLY A. WOODHOUSE
Queen's University

VICE PRESIDENT AND PUBLISHER	Don Fowley
EXECUTIVE EDITOR	Dan Sayre
EDITORIAL ASSISTANT	Francesca Baratta
EXECUTIVE MARKETING MANAGER	Chris Ruel
DESIGN DIRECTOR	Harry Nolan
PRODUCT DESIGNER	Jennifer Welter
COVER & TEXT DESIGNER	Thomas Nery
ILLUSTRATOR	Norm Christiansen
SENIOR CONTENT MANAGER	Karoline Luciano
PRODUCTION EDITOR	Sandra Dumas
MEDIA SPECIALIST	James Metzger
PHOTO EDITOR	Billy Ray
PHOTO MANAGER	Malinda Patelli
PRODUCTION SERVICES	SPI-Global
COVER CREDIT	Bertlmann/E+/Getty Images

All Gantt charts are provided as screenshots from Microsoft® Project and used with permission from Microsoft.

This book was set in 10/12 Kepler Light by SPI-Global and printed & bound by Courier(Kendallville). The cover was printed by Courier(Kendallville).

Founded in 1807, John Wiley & Sons, Inc. has been a valued source of knowledge and understanding for more than 200 years, helping people around the world meet their needs and fulfill their aspirations. Our company is built on a foundation of principles that include responsibility to the communities we serve and where we live and work. In 2008, we launched a Corporate Citizenship Initiative, a global effort to address the environmental, social, economic, and ethical challenges we face in our business. Among the issues we are addressing are carbon impact, paper specifications and procurement, ethical conduct within our business and among our vendors, and community and charitable support. For more information, please visit our website: www.wiley.com/go/citizenship.

This book is printed on acid-free paper. ∞

ISBN 13 978-0-47093949-9

Printed in the United States of America.

10 9 8 7 6 5 4 3 2 1

Preface

This book provides an introduction to the basic principles of engineering design in a way that is accessible and appropriate for first- or second-year undergraduate students, who do not yet have any discipline-specific expertise. It uses a first-principles approach, with an emphasis on problem definition and scoping, and a creative solution process, and is less concerned with rigorous, code-based design practice. However, it is also quite comprehensive, and includes significant material on communications, team strategies, and project management. This should allow instructors to pick and choose the components that they want as a custom print or digital text. The approach is based on the premise that "design thinking" can and should be introduced from the very beginning of an engineering education in order to frame the remainder of the subject area studies and motivate the students.

The past 15 years have seen a transformation of engineering design instruction. Historically, engineering education often consisted of two fundamental parts: a strong foundation of math, physics, and chemistry and a technical depth in one of the traditional disciplines. Although many programs included a final-year "capstone" design course, there was little formal education in design thinking and methodology leading up to this. Students were expected to learn how to design things during the often short and intense capstone experience.

Two forces have fomented change in this traditional approach. First, industry and accreditation groups have strongly indicated that this was not enough: That engineers needed to be further along in a full set of design skills. Second, design itself became recognized as a basic and significant engineering skill, with many commonalities across disciplines.

In recent years, a more formal introduction to engineering design thinking has been introduced in many undergraduate programs, often beginning in the first year of studies.

At the University of Toronto, Engineering Strategies and Practice (ESP) was introduced on a pilot basis in 2003 and rolled out to the full class of about 900 first-year students in 2005. We incorporated extensive material on communication, team, and professional developments into this first-year design course. We have now delivered this course to more than 8000 students, and we have been able to convey the important aspects of engineering design using a combination of lectures and progressively more complex design tutorials, leading up to community-service design projects in the second term.

There are many textbooks on engineering design, and they span a wide range of approaches. At one end, there are simplified, generic, and often shorter books on the basics of problem definition and creative solution methods. At the other end, there are comprehensive discipline-specific texts based in industry practice for individual fields. After more than a decade of teaching freshman design, we see a need for a text that sits between these extremes: a more comprehensive and flexible text that spans more topics, in more detail than some of the generic introductory texts, but does so in a way that is accessible and therefore useful for first- and second-year students. The result is this text, *Designing Engineers: An Introductory Text.*

Our course, and this text, strives to teach a structured, planned approach to the design process, but one that is flexible based on results, insights, and reflection. It is an iterative approach that considers the understanding that design solutions are never perfect, but engineers work to provide a solution that best balances various aspects in the context of the design, including emerging technology, economic and environmental concerns, and the interests of the client, team, users, and other members of society who may be affected by the design.

This text is organized as a set of small relatively independent modules, organized into clusters, which are further organized into sections. The sections differentiate major areas of engineering design knowledge, such as "Implementing a Project." Clusters associate information about specific areas of the section, such as "Working on a Team." The modules are the smallest text unit, covering a specific topic, such as "Organizing Teams." While the modules are not generally intended to be read sequentially, links at the start of each module indicate which modules would normally be read before the module, which modules can be read along with this module to supplement the information, and which modules would normally follow this module.

The text is conceived primarily as an electronic resource, something that an individual instructor can customize for a particular course and something that an individual student can read in a nonsequential manner, considering the student's individual needs. Although you may be holding in your hands a print form, you should look at it as you would a Web page or online encyclopedia. There are multiple focal points and most of these can be taken in any order. We strongly recommend that you use most of the core design process modules because these cover the essential foundation of the engineering design process, but after that, the text should be customized as needed, according to the demands of the design course and your own design projects.

There are six types of modules:

- narrative modules, which explain concepts and their context;
- navigation modules, which link modules and help the reader move from one to another;
- process modules, which enable application of a systematic design process;
- skill/tools modules, which provide specific techniques that support a systematic design process;
- resource modules, with complementary material to back up design learning; and
- review/reflection modules, which provide a summary and support reflective practice.

Some modules give the basic information needed to proceed with a project or a part of a project and that may be enough information for you at that given point. Other modules provide more detail and can be easily found, when you want to fill out your knowledge.

The material in the module is followed by a list of key terms and questions and activities that help you determine how well you have grasped the concepts in the module. In an electronic version, the key terms may be linked to the definitions in the glossary, and hashtags can take you to the sections of the text where a particular concept is dealt with in some depth.

Acknowledgments

Many colleagues have contributed to the development of our ideas over the years. We are particularly indebted to

- Members of the Engineering Strategies and Practice teaching team over the years. This dedicated team of instructors, staff, and graduate students has contributed enormously to our understanding of engineering design teaching.
- Our students, who provide invaluable feedback and inspiration.
- Maegan Chang, who developed the case studies for the text.
- Chirag Variawa, who developed a bibliography that identified the need for this text.
- Our editors at Wiley.

In addition, we would like to acknowledge the valuable feedback we received from the many people who reviewed early versions of the manuscript for the text:

Dr. Nadia Bhuiyan—Concordia University
Dr. Peter Byrne—University of South Alabama
Dr. Mauro Caputi—Hofstra University
Dr. Glenn Ellis—Smith College
Dr. Robert Fleisig—McMaster University
Dr. Robert Gettens—Western New England University
Dr. Hayden Griffin—East Carolina University
Dr. Margaret Harkins—University of North Carolina at Charlotte
Dr. Allen Hoffman—Worcester Polytechnic Institute
Dr. Jean Kampe—Michigan Tech University
Dr. Paul Kurowski—University of Western Ontario
Dr. Pierre Larochelle—Florida Institute of Technology
Dr. Stephanie Ludi—Rochester Institute of Technology
Dr. John Meech—University of British Columbia at Vancouver
Dr. J. Carson Meredith—Georgia Institute of Technology
Dr. Janice Miller-Young—Mount Royal University
Dr. Ibrahim Nisanci—University of Arkansas at Little Rock
Dr. Arun Srinivasa—Texas A&M University
Dr. Marlee Walton—Iowa State University
Dr. William Wild—SUNY Buffalo
Dr. Yuelei (James) Yang—University of Ontario Institute of Technology
Dr. Yih-Choung Yu—Lafayette College

Contents

How Engineers Design

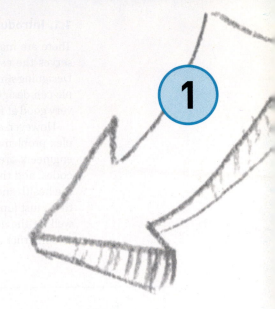

Introduction

1

#narrative module: #introduction

Learning outcomes

By the end of this module, you should demonstrate the ability to:
- Define what an engineering design process is
- Explain the purpose of engineering design
- Define requirements
- Define documentation and explain what it is used for

Recommended reading

After this module:
- **How Engineers Design > 2. Design Process Overview**

1. Introduction

Welcome to *"Designing Engineers: An Introduction."* This text, which is written in concise units, covers an engineering design process from developing an understanding of the problem to be solved, through idea generation, developing a detailed design, and implementation. The emphasis is on the first stages of a design process, in particular defining the project requirements, generating solution ideas, and evaluating the ideas. You will cover more specifics of the design process and the design of discipline-specific technologies in your upper-year courses. In this text we will discuss some of the common aspects of the design process across **disciplines** (e.g., the iterative nature of the process), and the need for modeling and testing. Also, we will compare and contrast the detailed design and implementation phases of the design process across project types.

The text is written in short units that are connected by recommended readings at the beginning of each unit, links, and a glossary. *Terms that are bold and italicized throughout the text are linked to definitions and can be found in the glossary.* Your instructor may use all the text or parts of it depending on how he or she wishes to teach design. Each unit will also contain the learning outcomes. These are things that you should be able to demonstrate by the time you finish the unit.

1.1. Introduction to the Engineering Process

There are many types of design. We will be focusing on engineering design, which serves the essential purpose of engineering: turning science into useable systems. Designing simple things generally does not require any special process, and many people can design simple things without learning how to design. In fact, we humans are very good at finding creative solutions to simple problems (see Figure 1).

However, as engineers you will generally be called upon to solve much more complex problems that require consideration from multiple perspectives. The problems engineers are asked to solve often involve specialized knowledge, regulations, or codes, and the resulting technology can have far-reaching consequences, including the health and safety of the public. Engineering work often involves finding solutions that must function well from many different points of view; the design must function well for the user, it must minimize impact on the environment, and it must be easy to construct and maintain. As the complexity of these problems grows, it becomes

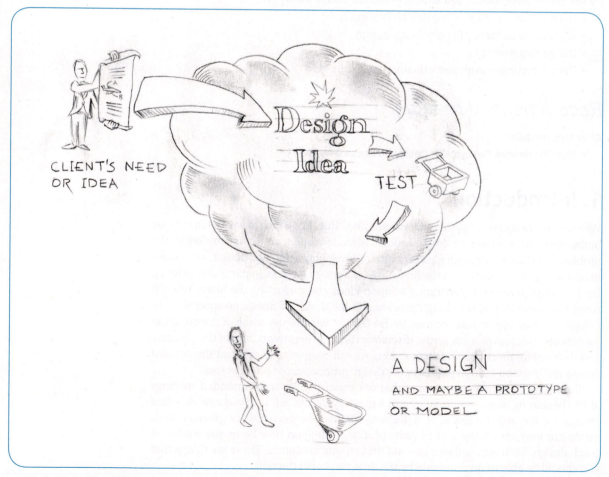

FIGURE 1 **The informal design process:** Designing a simple thing for your own use might not require any special design process.

increasingly difficult to organize all of the information, balance the trade-offs success-fully, and still find creative, effective solutions. As the complexity grows there is also a need for larger design teams.

A more formal *engineering design process* is used to help large teams manage the information that will be part of complex engineering problems, and to help engineers develop solutions that have the best chance of being effective from these many differ-ent perspectives. For example, the discipline of formulating the problem as a full set of *requirements* gives designers a means for comparing and prioritizing the differ-ent goals of the project. Requirements are a formal description of what is required of the design. Formulating a complete set of requirements is a way of making sure that as the work progresses, which may take years for a complex system, nothing essen-tial to the project is forgotten or missed. Furthermore, engineers as a profession are required to carefully and fully docu-ment a project. To *document* a project means to explain what has been done in writing, pictures (i.e., graphical communi-cation and drawings), and orally (in pres-entations, conversations, and meetings). The act of documenting the work also helps engineers develop ideas and clarify their thinking and it. This *documentation*, and the work it describes (the engineer's design work), will be the basis for deci-sions. People will decide whether to fund a project or not, whether to implement a design or not, based on the quality of the process that was used and the credibility of the documentation developed.

It may not seem necessary to use a formal process for the relatively easy design problems you will be given in the first few years of engineering school, or even for the more difficult term projects you will do for your courses in upper years. However, like learning any skill, it is valuable to learn the process in the context of simpler work so that as the complexity of the problems increases you have developed habits and learned tools that help you to be successful. In this regard learning design is much like learning to play a musical instrument or some other complex skill that combines thinking and doing. You start with simpler pieces of music, learn the basic process and techniques, so that you can successfully tackle increasingly complex works. Tech-niques for playing that work for a simple piece of music will not suffice for faster, more complex pieces later. The same approach applies to engineering problem solving and design. Being a good engineering designer requires both knowledge and practice.

There is no single universal engineering design process. Design processes vary from discipline to discipline and even from company to company. And design pro-cesses are changing as engineering tasks become more complex and more finely tuned with evolving technologies. However, they all share some basic similarities. In this text we will explain the common elements shared by virtually all strong engineering design processes. In the upper years of your engineering program you will probably learn about one or more design processes that are common in your discipline (civil, mechanical, electrical, chemical, etc.). These discipline-specific processes may use somewhat different terminology than what we use here, but the basic principles and methods will be essentially the same. Learning the basic techniques from this text will allow you to easily understand and adapt to the techniques prevalent in your engineering field and place of employment.

1.2. Navigating the Learning Modules

This text is organized differently from conventional textbooks. It does not have a sequence of chapters, but instead, it is made up of learning modules organized into sections that are grouped into clusters. Each cluster has a number of modules that you can read in any order, depending on your need at the moment. The following table lists and describes the kinds of learning modules in this text. If you are using a digital form of the text you can search on these by hashtag. You can also search hashtags for important concepts, such as #requirements, to find places in the text where these concepts are explained or examples are shown.

MODULE TYPE	DESCRIPTION
#narrative	Narrative modules discuss the ways engineering design happens in school and in industry. It explains the context of design projects and how they develop.
#navigation	Navigation modules help you find your way through the text modules so that you can plan your reading to fit your needs.
#process	Process modules enable you to apply a systematic design method. Each process module includes a section at the end that describes what you should have accomplished on your design project when you leave this particular process stage.
#skill/tool	A skill is a specific ability; a tool is a system, method, or technique that helps you perform a task related to a skill. Skill/tool modules may contain either or both of these related to particular parts of the design process.
#resource	Resource modules contain complementary information to back up process and skill/tool modules.
#review/reflection	More than just a summary, a review/reflection analyses the learning outcomes of a group of modules. It will help you put together and process ideas. These modules are a useful aid in studying for exams and provide a check on the quality of work you have done on a design project.

KEY TERMS

The key terms will be listed at the end of each chapter. Definitions for many of these terms can be found in the glossary.

discipline	engineering design process	requirements
document	documentation	

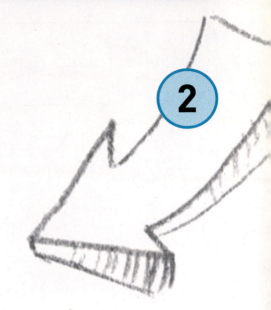

Design Process Overview

#narrative module: #processoverview

Learning outcomes

By the end of this module, you should demonstrate the ability to:
- Describe a design process
- Explain the concept of an iterative process
- Recognize good design process strategies including improvement through iteration

Recommended reading

Before this module:
- **How Engineers Design > 1. Introduction**

After this module:
- **How Engineers Design > 3. Project Phases**

1. Design Is Iterative

One of the essential similarities across all engineering design processes is the ***iterative*** nature of the process. Iterative means the steps in the process are repeated over and over, in a loop. With every iteration, the design team comes closer to an ideal solution. In the design process, iteration is used to enhance the design team's understanding of the problem they are trying to solve, to improve the proposed design, and to increase the quality and quantity of information generated in the process. Design without iteration (see Figure 1) may sometimes produce a good design, but more often produces a mediocre result that could be improved through iteration (#iterating).

The iterations in the design process always stop before an ideal solution is reached. The design team could always come up with a better design if they had more time, or more resources (people, money, space, and materials). And they could always find a better solution in the future, when science knowledge and improved technology make ideas feasible that are now only fiction. Computers in the 1960s were huge machines with relatively little computing power by today's standards, but they were not "bad" pieces of technology designed by poor engineers. They were the best that was possible

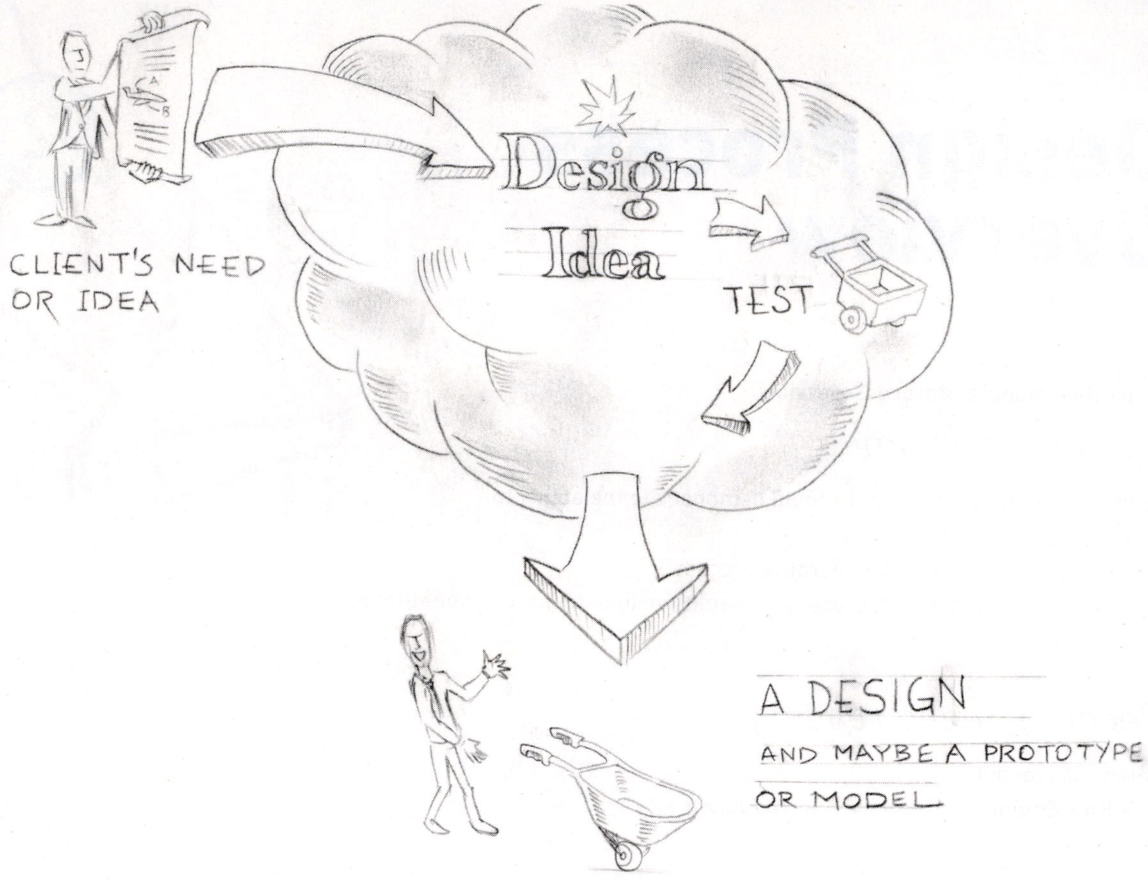

FIGURE 1 The informal design process.

at the time with the technology, funds, time, and other resources available. In order to bring a new piece of technology into being we must always stop our iterative design process and move on to actual implementation before the solution is the best it will ever be, or could ever be.

Novice designers may churn around in a hazy, fuzzy process exploring ideas randomly (see Figure 2, top). Sometimes this results in a good design but it is not an efficient use of time or resources. And the information that results from this type of process will be incomplete and not always well organized. After spending a month on an idea, for example, the novice designer may discover that the idea already exists, and is patented. Professional design engineers, therefore, use tools and strategies along with freeform creative and critical thinking to develop designs more efficiently (see Figure 2, bottom). A more rigorous process also produces more complete, well-organized information. This doesn't diminish the importance of creativity or innovation; tools and strategies channel this energy and effort to enhance the quality of the result. Like an artist who is taught how to see light and perspective in order to use these concepts intentionally (or choose to ignore them intentionally) in their work, your goal is to become an intentional designer.

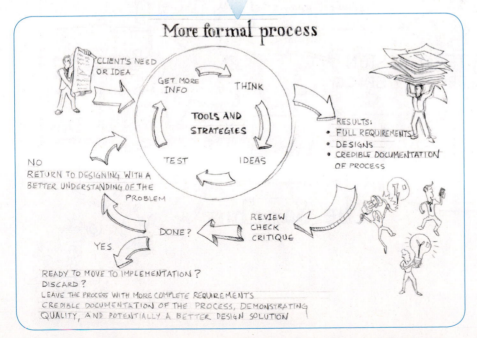

Using effective strategies and tools puts the design process in overdrive, increasing speed and creativity, and producing better results. Results from the design process can be measured in terms of quality and quantity of information.

FIGURE 2 **The formal engineering design process.** Using tools and strategies to improve the design process moves the process from a novice to an expert level.

The formal *engineering design process* shown in Figure 2 (bottom) relies on the use of tools and methods to develop the results. The results from this process are a full set of *requirements*; a high-quality design, or set of designs; and clear and complete *documentation* of the design and of the design process. We can envision this process as the activity that occurs in a design company (see Figure 3). The engineering design team gets a project to work on. They use investigation methods, creativity techniques, decision processes, and other tools to develop a set of requirements and design ideas. The team uses the requirements to test their ideas and choose between them. Then they bring their work to a manager or a project client to review and critique. As a result of the review they may go back into the project space to work on the design further, they may stop the design activities and move to the implementation phase, or they may stop the project altogether. Design requires iterating (looping) through these activities, usually multiple times.

Like levels in a video game, each time the team enters this project space they face new challenges. At first the challenge is just learning about the project, trying to figure out all of the aspects of the problem they have been asked to solve. As the team gathers more information about the design problem, they gain an understanding of how to navigate the challenges presented by this particular project. The team begins to

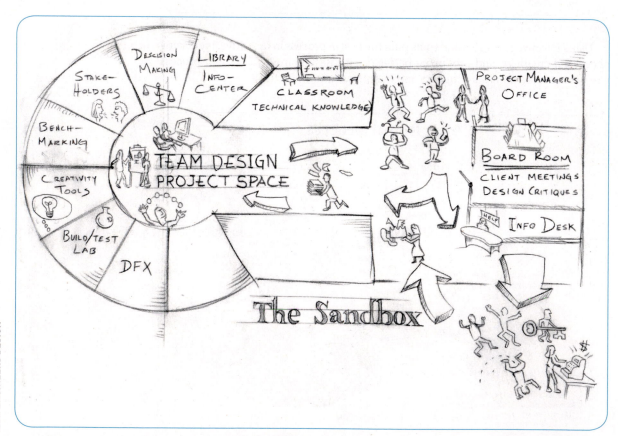

FIGURE 3 The activities involved in an engineering design process can be visualized as spaces in a design company.

find where the most intense challenges are going to be, and where there are already existing pieces of technology available that they may be able to use in the solution. The team gathers this information and organizes it into requirements. They can make use of *creativity methods* and other tools in their design toolbox to develop ideas for potential solutions. Each time the team completes a level they leave the project space equipped with a better understanding of the project requirements; more or improved design ideas; more information about the project as a whole (e.g., new sources of information, new contacts); and more tools in their toolbox—more experience, skills, and strategies to use in the next iteration. Some design processes are rapid. Others will go through years of iteration before the first version of the technology is produced commercially.

In a video game you know your score at the end of each level, but in a design project the score is more ambiguous. The measure of your accomplishments will come in the form of critiques from managers, clients, users, and others or the economic success of your design. No one has played this particular game before, each project is unique. The challenges, adventures, and ending are never predictable!

KEY TERMS

engineering design process   documentation   requirements
creativity methods   iterative

2. Questions and activities

1. Define "iteration" in your own words.
2. Give an example of an iterative process that you have engaged in (e.g., practicing the same piece of music repeatedly), and explain why the iteration was important in this process.
3. Explain why the engineering design process is necessarily iterative.

3

Project Phases

#narrative module: #projectphases

Learning outcomes

By the end of this module, you should demonstrate the ability to:

- Define the phases of an engineering design project
- Explain the primary purpose of each phase and its role in the overall design process
- Explain how each phase feeds into the next one

Recommended reading

Before this module:

- **How Engineers Design > 2. Design Process Overview**

After this module:

- **How Engineers Design > 4. Communicating throughout the Process**

1. The Phases of a Project

One of the essential differences between informal novice design and expert design is the development of *project requirements* (see Figure 1). Design problems are *open-ended problems* that could be solved in many different ways. The development of project requirements is crucial in an engineering design process. It is possibly the most important challenge in the process, and the one that is too often overlooked or treated superficially by novice designers. To develop the requirements the design team must research and document all of the key information that will guide and constrain the decisions that they make throughout the project. This phase of the process involves a lot of information gathering and analysis. The requirements phase results in as complete a description of the problem as the design team can formulate. The design team concurrently develops a *problem statement* that serves to define the problem and guide the process. Your design team will return to the requirements development process repeatedly throughout a project both to add to the definition of the problem and to remind themselves of the project goals. Each time your team makes a major revision to the project requirements you hit an important evaluation checkpoint where

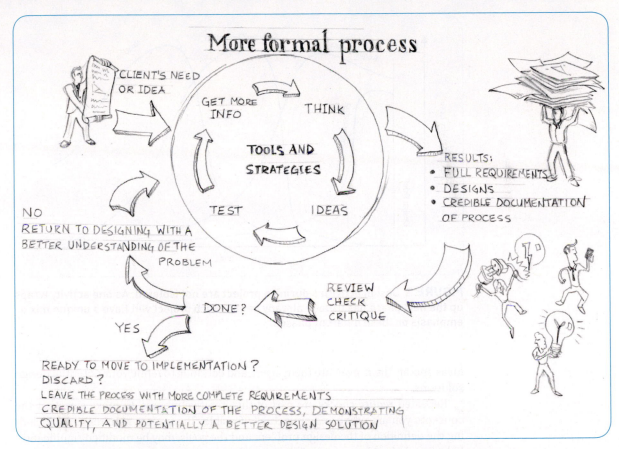

FIGURE 1 The formal engineering design process, which uses iteration with tools and strategies to improve results.

you must decide what direction to go next. The decision whether to continue with the project, and if you continue, what process to follow, will depend on this evaluation. Figure 2 shows the typical activity level for the different phases of a design project over time.

Once the problem begins to become defined the team will start generating ideas for possible solutions. These ideas are called *conceptual design alternatives*. This requires creativity and engineering work. Successfully generating good ideas is not just a talent; it can be learned and, like any ability, the more it is practiced the better you can get at it.

After generating a range of different possible solutions (conceptual designs) that the design team thinks might solve the design problem, the team needs to evaluate these ideas and decide which one (or ones) to pursue. This is a critical step in the design process and can be made more effective using *decision-making tools*. It requires comparing the proposed solutions against all of the design criteria you documented in the project requirements. It also requires getting input from a variety of people with an interest in the project, and using engineering tools and models to evaluate the possible solutions. This step often requires many iterations as you evaluate

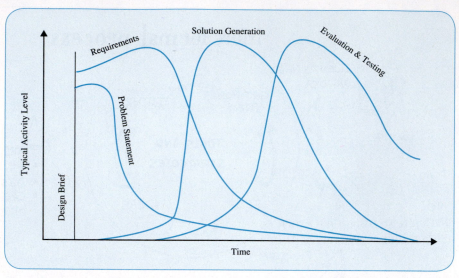

FIGURE 2 **Design activities during a project are not isolated.** As one activity wraps up the next phase has usually already begun. Each project will have a unique mix of emphasis on these different phases.

ideas, modify them, evaluate them again, and continue to refine your potential design solutions.

Repeated *reality checks* are part of this iterative process, to make sure that the concepts you are proposing are actually feasible. The project requirements, including the definition of the design problem and the goals, may be modified significantly. If the first design ideas you test fail to meet the requirements, you may need to return to the idea generation stage to develop more possible solutions. Each time you iterate you bring an increased level of understanding and expertise to the problem because of your experience with the project. This deepens your knowledge base and may allow you to conceive of solutions that were not apparent when you first approached the problem.

If the team is able to develop a feasible solution that sufficiently addresses the design problem then the project may move forward to the detailed design phase. The design may continue to evolve during this phase as you move toward implementation, construction, or production of the final design. In this stage the engineering team will continue to solve problems that arise as the details of the design are worked out. If successful, the system, process, or product that you have designed will be launched and ready for operation. However, the engineer's work does not stop there. Engineers are often involved in the operation of technology, and the issues that arise when a technology is decommissioned (i.e., disposed of).

KEY TERMS

project requirements	**open-ended problems**	**conceptual design alternatives**
decision-making tools	**reality checks**	**problem statement**

2. Questions and activities

1. What are the major phases in the engineering design process?

2. How do these phases relate to one another?

4 Communicating throughout the Process

#narrative module: #communicatingprocess

Learning outcomes

By the end of this module, you should demonstrate the ability to:

- Explain the purposes of documenting a design project
- List some of the users of the documentation produced during a design project
- Describe the essential features that differentiate engineering communication (e.g. engineering instructions) from casual communication
- Explain the purpose of an engineering notebook

Recommended reading

Before this module:
- **How Engineers Design > 3. Project Phases**

Alongside this module:
- **Design Process > Multi-use Tools > 4. Information Gathering**
  Implementing a Project > Communication > All Modules

After this module:
- **How Engineers Design > 5. What Engineers Design**

1. Communicating throughout the Process

> **The engineering design process is essentially an information process**
>
> An informal design process produces only a few design ideas and little information. Expert designers produce many ideas and extensive information while they are designing. The value that is produced is not just the end result, the design; it is all of the information that is created, gathered, and organized in the process. It includes documentation of all the important decisions that are made. For the engineer, the expert process results in information that enables significant learning and experience that can you take with you to your next project.

The engineering design process is essentially an information process. In the course of a design project you gather, organize, and manage information; you create ideas, examine them critically, make decisions, and formulate all of this into information that needs to be shared, transferred, and ultimately applied to implement your design. To be useful, all of this information has to be put in a form that can be understood and applied accurately both by yourself and by others. It has to be communicated through *documentation*.

The *documentation* of the design process can take many forms: written, graphical (drawings and pictures), and oral (presentations). It is an essential part of the design process and the documentation generally forms the bulk of the results. This type of writing is quite different than the essays, text messages, or other types of writing you may be used to. Engineers are required to write reports, memos, instructions, and so on. It requires a very clear, precise, and concise style (see the example in the box). The purpose of the documentation, which must be honest and transparent, is to communicate

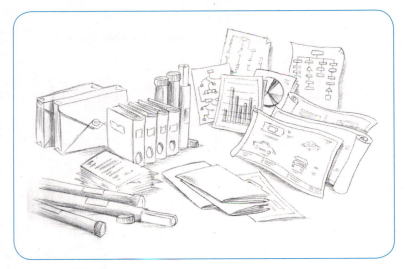

information in a way that is both credible and unambiguous. The quality of the documentation will often determine whether or not a design project is successful. Certainly, in proposals (as in the *Request for Proposals (RFPs)* process) and other key points of evaluation, the documentation will determine if a design is implemented or if it is discarded.

How Common Instructions Differ from Engineering Instructions

You might ask a friend, "Could you pick me up a coffee?" An engineer might more formally specify the request this way:

1. Could you go to the coffee shop at the southeast corner of College Street and St. George Street in Toronto, Ontario.
2. Go to the service counter.
3. Request a medium-size dark roast coffee.
4. Pay the cashier.
5. Go to the condiments counter and add 2 cc of whole milk and 1 gm of sugar.
6. Select correct size lid and snap lid into place.
7. Deliver coffee to me in 10 minutes or less from time of purchase with less than 1cc of spillage in transit.
8. Delivery location: Bahen building, 44 St. George Street, room 1007.

All values ±10% unless otherwise specified.

(continued)

Communicating throughout the Process

This may seem like it overly complicates the instructions. However, the probability that you will get your coffee exactly the way you want it (hot, with the right amount of milk and sugar) is drastically improved by this type of communication style. With a cup of coffee this may not matter, but if you are writing instructions for assembling a control system for an aircraft, it matters!

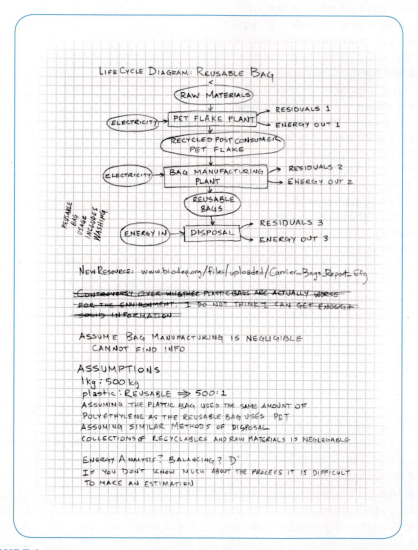

FIGURE 1 Engineers use notebooks, either electronic or on paper, to document their day-to-day work.

1.1. Communicating with Yourself

Communicating with yourself may seem easy and obvious. But when a design project goes on for months or years and involves thousands or millions of pieces of information, it is easy to forget facts and ideas or to remember them incorrectly. Documenting and organizing your work as you go is an essential information management skill in this process. It can save you from redoing work that has already been done, and keeps you from heading off in the wrong direction. The common method for keeping track of design work is an *engineering notebook* (see Figure 1). Your notebook is used like a journal to record your work as it progresses. Starting an engineering notebook is the first step to beginning a new design project.

KEY TERMS

documentation   **Request for Proposals (RFPs)**   **engineering notebook**

2. Questions and activities

1. Why is documentation of the engineering design process and the results important?

2. Practice writing up a list of steps needed to complete a simple procedure. For example:

   a. Opening a can

   b. Putting on a coat

   c. Filling a water bottle

   d. Measuring out sugar for a recipe

3. Now try handing your instructions from question 2 to a friend and see if following them exactly yields the desired result.

4. Set up your engineering notebook so it is ready for the start of a new project.

5 What Engineers Design

#narrative module: #whatengineersdesign

Learning outcomes

By the end of this module, you should demonstrate the ability to:
- Describe the different things that engineers design
- Explain how different disciplines in engineering contribute to a design

Recommended reading

Before this module:
- **How Engineers Design > 4. Communicating throughout the Process**

After this module:
- **How Engineers Design > 6. How Design Projects are Initiated**

1. What Engineers Design

While the basic aspects of the general engineering design process are similar among the various engineering disciplines (such as mechanical, electrical, civil, chemical, industrial, and biomedical), there are also some important differences among disciplines. Perhaps the most essential difference is in the types of things that engineers design. Nearly all engineering design projects are multidisciplinary (involving teams of engineers from various disciplines and people from outside the profession) but we can make some simplistic generalizations about the types of projects that are most commonly associated with the different types of engineering.

For example, civil engineering is most commonly associated with the design and construction of substantial structures and infrastructure such as water treatment and waste facilities, power plants, buildings, and transportation systems (e.g., bridges, traffic systems, subways, ships, airports), whereas mechanical, electrical, and computer engineers are more often associated with product or system design (e.g., medical equipment or consumer electronics). Computer or software engineers also design software or algorithms. Chemical and materials engineers may be involved in the design of products (e.g., pharmaceuticals, food, fertilizer), as well as the design of processes (e.g., chemical refining or production) and the design of plants. A ***plant*** can mean anything from a small production facility to a huge offshore oil platform. Industrial engineers design *processes*, such as manufacturing processes and products, and systems, such as information systems (e.g., financial systems, database systems, and web-based purchasing systems). Perhaps the most basic word we could use to describe what engineers design is *technology*; understanding that technology is anything real or virtual (e.g., a computer program) that does not occur naturally. So technology includes very simple products such as paper that were invented centuries ago, to high-tech processes under development such as quantum encryption. Look, for example, at all of the systems and devices used in a modern hospital, and you will see engineering of every kind in action.

What Engineers Design

Systems: A system is a set of organized components or elements that operate together as a unit. There are natural systems, such as the solar system, and engineered systems, such as a subway system or a communication network. The elements in an engineered system are chosen and arranged to perform a specified set of functions.

Plants: A plant is a structure, and its internal equipment, which is designed to produce a specified product. There are manufacturing plants that produce consumer or commercial products and power plants that produce energy products such as electricity and heat. Other common types of plants include water treatment  plants that produce clean water for people, and waste water plants that clean used water before it goes back into the environment.

Products: A product is a physical or virtual thing that is the result of a design process. There are consumer products, such cell phones, shampoo, and software packages; commercial products, such as oil drilling equipment or supercomputers; energy products, such as electricity or natural gas; and digital products, such as network services or online financial services. There are also agricultural products, such as corn, beef, or wheat, that are the result of a natural process that has been engineered to enhance the production of a desired product.

Materials: An engineered material is a substance that has been designed. Almost everything you have is probably made of an engineered material with the exception of plants (wood, natural food), animals, water (oceans, rivers), and rocks. Engineers design all kinds of gases, liquids, and solids, including alloys, ceramics, polymers (e.g., plastics) and combinations of substances.

Courtesy of the Authors

(*continued*)

Engineers can also design and produce some materials that are identical to natural products, such as diamonds or a cloned organism, or materials that can replace natural materials to some degree, such as artificial bone or skin.

Processes: A process is the sequence of operations needed to transform material, information, or energy from one form to another. Processes are designed to transform substances such as ore into useable forms, such as metals. Processes are also designed to transform energy from one form, such as gasoline (where the energy is in the chemical bonds), to another form, such as kinetic energy. And engineers design information processes, such as search engines. Process design generally involves not only the design of the operational steps of information, material, or energy transformation, but also consideration of the timing and scheduling, logistics, supply chain, and quality-control aspects of the process.

Courtesy of the Authors

Structures: A structure is an engineered product or system designed to support a load (i.e., counter a force). Structures include built systems such as bridges or buildings. There are also virtual structures, such as the structure of a network system, which are designed to support a virtual load (e.g., a large amount of network traffic).

Technologies: A technology is any product, system, or process that is designed and made by people.

There is considerable overlap among the disciplines. In fact, many engineers find that a single discipline does not really describe their work. An engineer's expertise is often better described by the industry they are in and the roles they have had (technical designer, **research & development** (R&D), manufacturing engineer, safety engineer, quality assurance, or project manager, as just a few examples). Your undergraduate degree will give you a good grounding in one field of engineering and get you started in your career. However, your subsequent experience and continuing education throughout your career will take you into areas well beyond your first discipline and ultimately define the type of engineer you become.

The differences in design across industries are in the design process itself, as well as the types of technology systems that result. For example, if a company is asked to design a large water treatment plant to be located in a big city, the project requirements stage comprises a large part of the design process and may take years to complete (see Figure 1). The proposed system will probably be thoroughly modeled and, as a public works project, will go through an extensive review by the government and other concerned organizations. However, there may be little really new technology that gets developed specifically for this plant, so the concept generation stage may be shortened and consist mostly of deciding between different configurations for locating the technology subsystems in the plant, and sizing the components. In contrast, a small product design company may decide to invent a new product just because they think it will be marketable. The engineers may spend some time during the project requirements stage satisfying themselves that there is no current product on the

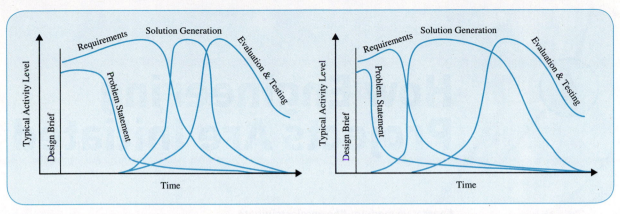

FIGURE 1 The emphasis on each phase in the design process will be different depending on the project.

market that fills the perceived market niche, and they may do some preliminary patent searches. But the requirements stage of a project like this is likely to be short. The engineering team will spend much more time on the concept generation step and may use a wide variety of creativity methods to develop a broad set of design alternatives. This is very different than the design process for the water treatment plant.

In every engineering design project all of the steps in the design process happen, but depending on the type of technology and the role of the engineer, each step will be more or less important to the project. Throughout your career as an engineer you will probably work on many different projects and will need to adapt to the unique design process and the job function you have on each project.

KEY TERMS

plant	technology	system
material	structure	process
product	research & development	

2. Questions and activities

1. Pick a technology you are familiar with (ATM machine, subway system, coffee maker, potato chips, etc.). Describe the types of engineering disciplines involved in designing and implementing (manufacturing or constructing) the technology.

2. Pick a company that has some engineering activities (most do). Look at the "careers" section of the company website. Find three job descriptions for engineering positions and identify the disciplinary background of the engineer they want to place in these positions.

What Engineers Design

6 How Engineering Projects Are Initiated

#narrative module: #howprojectsinitiate

Learning outcomes

By the end of this module, you should demonstrate the ability to:

- Explain the different ways in which engineering design projects are initiated
- Explain the difference between an in-house project and one that is contracted out
- Describe the basic function of consulting companies
- Describe what an inventor and an entrepreneur are

Recommended reading

Before this module:

- **How Engineers Design > 5. What Engineers Design**

After this module:

- **How Engineers Design > 7. Navigating the Engineering Design Process**

1. How Engineering Design Projects Are Initiated

To understand how engineering design projects come to be, you need to be aware of the two primary groups that are typically involved in a design project. One is the design team, which is made up of the engineers and other relevant individuals who will be tasked with designing the technology. The other is the *client*, which may be an individual, a company, or some other organization. The *client* is the person or organization that *commissions* the design; that is, they give the task of designing the system to the design team, and the client typically funds the project.

Some common modes for the initiation of a design project are described here. The list is in order based on the relationship between the engineering design team and the client, from projects where the client and the design team are very closely related or are the same people, to projects where the client and design team have a distant relationship (see Figure 1).

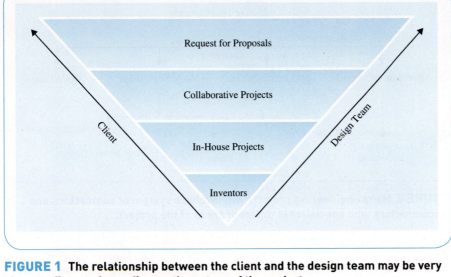

FIGURE 1 The relationship between the client and the design team may be very close or distant, depending on the nature of the project.

1.1. The Inventor and Entrepreneur

An *inventor* is an individual, or sometimes a team of people, who acts as both the client and the designer for a technology. They are people who identify a need for a new technology and they design it themselves. Some inventors will bring a design through the design process to the point where it can be patented, or where a working version has been built, and then look for a company to buy out or license the technology. Others will take a more *entrepreneurial* route and go into business, marketing their invention themselves and becoming the head of their own company. The initiation of a design project in this case requires only a good idea, and lots and lots of work.

1.2. In-house Projects

Companies that specialize in technology development employ *in-house design teams*. These companies serve as both the client and the design company. The design team members are employees of the company. Many of the "name brand" products and services you use are designed primarily by in-house teams that work at the company whose name appears on the product or system.

1.3. Collaborative Projects: Design Consultants, Contractors, and Custom Design Projects

In collaborative projects the client uses an outside engineering design team or company to accomplish the project. This is often done through a contract between the client and a design company or team. The client again can be an individual, a company, or another type of organization. They choose the engineering design company typically because the engineering group has a special expertise in the relevant field.

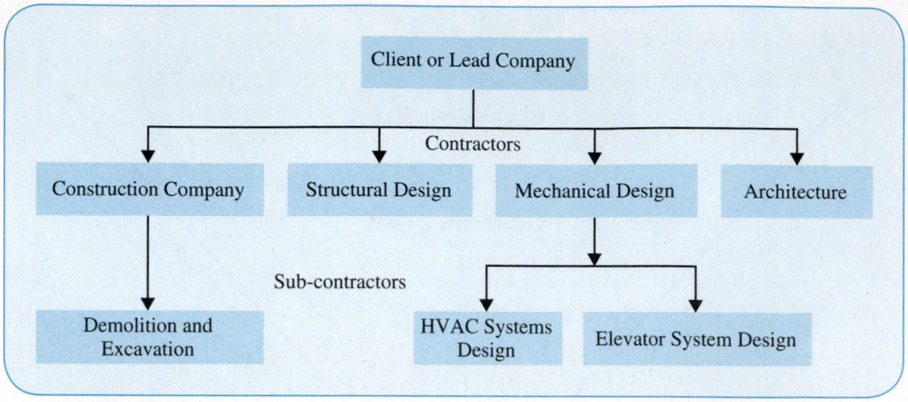

FIGURE 2 Many engineering projects will involve a system of contractors and subcontractors who specialize in different areas of the project.

However, they may have other reasons, including having worked with the engineering team before and having had a good experience. Even clients that are engineering design companies themselves often choose to contract or subcontract some of their design work out to other companies. In large projects there may be dozens of design teams from many companies working on the design of a system (see Figure 2).

1.4. Request for Proposals (RFPs)

A *Request for Proposals (RFPs)* is issued by a client organization, which is typically a company, some other type of organization, or a government agency, but could be an individual person. The RFP will usually state the purpose of the design project and explain the client's requirements. Anyone, or any company, can respond to an RFP by submitting a proposal for a design. In this type of project there may be no existing relationship between the client and the design company prior to the start of the project. The RFP will typically state what type of proposal is required; usually the proposal will need to be detailed enough, including estimates of the cost of the design and estimate of time for completion (this is called a *bid*), such that the client can make an informed decision about which proposal to choose. The engineering design team, or company, does not get paid for submitting a proposal. They will only be paid for their work if their proposal is accepted and a contract drawn up between the client and the engineering company.

The RFP method is used when the client wants to get a variety of proposals so they have a wide range of designs to choose from. Or, as is the case in most government work, the client is not allowed to preferentially choose an engineering company or group based on a prior relationship or other factors. An honest RFP system allows all engineering companies a fair chance at getting the contract.

2. Consulting Companies

One type of engineering company that works exclusively through collaborative projects, or RFPs, is an engineering *consulting company*. A consulting company usually takes on numerous projects for a wide variety of clients. They do not necessarily get involved in the full design process, or in the full implementation of the design. They are contracted to provide a specific outcome or set of deliverables. A *deliverable* is a piece of documentation or other result from the design process (e.g., a set of drawings, a working model of the design, a report, a cost estimate, a set of operating instructions, or construction of one part of the project).

The role of a consulting company is similar to that of a contractor. A client may decide to split up a project and contract out the work separately for each stage of the design and implementation of the project. Often this is done to take advantage of the different areas of expertise offered by different engineering design companies. For example, if Engineering Company 1 is designing a building for a client, they may have one of their in-house teams do the structural design. However, they may subcontract the design of the elevators to Engineering Company 2, a company that specializes in this type of work. In this case Engineering Company 1 now plays the role of client for the design of the elevator systems. Company 1 may also subcontract out the design

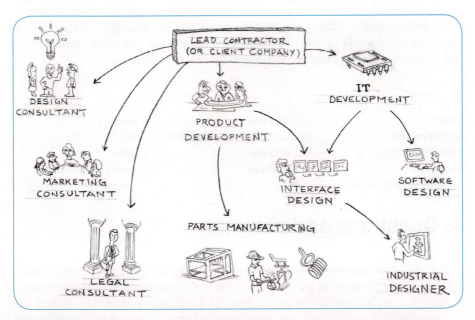

FIGURE 3 Large engineering projects will usually involve a complex network of contractors, subcontractors, consulting companies, and other professional services.

of the heating and air conditioning systems, the design of the electrical and lighting systems, and the design of some of the other specific components of the building. Complex engineering design projects may involve many different engineering companies who act as contractors and subcontractors and who all have to work together collaboratively.

3. Other People in the Design Process

In addition to the design team and the client, there will be other professionals involved in the design process. Engineering Company 1 may be an architectural firm (i.e., a group of architects), for example, that employees an in-house structural engineering group. In virtually every engineering design project of any magnitude there will be many different types of people involved in the design process. Some may work for the same company as the design team, and some may be contracted to work on a particular project.

There are also other groups of people that will be impacted by the design process and who are often involved in it. First, there are the **user** group and the **operators**. The users may be the operators (such as a person using and operating a laptop computer) or they may be different groups. For example, the users of clean water are a different group of people from the operators of the water treatment plant. There are also other **stakeholders** in the design process. Stakeholders are people or organizations who have an interest in the outcome of the design process. Stakeholders may include members of the public, government agencies, companies, and other organizations (which are called **non-governmental organizations**, or NGOs).

KEY TERMS

client	commission	inventor
entrepreneur	Request for Proposals (RFPs)	bid
consulting company	deliverable	user
operator	stakeholder	non-governmental
in-house design teams		organization (NGO)

4. Questions and activities

1. Suppose you have an idea for a new type of candy that combines fruit and caramel. You start your own candy manufacturing company to produce your product. To produce the candy in bulk you need manufacturing equipment that is custom designed to your product. What are some of the choices you have for getting the design work done? Explain how these choices work. For example: I could design the equipment myself or with a group of people (i.e., a design team). In this case I am the inventor because I am both the designer and the client. If I sell the manufacturing technology I invent to another customer, then I am also an entrepreneur.

Navigating the Engineering Design Process

7

#navigation module: #navigatingprocess

Learning outcomes

By the end of this module, you should demonstrate the ability to:

- List the essential process stages in engineering design
- Describe each stage briefly and identify where to find more information on part of the process

Recommended reading

Before this module:

- **How Engineers Design > Modules 1 through 6**

After this module:

- **Design Process > All Modules**

1. Design Process Skills and Tools

The engineering design process is iterative and nonsequential. We recommend that you start with requirements and finish at post-conceptual design, which includes implementation or production of the technology. However, where you start, which processes you use during a project, and where you end will really all depend on the nature and requirements of the design project you are working on. This is represented visually in Figure 1. It is normal during a design project to visit and revisit steps, such as idea generation or decision making, repeatedly and in no particular order. And there is no set of sequential recipes that will predictably work effectively for every design project. For this reason, we have organized the material in this text into clusters of modules and will provide some guidance on the ordering of the steps. However, your design team will need to consider whether the guidance provided fits the project you are working on, or whether a modified path through the steps would be more effective.

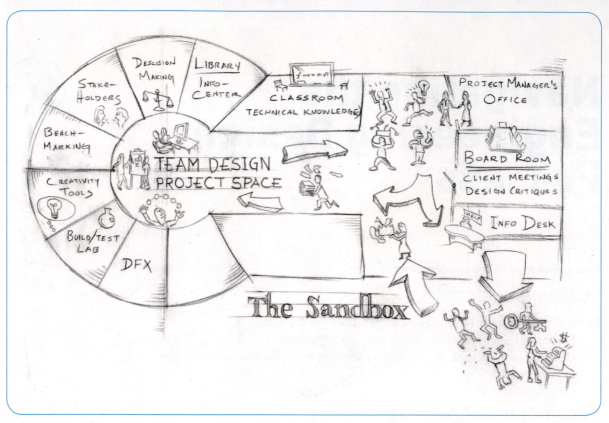

FIGURE 1 **The activities involved in an engineering design process can be visualized as spaces in a design company.** A typical design project will require visiting and revisiting these different activities before completing the project.

2. Map of the Design Process

The material in this text is organized into clusters of modules. All of the clusters are important to creating an effective, well-documented design. In this module we will focus on the clusters of modules which are central to the process (see Figure 2). These include the modules that describe the core engineering design process. These modules explain tools and methods to successfully work your way through a basic design project.

3. The Design Space

Fundamentally, the design process is about defining and reshaping the ***design space***. The design space is an abstract concept of a space that encompasses all of the design solution ideas that are being considered at any point in time during a project. Developing requirements defines the boundaries of the design space. Generating new solution ideas populates (expands) the space. Discarding ideas (i.e., removing solution ideas from consideration) shrinks the design space. Building and testing prototypes

Post conceptual design
– Leaving the design space

Requirements generation
– Defining the design space

Multi-use design tools
– Tools for characterizing/defining/
describing/detailing the design space

Idea generation tools
– Expanding the design space

Decision making
– Reducing the design space

Iterating
– Iterating in the design space

Investigation ideas
– Exploring the design space

Courtesy of the Authors

FIGURE 2 The typical engineering design method revolves around a few essential core processes: requirements specification, idea generation (ideation), decision making, and detailed design work. These processes are enabled and supported by using tools and strategies. The core process modules explain these activities.

or models explores (investigates) the design space. And creating a detailed design focuses the design space. The design process generally involves all of these processes. If you prefer to think about things from an abstract perspective, then the concept of manipulating, exploring, and clarifying the design space may be a useful way to envision the engineering design process.

4. Clusters in the Design Process Core

4.1. Requirements Generation: Defining the Design Space

It is generally recommended that the design process begin with a clear understanding of the problem that is being solved. This is one of the fundamental differences between a formal engineering design process and casual informal problem solving.

In engineering it is usually good practice to define a problem before solving it. In engineering terms a problem is defined by the **problem statement** and **requirements**. The requirements describe the **functions** that any possible solution must have in order to be considered a solution to the problem. Requirements also include the **objectives** for the design, and the **constraints**. In addition, the requirements describe the context in which the design solution will operate, including the environment, the users and operators, and other existing factors that will influence the design choices (#requirements).

The design team will usually revisit the requirements generation stage repeatedly throughout the design process to review the requirements and improve them.

4.2. Multi-use Design Tools for Characterizing/Defining/Describing/Detailing the Design Space

The multi-use design tools section includes a wide variety of techniques that can be used during the design process. These tools can be used at many different stages. All of the tools explained in this section can be used to generate requirements. However, they can also be used to generate solutions ideas, and evaluate possible solutions. For example, the tools in this section can be used as methods for rigorously comparing design ideas against each other or against existing technologies.

4.3. Idea Generation Tools: Expanding the Design Space

The idea generation tools explained in this section are used to create ideas and are sometimes called creativity methods. Like the multi-use design tools, idea generation tools can be used at many different stages of the design process. These tools can be used to generate ideas for requirements, but more often are used to generate ideas for design solutions. These types of methods are used substantially in the development of innovative technologies and may be the most important activity in some design projects. In other projects there may be very limited use of these tools. It is important that engineers be familiar with a wide range of creativity tools like these that can be used if the project requires it (#ideageneration).

4.4. Decision Making: Reducing the Design Space

The decision making modules explain techniques that are used to make well-grounded decisions; that is, decisions that are aligned with the requirements and goals of a project. The structured nature of these tools can help teams overcome indecisiveness and move forward when faced with complex and conflicting requirements or opinions. The tools covered in this section can be used to make all kinds of decisions throughout the design process. However, they are most crucial when the design team must decide which design solution ideas to pursue further and which to discard (#designevaluation).

4.5. Iterating: Iterating in the Design Space

The iterating section gives some guidance on how to iterate in the design process effectively. The modules suggest a cycle of iteration: generate → select → reflect. Following this process will assist you in developing a better design outcome. This section also explains the difference between an ideal solution and an **optimal solution**.

In design engineers strive to find the optimal solution that fits the functions and constraints of the project and this typically requires iteration (#iterating).

4.6. Investigating Ideas: Exploring the Design Space

The investigating ideas section covers tools such as developing metrics, modeling, and prototyping that are typically used to test and evaluate detailed design solutions. These tools can be used early on in the design process to give the design team a concrete understanding of the problem and the advantages and disadvantages of different possible solutions. Some design projects benefit from early prototyping. However, more often prototyping, modeling, and testing come later: after the requirements are well established, many solution ideas have been considered, a preferred solution has been chosen, and the team is working toward a detailed design.

4.7. Post-Conceptual Design: Leaving the Design Space

The post-conceptual design modules cover common activities that need to occur to bring a design into reality; that is, to implement, install, or produce the technology that results from the design process. The steps needed depend significantly on the type of technology that is being designed. Bringing a product to market is very different from constructing a plant or releasing a software package. This section covers some of the basic activities that are typical of many design projects. You will learn more about this step of the process, as it pertains to your discipline of engineering, as you move up through your program of study.

KEY TERMS

design space	problem statement	requirements
functions	objectives	constraints
optimal solution		

5. Questions and activities

1. Look at the concept map for the text. Identify each section. Try accessing the text material and navigating between modules.

8

Engineering School Projects

#narrative module: #schoolprojects

Learning outcomes

By the end of this module, you should demonstrate the ability to:

- Explain some of the differences between a design project for a course and an industry project
- In particular, explain the difference in the requirements stage of the process
- Identify essential steps in a design competition or school design project, for example, careful attention to the given instructions.

Recommended reading

Before this module:
- **How Engineers Design > Modules 1 through 7**

After this module:
- **Design Process > Requirements > 1. Introduction to Requirements**

1. Design Competitions and Other Types of Engineering School Projects

Many design projects in the first few years of engineering school are not done in service to an actual client. The client is an instructor who is asking you to design something for a course, or to produce a design for a design competition. The user of the design will be you, your classmates, or maybe the instructor, who will test the design once or twice to see if it works. The goal of this type of project is to give you a first experience with design to see how the process works, and the effort that is involved in getting something actually built and operating. The design work is very real, and the experience is an excellent first step toward being an engineer, but the project probably doesn't require fully documenting the project context in the depth that is necessary for an actual industry-based project.

In these types of in-school design activities you usually do not need to spend effort on developing detailed user, client, and stakeholder profiles. The emphasis in this type of course is on the build-test part of the design process. However, it is important to clearly define the problem you are intending to solve, and have an explicit statement of the requirements at hand. Many a design team has been disqualified from a competition because their design solution didn't meet the instructor's requirements exactly.

This careful attention to detailed instructions is essential for success in industry. For example, if your bid on a project (i.e., response to a Request for Proposals) arrives late or without the proper signatures, the bid—and all the work you put into it—will be immediately discarded. In a recent competition for the design of a new building on our campus, almost one-third of the bids were sent back unread because they arrived minutes late or were missing the signature of the company president. These "minor" errors cost these design companies millions of dollars. And a lack of exact adherence to safety standards can cost lives, and the company's reputation. Engineering school projects with detailed requirements and constraints that you must follow are good training for what is required in industry.

If you are asked to write up the requirements for any type of design project for a course, you should always check the assignment instructions carefully to see which requirement sections are necessary. Does the instructor want you to detail the stakeholder list and describe the client organization? Or is this the type of project that requires only a short list of the functions, objectives, and constraints? Make sure you know what type of project you are working on. As you move up through engineering school the projects will become increasingly authentic to real engineering practice. Eventually you will be a part of engineering projects in the field. The names of the categories (e.g., functions, objectives, or stakeholders) may change but the purpose of the requirements remains the same: to define the design problem. You will also find that the requirements, which might take only a page or so for a student assignment, will become chapters or volumes in a larger, complex project.

2

Design Process

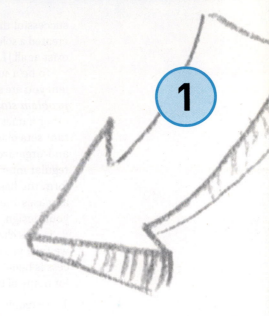

Introduction to Requirements

#process module: #requirements

Learning outcomes

By the end of this module, you should demonstrate the ability to:
- Define and explain the concept of requirements
- Explain the role of requirements in the engineering design process
- Define problem statement
- Analyze a design brief to identify the underlying need

Recommended reading

Before this module:
- **How Engineers Design > 7. Navigating the Engineering Design Process**

Alongside this module:
- **Design Process > Multi-use Design Tools > All Modules**

After this module:
- **Design Process > Requirements > 2. Functions**

1. Introduction

People often solve the wrong design problem. When faced with a simple math or technical problem it is easy to know what is being asked, and the kind of solution that is required. However, design is different. Even simple design problems can involve multiple competing objectives (e.g., light weight, fast, and low cost). For every one product or technology that is

successful there are dozens that never make it. Often this is because the designers created a solution that did not fit the problem, or a solution for a problem that doesn't exist at all [1,2].

To be a successful design engineer it is essential that you understand the problem you are solving. Here we will call the documentation of the design problem the **problem statement** and **project requirements;** however, this information goes by other names in industry (see Table 1). The **requirements** are one of the most important sets of information developed during the design process. They are a complete and organized documentation of the design problem and should be updated at regular intervals during the design process. In some projects the requirements will form the basis of the contract between your team (or company) and the client. In all cases they should describe what the client can expect, and not expect, from your design.

In this cluster of modules we introduce techniques for researching a problem and we take you through the process of developing a full set of requirements. This process is beneficial not only for professional designers, but you can use these methods for many of the projects you will work on during your undergraduate program.

These requirements modules cover:

1. The problem statement: A restatement of the design problem in terms of the need.

2. Developing the requirements: Defining the design problem in engineering terms.

3. Documenting the project context: Viewing the problem from multiple perspectives to more fully define the design problem. Documentation of the context will generate additional requirements.

4. Reflecting on the design problem: Critically reviewing the requirements.

TABLE 1 **Documentation of the design problem**

What the documentation of the design problem is called depends on the field. This material, or parts of the material, are called different things in different industries and sometimes different things in different companies in the same industry. This table gives a few examples. Whatever it is called, this information serves the essential purpose of describing the design problem that is to be solved.

INDUSTRY	DOCUMENTATION
Institutional building design	User specifications
Pharmaceutical industry	Discovery and screening
Software design	Requirements
U.S. Defense	**Specifications**
Aerospace	System Requirements
Chemical engineering	**Scoping** documentation

2. The Problem Statement

Gathering the information you need to understand a design problem starts by researching existing information. This type of research includes reading (both online and in print), but also will probably require listening, testing, and observing. The goal of the research is building your own information base. You will need to learn what the client wants; what the users need; and the technical facts, concepts, and methods that will be important in this design project. A good design team uses a wide variety of research methods, learns quickly and teaches one another, and they become project experts. Good designers are good learners.

A design project often starts with a statement from a client (i.e., *client statement*), sometimes called a *design brief*. In some cases a design team may seek out a problem to solve, but more often the team is given a problem to solve by a client or their own company (i.e., an in-house project). As an engineering student, you will often receive this design brief from your instructor in the form of project instructions. The design brief may be short (only a paragraph or two) and vague. Or it may be very long. Some request for proposals (RFPs) are thousands of pages long and can be confusing or contain conflicting information. If you are very lucky you will get a clear and complete client statement, but you still have to work through it to understand the problem. One of the first steps in the research process is digesting and investigating the design brief.

Digesting the design brief means making sure that your team understands the project *goals*. When dissecting a design brief you are trying to figure out what is the need. Sometimes clients will state what they want, not what they need. This is called a *solution-driven* statement. The statement tells you what the client thinks is the right solution. Part of your job is to analyze the statement to uncover what the need is, and decide if the solution the client has stated is really the right way to go. The engineer's job is to create a *solution independent* statement that preserves only the essential characteristics of the design problem and removes any unnecessary solution driven information. A solution independent statement of the client's need is called a *problem statement*.

Want versus Need

Client: "I *want* a bag that attaches to my wheelchair so I can carry groceries home from the store easily."

The engineer's job is to analyze this statement. The client statement can be broken into parts:

Which part of the statement is the real need here? Does the client only want to get the groceries home, or do they want to be able to carry other things? Do they want something they can operate alone, or would it be acceptable to be dependent

(continued)

on others for help? Is the real need here a secure attachment that can be used by the client, but will not allow theft? And how much volume and weight is the client thinking they need to transport? Does it need to protect the things being carried from rain, heat, dust, vibration, and other environmental factors? The engineer needs to begin to investigate the statement further to discover the need.

Engineer: "So what I am hearing is that you *need* a means for securely attaching things to your wheelchair. Is this correct? Or do you *need* a way to get your groceries home?"

Note: The client statement limits the solution to a narrow range of design ideas. The engineer's questions open up the range of possible solutions while still maintaining the essential function that meets the client's need.

But what if the client really does want a bag for groceries and will not be happy with any other solution? Then the engineering team needs to investigate this further with the client, if possible, to understand the issues better. It is appropriate for the engineer to ask the client what it is about a bag that makes it the solution of choice, and what features this bag would have beyond just attaching to the wheelchair. The engineer might also tell the client about some other ideas (a delivery service, or a mini-trailer that attaches to the wheelchair) to see if the client responds positively, or what their objections are. The engineer might go shopping with the client to measure the weight and size of a typical load of groceries. Part of an engineer's job is working with the client to explore alternatives and investigate the problem to better understand the client and their needs. Note that after this process, if the engineering team does end up designing a bag, then they will be able to provide detailed information about why they made this decision and why this is the design that best fits the needs of the client.

In addition to implicit solutions, design briefs may contain assumptions, conflicting information, or even errors. It is important that you investigate the essential facts to confirm what is presented in the design brief. This may include repeating some of the measurements if data is given or doing research online or in the library to confirm the facts. In the case of conflicting information it is appropriate to discuss the issue with the client so you are basing your design on the correct set of information.

Examples:

- "I want a hole drilled" → implicit solution. The hole is needed; drilling is a method (i.e., a ***means***).
- "I want a motor and the cooling fluid should drain out the bottom" → assumption – not all motors require cooling fluid.
- "It must hold 100,000 ping pong balls and fit into a subcompact automobile trunk" → conflicting information.
- "The camera must weigh no more than 2 ng" → likely a typo; the author meant 2 g maybe?

By the time you are done researching the client statement you should understand the statement well enough to be able to restate the problem in your own words

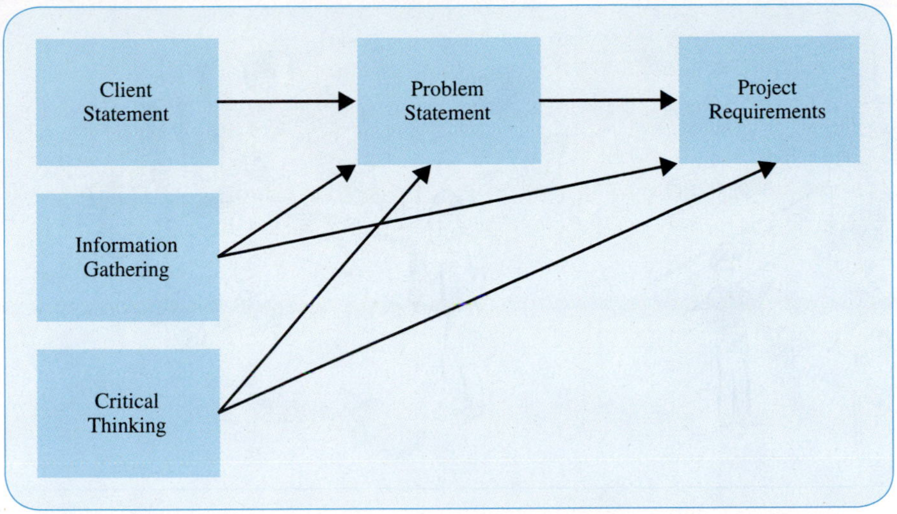

FIGURE 1 Moving from a client statement to requirements requires gathering more information about the problem and critical thinking.

(see Figure 1). The problem statement states what is *missing* or *lacking* in the world. It describes the gap or hole in the current reality that is motivating this project (i.e., what the need is). The design solution will fill this hole.

At this stage in the process, your statement of the problem will be incomplete. It might even be wrong in some respects. This is because the design brief represents only one perspective on the problem; it contains important information, but it is only a starting point. You will need to do more research to continue to evolve a more accurate and complete understanding of all aspects of the design problem. Your problem statement should evolve through the process as well; you may come back to revise the problem statement even after your client has examined some first design ideas you put forward. The problem statement is the first part of a complete set of requirements. It basically serves as the introduction to and motivation for the project requirements.

3. Developing the Requirements

Developing a clear set of requirements will help keep your design project manageable. Design projects are *indeterminate* (ill-structured or ill-posed), which means that design problems are inherently complex. They have many possible solutions. They often involve ambiguous or unknowable information. For example, what exactly do plant operators think about when sitting in the control room of the plant? How exactly will they interpret the information displayed on the screens around them? You could spend enormous amounts of time trying to get accurate information about every aspect of a design project, and still not be able to get exact information about everything you might want to know for the project. And designs must balance competing needs. For example, the needs of the plant operators, the needs of the public who live near the plant, and the needs of the plant owners. For these

reasons design projects can easily get out of control. They can become infinite time sinks, and the features that the client asks for as the project progresses can change and grow uncontrollably.

Your documentation of the project requirements will evolve with the project and with your understanding of the problem. However, having a clear, organized set of requirements for your team to use, and to share with your client at each phase of the project, is going to help keep the whole project under control and prevent you from becoming overwhelmed. Your team may never have all of the information you would want to get before designing a solution, but at least the information you have will be organized, and as complete as reasonably possible.

The engineering design process is essentially an information process

Developing and documenting a clear set of project requirements is part of the work you do for your client. This makes it part of the design. A professional engineering design is not just the thing you create; it is all of the information and documentation that backs up the process you follow and the thing you design. The requirements, and other documentation, are sometimes as valuable to the client as the design itself.

For engineering projects you do in school, the documentation may be an important part of the work that is graded. Your instructor is probably looking not just for a functioning design, but clear documentation of the thinking process you used, and the decisions that were made to arrive at the design. The quality of the thinking process, the methods you employed, and the documentation of this process may be important for your grade.

The project requirements can most easily be developed and managed by identifying categories of information that are necessary to define a design problem. In many industrial projects the requirements are complex and not neatly categorized. Or the headings in the requirements section are specified by the contract with the client or

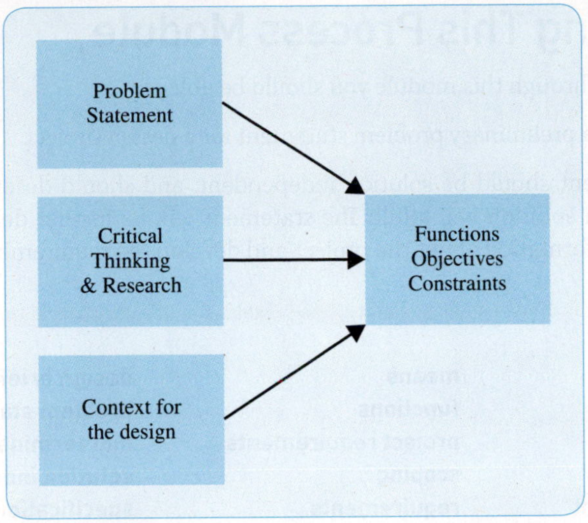

FIGURE 2 The requirements are developed based on the problem statement, the documentation of the context in which the design will operate, and critical thinking and research by the design team.

convention in a particular industry. Generally the requirements can be characterized as: *functions*, *objectives*, and *constraints* [3]. There will be overlap among these categories, and some important information may not neatly fit into any one of them. However, this structure will help you to organize the information and use it effectively in the design process and can help reveal major critical gaps in your understanding of the problem (see Figure 2).

We will start by discussing some general strategies for developing the basic requirements. In addition to these basic requirements, the definition of the design project will necessitate examining the design problem from multiple perspectives because the context in which the design will operate is important. Examining the problem from these perspectives will result in additional functions, objectives, and constraints (search hashtags functions, objectives, and constraints for more information).

Research

Research involves both investigation of existing knowledge and the development of new knowledge. At this stage of the design process most of the research your team will do is investigating existing knowledge: finding written material, talking to people and listening, and learning what you can about the project. You may also need to go to visit the site where the project will be installed (this is called a site visit), and take some preliminary measurements to help you understand the project.

Later in the project you will develop new knowledge: build models and prototypes to test; develop test methodologies; and collect and interpret data. Both of these research types are necessary in the design process, but they generally happen at different stages in the process (#informationgathering).

4. Leaving This Process Module

After working through this module you should be able to

- Develop a preliminary problem statement for a design project.

The statement should be solution independent, and should describe the "need" that the design solution will fulfill. The statement will be further developed as you gather more information about the project and develop the requirements.

KEY TERMS

constraints	means	design brief
objectives	functions	problem statement
goals	project requirements	indeterminate
solution driven	scoping	solution independent
client statement	requirements	specification
research		

5. Questions and activities

1. Given a design brief, analyze it for the underlying need.

2. Analyze an existing technology to determine the underlying need it fulfills.

   a. Write a solution independent problem statement that expresses this need.

3. Document your analysis of the design brief you have for a design project you will be working on.

   a. Identify questions that need to be researched.

   b. Write a preliminary problem statement for your design project.

6. References

[1] "The Gadget Failure Hall of Fame" by David Pogue, Scientific American, June 21, 2011, retrieved from www.scientificamerican.com/article.cfm?id=pogue-the-gadget-failure-hall-of-fame

[2] "Why Gadgets Flop", by David Pogue, Scientific American, June 21, 2011, retrieved from www.scientific american.com/article.cfm?id=why-gadgets-flop

[3] C.L. Dym and P. Little, *Engineering Design: A Project-Based Introduction*, 3rd ed. Hoboken, NJ: John Wiley & Sons, 2009.

Functions

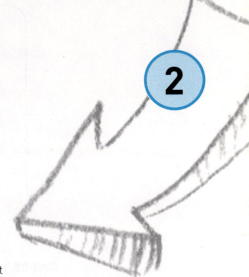

2

#process module: #functions

Learning outcomes

By the end of this module, you should demonstrate the ability to:
- Define and explain the concept of a "function"
- Explain how a function statement is different from a means statement
- Develop a set of functions for a design project
- Analyze an existing design with respect to its functions (i.e., list the functions of an existing technology)

Recommended reading

Before this module:
- **Design Process > Requirements > 1. Introduction to Requirements**

Alongside this module:
- **Design Process > Multi-use Design Tools**

After this module:
- **Design Process > Requirements > 3. Objectives**

1. Introduction

Functions are what the technology *must do* in order to work (i.e., in order to be functional). Basically any idea that meets the functional requirements could conceivably be a possible design solution for the problem. So it is important to explain all of the functions that the technology must fulfill in order to minimally be considered a possible viable solution (see Figure 1).

A function is a statement of exactly what the design must do to work. Suppose a team of chemical engineers is designing a new type of glue intended to be used for plastic parts (e.g., Plexiglas). The function of this product is to hold two or more plastic parts together. Function statements like this one typically include a verb ("hold") and the object of the action ("two or more plastic parts").

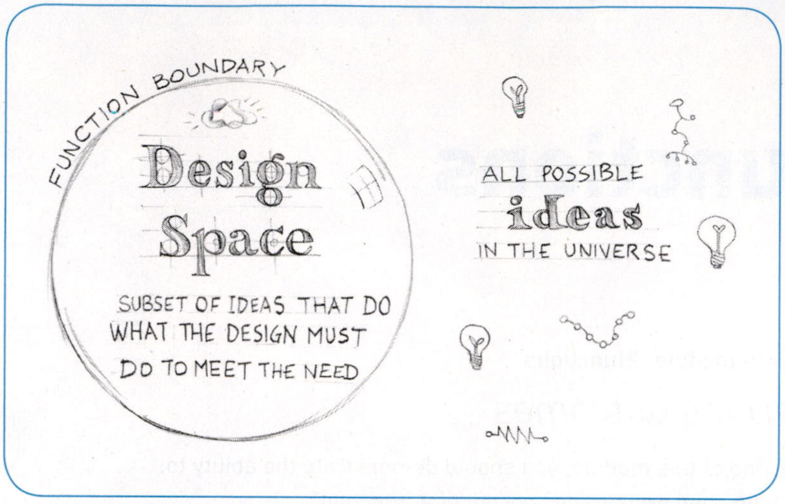

FIGURE 1 Defining the functions defines the design space. They draw the boundary between all possible ideas in the universe, and the subset of ideas that meet the functional requirements for what the design must do to address the identified need.

Note that the statement of the function "hold two or more parts together" is solution independent. It does not imply a specific solution, leaving open the opportunity for many design ideas. How well the glue should perform this task is not part of the statement. How well the glue works will be described as an objective, not a function.

The function statement does not overly constrain (unnecessarily reduce) the set of possible solutions that could be considered, but only rules out solutions that will in no way meet the basic functional requirement (see Figure 2). If the team includes this statement in the requirements then they are eliminating solutions such as vegetable oil, talc powder, air, aluminum foil, and a wide variety of other substances (and other technologies) that do not hold two or more plastic parts together.

This is exactly what a statement of a function should do. It should draw a boundary around the set of all possible solutions, and exclude ideas that do not function in a way that meets the needs of the design.

The functions statements must describe the boundary, not the ideas contained inside or outside the boundary.

1.1. Identifying the Obvious

In developing the functions and requirements in general, you are not yet trying to list or describe the ideas inside or outside the function boundary. You are trying to describe the boundary itself: trying to describe the design problem in engineering terms. Remember in your functions list to remove any mention of possible solutions, and just focus on the definition of the problem you are being asked to solve. If you do think of a possible solution (and you probably will), then think about the characteristics that make it a possible solution and write down those characteristics as requirements.

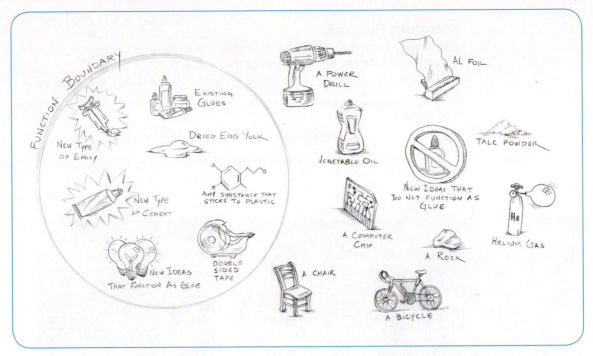

FIGURE 2 Visual example of the functional boundary between all possible ideas and the subset of ideas that serve the basic function of adhering plastic parts. Some of the ideas inside the function boundary (e.g., dried egg yolk) may be poor ideas because they are not good adhesives, but they minimally meet the functional requirements of holding plastic parts together so they are inside the boundary.

An important aspect of developing the requirements is making assumptions explicit and explicitly identifying aspects of the design that are so obvious you might assume that everyone is aware of them.

Let's consider an example. Suppose a team is designing a new bicycle security system (i.e., something that functions like a bike lock). At a minimum:

- A security system must secure a bicycle.

However, we can break this *primary function* down into sub-functions, sometimes called *secondary functions*, which describe what the design must do to enable the primary function.

- The design must secure a bicycle.
- The design must release the bicycle.

This may seem obvious; a bike lock that locks but won't unlock is an unacceptable solution. However, if the design team neglects to take into account the work the user must do to unlock the bicycle they may end up with a design that is awkward and difficult to use. You need to think through the user's need step by step so you make sure you are intentionally considering each functional element in your design.

Secondary Functions

Secondary functions result from, or enable, the primary function. Here are a few examples for the security system design:

- The design must secure the bicycle in place; this enables the primary function.
- The design must un-secure and release the bicycle; this enables the primary function.
- The design produces a secured bicycle; this results from the primary function.

Why is this last secondary function important for the design engineer to acknowledge? (See Figure 3.)

Courtesy of the Authors

Courtesy of the Authors

FIGURE 3 **Bike locks must be designed so they ultimately can be removed by someone other than the owner if necessary.** These bikes have obviously been abandoned by their owners and have now become public trash locked in place.

Courtesy of the Authors

FIGURE 4 **A concept for a keyless bicycle lock.**

Developing the requirements is a learning process, and intentionally considering every aspect of the functionality of the design teaches you more about the problem you are working on. In thinking through the bicycle security example there are actually many secondary functions to consider. For instance, the design requires user input in the form of information: when to lock the bike, when to unlock the bike, where to lock the bike, who is allowed access to the bike. Therefore the system must allow the user to input this information. This may seem like way too much thinking for a simple design, but considering all of these aspects of the problem allows the design engineer to consider the design from a much more creative perspective (see Figure 4).

Functions and Means

The statement of a function, or any requirement, must be solution independent. That is, the statement must not imply a specific solution because the requirements should define the problem, not the answer. A specific solution idea, or a *means* of solving the design problem, is called a *means statement*. Means can also refer to the means of implementing a solution. For example:

Functions: The bike security system must lock and unlock, or secure the bicycle and release the bicycle.

Means: This is accomplished using a key/lock system.

Any mention of a key or a physical lock does not belong in the function list and requirements for this design problem because it limits the solution ideas unnecessarily.

1.2. Generating Functions

A variety of methods for generating functions are explained in other modules (see Multi-use Design Tools modules and search hashtags such as ideageneration). Each entry in these module clusters will explain a method, what it can be used for, (e.g., generating functions or generating solutions), and its advantages and disadvantages. There are also examples given to help you understand how to apply each method.

It is recommended that you try using at least two or more creativity methods to generate a robust list of functions, and to get some practice using different techniques. Methods such as *brainstorming, 5 whys* (i.e., root cause finding), and the *black box method* are often used. *Means analysis, use case,* and *task analysis* can also be employed to assist in the generation of functions, helping you to understand the various human and non-human interactions with the design (search hashtags for more on these methods).

One very powerful method for articulating the functions of a design is called *functional basis*. The functional basis technique allows you to define the functions of a design in very simple engineering terms. This helps you to understand the essential nature of the technology you are designing. This method is particularly useful for new engineering students who are learning how to reorient their perspective to perceive the basic engineering functionality of technology (search hashtag functionalbasis).

Black Box Method

Example: Developing the functions for a bicycle lock system using the black box method

In this method you describe all of the mass, energy, and information that is going into the proposed design without actually describing the design itself. Then you describe all of the mass, energy, and information that you expect to result from the operation of the technology you are designing. For a bicycle security system the inputs are the bicycle (secured or unsecured), and information about when and where to lock the bicycle. There is also information about who is allowed to lock or unlock the bicycle. There may also be energy input. We assume here that mechanical energy provided by the user is available, and possibly electricity. The outputs are

(continued)

developed in the same way. Note that energy is a likely output because any energy input has to either be stored or exit the system (conservation of energy principle).

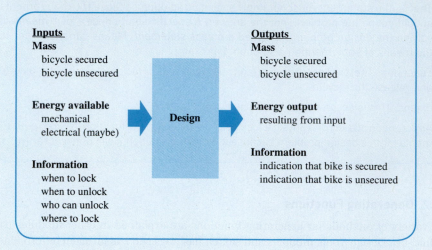

This type of diagram suggests functions of the design; it helps you discover what expectations you have for the functionality. In this example, we discover that we expect the bike security system will provide the user with a clear indication of when the bike is secured and when it is unsecured. The means for doing this is not specified; the system could indicate this information through visual, auditory, tactile, or other sensory input. The system could even send a text message to your phone telling you it is secured. Developing ideas for this functionality will be part of the solution generation process. For the requirements we would simply state:

- The design will produce a secured bicycle.
- The design will produce a released bicycle.
- The design will indicate when the bicycle is secured and when it is unsecured.

Additional pieces of information that are gleaned from this diagram will help us develop other requirements, notably objectives and constraints. For example:

Objective: The design should allow the user to secure the bicycle in a wide range of locations.

Constraint: The design must use only available energy as input if it uses energy at all.

2. Leaving This Process Module

After working through this module you should be able to

- Develop a full list of functions for a design project

It is important that you use several different methods to develop this list. The functions list should include both primary and secondary functions.

KEY TERMS

functions	primary functions	secondary functions
means	brainstorming	black box method
means analysis	5 whys	task analysis
functional basis	use cases	means statement

3. Questions and suggested activities

1. How are functions different from objectives or constraints?

2. Pick a design problem. Write a function for the design and write a means statement for the design.

3. What role do functions play in the design process?

4. Consider the following list of design needs.

  a. Write one or more primary function statements for each. Identify the verb and subject of the action in the function statements.

  b. Write two or more secondary function statements for each.

  - A user wants a means of carrying her soccer cleats without getting dirt on her clothes or other belongings.

  - A company wants a means of organizing their warehouse to make it as efficient as possible for assembling orders for customers from items on the shelves.

  - A company wants to design a web-based tool that suggests a car model to a customer based on their priorities and needs and the current available car models.

  - A user wants a means of moving their cello from their home to the location of their weekly cello lesson 10 km away.

  - A company want to create a new brand of shampoo that works without water.

5. Develop a complete list of the primary and secondary functions for the design project you are working on currently.

3

Objectives

#process module: #objectives

Learning outcomes

By the end of this module, you should demonstrate the ability to:

- Define and explain the concept of an "objective"
- Explain how objectives are used in a design project
- Explain the relationship between objectives and objective goals
- Develop a set of objectives for a design project
- Analyze an existing design with respect to its objectives (i.e., list the objectives that likely were considered in the design of an existing technology)

Recommended reading

Before this module:

- **Design Process > Requirements > 2. Functions**

Alongside this module:

- **Design Process > Multi-use Design Tools > 6. Pairwise Comparison**

After this module:

- **Design Process > Requirements > 4. Constraints**

1. Introduction

Objectives are what the design solution *should be*. Objectives are used to judge how *well* an idea solves the design problem. They are the evaluation ***criteria***. Some designers call objectives "measures of effectiveness" to emphasize this evaluative aspect of objectives. Unlike closed-ended problems that have one unique, exact solution, design problems are open-ended. Solutions to design problems are judged "better" if they do a better job at solving the design problem, or "worse" if they do a poor job at solving the design problem. Another way of saying this is that the quality of the design solution will be evaluated based on the criteria as documented in the requirements: Does the solution perform the necessary functions and to what degree does it meet the objectives? The objectives are the criteria that are used to make this judgment (see examples in Table 1).

TABLE 1 Examples of objectives.

DESIGN BRIEF	EXAMPLES OF OBJECTIVES
Lipstick: design a new type of lip-tinting product	• Long-lasting • Available in a wide variety of colors • Aesthetically pleasing • Won't come off on a glass, clothing, or another person • Won't chemically react with food or drinks
Building structure: design the support structure for a new building with a unique shape	• Inexpensive • Durable • Stable • Enhances architectural design (or at least doesn't impair it)
Course selection system: online system that allows university students to register for their courses	• Reliable • Easy to use • Easy access to information about the courses; such as course content and scheduling

If functions are the boundary that separates functionally appropriate solutions from non-functional solutions, then in this analogy the objectives describe the landscape inside the functional boundary (see Figure 1). The landscape has hills and valleys.

FIGURE 1 Functions define all acceptable solutions and objectives indicate which solutions are better or worse.

Objectives

51

When you begin generating solutions for the problem and evaluating them, you will use this "landscape" to judge whether a solution is near the top of a hill (meets the objectives for the design) or at the bottom of a valley (is a poor solution because it does not meet the objectives for the design).

A few notes about these examples:

1. These are partial examples; the actual list of objectives could be much longer.
2. Safety and other critical aspects of the design problem are not included here. Critical factors, such as safety, will be listed in the constraints section of the requirements.
3. Some of the examples (e.g., "reliable", "easy to use", "stable") are very general and could apply to many different design problems. You will need to add more specific objectives that explain what "easy to use" means for the specific design problem you are working on, and how it could be evaluated (measured).

2. Using Objectives

It is useful to set goals for the objectives. An ***objective goal*** is the minimum target level for an objective. Objective goals often inherently give measurement methods, called ***metrics***, for evaluating the objective. So in the case of "easy access to information" the metric could be the number of mouse clicks to get to the content (see examples in Table 2) (#metrics).

TABLE 2 **Examples of objective goals.**

OBJECTIVE	EXAMPLE OF A POSSIBLE GOAL
A wide variety of colors	Goal: at least 10 colors
Inexpensive	Goal: less than $10 million dollars
Easy access to information about the courses	Goal: no more than two mouse clicks to the course content or scheduling information from any point in the registration process.

The full requirements will be given to the client to check over. The ***objective goals*** help the client assess whether you are proposing to design a system that meets their needs, or whether your understanding of the design problem is very different from their expectations. For example, suppose you are designing the new type of lipstick and you set the goal for "wide variety of colors" at "≥10." The client may come back to you and say, "No, we were expecting more like a hundred different colors. You are off by an order of magnitude on this." That information is really important to know because it may later cause you to rule in or rule out some of the possible solutions that you devise for this problem.

However, objective goals are not absolute constraints. Suppose you develop an excellent lipstick product that, for some strange reason, can only be manufactured in 95 different colors. The objective goal you have listed in your revised list (e.g., "producible in at least 100 colors") would not cause you to throw out this design solution, even though it falls a little short of the goal. It would just motivate your design team to strive for a product that is as good in every other respect and is producible in even more colors. Or you might, in consultation with your client, settle for only 95 colors if the other attributes of the lipstick product were really great. Objective goals are meant to motivate the improvement of the design toward a higher peak in the landscape; set an approximate threshold for acceptable levels of performance or characteristics of the design solution. They are not constraints.

2.1. Generating and Managing Objectives

Examination of the design brief and team brainstorming are often used to generate an initial list of objectives. Also, many of the creative processes listed in the Multi-use Design Tools and Idea Generation sections can be used to generate objectives.

The most common method used to generate specific objectives and manage the list of objectives is an *objective tree*. Figure 2 shows a page from an engineer's notebook for an aerial photography system.

To build an objective tree, start with a list of objectives you have discovered through analysis of the client statement and other research you have done. At the top of the tree put the broadest, most general objectives. These objectives often come from the basic needs the client has stated and from the *DFX* (*Design for X*) considerations that are important in this project. These objectives are central to the project, but are often not directly measurable. *An objective must be measurable* so that later in the process you can use it as the basis of comparison between two possible design solutions. To make a fundamental, broad objective measurable you have to decompose it into measurable pieces.

To build the tree further, define the more specific objectives that are measurable. Under each general objective list the specific measurable objectives that, in combination, add up to the general objective. (see Figure 3). So, for example, a general objective for this system is "easy to set up." How will you know if the system is easy to set up? You can measure if it "requires few people to set up" with a goal of 2 people or fewer, and if it "requires a minimum amount of time to set up" with a goal of less than 10 minutes for set up. It will be easy to test a *prototype* (model of a design solution) against these

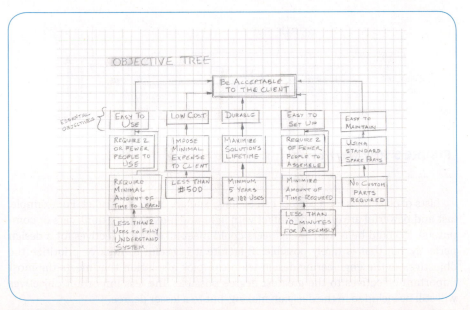

FIGURE 2 A page from an engineer's notebook shows an objective tree for an aerial photography project.

objectives. These specific objectives combine to allow the team to conclude which proposed system is easiest to set up.

The objective tree shown in the example notebook page is not complete. This is clearly just a first draft. There are many other objectives that should be added to this tree to fully define the problem. For example, under "easy to set up" in this project you would probably want to include statements about:

- The number of operations that are needed for set up, fewer is better.
- The tools required; no specialized tools are better.
- The amount of force (normal force and torque) energy necessary for set-up; less is better.

As engineers we mean energy in the physics sense: the force times distance, or torque times rotation, or energy of other kinds. This engineer would probably combine her work with the draft trees developed by other members of her team. Then they would review their research, and brainstorm further to create a more complete set of requirements.

How-Why Tree

An objective tree is an example of a **how-why tree**. The how arrow points down the tree: *How* will you make the design easy to set up? By making it fast to set up and require only a few people for set-up. The why arrow points up the tree: *Why* should you make it fast to set up and require only a few people? Because this makes the system easy to set up. How-why trees are used in many fields, including business, education, and government: basically, any field where the activities of the organization are motivated by performance objectives. They are used to organize information from high-level global objectives down to measureable local objectives.

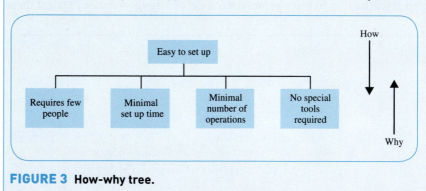

FIGURE 3 **How-why tree.**

Lists of objectives for a project can be long, and are often in conflict. For example, fast and powerful are often in conflict with light and inexpensive (think cars or computers). It would be terrific if engineers could meet all of the objectives for a design perfectly, but that is rarely possible. Therefore it is useful to rank or prioritize the objectives so the engineering design team can pick a solution that meets the most important objectives to the greatest degree possible. The less important objectives will still be included in the decision making, but to a lesser degree.

The easiest method for prioritizing a long list of any kind is using a *pairwise comparison* process. This allows you to compare just one item to one other item at a time, which reduces the complexity of ordering a long list. When using this method to order your objective list, make sure you use a list of objectives that are all at the same level on your tree (i.e., don't compare "easy to set up" to "requires no special tools for set up"). For the example tree shown in Figure 3, the prioritized list would rank from most important to least important: "easy to use," "low cost," "durable," "easy to set up," and "easy to maintain," along with the other essential objectives that are added during the process of developing the requirements.

2.2. Leaving This Process Module

After working through this module you should be able to:

- Develop an organized list of objectives for a design project
- Where possible, identify a goal or goals for the objectives

It is important that you prioritize the list (we suggest using pairwise comparison for this purpose), and that every high-level, abstract objective (such as "easy to use") have a list of sub-objectives that are measurable indicators of this intent.

KEY TERMS

objectives	**criteria**	**objective goals**
metrics	**objective tree**	**how-why tree**
design for X (DFX)	**prototype**	**pairwise comparison**

3. Questions and activities

1. How are objectives different from functions or constraints?

2. What role do objectives play in the design process?

3. Pick a design problem or existing technology. For this technology write an objective and an associated goal.

4. Consider the following list of design needs.

a. Write one or more objectives for each.

b. Identify a goal and metric to go with each objective.

- A user wants a means of carrying her soccer cleats without getting dirt on her clothes or other belongings.
- A company wants a means of organizing their warehouse to make it as efficient as possible for assembling orders for customers from items on the shelves.
- A company wants to design a web-based tool that suggests a car model to a customer based on their priorities and needs and the current available car models.
- A user wants a means of moving a cello from home to the location of weekly cello lessons 10 km away.
- A company want to create a new brand of shampoo that works without water.

5. Develop a complete list of objectives and associated goals for the design project you are working on currently.

a. Organize the objectives into one or more how-why trees.

b. Prioritize the high-level list of objectives using pairwise comparison.

Objectives

4 Constraints

#process module: #constraints

Learning outcomes

By the end of this module, you should demonstrate the ability to:

- Define and explain the concept of a "constraint"
- Explain how constraints are used in a design project
- Explain the relationship between constraints and objective goals
- Develop a set of constraints for a design project
- Analyze an existing design with respect to its constraints (i.e., list the constraints that likely were considered in the design of an existing technology)

Recommended reading

Before this module:

- **Design Process > Requirements > 3. Objectives**

Alongside this module:

- **Design Process > Multi-use Design Tools > 4. Information Gathering**

After this module:

- **Design Process > Requirements > 5. Documenting the Context**

1. Introduction

Constraints are what the technology *must be*. Constraints set absolute limits. If a potential solution violates a constraint, then it must be thrown out. In the analogy, where functions set the boundary and objectives describe the landscape, constraints define no-go areas in the landscape (see Figure 1). Constraints are areas of the landscape that are too hazardous, too expensive, or are fenced off for other reasons.

For example, constraints might result from:

- Federal aviation laws
- Maximum project or product costs

- Safety codes
- Interoperation (connection) requirements (e.g., must connect to or operate with an existing system)

In virtually every design project there will be constraints related to safety and reliability. Basically there are consequences to the failure of systems, and in engineering these consequences carry considerable risk to the health, safety, and security of people. Even if people are not physically hurt, if a design fails there may be other damage to property or economic value. It is very important, therefore, to list in the requirements the constraints the system must meet to manage the risks effectively. Typically this means identifying and adhering to standards, codes, and regulations.

FIGURE 1 Constraints identify sets of solutions that might otherwise meet the other requirements, but that are unacceptable.

What is the difference between an objective goal and a constraint? An objective goal sets an approximate level for the acceptable performance or characteristics that you want to achieve in your design (see examples in Table 1). A constraint is an absolute limit. Any design that does not satisfy the constraint, even by a little bit, will automatically be removed from consideration as a solution for the design problem you have defined. Some constraints will not be related to an objective, and you can have objectives that don't relate to a constraint.

TABLE 1 Examples of constraints.

OBJECTIVE	OBJECTIVE GOAL	CONSTRAINT
A wide variety of colors	At least 100 (increased from 10 to 100 based on client feedback)	Must be producible in at least 40 colors to be acceptable to the client
Inexpensive	Less than $10 million	Shall cost no more than $15 million maximum
Easy access to information about the courses	No more than two mouse clicks to the course content or scheduling information from any point in the registration process	<<No particular constraint associated with this objective>>
		Examples of other constraints that may be unconnected to the objectives: • Must meet all safety regulations • Shall conform to industry standards* • Must be legal to operate

*In some industries the use of the word "shall" is used exclusively to denote a constraint; the word "must" is not acceptable in the statement of a constraint. In a contract between the designers and the client the word "shall" indicates a contractual obligation that the design must meet. Sometimes the word "will" is allowed, but it is generally avoided because it may not be legally binding.

Constraints

1.1. Generating and Managing Constraints

Constraints are commonly generated in three ways:

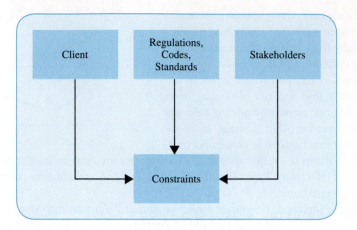

1. Consultation with the client

   The design brief may contain some constraints that you can echo in your project requirements. These include statements about the maximum cost of the design, or issues related to performance and marketing. It is important to dissect the design brief to find constraints, but also to consult with the client to uncover additional constraints that may not have been made explicit. Many times clients are so familiar with their own design problems that they forget to include basic assumed ideas in their statements. To create a design that is going to be acceptable to the client it is important to uncover these assumptions and make them explicit in the requirements.

2. Gathering information on *regulations*, *codes,* and *standards*

   Part of your job as an engineer is to identify the regulations, codes, and standards that pertain to the design problem you are working on (#informationgathering). Finding the codes and standards and understanding them for purposes of designing a system that meets these constraints is part of doing *due diligence* (doing your work responsibly). For example, building designs must meet the local building code. The code imposes constraints. Many of these codes, standards, and other information relevant to constraints can be found online, but you will probably also need to consult traditional sources (i.e., on paper). It can be difficult for a new engineering student to figure out which codes apply, understand the technical content of the code or standard, and figure out how to apply the code, standard, or regulation to the design problem. Not to worry. As your technical proficiency increases you will be able to grasp this material and use it. Once you become an engineer you must use codes properly, and get help when you are unsure of what to do. However, at the introductory level of engineering school you should be able to identify the appropriate codes and cite them in your constraints list. With some help, you might also be able to apply these in your design process, but at the very least you should show that you know which codes, standards, and regulations might apply to your design project.

There may be instances when it is not initially obvious which codes pertain to the project until you start developing some solution ideas. It is always worthwhile revisiting your requirements, particularly the constraints section, as you continue through the design process. You may find that it is necessary to add more information about codes, regulations, and other constraints to your requirements section as new ideas arise.

3. *Stakeholder*-imposed constraints and broader considerations

Within the client organization there may be other people who have ideas about the constraints for the project (#stakeholders). This includes the marketing and legal departments. The client organization may specify, for example, that the design should not make use of existing patented technologies because using this technology will require a license. Beyond the client there will be other stakeholders in your project who will suggest constraints. Constraints from stakeholders in the client organization and from outside the client organization often arise through DFX (design for X) considerations. It is important, therefore, to think through the DFX considerations and to seek out consultation with stakeholders beyond the client to uncover the requirements that arise from these people and organizations.

2. Leaving This Process Module

After working through this module you should be able to:

- Develop an organized list of constraints for a design project.
- State the justification for each of the constraints listed.

Adding constraints often adds cost and can also add legal liability. It is important that the constraints you specify for a project be justifiable based on the research you have done and the input from stakeholders.

KEY TERMS

constraints	stakeholder	regulations
codes	standards	due diligence

3. Questions and activities

1. How are constraints different than functions or objectives?

2. What role do constraints play in the design process?

3. Pick a design problem or existing technology. For this technology write a few constraints that pertain to this technology.

4. Consider the following list of design needs.

   a. Write one or more constraints for each.

   b. Identify the source of the constraint: design brief (client), code/standard/regulation, stakeholder, or other.

- A user wants a means of carrying her soccer cleats without getting dirt on her clothes or other belongings.
- A company wants a means of organizing their warehouse to make it as efficient as possible for assembling orders for customers from items on the shelves.
- A company wants to design a web-based tool that suggests a car model to a customer based on their priorities and needs and the current available car models.
- A user wants a means of moving their cello from their home to the location of their weekly cello lesson 10 km away.
- A company wants to create a new brand of shampoo that works without water.

5. Develop a complete list of constraints for the design project you are working on currently.

Documenting the Context

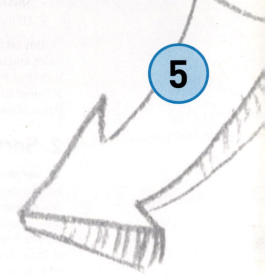

5

#process module: #serviceenvironment

Learning outcomes

By the end of this module, you should demonstrate the ability to:
- Define and explain the concept of a "service environment"
- Develop a complete description of a service environment for a design project
- Use the service environment description to generate additional requirements, particularly objectives and constraints
- Use the service environment description to add depth to the problem statement
- Analyze an existing design with respect to its service environment (i.e., describe the service environment of an existing technology)

Recommended reading

Before this module:
- **Design Process > Requirements > 4. Constraints**

After this module:
- **Design Process > Requirements > 6. Stakeholders**

1. Introduction

A substantial number of the functions, objectives, and constraints for a design project will result from the context of the design project. Initially the list of requirements will come from the design brief and your own thinking and research on the project topic. However, considering the project from multiple perspectives will add to the requirements list. For instance, documenting information about the users of the technology and the environment in which it will operate will give you additional insight into the problem and the types of solutions that will be effective.

There are many perspectives that could be investigated for any design project. We suggest that documenting a few essential aspects of the context will help you develop a more complete and successful project:

- *Service environment*
- *Users* and *operators*

- *Stakeholders*
- DFX considerations

This list is not meant to imply a sequence. The information can be developed in any order, and usually development is highly iterative. As your team does research using a variety of sources you will generate information for all of these categories. Your team will need to sort out the information and organize it to create a meaningful documentation of the design problem context.

2. Service Environment

The ***service environment*** describes the location where the design will operate. A location is more than just a latitude and longitude on earth or a point in space. The description of a location includes a description of all elements that might typically be present in that location (e.g., the amount of rain, the typical temperature range, the presence or absence of cell phone coverage, the types of animals and insects native to the environment, the noise level range, and the people). If the technology you are designing will be mobile, or will be installed in a number of different locations, then you need to describe the range of conditions typical for these different locations. If it will only operate outside, then you need to describe the outside conditions. And if it will only operate inside, then you need to describe the indoor conditions. You may think it is obvious what "indoor conditions" means, but indoor conditions in a yurt in Mongolia are very different than the indoor conditions in a gym locker room in Uruguay. Underwater, deep space, the Arctic, and many other situations will bring even more service environment factors into consideration if you are designing technology that must operate in these environments.

The description of the service environment is important when the conditions can cause the technology you are designing to fail for the user. Structural engineers always take into account the wind and weather conditions of the location for the design of a building structure. This should be obvious. However, service environment considerations are equally important for the design of a new type of snow tire that must operate in snow, but also safely on hot dry roads, or a new online banking system that must operate in a virtual environment accessed by people who speak a variety of languages and by people who want to try to hack the system.

It is sometimes easiest to develop this information as a diagram. Figure 1 shows some of the service environment considerations for a marine buoy (or channel marker). Note that the service environment should include systems available for installation and removal from service if this makes sense for the design problem.

The service environment documentation should *not* be written from the point of view of the design you are creating. That is, it is not *how* the device will be designed to operate within an environment, but instead a basic description of what the environmental conditions are. In other words, not "The design must be waterproof" but instead "The design will be used outdoors in Toronto, Canada, in all weather. Temperatures can be expected to range from 3.0°F to 85°F [1]. Average rainfall is 684.6 mm and snow is 115.4 cm [2]."[1] This statement about the service environment may then be translated into a constraint ("The design must be water-tight to a depth of 5 cm")

---

[1]Unfortunately, mixed units (SI and IP) are very common in engineering problems.

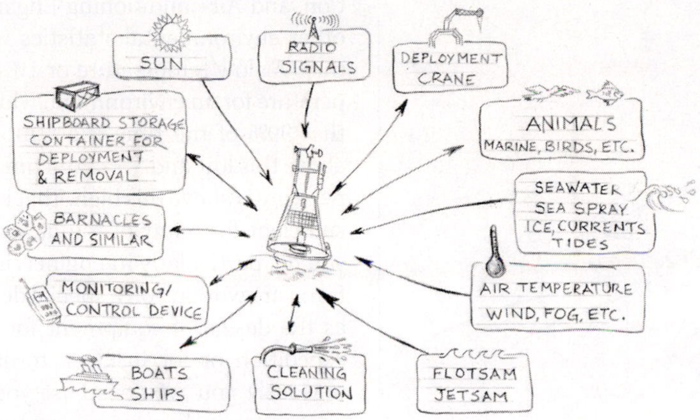

FIGURE 1 The service environment for a marine buoy defines all aspects of the environment that may influence the design.

and/or an objective and goal ("The design should be as water-resistant as possible, preferably water-tight to a depth of 10 cm").

2.1. Identifying and Describing the Service Environment

There are many aspects of the service environment that could be investigated and described. You should include all aspects that are potentially relevant for your design project, understanding that you can return to update and add to the service environment as needed. Developing a complete service environment description can prompt design ideas. For example, knowing that cell phone coverage exists in a location may prompt you to consider a wireless communication system that would augment your design in some way, or allow your users to use their smart phones with your system. However, it makes no sense to include an aspect of the service environment that clearly has no conceivable bearing on your design. For example, it makes no sense to specify the typical outdoor wind speed when you are working on the design of a new type of food processor, even if you might imagine that some user might take their food processor outdoors.

We suggest that you initially focus on three aspects that typically are important in many types of design projects. You can always add more characteristics to this list.

1. Physical Environment

Physical environment pertains to characteristics such as temperature range, pressure range, wind velocity, rain, snow, ice and salt water spray, sun conditions, humidity, dirt and dust, corrosive environments, shock loading, vibration, and noise level. Basically this includes the extremes of physical environment experienced where the design is operating. As a general rule, you don't need to choose the most extreme extremes (e.g., the record low temperature or high temperature). Engineers typically use ranges that are based on statistical percentages. For example, ASHRAE tables

(American Society of Heating, Refrigeration, and Air-conditioning Engineers) and other environmental statistics will report the 99% low temperature or 1% high temperature for an environment, which means that 99% of the time the temperature is above this low and 1% of the time the temperature is above this high. That is probably sufficient for many of the design projects you do, particularly in engineering school. If you are working on a special design, such as the design of equipment for an Arctic expedition or for inside a tornado, then obviously you would adjust your service environment description (and associated objectives and constraints) accordingly.

Some environmental factors work over time. Corrosion, such as rusting, is one of these. Other factors are more subtle; for example, you might design a product that survives a very high temperature, but will fail with repeated temperature cycles because parts are expanding and contracting with every cycle. Similarly bending a wire back and forth will break it. This is called fatigue. The reliability of a design over time is an important consideration. So the service environment description should document exposure to corrosive conditions or cyclic changes (e.g., temperature cycling) in the environment if this is a significant feature and likely to affect the design.

A description of the physical environment would also acknowledge other elements that may be active in the user's environment, including (but not limited to) appliances, machinery, or services (such as the availability of clean water). It could describe any assumptions you are making about other systems that the user has access to while using your design (e.g., the specifications of a computer necessary to run the software you are developing). Getting into the habit of explicitly acknowledging your assumptions about the operating environment and other aspects of the design problem is part of developing a design thinking perspective.

2. Living Things

Your service environment description should include the people and other living things that are in the environment. There will be special attention given to describing the users and operators of the design. In addition to the users there may be other people present in the service environment. You should try to broadly describe these people. This may be very general, but at least you are acknowledging the possible interaction between these other people and the design. For example, the design of a vacuum cleaner for home use must take into account the possible presence of children in the environment and should be, to the extent possible, child safe. On the other hand, the design of a vacuum cleaning system for an industrial machine shop would probably not mention children, and "child safe" would not enter into the constraints list.

In addition to people there are other living things in many environments. These include the possible presence of pets, livestock, wild animals, insects, microbes, and plants. These should be described to the extent that they may conceivably be important to the design project you are working on. In the example of the design of a vacuum cleaner for home use, the possible presence of pets, insects, and house plants should be acknowledged. You might also want to describe some common household microbes if you think you might want to design a system that removes these from indoor surfaces. You would probably not mention livestock or wild animals in your service environment description for a household vacuum cleaner project.

In addition to users, other people, and other living beings who have a positive or neutral interaction with your design, you also need to describe the presence of malevolent beings in the service environment. These include people who could vandalize the technology or use it improperly; Insects or animals that could damage the technology; and microbes, fungus, or mold that could grow in or on your system (if it is a product or installation). Many clients for technology that will be used in hostile environments (such as the military) will specify that the design surfaces be anti-fungal, and that any sensitive technology be enclosed in a vermin-proof casing. For virtual technologies, your service environment should mention the presence of hackers, particularly if the system might be making use of sensitive information (e.g., credit card numbers, banking information, or personal information).

3. Virtual Environment

The last category that we suggest you include in your service environment is *virtual environment*. Many design projects will not need to acknowledge or describe the virtual environment at all. If, for example, you are designing a new type of shampoo bottle, it is unlikely that information about the virtual environment will affect your design and you can leave this section out. However, if there is a good possibility that your design could make use of existing virtual infrastructure, such as wireless networks, satellite coverage (e.g., GPS), or cell phone signals, then you should explicitly state what your assumptions are. What type of

access to these systems are you assuming will exist in the environment where the design will be operating?

If you are not sure whether virtual systems will be important to your design, then leave this section out. You can always come back to add it if you find that the design process is leading you toward a solution that requires the existence and access to these types of systems. Remember not to describe the design solutions you are considering, or how they will function. The service environment description should just describe the characteristics of the environment where the system will be operating.

3. Further Considerations

3.1. What about Parts of the Service Environment that Don't Neatly Fit into One of These Categories?

There are plenty of characteristics of many environments that don't neatly fit into one of these categories. You can always create an "other" category for these or put them into one of the categories that you think fits best. For example, the availability of electricity in the location where the design will be operating, and the characteristics of the standard electrical *service* that is available (e.g., in North America, 110V). Is this a physical characteristic of the environment or a virtual characteristic? It doesn't matter which category you place it in, as long as you list is somewhere. Again, if you are designing a shampoo bottle, you shouldn't state information about the electrical service anywhere in the service environment, at least until you start down a solution path that requires electricity to operate your shampoo bottle. But if you are designing a water pump system for a remote location, it is worth noting that you are assuming electrical service is available (or not) because this assumption would change the constraints list for the design project. The categories we have suggested are only to help you get started in your thinking process. Feel free to add a "miscellaneous" category or sort your service environment information in any other way that makes sense for your project. The category names do not matter; it is the information about the environment that is important, because it is this information that you will use to shape your design process.

3.2. Using Your Service Environment Information

Once you have developed a clear description of the service environment, go back and revisit your functions, objectives, and constraints. Use your service environment information to create objectives and constraints for your system. For example, if you have uncovered the fact that there is no reliable access to electricity in this service environment, then you may want to add a constraint: "The design must operate without electricity input from a power grid." If you know that the design will require energy input (because the design must do work, and work requires energy input), then you might add to the functions list "Design will use available energy sources as necessary to provide functionality," without yet specifying the nature of the energy input (e.g., solar, manual, water power, wind). Or if, in the development of the service environment description you have realized that the environment may contain people who could vandalize your system, then you might want to add to your objectives list: "The design should be difficult to vandalize" (the more difficult, the better). This objective would have to be further developed to make it properly testable.

4. Conclusion

The reason for including all of these aspects in your service environment description is just to make sure that you make your assumptions explicit. This process reminds your team to do their homework (i.e., due diligence), research the environmental conditions, and gather information about the location of operation. It is part of being intentional as a designer. Numerous technologies have failed because designers have relied on the service environment having certain attributes, and their assumptions were incorrect. But they may never have checked their assumptions because they never explicitly stated them.

5. Leaving This Process Module

After working through this module you should be able to:

- Develop a complete description of the service environment for a design project.
- Use the service environment description to further develop the problem statement and list of requirements for a project.

The description of the service environment should contain all of the aspects that are likely to be important for the type of project on which you are working.

KEY TERMS

service environment	users	operators
stakeholders	virtual environment	service

6. Questions and activities

1. Define and explain the meaning of service environment.
2. What role does the description of the service environment play in the design process?
3. Pick a design problem or existing technology. For this technology, write a description of the service environment.
4. Consider the following list of design needs. Write a description of the service environment for each.
   - A user in North America wants a means of carrying her soccer cleats without getting dirt on her clothes or other belongings.
   - A company in Tokyo wants a means of organizing their warehouse to make it as efficient as possible for assembling orders for customers from items on the shelves.
   - A company wants to design a web-based tool that suggests a car model to a customer based on their priorities and needs and the current available car models.

- A user in New York City wants a means of moving their cello from their home to the location of their weekly cello lesson 10 km away.
- A company want to create a new brand of shampoo that works without water for use in arid environments and regions experiencing drought.

5. Develop a complete description of the service environment for the design project you are working on currently.

7. References

[1] *2009 ASHRAE Handbook: Fundamentals.* American Society of Heating, Refrigerating and Air-Conditioning Engineers, Knovel Publishing, 2009.

[2] Environment Canada (2011, May), *National Climate Data and Information Archive.* [Online] Retrieved August 21, 2011, from www.climate.weatheroffice.gc.ca/Welcome_e.html

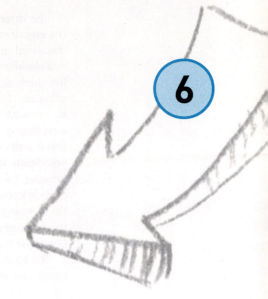

Describing Stakeholders

6

#process module: #stakeholders

Learning outcomes

By the end of this module, you should demonstrate the ability to:

- Define and explain the concept of a "stakeholder"
- Develop a list of stakeholders and their interests for a design project
- Explain how a marketing department or people contribute to the development of requirements
- Describe how to use community consultation to increase your understanding of a design project
- Use the stakeholder interests, community input to generate additional requirements, particularly objectives and constraints
- Use the stakeholder list and community input to add depth to the problem statement
- Analyze an existing design with respect to its stakeholders (i.e., describe the stakeholder interests of an existing technology)

Recommended reading

Before this module:

- **Design Process > Requirements > 5. Documenting the Context**

After this module:

- **Design Process > Requirements > 7. Describing, Users, Operators, and Clients**

1. Introduction

Stakeholders are people or organizations that have a stake or *interest* in the technology you are creating. A ***stakeholder interest*** is an aspect of an organization or person's ongoing welfare that might be affected by a design. It may be economic, physical, or psychological. An interest implies that there may be benefit or loss due to the design. Defining the design problem also requires identifying the various stakeholders in the project. We mentioned stakeholders when we discussed constraints and they play a part in virtually every step of the design process.

The three most obvious stakeholders are the client, the users (or operators), and the engineering design team. In fact, users and client are so obvious, and so clearly discussed in other parts of the design requirements, that we do not list them under stakeholders. This may seem odd, but it is a convention. It would be redundant to list them again in the stakeholder section after describing them so thoroughly in other sections of a design report. The engineering design team is also, obviously, a stakeholder. The quality of the work they do will impact their reputation, and possibly their paycheck. However, the design team is never listed as a stakeholder. Again, this is a matter of convention. Remember that the requirements are specified to help you define the needs of the design project explicitly and to articulate an agreement between the engineers and client about what is being designed. It is expected that you already know your own needs, and it isn't necessary to explain them to your client. So these most obvious stakeholders are not actually listed in the stakeholder section of the requirements. This section details the interests of the other people and organizations that have a stake in the project.

The kinds of interests that the stakeholders have in the project generally fall into categories related to economics, ethics or morality, legality, human factors, social impact, and environmental impact. The stakeholders can be individuals, organizations (companies, non-profit organizations, or professional organizations), or government agencies. In this context non-profit organizations (such as charities) are often referred to as ***non-governmental-organizations (NGOs)***. The stakeholder section of the requirements should list each of the stakeholders that should be taken into account in the design process, and the interests that they represent. Plants, animals, and the environment are not stakeholders, but are represented by NGOs with an interest in them (e.g., the Humane Society, World Wildlife Fund, Sierra Club, and similar organizations). The point of this list is to alert the client to the existence of these interests and remind the design team that there are other perspectives and people who will have an impact on the success, or failure, of a technology.

Certain disciplines of engineering are particularly sensitive to stakeholder concerns. Civil engineering, for example, often deals with large infrastructure projects that will impact a whole community or region. Often these types of projects require environmental impact statements and community consultation. Insufficient attention to stakeholder interests on these types of projects can result in the project being delayed for decades or even cancelled [1]. Gathering information to develop a complete list of the stakeholders early in the project, and following this up with ongoing consultation with these groups, can significantly improve the chances of success for projects of this nature.

Once you have identified the stakeholders and their interests, these can be used to add to the list of functions, objectives, and constraints in your project requirements. For example, the economic interests voiced by the chamber of commerce may be used to create a requirement that the plant be expandable to handle additional load resulting from future demand. A requirement of this nature would require consultation with the client before including it. Or the interests of the local community organizations might result in the addition of an objective to maintain some park land area at the site with an improved picnic area. Listening to the residents and taking their concerns into account can result in a set of requirements that motivates a more innovative and improved design solution than would have resulted otherwise.

TABLE 1 Example of stakeholders and their interests for a municipal water treatment plant.

Example: A new water treatment plant is being planned. It will be located on a lake shore. The plant is intended to deliver clean drinking water to a community of 30,000 people, replacing an aging plant that is no longer able to meet the growing demand from the community.

Users: Public that will be getting the clean drinking water.

Operators: The people who will be running the plant.

Client: The town or municipality if this is a public water system.

Table 1 shows just a small sampling of some of the other people and organizations who would have an interest in this project and would probably want their concerns taken into account by the design engineers.

STAKEHOLDER	INTEREST
Community organizations	There may be special interest groups in the town (e.g., people whose homes are located near the planned plant site), who will voice concern. Their concerns may be: – Economic: impact on their house and land values – Economic and social: job opportunities as plant operators – Environmental: impact on their local natural environment – Ethical and social: concerns about impact on local historical sites, proximity to religious sites (a nearby graveyard), impact on traditional sites used for social activities (e.g., removal of a park and picnic area), and aesthetic impact
Regulatory agencies at the state or federal level	– Legal: compliance with state and federal regulations
Environmental NGOs	– Environmental: concerns about the plant, business development and population growth, and the pressures of these on the natural environment around the community including flora and fauna – Ethical: concerns about the trade-off between environmental impact and public health.
Local sporting clubs	– Social and environmental: concerns about impact on fishing, hunting, or water sport activities
Chamber of commerce or other local business associations	– Economic: interest in having clean water available for business development and population growth
Local health agencies	– Social and ethical: interest in the health and welfare of the community, which depends, in part, on the availability of clean drinking water
Tim and Sara Murphy	– Economic and human factors: Mr. and Ms. Murphy's house is going to be seized by eminent domain because it sits on the planned site for the plant. They will be compensated. However, they obviously have an interest in fair compensation, and the physical and psychological upheaval that will result from having to move.

Having details about the stakeholders in your design documentation will help the readers understand the sources of the various requirements you generate, and later the reasons behind the design choices you make.

2. Design for the Community and Community Consultation

In addition to the natural environment, an engineering project or design can impact human society. In particular, the local community may be affected by the construction or implementation of a technology. This is especially the case for infrastructure, plants, and facilities development projects (e.g., roads, power stations, pipelines). While technology can bring positive changes to a community, it can also have negative impacts and generally there are trade-offs associated with every project. It is ethically appropriate to inform people about plans that may impact their community, and it is advisable to involve the community in the planning process when the project will directly impact the local environment. In many countries, a lack of community involvement can result in a project being delayed by legal action, or cancelled. This is because many jurisdictions require community consultation as part of the engineering process for projects that have the potential for impact.

Community consultation is also called public or stakeholder participation, or community involvement. A number of organizations offer guidelines and practical advice on community consultation:

- International Association for Impact Assessment (IAIA): www.iaia.org/publications/
- The International Association for Public Participation (IAP2): www.iap2.org/

Codes of practice can also be found at government websites (the Environmental Protection Agency or Ministries of the Environment).

The *stakeholders* that may be interested in participating in the decision-making process that goes into the design and implementation of a project can include the public, local companies and organizations, government agencies, and *non-governmental organizations* (NGOs). All of these groups should have the opportunity both to learn about the project as it goes through the design process, and also to have their views and perspectives on the project heard and used as input to the design process. Public consultation can help the engineering team gather information about the service environment and other aspects of the problem requirements. Consultation can also help them identify policies, standards, licences, permits, and other approval processes that they will need to complete the project. Community input can aid in the development of impact plans, and improve the quality and effectiveness of a design.

Working with the community through all stages of the project may improve the community support of the project. However, even if full consensus is not achieved, an open, honest, and transparent process can assist in the building of trust and improve understanding on all sides of the issues involved. Communication with the public should begin early in the design process, and continue through *commissioning* of the technology. Commissioning means getting a plant or system up and running after it has been constructed. Ideally the engineering team will follow up with the stakeholders after the design is implemented to assess the impact and address any lingering issues. In a sense, the community, especially in *public works projects* or plants, should be part of your team.

Some best practices for community consultation (adapted from [1] and [2]):

- Welcome and encourage appropriate community participation
- Consultation is a two-way communication process; it involves teaching and learning, telling and listening
- Consultation is useful at every step in the design process
- Consultation should be open to everyone and honest
- The more significant the potential impact, the greater the need for more frequent, and deeper participation
- Don't "dumb down" information for people, but communicate at a level and in terms that are understandable and respectful of your audience
- Address conflicts early and use mediation if possible to resolve these
- Include the cost and time for consultation in your project plan

---

Advanced Material

Methods for encouraging participation (adapted from [1] and [2]):

- Open house or exposition: show posters and exhibits, have information flyers or brochures available, and people there to answer questions, listen and take notes, and explain the project. This can be as simple as a table set up in an effective location (e.g., near the site of the planned design installation or at a local gathering place such as a shopping center).

- A public forum or meeting: usually involves a presentation by the project team, followed by a question and answer period. A public forum can also be done virtually (e.g., using a tool such as Textizen).

- Discussion groups provide an opportunity for a smaller group of people to sit down together and discuss some aspects of the project in more depth. Types of discussion groups can include advisory groups, task forces, consultation groups, focus groups, and others.

- Workshop is a short course that also actively involves the participants in the planning process.

- Surveys and interviews are particularly useful in the post-implementation phase to get feedback on how well (or not) the technology is functioning for people. Surveys and interviews are usually carried out by professionals in the social sciences and business, not by engineers.

- Newsletters

- Newspaper stories, public notices, and websites

- Bargaining, mediation, or negotiations are generally used to resolve conflicts or disputes.

To be effective, a combination of methods should be used. Also, if it is feasible, they should be repeated more than once to reach and get feedback from as many people as possible.

3. The Role of Marketing

For projects in your first few years of engineering school the categories discussed in this chapter are probably sufficient for you to form an understanding of the problem and document it. However, for advanced projects and projects in industry, there will be additional areas that will need to be researched and expressed to define the design problem.

One of the most common considerations that occurs in industry, but not often in university projects, is marketing. The marketing department in a consumer or commercial product or services company will play a significant role in the development of requirements for a new design. Typically the marketing department will contribute objectives for the project. However, the marketing department may also insist on the addition of functions, and may apply constraints. The most common constraints will be in regard to cost, and the relation of the features of the technology you are designing relative to current technology on the market. They may insist, for example, that the new product or service be faster than the competing technology already on the market. If your design does not meet this requirement, then the marketing department may decide not to support the continuation of the design project. Marketing can also be a major source of requests for design changes later in the design or implementation process. This is one source of **scope creep**, which is when the requirements for a project are changed or the functionalities increased while the engineers are in the process of designing or building a system. It is easier to cope with these requests if your team has a clear set of requirements and uses some design for flexibility strategies.

4. Leaving This Process Module

After working through this module you should be able to:

- Develop a list of stakeholder interests for a design project.
- Use the stakeholder interests to further develop the problem statement and list of requirements for a project.

The description of the stakeholders should contain the people and organizations that are most likely to be important for the type of project on which you are working.

KEY TERMS

stakeholder	stakeholder interest	non-governmental
commissioning	public works projects	organizations (NGOs)
community consultation		scope creep

5. Questions and activities

1. Explain what a stakeholder is and what a stakeholder interest is. Give some examples.

2. What role do stakeholder interests play in the design process?

3. Pick a design problem or existing technology. Identify stakeholders in the design of this technology and their interests.

4. Consider the following list of design needs.

a. Write a list of stakeholders for each.

b. Identify the stakeholder interests in each case.

- A user wants a means of carrying her soccer cleats without getting dirt on her clothes or other belongings.
- A company wants a means of organizing their warehouse to make it as efficient as possible for assembling orders for customers from items on the shelves.
- A company wants to design a web-based tool that suggests a car model to a customer based on their priorities and needs and the current available car models.
- A user wants a means of moving their cello from their home to the location of their weekly cello lesson 10 km away.
- A company want to create a new brand of shampoo that works without water.

5. Develop a complete list of stakeholders and their interests for the design project you are working on currently.

6. References

[1] "Paradise Crossed" by Ray Bert, Civil Engineering Magazine - ASCE, Vol. 68, No. 7, July 1998, pp. 42–45. http://cedb.asce.org/cgi/WWWdisplay.cgi?112937

[2] International Association for Impact Assessment (IAIA) (2009), publications [online]. Retrieved August 22, 2011, from www.iaia.org/publications/

[3] International Association for Public Participation's (IAP2), homepage [online]. Retrieved August 22, 2011, from www.iap2.org/

7

Describing Users, Operators, and Clients

#process module: #users&operators

Learning outcomes

By the end of this module, you should demonstrate the ability to:

- Define and explain the concepts of a "user" and an "operator"
- Identify the users and operators for a design project
- Describe the users and operators in a manner that is respectful, and identifies characteristics of these groups that are important to the development of the design project requirements
- Use the identified characteristics of the users and operators to generate additional requirements, particularly objectives and constraints
- Use the user and operator characteristics to add depth to the problem statement
- Analyze an existing design with respect to its users and operators (i.e., describe the characteristics of the users and operators that the designers of an existing technology likely took into account in the design process)

Recommended reading

Before this module:
- **Design Process > Requirements > 6. Stakeholders**

Alongside this module:
- **Design for X > Human Factors > 1. Design for Human Factors:** Introduction

After this module:
- **Design Process > Requirements > 8. Characteristics of Good Requirements**

1. Introduction

The terms **users** and **operators** describe the people who will be using and operating the technology you are designing. The term **client** describes the person or organization that commissioned the design. The users, operators, and client for your design play an essential role in the design problem and should be described explicitly. This is an especially sensitive section of the problem description because you must describe the client back to themselves. The descriptions of the users, operators, and client must have a tone of respect; this is an important aspect of the professionalism exhibited in your work. As much as possible, it should use the terms that the users and operators use to describe themselves.

> **Example: Using Unbiased Language that Values People**
>
> Suppose you are designing an online system that is being used by an organization that provides counseling for people being treated for cancer. They want to provide a confidential chat system for their clients. If the organization refers to their clients (the users of your system) as "cancer survivors" then that is how you should describe them—not, for example, as "cancer victims." It may seem like a small difference to you, but for the organization that is reading your proposal and deciding whether to hire you to design their system, it is a big difference.

1.1. The Users and Operators

The description of the users and operators should include as much detail as possible about the people (or other beings) who will be using the design. This should not be overly narrow, focusing on only a single culture, for example. It can be a difficult task to avoid being too narrow while being as detailed as possible, but try to think about it in terms of what all of the users or operators have in common. Often the users and operators are the same people, but sometimes they are different groups of people who should be described separately (#humanfactors).

For example, consider a new piece of surgical equipment, such as a new type of surgical staples used to suture a laceration or incision. The medical staff (i.e., doctors and nurses) is operating the technology and therefore using it. However, the patient is also a user, so that perspective should also be considered in the design process.

The description of the users and operators should include physical, psychological, social, and organizational factors [1]. You also must show respect for the users, no matter who they are or how you personally feel about them. Consider the following questions.

Physical

- Who are the users? What do they have in common?
- What is their height range, mobility, and other anthropogenic data? These data are available in handbooks and online.
- What are their physical characteristics? This is important because many designs rely on some aspect of the users' physical characteristics.

You will need to comment on the assumed physical traits of your users and operators further once you have begun to develop a design solution, but this information will come in your conceptual design documentation.

Psychological

- What languages do the users speak?
- What level of education do they have (e.g., reading ability, computer or math skills if these might play a role in your design)?
- If your design will be operating in a special environment (such as an industrial workplace), do the users have a special set of skills or special knowledge? Do they receive special training?
- What is the cognitive ability range of your user group? This is particularly important if your system is going to rely on the cognitive ability of the user or if the design is for a set of users with different cognitive abilities from your own (such as children).
- What are some of the other "psychological norms" that may play an important role in the design problem you are working on? Is there an expectation of a design based on the normal operating environment, such as large buttons for equipment in a steel mill where the steel rolling equipment is huge and heavy?

> There is a temptation to send out a questionnaire or do a poll to get information about users. However, designing questions to collect good information that do not introduce bias is difficult, and requires specialized training. Except for simple cases, it is likely beyond the expertise of most students and many engineers.

Keep in mind that psychological norms are often culturally specific. For example, in North America it would be "normal" to design a switch where the up position is on and the down position is off. But in other parts of the world up is off, and down is on. Imagine how frustrating it would be to use a device that violates the basic psychological norms you are used to.

Social

- What are the social norms of your user or operator group?
- Do they share a common culture? If so, what are some of the aspects of that culture? Keep in mind that culture does not need to be based on the culture of a country or location; it can be based on a community.

For example, there is the "culture" of university students around the world who share some commonalities (e.g., exams, courses, dealing with professors, social norms). These things may be somewhat different from place to place, but many of the aspects of the culture are common. In fact, Facebook is an example of a technology that was designed explicitly for the common social culture of university students worldwide. The description of the social aspects of your user or operator group can be essential to your design depending on the nature of your project.

Organizational

- Do your users or operators all belong to the same organization? For example, do they all work for the same company? Or are all citizens of the same country? Or belong to the same profession?
- What are the values and goals of this organization (if the design project is focused on users in a particular organization)?

For example, suppose you are designing a system that will be used by doctors and nurses. You would want to make sure that your design fits the values and goals of the medical profession. Here, in the users section, you would briefly describe the organization from the perspective of the users. In this example you would briefly describe the medical profession (values, culture, and ethics) from the perspective of medical practitioners (i.e., doctors and nurses). This would help you keep in mind the users' perspective when you begin to design the system.

These organizations do not have to be formal. For example, if you are designing a bicycle accessory, you will be dealing with the biking community. In general their aims and attitudes may differ somewhat from those of other users of the roadways, such as drivers and pedestrians. Your design choices may be directed by those differing aims and attitudes.

1.2. The Client

If the client is different from the users then you need to also describe the client. That is, if the client is an individual or organization who is commissioning the design but will not directly be using the design, then you need to describe the client. The description of the client is typically briefer than the description of the users and operators. It describes only the essential features of the client, and those aspects that pertain directly to the project. We recommend that the client description include a statement about the client's ethical beliefs and values in your own words. This is because whatever you are designing must fit with the client's ethical beliefs and values to be adopted.

1.3. Getting to Know the Users, Operators, and Client

Even for professional design engineers it is very difficult to design systems that will be used by people different from themselves. And the more different the user and client are from the design engineer, the more difficult it is for the engineer to imagine how the user and client will perceive and make use of the design. It is therefore critical for the design engineer to investigate who the users and client are and take the time to get to know their needs. It is not unusual now, for example, to find engineers in the operating room with surgeons observing how medical instrumentation is used; on the floor of a manufacturing plant talking to assembly line workers; or working with development groups in rural villages to understand the perspective of the people who will be using the technology.

Equally, it is easy for an engineer to misjudge a design project if the users are very similar to the engineer. This is because the engineer might assume that all of the users will approach the design, and will perceive it, the same way they themselves do. They make the mistaken assumption that all of the users are exactly identical to themselves.

FIGURE 1 Interviewing users of a potential design or observing them using a related technology is a method for gathering information.

To avoid these mistakes, the design team must take the time to understand the users, operators, and client. There are several strategies for doing this:

1. *Try it yourself.* If you are designing a new system to replace an existing technology, try using the old technology yourself. Become the user. Observe what you do and how you do it. Talk out loud and make lots of observations about the experience. Sometimes an experienced operator will neglect to tell you the lessons they learned when they first started using the system, habits that are now routine for them.

2. *Gather information.* Use standard resources to gather information about the client and users.

3. *Live it.* It helps if you have lived with the users and worked with them. A design fits into an operational culture, and the better you know the culture, the better your design.

4. *Interviews.* Interview the client and users about their perception of the existing systems and the need for a new technology (see Figure 1). And ask them about themselves. You need to know, however, that many users will report that an existing technology is "just fine" but if you observe them using it you might see a very different situation. Talk to the people who are very hands-on with the technology. Maintenance people often will be able to tell you more about an existing design than top-level managers (but don't tell the managers this, of course!).

5. *Observation.* Watch the user using an existing system and ask them to tell you about it while they are using it (i.e., in situ observation). This is part of what is called **task analysis**. Be sure the users know that you are not judging their operation or jobs but are just there to learn.

Use Cases, Task Analysis, and Benchmarking

Here is a brief overview of more methods for understanding the user's and operator's perspective (search #usecase, #taskanalysis, #benchmarking for more information).

A **use case** is a description of how a person or system achieves a goal using the technology you are designing. A use case describes the interaction.

A **task analysis** is an analysis of the way a user performs a task. This is usually done by observing or videoing a user performing a task. Then the engineer writes up an analysis of what the user did and how they did it. This is especially useful if you are designing new technology to replace an old way of doing something.

Benchmarking is the analysis of an existing technology. This involves the engineer, or others, using an existing piece of technology that has similar functionality to the design project they are working on. The engineer observes the interaction and analyzes it (like a task analysis); how long does it take to do something, how

(continued)

difficult it is, what are the relevant performance features of the technology. Then the engineering team dissects the existing technology to understand why and how it was designed the way it was.

Increasingly companies that have a design focus are employing people who specialize in understanding other people's perspectives. These include psychologists and cultural anthropologists who might work with the design team to help the team understand their users.

What you and your team experiences, measures, and sees themselves is often more reliable and more revealing than information that comes second or third hand, relayed (and interpreted) by others. This list probably covers more than you might be expected to do for a design project early in your engineering schooling. However, understanding the users, operators, and clients for your design is a critical step to creating a technology that will be accepted and used.

Many designs are unsuccessful because they solve the wrong problem. Sometimes it is because the technology is designed for the engineers themselves rather than the user. What engineers think is cool, usable, or necessary frequently isn't for other people. A good design engineer listens to the client and the users, and constantly keeps their perspectives in mind while working through the design process.

Often information about the operators, users, and clients is included in the design documentation. This helps the reader understand why certain requirements were included. Inclusion of use cases, task analysis, and benchmarking information can also help to deepen your understanding, and your client's understanding, of the design problem.

Once you have described the users, operators, and clients for the technology you are designing, you then need to use this information to further develop your requirements. In particular, the information about the people and organizations that will be interacting with the technology should enter into the objectives and constraints for the design problem. Think through the technology from their perspective and note what features and hazards are associated with the use or misuse of the technology you are designing. For example, if your user is a child (e.g., you are designing toys for children) then you need to take into account that the user may not be able to read warning labels, or be responsible for following a warning. Therefore, your constraints list should require removal of all possible hazards (include a specific list) and compliance with appropriate regulations regarding children's toys. In addition, the psychological profile and developmental information about children should guide the development of objectives that would improve the potential success of your design. You might also note unintended uses of the toy (eating it, throwing it, pounding it, and putting food into it) that need to be taken into account in the design.

2. Leaving This Process Module

After working through this module you should be able to

- Describe the characteristics of the users and operators for a design project
- Use these characteristics to further develop the problem statement and list of requirements for a project

Describing Users, Operators, and Clients

The description of the users and operators is a highly sensitive part of an engineering report. Care should be taken to use fair, unbiased language that values the diversity of people who may be interacting with the technology you are designing.

KEY TERMS

users	operators	client
use case	benchmarking	task analysis

3. Questions and activities

1. What role do users and operators play in the design process?

2. Pick a design problem or existing technology. For this technology identify users and operators of the technology. Be sure to include all of groups that use and operate technology (e.g., maintenance and repair people). Describe these people at all levels (physical, psychological, etc.)

3. Consider the following list of design needs. Write a list of users and operators for each.

- A user wants a means of carrying her soccer cleats without getting dirt on her clothes or other belongings.
- A company wants a means of organizing their warehouse to make it as efficient as possible for assembling orders for customers from items on the shelves.
- A company wants to design a web-based tool that suggests a car model to a customer based on their priorities and needs and the current available car models.
- A user wants a means of moving a cello from home to the location of weekly cello lessons 10 km away.
- A company wants to create a new brand of shampoo that works without water.

4. Develop a complete description of the users and operators for the design project you are working on currently.

4. References

[1] K. Vicente, *The Human Factor: Revolutionizing the Way We Live with Technology*. Toronto: Vintage Canada, 2004.

Characteristics of Good Requirements

(8)

#review/reflection module: #goodrequirements

Learning outcomes

By the end of this module, you should demonstrate the ability to:

- Identify the DFX considerations that are likely to be important in a given design project
- Review a set of requirements for essential characteristics such as completeness, independence from implied solutions (i.e., solution independent), and testability
- Define and explain "project scope"
- Apply the concept of "project scope" to describe the scope of a given design project
- Perform a preliminary evaluation of a design problem to identify if it can be broken apart into pieces and whether the some parts of the problem can feasibly be solved, are routine, and/or can be solved with "off-the-shelf" technology
- Also identify which parts of the design problem are closed and can be solved definitively, and which are open and require a creative design solution

Recommended reading

Before this module:
- **Design Process > Requirements > 7. Describing, Users, Operators, and Clients**

After this module:
- **Design Process > Requirements > 9. Summary; Putting It All Together**

1. Introduction

Development of the project requirements involves researching information, analyzing it, and organizing it. This process helps the team learn about the project and familiarize themselves with the specialized knowledge they will need to solve the problem. This is essential because of the complex and open-ended nature of design problems. It is also essential because the design process is an information and communication process. As a design engineer you need to make sure that your whole team has a clear, common understanding of the design problem; that you can communicate this understanding to other teams that you will work with on the project; and that you are in agreement with the client about the problem you are solving.

Once the requirements have been documented the design team should reflect on the design problem before moving forward. This *reflection* is part of the critical thinking that goes into a design project. The team needs to:

- Come to agreement on the problem that is being solved (i.e., agree with the requirements and that the requirements are sufficiently well formulated and sufficiently complete to move forward), recognizing that the requirements will be revisited as the process continues.
- Come to agreement on the *scope* of the project.
- Do a preliminary evaluation of the nature of the problem, which will help to determine the solution methods that are used.

2. Project Scope

One outcome of defining the design problem is the setting of the project *scope*. The scope is the definition of the breadth and depth of the problem you are proposing to solve. It defines the limits or boundaries of the problem. This is why defining a design problem is sometimes called a scoping activity, and in some industries the project requirements document is called a scoping document. It is important to review the requirements list you have written and make sure that the team agrees on the scope: what is within the scope of the project, and what is beyond the scope of the project. In industry you would also want your client, manager, and some of the other departments at the company (e.g., marketing) to sign-off on the scope before proceeding. This can help to defend against *scope creep*.

For example, your project requirements indicate that you will be developing a design that will take aerial photographs at social events (see #aerialphotography case study). You have worked up a list of functions, objectives, and constraints for this project. A device that will work with a Canon EOS-5D camera equipped with a variety of lenses is within the scope of this project. This is explicitly written into the requirements. The requirements written up for this project make it clear that you are intending to design a device that reliably works up to 500 m^2. Designing a device that must be capable of taking pictures that cover 750 m^2 is beyond the scope of the project. You are not promising to design a device that works reliably for areas greater than 500 m^2. The scope sets the expectations for the project.

3. DFX Considerations

The requirements are incomplete without taking into account DFX considerations (#DFX). Like the stakeholders section, information about DFX considerations will feed back into the lists of functions, objectives, and constraints.

DFX means Design for X, where "X" is a design consideration that is of particular importance to the specific project you are working on. Some common DFX considerations are:

- Design for human factors
- Design for the environment and community considerations
- Design for manufacturability
- Design for safety
- Design for durability
- Design for maintainability
- Design for intellectual property
- Design for aesthetics
- Design for flexibility

Look down this list carefully and identify whether any of these DFX considerations particularly apply to the design project you are working on. Think about any other DFXs not listed here that might be important in your project. Explain briefly why each of the DFXs will apply particularly to this design problem. Remember that at this point you are explaining how the DFX pertains to the design problem, not any particular solution. If there is one DFX topic that is especially applicable to the project then you may want to research this subject further to learn methods and standard practices that you can implement to create a more effective design solution. You should also use the DFX topics you have identified to add to your lists of objectives, functions, and constraints.

For example, suppose you are designing a new highway. Safety will probably be a key concern. You would add "design should be safe" to your list of objectives, if it isn't already there, later expanding this into testable requirements. There will also be a whole set of special codes and standards for the design of a highway that must be followed to improve safety. However, in researching design for safety you will also find that there are various common ways of approaching this DFX topic. You can remove hazards, guard against hazards, and warn against hazards. Knowing this information, you might then add further to the list of constraints: "specific exceptional hazards shall be removed, guarded, or warned against" (followed by a list of the specific hazards). So identifying this DFX consideration allows you to make your requirements list more complete and specific. It also reminds you later to specify the addition of the "Beware, moose crossing ahead" sign in the appropriate location.

After working through this chapter you should have at least a first draft of a problem statement and requirements for your project. You and your team should have learned much of what you will need to know to solve the design problem, and should have identified any additional knowledge you will need to learn. Defining the problem is an essential step in the design process and it is an iterative exercise. You will continue to revisit this definition repeatedly as you work through the project both to update the requirements and to remind yourself about the problem you are working to solve. A well-written set of requirements will communicate to your manager and your client the goals and scope of the project.

If this is a project for one of your engineering courses, the requirements document will demonstrate to your instructor that you understand the problem you are working to solve. It will tell your instructor or client what they can expect from the solution you are creating, and what not to expect. It will give you and them a way of measuring your success. The requirements will help keep your whole team, and other professionals that you are working with, on track.

4. Characteristics of Good Requirements

Good requirements have many characteristics. They are clear and unambiguous. They are relevant to the project. They have three essential characteristics that are evidence of good engineering practice: being complete, being solution independent, and being testable.

4.1. Completeness

The more complete your requirements are, the easier your solution process will be. Ideally, you would want perfectly complete requirements before moving to the generation of solution ideas. However, this is not only impossible to achieve, but the time it would take to reach this ideal is not worth it. At some point there is diminishing return on the time invested in development of the requirements. You just need to start working on the solution.

There are many things in engineering (and other fields) that seem to follow an approximate 80% to 20% rule, and a compromise on ideal requirements is probably one of these things. This means that the last 20% of the work you could do to achieve ideal requirements (if this is even possible) would probably take 80% of your time (i.e., four times longer than it takes to generate the first 80% of your requirements). This is a lot of time and effort for marginal improvement. So make sure your requirements are as complete and clear as possible, and really push toward the ideal of completeness, but don't expect to have a perfect set of requirements before you move on to idea generation.

> ### Other examples of the 80/20 rule
>
> In construction: the last 20% of the construction of a building will take 80% of the time.
>
> Working with people: 20% of the people you work with will cause 80% of your headaches.
>
> In fundraising: 20% of donors give 80% of donations during a typical fundraising campaign.

4.2. Solution Independent

The definition of the design problem (i.e., the requirements) must be as *solution independent* as possible. This means that the requirements should not imply a specific design solution, or be written with any one solution in mind. The requirements must leave open the possibility of solving the problem in as many different ways as possible. This is very difficult because in many ways defining the problem defines the solution.

In fact, some engineers argue that the effort and thinking needed to solve engineering problems well is often 80% defining a problem effectively and only 20% solving it (which is another example of the 80/20 rule). Creating solution independent requirements is a balance between a strategy of least commitment and a need for determinacy.

The ***strategy of least commitment*** states that at any point in the design process you should only make decisions that are "forced" upon you, which means don't be lazy in your decision making, be intentional. The strategy of least commitment is a strategy of leaving every possible design idea on the table for discussion until you rule it out through deliberate decision making. Good designers do not exclude an idea because they assume it won't work; they think it through carefully. They ask themselves, "Why not?" In the development of the requirements, the strategy of least commitment advises that you carefully write the requirements to allow the widest range of solutions possible.

Carefully examine each requirement you have written. What does the requirement assume? For example, if the requirement states, "The memory size must be at least 1 MB (one mega-byte)," you should be asking yourself:

- Why 1 MB? Why not less, or more? This is a specific number—what should be specified is what the memory has to do; 1 MB is an implementation decision (i.e., a ***means*** statement).
- Why a memory is specified at all? Is this also a means statement?

However, there is also a need for ***determinacy***. The problem with leaving your requirements too broad or general is that you may never converge on a solution. The requirements must be specific enough to guide you toward a solution, preferably a really good solution. The point of defining the design problem is to start to move an indeterminate, open-ended problem toward being determinate and closed-ended: that is, to actually begin to converge toward a solution. The skill in defining a design problem is putting together a thorough set of information about the problem that both clearly defines the problem and also leaves open a wide variety of potential solutions. A skilled design engineer balances the strategy of least commitment with the need for determinacy.

4.3. Testable

Read through your list of requirements again and make sure that they are testable using a ***metric*** (#metrics). If you have general functions, objectives, or constraints, make sure they are decomposed (detailed out) into a set of more specific statements that are testable.

For example, what does "user friendly" really mean? To an experienced photographer this may mean that it is easy to precisely set the shutter speed, focus, and aperture on a camera. To a less experienced or casual user, it may mean that all the settings on the camera are automatic; they can just press a button to get a good picture. There are no specific camera attributes that are denoted by the term "user friendly," nor are there for any other design. The general objective "user friendly" would need to be broken up into a set of specific measurable objectives to be useful. (See the section on objective trees.)

On the other hand, the statement "The design should be blue" is easy to determine. Blue is a specific wavelength of light. It can be measured. Requirements should have obvious, realistic methods of being tested. Some examples are shown in Table 1.

TABLE 1 Examples of requirements that are well formulated (good) and poorly formulated (bad).

GOOD/BAD	EXAMPLE	REASON
Good	The exterior should be yellow.	The color can be seen to be either yellow (pass) or some other color (fail).
Mediocre	The glue should hold the plastic parts together.	How well? Double-sided tape will hold the parts together, but not as well as most plastic glues. This statement doesn't give the engineer much to measure in terms of performance.
Bad	The interface should be user friendly.	Some people's user friendly is other people's boring and still other people's frustrating. There is no method of measuring "user friendly." See the comments preceding.
Good	The filter shall pass signals above 3 kHz with under 2% attenuation and reduce all signals 2 kHz and lower by at least 10 dB.	An engineer can run tests or simulations on the filter that will check to see if it meets these criteria.
Bad	The design should respond quickly.	Is this hours, milliseconds, or nanoseconds? How quick is quickly? This is not testable without further detail.
Bad	The product should be light and portable.	Light is a relative term. Portable just means movable. There is no clear pass or fail.
Good	The battery must perform for a minimum of 5 days when device is in standby mode.	There is a simple test (*metric*) to see if the device meets this requirement. The engineer can run the device on standby for 5 days on a battery and test if the device is still operable.
Good	The system should be down for no more than 4 h in a year.	While there is no simple test for this, there is a hard target given. Careful failure analysis done using a method agreeable to the client would be needed to determine if the system met this requirement.

5. Evaluating the Problem

Once your team has defined the design problem you should stop and ask yourself some questions that will help you decide where to go next. Going back to the Sandbox (see Figure 1), the office space of our fictional design company, you have just spent a lot of time in the library gathering information and in the boardroom talking to the client. You have also been out of the office observing users, and talking to operators and stakeholders. Now you have a much more thorough understanding of the project.

Your team should spend some time in the project space discussing the problem before you proceed. The natural tendency at this point is to want to start right in on solving the problem. But you will save yourself enormous amounts of time, effort, and frustration if you first reflect on the problem you are solving, and critically analyze the nature of the project.

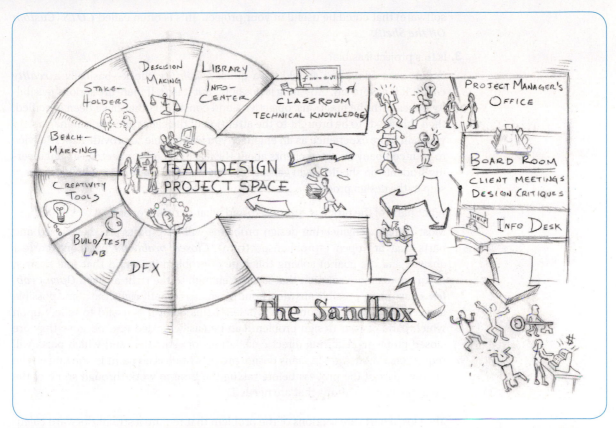

FIGURE 1 **The activities involved in an engineering design process can be visualized as spaces in a design company.** Where your team goes after defining the requirements of a design problem will depend on the problem you are solving.

1. Is this a *routine design* problem?

   Routine design means that there is a commonly accepted method for solving the problem (#routinedesign). Routine design problems occur frequently in industry. As you take more courses in your discipline you will be introduced to the routine design methods typical in your field. Examples of routine design include structural design for routine buildings (heavily driven by the building code and standard practices), minor modifications to an existing design, or algorithmic design (e.g., layout of a printed circuit board; component descriptions are entered and computer software does the layout for you, which may require minor tweaking after the software completes the job).

2. Is there an *off-the-shelf* solution, or components that could be used?

   "Don't reinvent the wheel" is a common cliché, but it is really true in engineering design. In examining your design problem, you may find that there are elements of the design that can be satisfied using off-the-shelf technology. It is worthwhile familiarizing yourself with commercial materials and components (hardware and software) that could be useful in your project. This is often called **COTS** (*Custom Off the Shelf*).

3. Is this project feasible?

Before proceeding you should do a quick *feasibility* check—basically a *reality check*. It is easy to say, "Anything is possible," but it isn't. If there are serious feasibility issues with the functions, objectives, and constraints that have been specified, then this needs to be brought to the attention of the client and your managers. As you gain more experience as an engineer you will be able to provide expert advice to your company and your clients about the feasibility of a project. As an engineering student you should begin getting into the habit of assessing feasibility at every step in the design process.

4. What parts of the problem are closed, and what parts are open?

Most complex engineering design problems will have parts are that closed and parts that are open (#problemspectrum). **Closed problems** have one correct answer, and the goal of solving this type of problem is to find that one answer, or an approximate solution that is close enough to the right answer. **Open problems** have many possible answers, none of which is exactly right, but some possible answers are better and some are worse. At this point it is useful to reflect upon which parts of your design problem can be easily tackled first because they are closed problems requiring direct calculations or estimates, and which parts will require creative design. In many design projects there is no point in starting on the creative part of the process before taking the time to work through some of the preliminary calculations that are needed.

The closed parts are sections of the problem that require a straightforward calculation process, or *sizing*. Sizing means doing a calculation to determine the capacity of the equipment or system needed for the job. For example, suppose you are designing a ventilation system for a building. You will need to estimate the volume of the building for this project so you can appropriately size the fans and other equipment that will be part of the ventilation system. Calculating the volume of the building is a closed-ended problem; it should have a correct answer, even if you are just making an estimate, which isn't exact. Estimating the volume is one of the closed-ended parts of this design problem. Deciding where the ducting should go, the types of equipment to use, balancing the trade-off between duct size and noise, and other aspects of the solution are open-ended problems. There are many possible choices and no definitive solution. Some choices yield a better solution and some choices will produce a solution that isn't as good. Once you have designed the duct system (this usually requires iteration and other procedures typical of an open-ended design process), and selected the air exchange rate, then sizing the fans is a straight-forward closed-ended calculation.

As you reflect on the four preceding questions, document your thinking in your engineering notebook. You may think of ideas that are not immediately useful at this stage of the process, but will become useful later. Make sure you also have adequate discussion with your team about the design problem; this is called ***group processing***, and it is essential for good team performance.

KEY TERMS

scope

strategy of least
  commitment

metric

off-the-shelf

feasibility

sizing

scope creep

means

reflection

custom off-the-shelf (COTS)

closed-ended problems

group processing

solution independent

determinacy

routine design

reality check

open-ended problems

6. Questions and activities

1. Define and explain the concept of "scope."

2. Describe the scope of the design project you are working on currently.

3. Pick a design problem or existing technology. For this technology, identify specifically what DFX considerations went into the design of the technology and how are they manifested in the design.

4. Identify the DFX considerations that are important in the design project you are working on currently.

5. Review a given set of requirements for completeness, solution independence, and testability.

6. Review the requirements you have written up for the design project you are working on currently. Are they complete and solution independent? Are all of the requirements testable?

7. Review the design problem you are working on currently.

   a. Can it be broken up into smaller sub-problems? If yes, which of these sub-problems are open and which are closed-ended?

   b. Is this whole design problem routine? Are any of the sub-problems routine?

   c. Is the project feasible?

9

Summary:
Putting It All Together

#review/reflection module: #requirementssummary

Learning outcomes

By the end of this module, you should demonstrate the ability to:
- Summarize the elements that constitute a complete set of requirements
- Identify whether you have completed the requirements for your design project sufficiently to move on to the next phase of the design process

Recommended reading

Before this module
- **Design Process > Requirements > Modules 1 through 8**

After this module
- **Design Process > Idea Generation > 1. Introduction to Idea Generation**

1. Introduction

Developing the lists of the functions, objectives, and constraints for the design project is an essential step and will form the basis for generating solution ideas, evaluating those ideas, and eventually choosing an idea or ideas to develop into a detailed design. As with other parts of this process, the development of the functions, objectives, and constraints is highly iterative. These sets should be revisited and refined throughout the design process. What these lists are called, the terminology used (i.e., functions, objectives, and constraints), matters less than understanding the purpose that these statements serve. They serve to clarify the problem as much as possible for the design team, the client, and other stakeholders in the process. They set a target that everyone can work toward. Requirements that are ambiguous or incomplete can lead to inefficiency in the process and to frustration, as people try to figure out what is being designed as they go along (see Figure 1).

It is worthwhile including in your documentation the reasoning behind your stated requirements. At a minimum, the reasoning should be recorded in your engineering notebook. It is our experience that requirements may be dropped later as being unnecessary restrictions on the design if there is no supporting documentation to remind the team why the requirement was included. This can result in the failure of the design if, in fact, the dropped requirement was essential.

TABLE 1 **Summary: Project requirements**

Problem Statement: A clear statement that defines the design "need" and usually encapsulates the requirements and context of the design project.

REQUIREMENT TYPE	DEFINITION	NOTES
Function	What the design *must do*	Active statements. Functions are necessary for any acceptable design solution
Objective	What the design *should be*	Descriptive statements. Objectives help differentiate and rank the various design solution ideas
Constraint	What the design *must be*	Constraints are required by society or by the client. Legal requirements; cost requirements; overall system or situation requirements (such as "must work with a particular operating system")

Project Context: Informs the design requirements. The context adds information about the circumstances in which the technology will operate. This information is used to further develop the functions, objectives, and constraints

CONTEXT FACTOR	NOTES
Service environment	Describes the environment in which the technology will operate, including the physical and virtual aspects of the environment
Stakeholders	Describes the people and organizations with a stake (called an "interest") related to the technology that is being designed
Users and operators	Describes the characteristics of the people and groups who will be using and operating the technology that is being designed

FIGURE 1 **When specifications are not clearly stated, the team may not be working toward shared goals.**

Other ways to categorize requirements: *goals*, **constraints, and** *criteria*

The use of the categories "functions, objectives, and constraints" is just one of many different ways to organize the requirements. Another organization that is used in product design is *goals*, constraints, and **criteria** [1]. In this framework constraints are defined in the same way as we have defined them here. Goals include both functions (called functional goals in this framework), but also include objectives (called non-functional goals). And criteria in the goals/constraints/criteria framework describes the directionality of the goals (e.g., in cost *less* is usually better, but in speed *higher* is often better, depending on what you are designing).

This is only one example. Different disciplines and industries will have their own way of categorizing the requirements for a design project, and these categories are evolving. However, the purpose of the requirements documentation is always the same: to clearly define the design problem.

2. Leaving This Stage

Leaving this stage of the process you should have:

A well written problem statement that clearly explains the need and scope of the project.

A list of requirements with supporting documentation[1] and explanations:

- Functions: a well-organized list
- Objectives: a tree and a prioritized list
- Constraints: a list with supporting references

Supporting documentation and explanations for the requirements are typically included in the design reports you prepare for your instructor (and manager and client).

For projects that have a known context:

- A complete description of the service environment
- A clear description of the users, operators, and client for your project
- A list of the known stakeholders and their interests
- A list of the DFX considerations

Every element in the context should include supporting documentation. This supporting documentation is typically included in the design reports you prepare for your instructor (and manager and client).

Reflection on the project:

- Ideas on testability of the requirements
- Notes-to-self about the nature of the design problem: the feasibility, sub-problems that need to be solved, off-the-shelf materials, and so on

---

[1]Note: Supporting documentation means the evidence or information that you have gathered that supports your ideas. The explanation explains your thinking process: the logical steps you took to get from the information to your ideas (i.e., your requirements and conclusions about the context).

- Notes-to-self about further research that needs to be done on the problem to enhance the solution generation process

The reflection on the project is typically not included in design reports. It takes the form of notes in your engineering notebook, or other team documentation to remind yourself of things that will be important as you move through the process, and for when you revisit the requirements and problem statement.

KEY TERMS

goals   **criteria**

3. Questions and activities

1. Review the project requirements for the design project you are currently working on. Do they include all of the necessary components and fit the criteria listed above?

4. References

[1] P.H. Roe, G.N. Soulis, and V.K. Handa, *The discipline of design*. Boston: Allyn and Bacon, 1967.

1

Functional Basis

#skill/tool module: #functionalbasis

Learning outcomes

By the end of this module, you should demonstrate the ability to:
- Apply the functional basis technique to identify functions for a design project
- Apply the functional basis technique to analyze an existing technology and compare/contrast technologies and how they operate

Recommended reading

Alongside this module:
- **Design Process > Requirements > 2. Functions**
- **Design Process > Idea Generation > 4. Morphological Charts, Analogy, and TRIZ**

1. Functional Basis

A number of people who study design in engineering have noted that the functions of virtually all products can be described using a small set of terms [1]. This idea began to be developed nearly a century ago and has been an active area of study in the last few decades as people seek ways to model design processes computationally. It is a powerful technique for dissecting an engineering design problem and understanding it from a functional perspective. The ***functional basis*** method is generalized here to be applicable to all types of engineering design projects.

Everything in the world can be described in terms of its mass, energy, and the information it carries. This perspective on ecological systems was first developed by biologists in the 1970s. More recently engineers have made use of this perspective in design. Technology can be described as systems to control and manipulate mass, energy, and information for a specified purpose. Generally the desired effects of technology are safety, keeping people healthy and well, and enabling people or living things to pursue the lives they want to live. Not just products, but all technologies can be analyzed using this framework. Mass, energy, and information are sometimes called ***flows*** but they do not always move. So instead we will refer to these as ***components***. All technologies are made up of these components and operate on

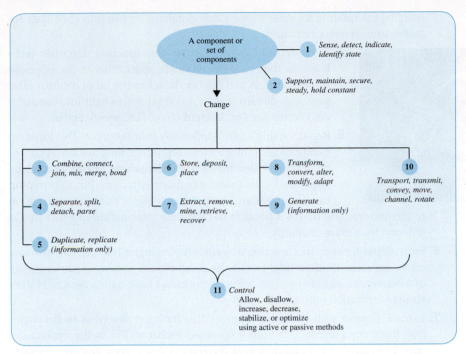

FIGURE 1 Functional basis: The essential functionality of virtually every technology can be described in terms of these essential actions.

components of mass, energy, and information. You can use this framework to describe the functions of any design project.

There are basically 11 types of operations (i.e., functions) that a technology can perform on a set of components. Nine of the operations are applicable to mass, energy, and information, and two apply only to information. The overarching goal of these operations is control, which is itself a function. These 11, including control, are shown in Figure 1. For each function several synonyms are given to help fully describe the meaning of the function.

1. Sense, detect, indicate, or identify state: This function describes technologies that do not act on mass, energy, and information, but are only intended to indicate the state of the components. Examples include sensors, detectors, gauges, meters, scanners, indicators, and test equipment or algorithms. These technologies assess some aspect of the components. Usually the output is information about the state of the system.

2. Support, maintain, secure, steady, hold constant: This function describes technologies that act to prevent change. Technologies of this type are intended to keep information, mass, or energy from moving or changing; for example, preservatives, structures, or secure storage systems.

3. Combine, connect, join, mix, merge, bond: This function describes technologies that act to assemble separate components into a single unit. This could include

Courtesy of the Authors

Functional Basis

97

merging information together into a single database, or reacting chemicals together to get a new compound.

Courtesy of the Authors

4. Separate, split, detach, parse: This function describes technologies that take components apart. This is the opposite of "combine". A can opener detaches the lid of a can, a filter separates dirt from air, a prism separates light into constituent components of different colors (i.e., wavelengths).

5. Replicate, duplicate: *Applies only to information.* This function describes technologies that reproduce information. A photocopier duplicates information. Mass and energy cannot be duplicated, only transformed or rearranged, but information can be duplicated. This process does not add new information to the universe, but makes it possible to distribute information in a way that is different than mass or energy.

6. Store, deposit, place: This function describes technologies that put components into a specified location (real or virtual). Examples: A recharger stores energy in a battery; a hair care product deposits chemicals onto hair; a backup system stores information onto a hard drive.

7. Extract, remove, mine, retrieve, recover: This function describes technologies that bring components out from a specified location. This is the opposite of "store." Oil drilling, data mining, heat recovery systems, recycling are all examples of extraction or recovery technologies.

8. Transform, convert, alter, modify, adapt: This function describes technologies that change the component, or components, from one form to another. Clothing dye transforms cloth from one color to another color, speakers convert electrical information and energy into sound waves, a drill press modifies the shape of an object. All of these are technologies that change the form of the component they operate on.

9. Generate: *Applies only to information.* This function describes technologies that put mass, energy, and/or information together to generate new information. What is created is not simply a sum of the parts, but actually contains new emergent information. An expert system in a nuclear plant combines information from many different sensors to reach a conclusion about the plant's operating mode. The system is not simply putting the information together, but actually using that pattern of information to draw conclusions and make decisions. Similarly, a prototype (such as a collection of mass put together in a new way) can generate information about a design.

Can Information Actually Be Created or Destroyed?

Information theory says that information cannot be created or generated, which we agree is true in an absolute sense. However, from a practical perspective, technology can generate new understanding and meaning for the user. So from this perspective technology can generate meaning and understanding that would not be readily evident to someone looking at the pieces of information together. We can say that information has functionally been created from the user's perspective. This is related to the concept of emergence.

(continued)

Similarly, information cannot be destroyed. However, practically, technology can be used to render information unperceivable: unreadable, or impossible to recover. If a unique manuscript is incinerated, it is not humanly possible to recover the information it contained. From a functional basis perspective, the information has been *transformed* into an unusable state.

In this text we have chosen to add "generate information" to the list of functional basis terms recognizing that this violates information theory, but is practically important for describing expert systems and other technologies. We have also chosen not to include "destroy information," but instead to view this as a transformation of information to an unusable form. This is an artificial but practically useful way to describe kinds of technology design that are currently important.

10. Transport, transmit, convey, move, channel, rotate: This function describes technologies that move components through space and time. Radio waves, trucks, and power lines are all examples of transport technologies. These technologies carry or convey mass, energy, or information from one place to another.

11. Control: Control is an essential function of virtually all technologies. The goal of controlling the functionality can be to allow or disallow (prevent) change or transport, to increase or decrease change or transport, to stabilize a process or component, or to optimize a process or functionality. And control systems can be active (requiring energy) or passive.

Examples

Pedestrian crossing: a stoplight actively *controls* the flow (*transport*) of people across a street, a pedestrian bridge passively *controls* the flow of people.

Heating a building: A window facing south passively increases solar energy *gain* in a building. A solar collector that uses pumps to carry heated water through the building is actively increasing the *transport* of solar energy into the building.

A computer mouse: *transforms* information from hand movements to an electronic signal

The structure in a building: *supports* mass, holding it in place. It prevents the building from collapsing.

A barcode scanner: *detects* information. It reads the barcode on a box that identifies the contents and *transforms* that information into a digital signal.

A wind turbine: *extracts* energy from the wind and *transforms* it into electricity. It is designed to *optimize* the extraction process.

BHA, a common food preservative: *maintains* mass; it prevents the food from changing (i.e., spoiling). It *controls* (tries to prevent) mass transformation.

Lipstick: *Deposits* material onto the lips to *transform* the look and feel of the skin. It *controls* the appearance of the lips.

Engineers also design models, which is a form of information generation. Models are representations of real systems. They put ideas together in unique ways to improve understanding. Engineers try to capture the aspects of the behavior of real systems in equations or other types of models. Engineers can then solve the equations or explore the behavior of the model and apply the results back to the real system.

Courtesy of the Authors

Courtesy of the Authors

Courtesy of the Authors

FIGURE 2 The same functional basis can be embodied in many different design forms.

The functionality of a model is to turn the behavior of a real system into a set of information that can be understood and used for prediction purposes. Modeling is a process of transformation, transforming real behavior into information that can be analyzed.

Functional basis is useful in several ways. First, it can help you identify the underlying functionality of the system you are designing. This is an important aspect of defining an engineering design problem. You can also use this framework to compare technologies that serve the same functional basis (see Figure 2). You can use these other technologies as analogies; are there aspects of these other designs that could be used in your design? Why is a dump truck different from a fork? This perspective may assist you in gaining deeper insight into the design problem you are working on.

KEY TERMS

functional basis   components   flows

2. Questions and activities

1. Pick three pieces of technology you have used today. Identify their functional basis.
2. Consider the following list of technologies. Identify the functional basis for each and identify whether there is a control process that is part of the functionality.
   a. A subway system
   b. A bag of potato chips (just the bag)
   c. A stapler
   d. A computer game controller
   e. Toothpaste
   f. A bicycle
   g. A coffee maker
   h. An online banking system
   i. battery
3. Identify the functional basis of the design problem you are working on for your current design project.

3. References

[1] J. Hirtz, R.B. Stone, D.A. McAdams, S. Szykman, K.L. Wood. "A functional basis for engineering design: Reconciling and evolving previous efforts," Research in Engineering Design 13(2):65–82, March 2002.

DESIGN PROCESS

Black Box Method

#skill/tool module: #blackbox

Learning outcomes

By the end of this module, you should demonstrate the ability to:
- Apply the black box technique to identify requirements such as functions and objectives for a design project
- Apply the black box technique to analyze an existing technology and compare/contrast technologies and how they operate

Recommended reading

Before this module:
- **Design Process > Multi-use Design Tools > 1. Functional Basis**

Alongside this module:
- **Design Process > Requirements > Modules 1 through 4**

1. Black Box Method

The **black box method** is a method for exploring the function space of a design problem. It can also be used to generate objectives and constraints, and trigger design solution ideas. This method can be used as a first step in the decomposition process (#decomposition).

1.1. When to Use

In this method you describe all of the mass, energy, and information that is going into the proposed design without actually describing the design itself. Then you describe all of the mass, energy, and information that you expect to result from the operation of the technology you are designing. The design solution is treated as a "black box." This allows you to create ideas for what the design will do, or should be, without specifying a solution (i.e., creating solution independent requirement statements).

1.2. How to Use

In this method you create statements that connect the input to the output. In the example for a bicycle security system, the design takes an unsecured bicycle as an input and produces a secured bicycle as an output, or takes a secured bicycle as an input and produces an unsecured bicycle as an output. The primary function of the design is to change the state of the bicycle. As a secondary function the design takes information as an input and utilizes this information in its operation. We also expect the design to produce information.

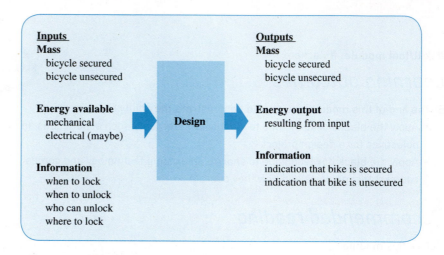

This type of diagram decomposes the functions of the design; it helps you discover what expectations you have for the functionality. In this example, we discover that we expect the bike security system will provide the user with a clear indication of when the bike is secured and when it is unsecured. The means for doing this are not specified; the system could indicate this information through visual, auditory, tactile, or other sensory input (smell? taste?). The system could even send a text message to your phone telling you it is secured. Developing ideas for this functionality will be part of the solution generation process. For the requirements we would simply state:

- The design will indicate when the bicycle is secured and when it is unsecured
- The design will produce a secured bicycle
- The design will produce a released bicycle

Additional pieces of information that are gleaned from this diagram will help us develop other requirements, notably objectives and constraints. For example:

- Objective: The design should allow the user to secure the bicycle in a wide range of locations.
- Constraint: The design must only use available energy as input if it uses energy at all.

If you use the black box method for solution generation, then after constructing the diagram you would use brainstorming to develop a long list of means for each of the functions and subfunctions revealed by the diagram.

KEY TERMS

black box method

2. Questions and activities

1. Pick a piece of technology you have used today. Develop a black box analysis of this technology to identify primary and secondary functions of the technology.

2. Consider the following list of technologies. Develop a black box analysis of one of these. Use this analysis to identify for primary and secondary functions of the technology.

    a. A subway system

    b. A bag of potato chips (just the bag)

    c. A stapler

    d. A computer game controller

    e. Toothpaste

    f. A bicycle

    g. A coffee maker

    h. An online banking system

    i. A battery

3. Apply the black box method to the design problem you are working on for your current design project.

3 Decomposition

#skill/tool module: #decomposition

Learning outcomes

By the end of this module, you should demonstrate the ability to:

- Apply the decomposition technique to identify requirements for a design project
- Apply the decomposition technique to analyze an existing technology and compare/contrast technologies and how they operate.

Recommended reading

Before this module:

- **Design Process > Multi-use Design Tools > 1. Functional Basis**

Alongside this module:

- **Design Process > Idea Generation > 4. Morphological Charts, Analogy, and TRIZ**

After this module:

- **Design Process > Multi-use Design Tools > 5. Benchmarking**

1. Decomposition

Complex problems can overload our ability to generate comprehensive requirements or solutions. A complex problem may have many requirements, and you are simultaneously thinking about many different parts of the problem and trying to come up with an integrated solution all at once.

You can do this process more efficiently by breaking the challenge into smaller parts while generating requirements or before you start looking for a solution. This method is called ***decomposition***. After the problem is decomposed you can then generate requirements such as primary and secondary functions for each part. Or generate solution ideas for each of the structural or functional elements. This produces a much wider range of ideas than you could produce by considering the whole system all at once. Also, by disassociating each element from the overall system you remove mental blocks or filters that might impede your thinking. You are better able to divorce a single element from any preconceived notion you have about what the whole system should be like, and consider a wider range of ideas.

Structural decomposition divides the problem up into the various structural elements, which are discrete physical units or subsystems. In the example we show the decomposition of a vacuum cleaner (see Figure 1).

Functional decomposition divides the problem up into its functional units, and this approach is generally more useful than trying to solve the problem otherwise. A vacuum cleaner needs to suck air and dirt from the ground, separate the air and dirt, store the dirt, exhaust the air, and so on. By separating the various subfunctions, we can focus our attention on small subsets of the problem, and examine these subsets at their most basic level. Let's work on the vacuum cleaner, and focus on the key problem of separating dirt from air.

In Figure 2 we have identified a few different ways we could separate dirt from air. Until about 1990, the household vacuum industry was entirely focused on the use of a filter medium. Vacuums all used bags or porous filter plates. (A third method would be to pass the air through a loose medium such as sand. This is used for swimming pool filters, but there are obvious problems in applying this to household vacuums.)

However, there are many other suggestions here. We could put the dirty air in a large tower, and wait for the dirt to settle out while removing clean air from the top. Not very practical for a portable vacuum cleaner that must be small and light, but this principle is used in large settling ponds created to treat liquid waste. We could speed up the settling process with a centrifuge. Or we could speed it up using a cyclone, where air is pumped at high velocity around a cylinder or cone, and dirt is forced to the periphery and drops to

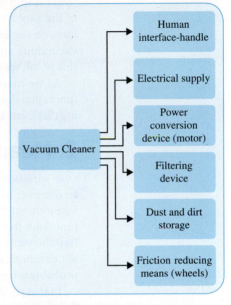

FIGURE 1 Example of the structural decomposition of a vacuum cleaner.

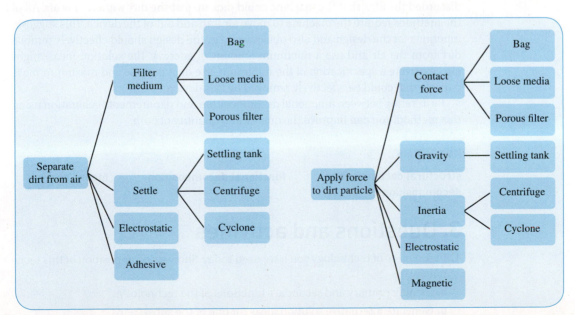

FIGURE 2 Two alternatives for the functional decomposition of the function "separate dirt from air."

the bottom. We could charge the particles with ions, and then get them to stick to charged plates. We could blast the air against a piece of adhesive tape. And we could combine these methods.

An even more fundamental way of doing the functional decomposition in this case is to think of the different types of forces that need to be applied to a particle of dirt to get it to move in a different direction from the surrounding air, which is the heart of the problem. In Figure 2 we realize that we missed the possibility of a magnetic force on our first pass through. Unfortunately, this is only going to be helpful if we are vacuuming up iron filings in a steel mill, but remember that we just generate ideas at this point, we don't filter them.

As we mentioned earlier, James Dyson became a billionaire based on this simple functional decomposition, realizing that there were other possibilities for of separating dirt from air in a household vacuum besides filtration.

2. Means Analysis

One variation on functional decomposition is **means analysis**. This can be used to discover or generate additional functions, objectives, or constraints to add to a requirements list. To perform a means analysis you decompose the problem into sub-functions, then you generate as many ideas as possible that satisfy one of the functions (see the vacuum cleaner example). Now examine the means: are there any common-alities among these solutions? Functions that all of the solutions have in common are probably integral to the design problem.

In this particular example all of the means require the user to remove the dirt from the machine, so the functional requirements must allow dirt to be moved to a loca-tion chosen by the user (typically the trash). The means shown also all presuppose the movement of air with the dirt through the machine. In this case the designer has discarded the idea that the machine could pick up just the dirt without any air. All of the methods require the machine to move air into and out of the device. This suggests functions for the design and also objectives (e.g., the design should effectively remove dirt from the air and use a minimum amount of energy). The solution ideas might also motivate a specification of the minimum size dust particle and maximum mass particle that could be effectively removed by this device.

By iterating between functional decomposition and requirements generation using this method, you can improve the quality and quantity of both.

KEY TERMS

structural decomposition   **functional decomposition**   **means analysis**
decomposition

3. Questions and activities

1. Pick a piece of technology you have used today. Show a decomposition of this tech-nology to:

   a. Identify primary and secondary functions of the technology

   b. Generate alternative solution ideas for one of the subsystems

2. Consider the following list of technologies. Show a decomposition of one of these technologies. Use this analysis to identify primary and secondary functions of the technology. Then generate alternative solution ideas for one of the subsystems.

a. A subway system

b. A stapler

c. A computer game controller

d. A tube of toothpaste

e. A bicycle

f. A coffee maker

g. An online banking system

h. A battery

3. Apply decomposition to the design problem you are working on for your current design project. Identify additional requirements using this method, and generate a wide set of solution ideas for the subsystems.

Information Gathering

#skill/tool module: #information gathering

Learning outcomes

By the end of this module, you should demonstrate the ability to:

- List sources of information that will be useful for research during a design project
- Evaluate a source of information for credibility
- Use a source of information properly, including taking notes from the source and recording citation information

Recommended reading

Alongside this module:
- **How Engineers Design > 4. Communicating throughout the Process**

After this module:
- **Design Process > Multi-use Design Tools > 5. Benchmarking**

1. Introduction

Engineers gather information for a multitude of purposes during a design project. Information gathering is used to:

- Add detail to the requirements
- Investigate *codes*, *regulations*, and *standards* that pertain to the design
- Accurately describe the context of the design (i.e., *service environment*) and *stakeholder interests*
- Inform the idea generation process; take into account the *state of the art* through *benchmarking* and review of the literature on existing technologies
- Test design alternatives by gathering information through *metrics*
- Specify the detailed implementation of a design using existing technologies such as parts, materials, systems, and manufacturing or building processes

In this module we will focus on the gathering of information using literature sources, both traditional and online, for the purposes of adding accurate information to the requirements and informing the idea generation process. There is a special emphasis on critically thinking through the credibility of the sources you use so your design process is built on reliable, valid information.

2. The State of the Art

It is important for engineers to know about existing technologies related to all or part of the design project they are working on. The sum total of the existing knowledge is referred to as the *state of the art* or *prior art* (#priorart).

Searching for existing solutions is a mandatory part of any design process, but the timing should be chosen strategically. Some people may feel that the creativity process may be hindered by preconceived ideas based on existing solutions to similar problems. In fact, novice designers might well benefit from some *blue sky thinking* about the problem before doing the necessary investigation of the literature and reading. On the other hand, it is inefficient to generate design solutions that are already well known by others. There are many possible alternatives for most design problems, and if you are prepared to search for these in a systematic way, without constraining your creativity, then you can save time by first reviewing the state of the art.

In order to really save time and produce high quality work, you must bear in mind that although more information is more easily available than ever before in human history, not all of the information that comes to you quickly is useful. Information must be analyzed for its credibility and relevance before it is used.

3. Gathering Information

To find out what is already known, you need to know where to look. The *state of the art* may take the form of commercial products or services, patents, standard industry practices and codes, and the personal knowledge of experienced engineers in the field.

3.1. Published Sources

Searches should always start with the most general *sources*, and then drill down to more and more detail and specificity. If your task is to design a new filtering sequence in a chemical plant, you do not start with a detailed scientific reference on the effect of pore size on membrane efficiency. Instead, you would begin by using a general reference, such as an encyclopedia article on filtration. If you are working in an engineering firm, you might try to find a knowledgeable senior engineer to chat with initially, just to get a feel for the general area. Talking to people is a very important part of information searching. As you build up expertise of your own, you would move to more and more detailed and specific sources of information.

Here is a simple bullet list of some places you might look for printed literature and a suggested order (from general to more specific). If you are already familiar with the general details concerning your design, it may be appropriate to start part way down this list.

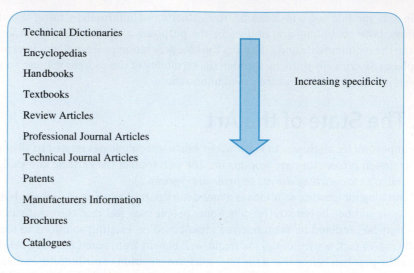

Technical Dictionaries

Encyclopedias

Handbooks

Textbooks

Review Articles

Professional Journal Articles

Technical Journal Articles

Patents

Manufacturers Information

Brochures

Catalogues

Increasing specificity

FIGURE 1 Search priority when researching information for a design project using printed or online material [1].

Note also that many industries in engineering have their own publications (magazines and handbooks), which are often published by professional trade organizations (IEEE, ASME, CSME, AIChe, etc.). These are usually available from libraries or can be borrowed from professionals in the field. Also, many catalogs from engineering companies contain more than just the description of the items for sale and prices (see Figure 2). They may also contain information on the technology more generally and the way to properly use it in applications (much like a handbook).

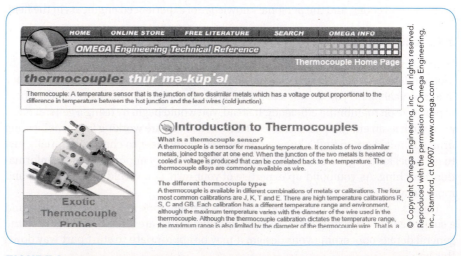

FIGURE 2 Many companies that manufacture components used in design will provide information to help engineers select the right component and utilize it safely.

3.2. Benchmarking: Reverse Engineering

A survey of existing designs is a critical part of investigating the state of the art. The process of dissecting and understanding an existing design is known as **benchmarking.** If you want to design a new hair dryer, it would obviously be very instructive to buy a few of the better ones already on the market and to disassemble them to discover how they are designed. It might even be a good idea to buy some bad ones to find out what is wrong with them. Benchmarking is common practice in large industries such as the automotive and home appliance industries [#benchmarking].

3.3. Prior Art in Other Domains

It is worthwhile spending some effort to think about where you might find relevant background information, aside from the obvious places. In particular, it is important to recognize that prior art may not be restricted to the area of application of the current design. In other words, it is a good idea to *plan* your search! For example, suppose you have to design a nozzle for a high-pressure steam jet to be used to remove unwanted deposits inside an industrial furnace. Have other engineers worked on similar problems? The answer is yes: Sir Frank Whipple spent a great deal of time on high-speed flow when he was designing the original jet engine in the 1930s and 1940s. In fact, modern jet engines have very efficient "nozzles" to direct high-velocity air in one direction. The high-efficiency nozzles now widely used in industrial boilers in the pulp and paper industry are based on these principles.

3.4. Researching the Context

Beyond the state of the art, research is necessary when developing other sections of requirements, which provide details of the context of the design, the service environment, operators and users, human factors as well as environmental, social, and economic concerns. The details of these concerns may well be outside of your own experience and they have to be accurate and credible. So the research sources for context must be analyzed as carefully as those for state of the art. The strategies for information searching shown in Figure 1 are appropriate for gather context information also. The goal is to gather, use, and document accurate information at every stage in the design process, from requirements development to details of implementation.

4. Analyzing Information

Gathering information and putting it into a document is not enough; in fact, it is dangerous. If the information does not come from a reliable source and contains errors and inaccuracies, you run the risk of misinforming your reader or client or worse, creating physical, economic, or environmental harm.

When you are doing research, you have to ask questions about the source of information and about the information itself. Given the amount of information that is now available, gathering and analyzing information is probably more challenging today than it ever has been, and when you are assessing information it is important to remember the key critical thinking concepts of *frame of reference*, *bias*, and *purpose* (#criticalthinking).

Information you use should be rigorously validated. You should assume there is no such thing as "general knowledge." If you know something because "everyone knows" it, then your client knows it too and you are not adding any value by citing it. There-fore, to support your claims you have to use data you have generated in a reliable manner or located from reliable sources.

5. Evaluating a Source of Information

You may not be used to questioning research material. That is, up until now, you may have searched for a piece of information you needed for a project you were working on and when you found something that seemed to fit, you used it. Critical thinkers, however, do not utilize information without analyzing it first. Just as they try to be aware of the factors that affect their own objectivity and credibility, they also actively examine information from others (#skepticalthinking). It is a best practice to analyze any information before accepting it. The best way to do this is to formulate a series of questions and try to answer them before using the information. Questions you might want to consider include:

1. What is the authority of the source of information? If it is a person, on what basis does the person know about what they are telling you?

   a. Is the person an expert? Expertise can be determined by a professional designa-tion. Engineers, doctors, and lawyers are examples of regulated professions. They cannot practice unless they are authorized to do so by a professional associa-tion. Professional associations, job designations (such as professor, counselor) are external designations of authority.

   b. Expertise can also be determined by less obvious factors. One is experience; a person who has used a particular program for a long time probably has some good information about it. You would look for informal signs of authority such as experience in online reviews or blogs. However, when dealing with non-obvious measures of authority, you must be aware of bias and purpose.

   c. Print or web published sources such as scholarly or academic journals use pan-els of objective experts to review their material. These help ensure the authority of the content. Many encyclopedias, too, have boards of scholarly editors. But encyclopedias such as Wikipedia, which invite the democratic participation of the public, must be much more closely evaluated for bias and validity. Fortu-nately, Wikipedia, for one, is self-regulating and posts warnings when informa-tion lacks sufficient referencing. It also publishes a history of the development of each page. Examining the history helps to analyze the validity of the information.

2. Is there evidence of bias? Bias does not invalidate information; it only decreases the objectivity of information. A nonobjective source of information may be valuable *because* of its point of view.

   a. Does the source already have a reputation for its political or social viewpoint? Sometimes newspapers, for example, are known for their political view-point. Some are known to be conservative, while others have a liberal point of view. Some newsletters include this information in their titles; socialist newslet-ters frequently do this. A communication from an organization with a particular political viewpoint as part of its name is certain to have a particular point of view or bias.

b. Can you tell, from the details chosen or from the tone, what kind of values the writer or organization may have? Does it use value-laden terms, or express qualitative judgments? Does it only give one side of a multi-sided situation?

3. What are both the explicit and the implied meanings in the language used?

a. Language communicates in two distinct ways—explicitly and implicitly. The first term refers to the obvious meaning of the words themselves. The statement "To get out of full screen view, press the 'Esc' key" has a simple, clear and obvious—or explicit—meaning. If something is enlarged on your screen and the program allows you to use "Esc" to reduce it, then pressing the key marked "Esc" will have the expected result. That's it. There is no other "hidden" meaning.

b. Implicit meanings, on the other hand, are not so obvious; they are not on the surface of the words and cannot be grasped easily or with intellectual certainty. Rather, they are felt or intuited. They may be embedded in the particular words chosen. They carry double meanings. Take, for example, the sentence "You deserve some credit" on a credit application in a car advertisement (www.ford .ca/app/fo/en/cars/focus.do#). The explicit meaning is that you can apply for a loan. The implied meaning comes from the normal use for this sentence—it refers to someone getting recognition for some achievement. Thus, the idea of taking out a loan is equated with deserving something because of a particular achievement, as if a loan is an award.

4. What is the purpose of the information? In some sources, such as encyclopedias, the purpose is to give the reader a basic introduction to a topic. Textbooks are similar. The purpose of certain handbooks may be to ensure that a procedure can be followed successfully and safely. Other kinds of purpose, however, can considerably affect the usefulness of information. Take advertising, for instance. The goal of advertising is to get someone to buy something; it therefore must interest and excite the buyer. It cannot lie in doing this; there are organizations that ensure it does not. Therefore, specifications have to be accurate. However, advertising often operates in a way that allows it to excite without lying; it uses emotional language and emotional situations that excite the buyer without actually saying anything about the product. Understanding purpose allows you to use the information appropriately.

5. Who is the intended audience? The intended audience will affect the degree of complexity and the type of the information. Only certain things will be reported to certain audiences. When answering this question, factors to consider include the age of audience, amount of education expected, and their political viewpoint. Ask yourself how well you match up with the intended audience or how well it matches up with the audience you are writing for or speaking to. Often, in order to understand a concept, we will begin with an article in Wikipedia. The article may be intended for an audience with far less education than we have; the concept will be explained fairly simply. If we stop there, though, we may be underestimating our needs as well as our audience's needs.

6. Documenting What You Learn

As you gather information it is very important to record not only the information, but also to record the *source* of the information. Do not simply cut and paste material from the Internet into a file to keep for use later. You will eventually need to explain the information in your own words, so it is good to make your own notes as you find information. If you do not understand the material well enough to rewrite it in your own words, you shouldn't be using it in a document! Make sure that as you take notes on the information from an information **source** you also write down, absolutely accurately, the **citation information** for the source so you can properly cite the ideas when you use them in your design reports and other documentation.

KEY TERMS

requirements	codes	regulations
standards	service environment	stakeholder interests
state of the art	benchmarking	metrics
blue sky thinking	bias	frame of reference
citation information	purpose	source
prior art		

7. Questions and activities

1. Find an example of each of the sources listed in Figure 1 that has some relevance to your field of engineering. Correctly document the citation information for each of the sources.

2. Find three sources of information authored by one person. Referring to the section in this module on Evaluating a Source of Information, item 1: For each source: on what basis does the person know about what they are telling you? (See section 5, item 1.) Explain their credibility using the three areas identified in section 5, item 1. Make sure you also cite the examples. (Provide full citation information for the three sources).

3. Find three sources of information related to your discipline in engineering from any source. Is there evidence of bias? Explain. Use the criteria in section 5, item 2 to judge bias. (Provide full citation information for the sources.)

4. Find three examples of information where the explicit and implied meanings are different. These examples can be short, such as just one sentence. What are both the explicit and the implied meanings in the language used in these examples? Explain. (Provide full citation information for the examples of information used.)

5. Find three examples of information. For each example: what is the purpose of the information. Can you identify a purpose based on the ideas in section 5 item 4? Explain your reasoning for the purpose you identified. (Provide full citation information for the examples of information used.)

6. Find three examples of information. For each example: who is the intended audience for the information. Can you identify the intended audience based on the ideas in section 5 item 5? Explain your reasoning. (Provide full citation information for the examples of information used.)

8. References

[1] Adapted from Dieter, G.E. *Engineering Design*, 3rd ed. McGraw Hill, 2000, p. 125.

[2] www.ford.ca/app/fo/en/cars/focus.do#

Benchmarking

#skill/tool module: #benchmarking

Learning outcomes

By the end of this module, you should demonstrate the ability to:

- Explain the process of benchmarking and how it is used in the design process
- Perform a benchmarking analysis of an existing technology

Recommended reading

Before this module:

- **Design Process > Multi-use Design Tools > 4. Information Gathering**

Alongside this module:

- **Design Process > Requirements > Modules 1 through 4**

1. Introduction

A survey of existing designs is a critical part of investigating the state of the art. The process of dissecting and analyzing an existing design is known as ***benchmarking*** or ***reverse engineering.*** If you want to design a new coffee maker, it would obviously be very instructive to buy a few of the good ones already on the market and to disassemble them to discover how they are designed. It might even be a good idea to buy some bad ones to find out what is wrong with them.

Courtesy of the Authors

The term "benchmarking" is also used to refer to the practice of using the analysis of an existing design to set goals for a new design. For example, the new coffee maker should have the same energy usage or less than models currently on the market. In this way the existing design serves as a benchmark or target for the new system under development.

1.1. Gathering Items to Benchmark

Start by gathering as many examples as you can of the technology you are analyzing. Ideally you want to include a wide variety of different approaches that have been developed to solve the same problem (e.g., a broad range of different coffee makers if you are designing a new coffee maker, or several different software packages that are related to a new piece of software you are designing). In industry you would have a budget for benchmarking, but in school this may be more difficult.

There are several lower-cost or no-cost options. Try looking around for technology operating in your environment (e.g., the coffee maker in your residence). You might not be able to take it apart, test the construction materials, or see the source code, but you can observe it operate. This will allow you to measure the operation

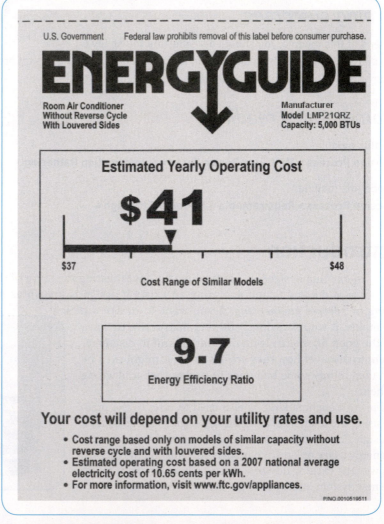

FIGURE 1 Published specifications for products and systems already on the market can be used to benchmark your design project.

and performance *metrics* for the technology. If you are benchmarking a larger installed piece of technology (e.g., a bridge, or a subway fare system), try visiting as many instances of the technology as you can; take pictures and make observations about how the technology is working. You can also visit sites virtually, online. For consumer items try going to a store (real or virtual). You can't always see the technology operate in a store, but you can often get the manufacturer's specifications from the packaging or online catalog (see Figure 1). If you have a small budget you could try looking at a second-hand retailer such as a thrift shop, junk yard, or online resale site for some used items. This is particularly useful if you want to disassemble the items to understand better how they work.

2. Using a Performance or Task Analysis

There are a number of different ways to analyze the design of an existing system for the purposes of benchmarking. *Performance analysis* and *task analysis* are related approaches that do not, generally, require dissecting the target item.

2.1. Performance Analysis

Select the target technology that you are going to analyze. Identify the different subfunctions of the item and *metrics* that can be used to measure the performance of these subfunctions. For example, there are many possible subfunctions that can be identified for a vacuum cleaner (see Table 1). The table lists a few examples and possible metrics that could be used to measure the performance of the particular subfunction (#metrics).

This type of analysis is much like running a set of lab experiments and is often carried out in a lab (when there is the budget available to do this in industry). The methodology is carefully controlled so the results from the investigation of one target item can be compared to another. It is important to use good experimental practices when carrying out this type of analysis, for example: running the experiment multiple times, which improves the reliability of the results, or controlling for other factors such as the quality of the dirt, the surface, and other variables.

TABLE 1 Analysis of a vacuum cleaner identifying a few of the subfunctions and possible associated metrics.

SUBFUNCTION	POSSIBLE METRIC
Sucks dirt from surfaces	1. Put known quantity of dirt on a surface (e.g., 1 kg) 2. Sweep over the surface with the vacuum 3. Measure dirt collected or dirt left on the surface
Separates the dirt from air and other items (e.g., large toys)	1. Feed known quantity of dirt into the vacuum 2. Measure amount captured by vacuum and/or amount exhausted to the environment (i.e., uncaptured)
Stores the dirt	1. Measure the size (volume and/or weight) of dirt that can be stored in the vacuum before emptying is required
Collapses or closes up to store neatly and efficiently in a closet	1. Measure the volume required to store the vacuum cleaner 2. Possibly also measure the appearance of "neatness" using a focus group or the design team's opinion

Courtesy of the Authors

However, it does not always require that you have a lab. There are many performance measurements that can be gotten, measured, or inferred without taking the item to a lab. The volume of dirt a vacuum can hold before emptying is required, for example, is a value that is probably available from a catalog, observing the model in a store, or measuring it at home (outside a lab). Other performance measures can be made in the course of ordinary usage of the target item. For example:

- How long does it take to roll up the cord and store the vacuum?
- How many different attachments or settings does it have and what are they used for? (Read the owner's manual.)
- How long a reach does it have? How small a space can it reach into?
- How long does it take to empty the dirt? And is it easy to do, or does some of the dirt fall back onto the floor?
- How much energy does the vacuum cleaner use? (Check the electrical rating, usually found in the owner's manual or specification information on the machine.)

All of the results and observations should be recorded in your **engineering notebook**. Once you have measured the performance of several different systems you can make more informed choices for the **objective goals** and other **requirements** for the design project on which you are working. You may also gather ideas about which designs are most effective, which are less effective, and why. Make sure this information is documented so you can use it later to inform your idea-generation and decision-making processes.

2.2. Task Analysis

Task analysis is similar to performance analysis in that it tests the way the target item operates (#taskanalysis). However, in task analysis the view is entirely from the operator's or user's perspective:

- What steps does the user follow to perform an operation?
- How many steps?
- How much effort and time is required for each step?
- Is the operation intuitive or does it require training?

Task analysis can be performed through self-observation (e.g., talking out loud or recording the steps in a process while you personally are performing them). Or it can be done through observing another person interacting with the target technology via video, focus group, or other types of reconnaissance. In either case the goal is to document the characteristics (time, effort, and level of difficulty) of the operation of the existing technology so its operation can be compared and contrasted with other existing technologies and to the new technology you are designing. Like performance analysis, task analysis is used to set requirements, such as objectives and objective goals. It is also used for idea generation and decision making because it allows you to compare the characteristics of the new design solution you are proposing to the existing technologies that are already available.

3. Using a Form, Flow, Function Analysis

A second benchmarking approach is to identify and document the functions of the existing system and how these functions are enabled by the ***form*** of the design and the ***flows*** of mass, energy, and information. This type of analysis is related to functional basis, black box method, and functional decomposition. However, instead of using the flow analysis to generate ideas for a new design, in benchmarking we use flow analysis to understand and quantify an existing technology.

This type of analysis is best performed when it is possible to dissect the existing technology, but a rough analysis can be done without taking the target item apart or accessing the inner workings. We suggest that if you are doing both a performance analysis and a form, flow, function analysis that you do the performance measurement first, particularly if the form, flow, function analysis is going to require dissecting the target technology such that it is not going to be easy to put back into its original condition.

To do this type of benchmarking, pick one of the items you have collected for benchmarking. Perform a functional decomposition on the target item also identifying key features (objectives) that are included in the design; that is, identify the functional subsystems. For example, a vacuum cleaner typically:

- Sucks dirt from surfaces
- Separates the dirt from air and other items (e.g. larger toys)
- Stores the dirt
- Exhausts the air and other items
- Provides a means for deposing of the dirt; that is, emptying the dirt out of the machine
- Provides a means for controlling the machine: off/on and where it is directed
- Functions effectively on different types of surfaces
- Agitates the surface to loosen dirt
- Allows the user to move around a room easily
- Collapses or closes up to store neatly and efficiently in a closet

Note that the ability to store neatly in a closet is not related to the primary function of a vacuum cleaner, but is an essential feature that relates to an important objective.

For each of these subsystems, identify the way the designer has chosen to address the requirement. This is done by documenting the subsystem function, the form used to fulfill the function, and the flows of mass, energy, and information utilized. Table 2 shows how to organize the information for this type of analysis. The analysis should include a complete description of the subsystem and how it operates to achieve the function or goal. It also includes a section (titled "Analysis" in the table) that records your observations about the operation, performance, and characteristics of the subsystem—in particular, how the form and flows relate to and enable the function.

Benchmarking

119

TABLE 2 Example of a table to document this type of benchmarking analysis

Function 1: *Describe subsystem being analyzed*

FORM	FLOWS
Describe the form this takes in the target item and how it operates to perform the function in detail. Explicitly detail measurements of size, shape, structure, material properties, algorithm strategy used, and so on.	Mass: *Describe any mass flow related to this subsystem in detail—type and quantity* Energy: *Describe any energy flow related to this subsystem in detail—type and quantity* Information: *Describe any information flow related to this subsystem in detail—type and quantity*

Analysis: Observations about this subsystem, the quality and characteristics of the solution that the designer implemented in this particular design solution.

• *How does the form relate to the flows to perform the required function?*

Function 2:

FORM	FLOWS
	Mass:
	Energy:
	Information:

Analysis:

Ideally you would complete a table like this for several different target items in your investigation (e.g., a wide variety of different vacuum cleaners). After completing this work, you would have a deep understanding of how existing technologies work and how different designers have chosen to address the same problem.

4. Benchmarking Tangentially Related Designs

Innovation often comes from combining existing ideas that have never been put together before. In addition to benchmarking systems that are close in function to your design project, it might also be useful to benchmark systems that are only tangentially related if you are trying to develop an innovative technology (see Figure 2). Using functional basis to find systems that are superficially unrelated, but essentially serve the same purpose is one method of identifying non-obvious technologies for benchmarking.

5. Advantages and Disadvantages of Benchmarking

Benchmarking is a very effective tool for investigating existing technologies. It helps you to uncover requirements that you may not have previously considered. And it can help you think of new ideas or position the technology you are designing in a new design space that is not covered by currently available technologies. It is also important to prevent you from "reinventing the wheel" or accidently infringing on other people's intellectual property (i.e., patent infringement). It is often an essential process because it helps you learn about a field in depth in a way that is difficult to achieve using other methods.

However, benchmarking can impede creativity. Once you have immersed yourself in the existing design solutions it becomes difficult to imagine other ways to solve the problem. In this way, benchmarking pollutes your design space. Expert designers are practiced at using benchmarking for learning, but separating their own ideas from what they have seen. Another strategy experts use to keep benchmarking from polluting their idea set is to use structured idea generation or creativity methods (see idea generation) to develop innovative solutions. But for a new designer this is less effective. So while it is imperative that you be familiar with the existing systems that are already available, we recommend that you not engage in the benchmarking process until you have gone through at least one round of idea generation.

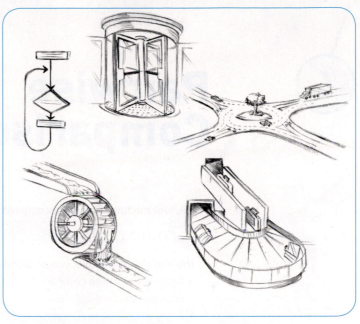

FIGURE 2 Benchmarking related systems can assist in the generation of requirements and solution ideas.

KEY TERMS

benchmarking	reverse engineering	metrics
performance analysis	engineering notebook	objective goals
requirements	task analysis	flows
form		

6. Questions and activities

1. Develop a benchmarking plan for a technology or system.

2. Perform a benchmarking analysis of a technology or system.

3. Perform a benchmarking analysis of target items related to the design project you are working on currently.

    a. Use this analysis to add to your list of requirements, in particular your objective goals.

    b. Use this analysis to add to your list of potential solutions for the subsystems of the design.

Benchmarking

6

Pairwise Comparison

#skill/tool module: #pairwisecomparison

Learning outcomes

By the end of this module, you should demonstrate the ability to:
- Apply the pairwise comparison technique to prioritize a list, such as the list of objectives for a design project
- Apply the pairwise comparison technique to prioritize a list for the purpose of reducing the number of items or ideas under consideration (e.g., narrowing the design space)

Recommended reading

Alongside this module:
- **Design Process > Requirements > 3. Objectives**

1. Pairwise Comparison

The ***pairwise comparison method*** is best used for prioritizing a list. It can be used in the design process for prioritizing the list of objectives for a design project. This helps the team focus on the most important objectives that the design must fulfill in order to meet the needs of the client, users, and operators.

However, pairwise comparison can be used in a much broader range of applications. It can be used in any application where there is a need to prioritize a list (e.g., to prioritize budget items when there are limited funds available). It can also be used to eliminate ideas or items from consideration (i.e., reduce the design space).

What it does: Pairwise comparison is a simple method for ranking a list. Ranking a list can be difficult because you are trying to compare many items to many others at the same time. A pairwise comparison simplifies the task by breaking one complex decision up into a set of decisions; each decision in a set compares just one item to one other on the list.

How to do it: In each cell of the table you only need to compare one item on the list to one other and decide which is better, has a higher priority, or is more important. This can be used for ranking, or prioritizing, any list of items: objectives, design solutions, items you want to buy, things you need to do, and so on. For this method you put your list along both the first row of a table, and repeat the list down the first column (see table). Then you compare each item in the first column to the each item in the top row, one at a time. Each decision results in a 1 or a 0; 1 if the item in the column is ranked higher and 0 if it is ranked lower than the item in the top row.

Consider the first row in the table. In this example the team has decided that item #1 (for example, objective #1) is more important than item #2, item #3, and item #5, but less important than item #4. Overall item #1 gets a score of 3 by adding up the score across the row. We repeat this for the other rows to get the scores shown.

	ITEM #1	ITEM #2	ITEM #3	ITEM #4	ITEM #5	SCORE
Item #1	— — —	1	1	0	1	3
Item #2	0	— — —	1	0	0	1
Item #3	0	0	— — —	1	0	1
Item #4	1	1	0	— — —	1	3
Item #5	0	1	1	0	— — —	2

In this example items #1 and #4 have tied for high score and would be moved to the top of the list because they appear to be the most important, or top alternatives. Items #2 and #3 have the lowest scores and would be moved to the bottom of the list. You could then use team discussion to decide which item (#1 or #4) should be put at the top of the list, and whether to drop items #2 and #3 off your list.

Advantages: Pairwise comparison is a very fast method for ranking a list or eliminating some alternatives.

Disadvantages: Pairwise comparison obviously yields approximate results. The scores in the example do not mean that item #1 is three times as good as or three times more important than item #2. The results should always be discussed. Also, even if an item receives a score of zero it does not mean that it has no importance. It just means that it has a lower relative priority.

Example of Pairwise Comparison to Rank a List of Objectives

Suppose we are given a design project that has three essential objectives: the design must be easy to use, reliable, and have a low environmental impact. It might be difficult to meet all three of these objectives with one solution. So through discussion and consultation we try to determine how the client, users, and other stakeholders value these objectives relative to each other. The team then translates this information into a pairwise comparison chart, as shown.

	EASE OF USE	RELIABLE	LOW ENVIRONMENTAL IMPACT	SCORE
Ease of use	— — —	1	1	2
Reliable	0	— — —	1	1
Low environmental impact	0	0	— — —	0

In this example the objectives would be listed, from most important to least important:

1. Ease of use
2. Reliable
3. Low environmental impact

The ranking does not mean that environmental impact is unimportant. It should still be included in the list. This simply suggests that given the choice between a design solution that is easy to use and one that has a low environmental impact, all else being equal, we would choose the one that is easy to use in this case. Obviously, it would be preferable if we could find a solution that satisfies all three objectives well.

KEY TERMS

pairwise comparison method

2. Questions and activities

1. Specific example of pairwise comparison: Suppose you are given money to spend for your birthday. What do you buy?

	EVENING OUT WITH FRIENDS	BOOKS	MUSIC, APPS, OR GADGET	CLOTHES OR ACCESSORIES	FOOD	SCORE
Evening out with friends	— — —					
Books		— — —				
Music			— — —			
Clothes or accessories				— — —		
Food					— — —	

Try filling in the table to see what your priorities are, or add you own items to the list and then fill it out.

2. Perform a pairwise comparison to rank the objectives for the design project you are working on currently.

Introduction to Idea Generation

1

#process module: #ideageneration

Learning outcomes

By the end of this module, you should demonstrate the ability to:
- Explain strategies for developing a wide range of solutions for a design problem
- Describe the role of idea generation in the design process

Recommended reading

Before this module:
- **Design Process > Requirements > 9. Summary; Putting It All Together**

After this module:
- **Design Process > Idea Generation > 2. Brainstorming**

1. Introduction

Once the design problem is defined in sufficient detail, the engineering design team can begin to generate ideas for designs that satisfy the requirements (***idea generation*** also called ideation). This is the heart of the design process, and for many design engineers, it is also the fun part. It is a ***divergent process***, where the ***design space*** is deliberately expanded to provide the widest possible spectrum of solutions from which to choose. The word "creativity" conjures up images of artists and poets, but the invention of new and diverse ideas to solve an engineering problem is very much a creative endeavour in which engineers create solutions to both routine and innovative design challenges.

The process of generating creative design solutions is fundamentally different than the familiar process of finding a solution to a closed-ended analytical problem. In a typical engineering exam question in a university course, the goal is to understand the given problem, recall the pertinent equations (which represent physical relationships), and to transform the given information into the required solution using these equations. Such problems almost always have single correct solutions. A design problem is characterized by many possible solutions, and a talented designer is one who develops the skill of finding the *optimal solution*. It would be unusual if this were the first and only solution developed, so a key to the process is an ability to generate many different solutions so that the optimal one can be chosen. This is referred to as populating or expanding the *design space*.

2. Building Skills in Solution Generation

When faced with a design problem, most people immediately begin to dream up solutions. Weak designers go with one of their first ideas and develop it without fully considering the alternatives. Strong designers recognize that there are likely to be many solutions and that they will have to work hard and use *creativity methods* and structured thinking to generate a complete set of feasible solutions from which to choose. Practicing these creativity methods will help you develop the ability to generate many diverse solutions to a problem.

You are already familiar with the most widely used method for finding creative solutions to a design problem: just thinking about it. If you are presented with a design challenge, you are probably quite capable of coming up with a variety of feasible solutions. But are you capable of coming up with the optimal solution? Adding some discipline and strategies to your thinking can greatly enhance your engineering creativity and chances of success.

You should end up with a long list of ideas, only some of which would fully satisfy the functions, objectives, and constraints. After the initial list-building exercise is well and truly exhausted, you would combine elements, examine the nonsense answers to see if anything useful can be extracted from them, and generally try to build a few more realistic design alternatives from which to choose.

You might want to go through the idea generation process several times, as you discover new areas for ideas while working with the older ideas. The idea generation process will inevitably continue even as you converge on a single solution that will be implemented. This is the reality of the design process.

Of course, like any activity that has commercial and practical significance, there have been many books, articles, and software programs written to promote and facilitate creative idea generation. Some of the methods that are widely used in professional engineering design for creative solution generation include free and structured *brainstorming, lateral thinking, SCAMPER, morphological charts, analogy,* and *TRIZ*. These methods are used not only for generating creative solutions to the design problem, but can also be used for other parts of the design process, such as for generating the design requirements. We encourage you to try using several of these methods to fully populate the design solution space.

3. Leaving This Process Module

After working through this module you should have:

- A long list of design solution ideas
- Documentation of the idea generation activities you have used:
  - Documentation of the creativity methods your team used to generate solutions
  - A much longer list that includes all of the ideas you generated before you eliminated duplicates and combined ideas

KEY TERMS

divergent process	design space	optimal solution
creativity methods	brainstorming	lateral thinking
SCAMPER	morphological charts	analogy
TRIZ	idea generation	

4. Questions and activities

1. Explain the role of idea generation in the design process.
2. Explain three methods that can be used for idea generation.
3. Pick a design problem and try using a variety of creativity methods to generate a large number of potential solutions.
4. Generate potential solution ideas for the design project you are working on currently. Document your thinking process and results in your engineering notebook.

2

Brainstorming

#skill/tool module: #brainstorming

Learning outcomes

By the end of this module, you should demonstrate the ability to:

- Explain how to run a brainstorming session
- Describe the strengths and weaknesses of brainstorming
- Participate appropriately in a free or structured brainstorming session
- Determine if free brainstorming is a comfortable method for a team

Recommended reading

Before this module:

- **Design Process > Idea Generation > 1. Introduction to Idea Generation**

After this module:

- **Design Process > Idea Generation > 3. Creativity Methods**

1. Introduction to Brainstorming

The term **brainstorming** is usually used to describe the process in which a group of people have an intensive session of idea generation. Brainstorming is used to generate a large number of ideas, and if successful, unusual ideas, in a relatively short time. The concept of brainstorming was developed by Alex Osborn and is used to generate ideas to address a specific question or defined problem [1]. The main focus of this technique is to generate a large quantity of ideas without worrying about their quality.

Many people have come to believe that the quantity and quality of ideas generated in a group is much better than what would be generated by the individuals in the group operating alone. In fact, extensive experimental research demonstrates that the opposite is true: individuals significantly outperform groups [2]. As Osborn himself recognized, "The creative power of the individual counts most" [3] and

recommended group brainstorming "solely as a supplement to individual ideation." [4] Stroebe and colleagues discussed three possible reasons for the relatively poor performance of groups [2]:

- *Free riding* (some members don't participate)
- *Social Inhibition* (members afraid to put forward their ideas)
- *Production blocking* (time waiting for others to speak is unproductive and ideas are lost)

Of these three, research suggests that production blocking is the primary cause of the loss in productivity in a group session [2]. The necessity to wait for others to speak reduces the productivity of an individual in a group setting. The longer the average wait time is, the greater the effect.

There are a number of variations on rules and variations devised for brainstorming that attempt to mitigate these issues. Free brainstorming may be the concept with which you are most familiar. In free brainstorming, all the participants generate ideas as a group and someone records the ideas. People often talk across one another, and shout out their ideas. It is fast paced but sometimes quieter people or people who need to take a little more time are left out in this type of brainstorming. Structured brainstorming offers an alternative and can be done in a number of ways. The underlying premise is to provide a method where people can sit quietly for 5–10 minutes and generate ideas on their own before they suggest them to the group.

In order for brainstorming of any type (and indeed most idea generation techniques) to be successful, there are some basic guidelines to follow. These were originally described by Osborn and have been modified somewhat since that first publication.

1. **Seek quantity:** The idea here is that by generating a large number of ideas, some high-quality, innovative solutions will come through.

2. **No criticism:** When there is no criticism, people feel freer to generate more ideas. You should not make either positive or negative comments about the ideas. Even positive comments indicate judgment.

3. **Get all the ideas out:** Encourage unusual ideas and freewheeling. Team members should not be inhibited about providing a "wild" or "unrealistic" idea. These ideas often form the basis for very original solutions. They can also provide a trigger for another new idea from someone else. Every ideas counts and when you are generating your list, duplicates are allowed. You will remove them later.

4. **Combine and improve ideas:** Good ideas may be combined to form a single better idea. It is helpful to pick up on the ideas of others and to look at combinations or associations. This helps build more ideas.

The ability to build on others' ideas is an obvious advantage to the group setting. It is possible for an individual to miss an entire line of thinking if working on his or her own, and yet still have useful contributions to that line once exposed to it. One method of combining the advantages of the group and individual modes is to link participants anonymously by computer. This eliminates social inhibition, and if more than one person can post to the communal board at the same time, production blocking is also eliminated. Gallupe and colleagues found that the production of electronically linked groups increased with group size, whereas the production of physical groups did not [5]. Surprisingly, in an anonymous computer-based platform, studies have shown that critical comments were actually beneficial and improved the production of the group [6].

What does all this mean for your brainstorming activities? You need to be careful in depending solely on group brainstorming. When conducting a session, it is useful to observe Osborn's rules: encouraging quantity, building on the ideas of others, and suppressing criticism. If you are running a face-to-face session where production blocking is going to be an issue, you could try a dual recording system, in which participants record their own ideas individually, then pass them along to the group when the opportunity arises. You could also use the structured brainstorming approach, in which you do some individual work prior to coming together as a group. Finally, you could try an online collaboration system, preferably one that is anonymous.

You should always add time pressure when brainstorming. People will respond if you limit their time and a bit of competition to the brainstorming session. Groups that are highly practiced in brainstorming can generate hundreds of ideas in less than 15 to 20 minutes (groups of 5 or so). Many of the ideas will not be appropriate but some will inevitably be useful and creative.

1.1. Post-it® Notes

Post-it® Notes can be used to great advantage in brainstorming. There are generally two techniques that are used, and both utilize the large 4-by-6-inch notes. In the first method, people write their ideas on the notes and as they write they read out their idea (this is closer to "free" brainstorming). In the second method, the team members write on their Post-it® Notes and then at the end of the idea generation they come up and place the notes on flip charts or on the wall (this is more like structured brainstorming).

Post-it® Notes can also be very useful for moving to idea selection. You can group the ideas under themes, eliminate duplicates, and use the notes for ranking as well. In addition, placing the notes on flip charts or a wall helps to get people up and moving. This encourages creativity and team building.

2. Free Brainstorming

2.1. When to Use

Brainstorming should be used with a well-defined question or problem. It is strictly for uninhibited idea generation, not for selection. Free brainstorming generally works well if everyone on the team knows one another and they are comfortable together. If you start to free brainstorm and find that one or two people are dominating, then it is a good idea to move to a more structured style so that all members of the team can participate.

2.2. How to Use

In order to free brainstorm, you will need someone to lead the session and to record and show the ideas generated. You may wish to use flip charts or a computer and projector. In free brainstorming, one person records the ideas as they are generated by the group and leads or facilitates the brainstorming. The person who is recording will need to be able to record quickly and to respond when the ideas start to slow down. Your brainstorming sessions should always have a set time limit and the leader should indicate the time to add some competitive pressure to provide more ideas. The first round of brainstorming should be about 15 to 20 minutes. You should then brainstorm again on the same topic after a break. When people are required to brainstorm more ideas, they become more creative.

2.3. Steps

1. Set up a comfortable atmosphere for the brainstorming.
2. If the team does not know one another, use a fun 5-minute "warm-up" to brainstorm around something silly—for example, all the uses of brick or of a pink balloon. Remember that laughter is helpful for the creative process.
3. Generate ideas out loud in a group for 15 to 20 minutes, recording all ideas.
4. Eliminate the duplicates and then move to narrow the number of ideas through multi-voting.

If you have time:

1. After step 4, take a break.
2. Reconvene and generate ideas for 5 to 10 more minutes.
3. Collect all the ideas and go back to step 4.

2.4. Strengths

Free brainstorming allows the team members to build on each others' ideas and to generate a large number of creative ideas in a relatively short space of time.

2.5. Challenges

The chaos of free brainstorming can make some members of the team uncomfortable. They may be quieter, or think in a less lateral fashion, or may be less practiced than other members of the group. If this is the case then you can end up in a situation where only a few members of the team are generating the ideas, while others do not participate. It is important to change the style at this point. Everyone has good ideas and they need to be able to express them.

If you are not brainstorming around a relatively well defined problem, free brainstorming can sometimes be so unfocused it is useless.

3. Structured Brainstorming

3.1. When to Use

Structured brainstorming can be used for requirements or concept generation. Structured brainstorming is often used when the team is new and people don't know one another, when there is the potential for conflict, or when it is difficult to get ideas from all the team members. It is a technique that reduces production blocking and free

riding. It is often also used as part of the nominal group technique (NGT), which combines idea generation with idea selection or decision making.

3.2. How to Use

In structured brainstorming, you follow all the same concepts as for free brainstorming but you make a small modification.

3.3. Steps

1. Set up a comfortable atmosphere for the brainstorming.
2. If the team does not know one another, use a fun 5-minute "warm-up" to brainstorm around something silly—for example all the uses of brick or of a pink balloon. Remember that laughter is helpful for the creative process.
3. Provide 15 to 20 minutes for everyone to generate ideas on their own by recording their ideas on a piece of paper or Post-it® Notes.
4. The leader then goes around to each member of the group and asks for the person's first two ideas. This is repeated until all the ideas that people have recorded are up on one list. Duplicates will be automatically eliminated.
5. Move to narrow the number of ideas through the techniques described under idea selection.

If you have time:

1. After step 4, take a break.
2. Reconvene and generate ideas for 5 to 10 more minutes.
3. Collect all the ideas before going to step 5.

3.4. Strengths

Structured brainstorming eliminates the chaos of free brainstorming and it is therefore generally easier for everyone to take part, particularly if the team is relatively new and the team members do not know one another very well. It is also effective when the topic of the brainstorm session may cause conflict.

3.5. Challenges

Structured brainstorming makes it more difficult to build on other people's ideas. This is why it is recommended that you do several rounds of structured brainstorming in a session; this gives people an opportunity to build on one another's ideas.

KEY TERMS

brainstorming	**free riding**	**social inhibition**
production blocking	**free brainstorming**	**structured brainstorming**

4. Questions and activities

1. Try using structured and free brainstorming using the following triggers. Which method works best with your team? When brainstorming it is very helpful to have an example of the trigger (e.g., an actual coffee cup).

a. Brick

b. Pink balloon

c. Screwdriver

d. Coffee cup

e. Nail polish

f. Ping Pong ball

g. Pick something out of your backpack or from around your room to use as a trigger. If you have access to a kitchen, items like spoons, sieves, pot lids, and the like work well.

2. Brainstorm a list of ideas triggered by putting together two unrelated ideas. To do this, write down a list of ideas on separate small pieces of paper (see examples below, and add your own ideas). Put them in a container and shake. Draw two out and brainstorm ideas triggered by this combination of items. Reverse the order of the items and see if this triggers more ideas.

Family	Red	Triangle
Chair	Purple	Circle
Grapes	Smart	Water
Smooth	Guitar	Paper
Malleable	Belt	Desk
Fish	Pen	Light

3. Conduct a brainstorming session for the design project you are working on currently to identify additional requirements or to develop a list of solution ideas.

5. References

[1] Osborn, A.F. *Applied Imagination: Principles and Procedures of Creative Problem Solving,* 3rd rev. ed. New York: Charles Scribner's Sons, 1963.

[2] Stroebe, W, Nijstad, B.A., and Rietzschel, E.F. "Beyond Productivity Loss in Brainstorming Groups: The Evolution of a Question." In *Advances in Experimental Social Psychology,* vol. 43, San Diego: Elsevier, 2010, pp. 157–203.

[3] Osborn, p. 139.

[4] Osborn, p. 142.

[5] Gallupe, R.B., Dennis, A.R., Cooper, W.H., Valacich, J.S., Bastianutti, L.M., and Nunamaker, Jr., J.F. "Electronic brainstorming and group size." *Academy of Management Journal* 35, 2 (June 1992):350–369.

[6] Connolly, T., Jessup, L.M., and Valacich, J.S. "Effects of anonymity and evaluative tone on idea generation in computer-mediated groups." *Management Science* 36, 6 (June 1990):689–703.

3 Creativity Methods

#skill/tool module: #creativitymethods

Learning outcomes

By the end of this module, you should demonstrate the ability to:

- Describe several *creativity methods* for generating new ideas and explain how to make use of these methods
- Apply the methods to generate new alternative ideas
- Foster a creative atmosphere with your team

Recommended reading

Before this module:

- **Design Process > Idea Generation > 2. Brainstorming**

After this module:

- **Design Process > Idea Generation > 4. Morphological Charts, Analogy, and TRIZ**

1. Creativity

The work of an engineer is to create new technology from existing science to create solutions to problems; to create new products and processes. Standard work is done primarily by technicians; engineers are in place where the work may generate situations that require original thought, solutions that are not in the manual, or choice before application. When Chester Carlson set about to produce a method of copying patents, his explorations took him off in a new direction, an exploration that eventually netted him the patent on xerography, the basis of modern photocopiers [1].

Many students, even ones doing very well at their engineering schoolwork, and even many graduated engineers, will latch onto the first workable idea that solves a problem. Good design engineers will look more broadly than the first solution that seems feasible, even if it is the standard industry solution, just in case a better solution exists for this particular situation. With the advances in technology and products, new opportunities are constantly being presented. To jump at the obvious or the first idea is to miss these opportunities.

The Basic Key to Creativity

As hard as it may be to grasp, the key to creativity is mistakes! For every great idea there are dozens, maybe hundreds, that are silly, demonstrably incorrect, infeasible, or inferior. The most creative among us will go through a tremendous number of ideas in order to find the different, most wonderful new idea that solves a problem better. The key to creativity is to be open and fearless!

With this in mind, we will look at ways to increase your creative abilities. These techniques can be used most effectively when you are brainstorming (#brainstorming).

2. Expanding Creative Thought: Lateral Thinking

Edward de Bono is a creativity guru who coined the term *lateral thinking* in 1967 [2]. The underlying motivation for lateral thinking is summed up nicely in this quote: "You cannot dig a hole in a different place by digging the same hole deeper" [3]. Thinking about a problem when you already have a solution in mind is like standing in a hole with a shovel. The temptation is to keep digging down, to make more and more excuses for a solution that doesn't work well. Lateral thinking means moving to the side, starting a new hole in a new place.

> **Put off your Review of the State-of-the-Art**
> Because of the tendency for novice designers to jump at an early feasible solution, we recommend that they start with some idea-generation before reviewing the state-of-the-art since the state-of-the-art is nothing more than a hole dug for you by someone else. While you should strongly consider any available solution, this should never be done without some thought about alternatives.

De Bono and others have described a number of techniques for stimulating lateral thinking. Just a few are presented here.

- *Challenge assumptions:* Assumptions that are generally accepted may not be right, merely passed on from person to person. James Dyson, who invented a new way to vacuum dirt, asked the question, "What if you don't need a filter to separate dirt from air?" [4]
- *Ask "Why, why, why?"* A child sometimes asks "Why?" over and over again. Doing this as an adult can help you challenge your assumptions.
  - Why do people use money to buy things? Why do people carry several different credit cards? Why doesn't the debit or credit card machine just know who someone is?
  - Why are drink bottles cylindrical?
  - Why can people make fuel from corn (ethanol) but not from other plants, like grass (hay)?

This method is also called *5 whys*, because of the typical number of "whys" you need to ask to get to the root cause of a problem.

Creativity Methods

135

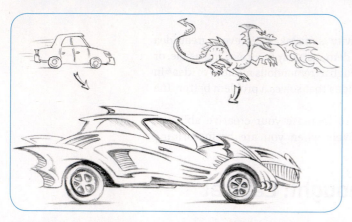

FIGURE 1 A randomly chosen concept can be used to stimulate new ideas or as the basis for a warm-up exercise before a brainstorming activity.

- *Reversal method*: Reverse the direction or sequence of things. The reversal may not lead to a valid solution, but will stimulate divergent thinking. Rather than worry about how a gasoline engine is going to drive the tires in an automobile, think about what the tires have to do and what that could mean about how they are driven.

- *Random stimulation*: Seeking some random word or idea and incorporating it in your line of thinking (see Figure 1). What if your solution had to include a monkey? Saltwater? A triangle? Music? This may not give you a solution, but it may open your mind to many divergent solutions that don't follow standard-practice thinking. This will take your creative thoughts away from the ordinary and get you started on the truly different.

- *Blue sky thinking*: Imagine what is possible with no constraints. In blue sky thinking you try to ignore any preconceived notions about what is possible or impossible and try to imagine ideal solutions if there were no limits.

2.1. A Few Other Strategies

Here are some other basic strategies used by professional designers when they are trying to generate ideas—for example, securing a horse to a tree:

- Be a bit crazy, off-the-wall, and absurd. Think about solutions to the problem that are not feasible or practical: skyhooks, teleportation, thousands of insect slaves. Suspend judgment. You need to secure a horse to a tree. Use a force field!

- Be skeptical whenever anyone makes a statement about the problem. Is the statement always true? Can you think of any circumstance or part solution that will do the opposite of the statement? Can you add to the design to remove an objection? You need to secure a horse to a tree. Or do you? Maybe the real problem is just keeping the horse around so it's available when you want it again. This is a version of the "why-why-why" strategy.

- Remove the possibility of the standard solution. You need to secure a horse to a tree, but can't use the standard solution of a rope. How do you then solve the problem?

- If one or more of the design conditions seems to be a sticking point, throw it or them out and create solutions for the remaining problem. Then try to adapt your ideas to cover the removed conditions. Throw out the horse idea and call a cab!

- Understand, tolerate, and encourage all of these behaviors in your teammates. Often great ideas come from the interactive, off-the-wall banter between active minds.

3. Using One Idea to Generate Others

One way to generate new ideas is to look at existing solutions and try to expand or modify them to fit the new situation. Take a solution and an objective and extend the solution to perform better in terms of that objective to create another solution.

3.1. *SCAMPER*

SCAMPER is an acronym for a set of suggestions proposed by Robert Eberle to trigger novel design ideas from existing solutions [5]: Substitute, Combine, Adapt, (Modify, Magnify, Minify), Put to other uses, Eliminate, Rearrange or Reverse. SCAMPER is one of the techniques referred to as ***force-fitting***; it forces the mind to make creative jumps. This means it is particularly useful as a creative tool when a team is stalled finding solutions, since it can trigger fresh trains of thought.

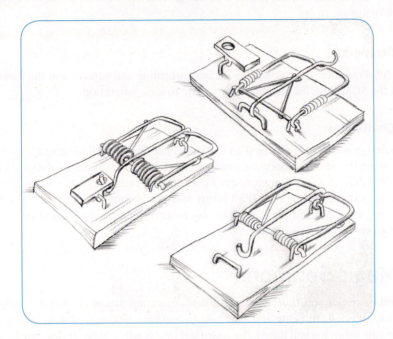

3.2. How to Use [5]

Substitute: What might you replace or exchange for something else. What if you used a different person, place, material, ingredient, process, or approach?

Combine: What could you put together to solve your design or problem? Can you combine material, products, or purposes?

Adapt: How might you change a part of the design or problem to solve it? What might you do differently? What can you copy? What other idea is it like?

Modify, **M**agnify, **M**inify: Can you add or subtract something to solve your design question, can you make something larger or smaller or increase or reduce something?

Put to other uses: How might you use something in a different way to solve the design or improve it? Change the context of the problem.

Eliminate: What might you remove to improve or solve your design problem? Can you remove a function or a unit without affecting the result or would it improve it?

Rearrange or **R**everse: Can you interchange components within the design? Could you reverse it, turn it upside down or look at backward?

3.3. Steps

1. Set up a comfortable atmosphere.
2. Read the SCAMPER question out loud in your team (you can go in any order).
3. Write down the ideas that are generated by the team (this can be done as a group or individually).
4. Continue to apply the questions until no ideas flow. Note that some questions may not be applicable to your problem or design and therefore no ideas will flow. It is fine to just move on.

3.4. Strengths

SCAMPER questions act as triggers for brainstorming and can expand the ideas. The use of the SCAMPER list also brings structure to idea generation.

3.5. Challenges

Some design problems may need an expanded list of potential solutions to generate ideas using this method because you need to start with potential solutions that fit the SCAMPER structure. For some projects you may need to create your own trigger words, again because the SCAMPER terms may not be applicable to a particular project. Using the SCAMPER structure can also sometimes limit the creative energy that comes from the free flow of ideas.

4. Magic Solutions

Generating *magic solutions* during the idea generation process is useful. It helps the team expand their thinking and may lead to feasible solutions. At some point, however, magic solutions will have to be reworked into feasible ideas or discarded.

What Is a Magic Solution?

Suppose your project includes the requirement to count red blood cells in a blood sample. A solution with "magic" in it might say "Somehow this device will do the red blood cell count" or "Red blood cells will be counted by getting the red blood cells to count off."

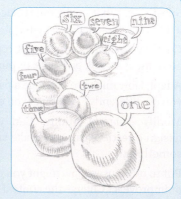

A derived solution without magic might be "The device will force the samples through micro-channels in a plastic substrate where a camera mounted on a microscope will relay a video picture to a processor that will recognize and count the red blood cells." This method may or may not work as described, but it is at least one possible idea that employs realistic technology. The magic idea can be a stepping stone to workable ideas that otherwise might not occur to your team.

lateral thinking	why, why, why (also called 5 whys)	reversal method
random stimulation	SCAMPER	force-fitting
magic solutions	blue sky thinking	creativity methods

5. Questions and activities

Here are some design creativity exercises for your team:

1. What features would you design into a table-sized tablet-like computer that was the top of someone's desk?

2. What features would you design into a table-sized tablet-like computer that was a conference table?

3. What could a walking-stick do?

4. Puppies often will not return when called. What device could you design that would encourage the proper behavior?

5. Use the techniques discussed in this module to generate potential solutions for the design project you are working on currently. Document your thinking process and results in your engineering notebook.

6. References

[1] *The Great Idea Finder: Chester Carlson* [Online]. Retrieved from www.ideafinder.com/history/inventors/carlson.htm

[2] De Bono E., *The Use of Lateral Thinking*. London: Cape, 1970, 1967.

[3] De Bono E., *Lateral Thinking* [Online]. Retrieved from www.edwdebono.com/lateral.htm

[4] Dyson J., *Against the Odds: An Autobiography*. New York: Texere, 2003.

[5] Adapted from Eberle, R. "SCAMPER: Games for Imagination Development." Buffalo, NY: D.O.K. Press, 1990.

Creativity Methods

Morphological Charts, Analogy, and TRIZ

#skill/tool module: #morphchart, #analogy, #TRIZ

Learning outcomes

By the end of this module, you should demonstrate the ability to:

- Explain how to use morphological charts, analogy, and TRIZ as idea generation methods.
- Apply your knowledge of these methods to effectively develop new alternative solutions to a design problem.

Recommended reading

Before this module:

- **Design Process > Idea Generation > 3. Creativity Methods**

Alongside this module:

- **Design Process > Multi-use Design Tools > 3. Decomposition**

After this module:

- **Design Process > Decision-making > 1. Design Evaluation and Selection**

1. Morphological Charts

A ***morphological chart*** is a simple graphical representation of the process of putting ideas together to make some realistic integrated solutions for a whole problem. This method builds on decomposition, which is just a more structured and effective way of thinking about the design. Regardless of which method you used, you need to build integrated solutions from the various pieces.

In a ***morphological chart***, or ***morph chart*** as it is commonly known, row headers represent the subfunctions, and columns contain ***means*** of accomplishing the subfunctions. Ideas are added to the rows in no particular order. An integrated design is represented by a path through all subfunctions from top to bottom (see example in Figure 1).

1.1. When to Use

Morphological matrices, as they are sometimes called, or morph charts are used to find new combinations of functions and means. They can, therefore, be used to help populate the design space. They are used after the functions and specifications have been determined and you are now looking at generating alternative designs.

1.2. How to Use

In using a morph chart for generating alternative designs or solutions, you list the subfunctions or features in the left column of a chart and then the possible ways in which the functions or features can be implemented across the rows. (See the example chart.) You then "mix it all up" to get multiple designs or solutions. The easiest way to do this is to number the ideas across a row and then randomly put them together. You can use dice or a random number generator to force you to think about unusual combinations.

1.3. Steps

1. List your subfunctions or features down the left-hand column.

2. List all the means or methods of achieving each subfunctions across each row.

3. Now connect the means or methods in different patterns randomly.

4. Continue until you have explored a wide variety of combinations.

1.4. Strengths

Morphological charts expand the design space and can lead to creative solutions to well defined problems.

1.5. Challenges

These charts can get very large and unwieldy very quickly.

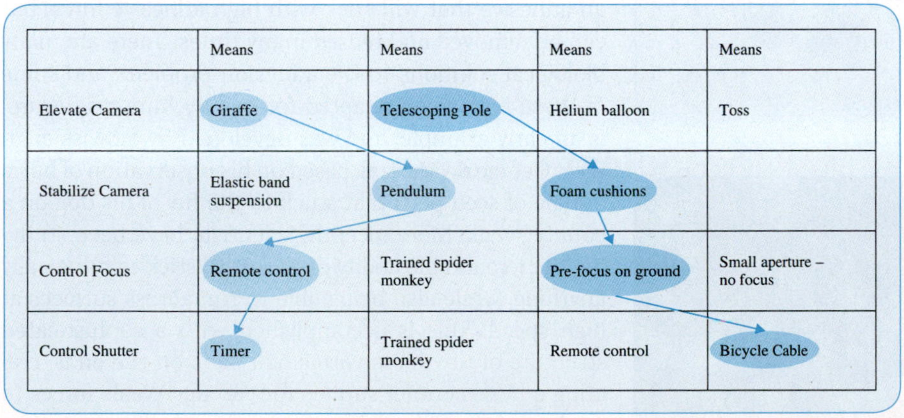

	Means	Means	Means	Means
Elevate Camera	Giraffe	Telescoping Pole	Helium balloon	Toss
Stabilize Camera	Elastic band suspension	Pendulum	Foam cushions	
Control Focus	Remote control	Trained spider monkey	Pre-focus on ground	Small aperture – no focus
Control Shutter	Timer	Trained spider monkey	Remote control	Bicycle Cable

FIGURE 1 Example of a morphological chart for elevating a camera to get a wide-angle picture at a wedding (#aerialphotography case study).

Morphological Charts, Analogy, and TRIZ

141

2. Analogy

Analogy is a powerful inventive tool in which solutions to similar problems in other fields are used. It has been discussed by De Bono and many others. To use a **technical analogy**, you must ask, "What other technical problems are similar to the current problem, and how were they solved? Consider, for example, the adaptation of jet engine technology to high-speed steam nozzles used for cleaning industrial boilers. James Dyson (the inventor of the cyclone vacuum cleaner) also developed a commercial hand dryer known as the Dyson Airblade™. We don't know how the solution was developed, but it might have been by technical analogy. Conventional washroom hand dryers use hot air to evaporate the water from hands, a very energy-intensive method. To use analogy, an engineer might ask, "What other things are dried industrially, and how is it done?" In a car wash, there is no way to economically evaporate all the water used. Instead, the water is blown off with an air knife. The Dyson Airblade™ uses this idea to dry hands.

Another type of analogy that has become an important field of academic and industrial research is analogy to the living world. The process of adapting solutions from the biological world to engineering design problems is called **biomimetics**. It is based on the knowledge that evolution has, in many cases, produced highly optimized solutions for problems similar to those that must also be solved by engineers for man-made objects and systems.

Scientists routinely study the structure of materials such as spider silk, rams horns, and oyster shells for clues about how to make strong and tough materials. They study the many chemical processes used by life to process food. They study the decision-making processes underlying the behavior of a flock of birds or school of fish to create simple swarming robots for large-scale tasks such as cleaning up oil spills. Many fields of technology have a biological analogy.

To make use of the **biological analogy**, you must simply ask, "How does nature solve this problem?" As an example, assume that your task is to come up with an adhesive that will stick with high adhesive force, but can be removed and reused many times. There are many biological solutions to the adhesion problem, and some of these have been adapted for use by humans. Velcro® is an early example, and was developed by a Swiss engineer, George de Mestral, based on his observation of burrs (a type of seed pod) that stuck to the fur of his dog on a country walk. More recently, scientists have been studying the remarkable ability of geckos to stick to practically anything, while also being able to run across surfaces at high speed. This is accomplished with a sophisticated structure of tiny hairs with small pads on the ends, creating a huge contact surface for van der Waals forces to work on. Scientists and engineers are searching for ways to recreate this micro-structured surface to make super adhesives. A football wide receiver with "gecko gloves" would have a serious advantage.

Courtesy of the Authors

Courtesy of the Authors

DESIGN PROCESS

3. TRIZ

TRIZ is an acronym for the Russian phrase "Theory of Inventive Design," developed by Russian inventor and author Genrich Altshuller [1]. Altshuller spent a great deal of time reviewing patents for the Navy, and in 1946 began to develop a system of invention based on the recurring elements he found. He came to the conclusion that the need for an inventive design arose when there were technical contradictions inherent in a problem. For example, if you need to create a perforated rubber hose to water a garden, you need a hose that is flexible for use, but it is difficult to drill holes in a flexible tube [1]. A typical engineering compromise would be to make a hose with intermediate flexibility. In contrast, an inventive solution for this problem is to freeze the hose before drilling the holes.

Changing the state of the physical property is one of the 40 inventive principles that Altshuller identified as appearing over and over in the solution of inventive problems. It is this repetition that allows technical analogies, described earlier, to be extremely useful. Other principles include things such as separate in time or space, or nesting one thing inside another. For a list of all 40 key inventive principles with examples, refer to www.triz-journal.com/archives/1997/07/b/index.html.

KEY TERMS

morphological chart (morph chart)	technical analogy	biomimetic
biological analogy	TRIZ	analogy
means		

4. Questions and activities

1. Practice applying these three techniques to a design problem. Pick one of the following, identify the essential subfunctions of each, and apply the creativity methods explained this section to develop alternative solution ideas.

   a. A pencil sharpener

   b. A hairbrush or toothbrush

   c. A watch

   d. A fork

   e. A computer mouse

   f. Another simple, ordinary piece of technology that you use

2. Try using these three methods (morph charts, analogy, and if it is applicable, TRIZ) to the design project you are working on currently. Document the process of applying the technique and the results in your design notebook.

5. References

[1] Altshuller, G. *And Suddenly the Inventor Appeared: TRIZ, the Theory of Inventive Problem Solving*. Worcester, MA: Technical Innovation Center, 1996.

Morphological Charts, Analogy, and TRIZ

1

Design Evaluation and Selection

#process module: #designevaluation

Learning outcomes

By the end of this module, you should demonstrate the ability to:

- Take a set of potential solutions and from them select an optimal solution for implementation

Recommended reading

Before this module:

- **Design Process > Idea Generation > Modules 1 through 4**

After this module:

- **Design Process > Decision-making > 2. Selecting a Design Solution**

1. Introduction

The design process includes generating and selecting ideas repeatedly before proceeding to detailed design and execution. Coming into this module you should have already generated a list of potential design solutions. In this module we focus on the evaluation that results in design selection.

You may perform the evaluation and selection process with only one or a few potential solutions. This can be important as a test and generator of your project requirements, and to test your selection methods. However, ultimately it is very important to explore the design space well in the solution generation and evaluation processes. You do not want to miss good potential solutions because you were trying to be quick, or you too quickly closed your mind because you have a solution idea that you think "works."

Enter this process with the following:

- *A set of well-constructed requirements.* Earlier you generated the project requirements, your goals for the project. Your requirements should be solid at this point. These are your ultimate target and every decision you make during this process will reflect this. If the target is wrong, your analyses will be incorrect.

- *A list of potential solutions*. Ideally, many diverse potential solutions. These could be solutions to the whole problem, or part of the problem. These potential solutions should have enough detail such that implementation methods seem within the realm of current practice, although it may require some extensions of current practice. This means there is no "magic" in the solutions on your list, although there can be methods that require analysis to determine if they actually will perform as required.

1.1. Goal of the Evaluation and Selection Process

The requirements generation process created a set of functions, constraints, and objectives. Any solution must meet the functions and constraints, and will be compared to other solution candidates using the objectives as the evaluation criteria.

You may find the evaluation and selection process indicates you should be revisiting your requirements or your idea generation process. Even moderately involved projects are not easily fully understood on the first attempts and experienced engineers will expect to reach a good solution only through iteration of these design steps.

1.2. Inside the Evaluation and Selection Process

The evaluation and selection process is conceptually fairly simple. Potential solutions are compared to the functions, constraints, and objectives, and compared to one another. The potential solution that best fits the requirements (criteria) is selected. This is illustrated in Figure 1.

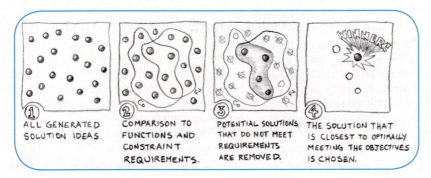

FIGURE 1 The best solution is the one that meets all functional requirements and constraints, and optimally fits the objectives.

2. Steps of Evaluation

When you were generating solutions it was important *not* to discard potential solutions too soon. This allowed a wide range of potential solutions to be generated. Now you need to change your perspective; you must judge your ideas.

There are 3 steps in evaluation:

Step 1: Assess each design idea against the functions and constraints in the requirements, and against practical implementation limits. The functional and constraint boundaries are shown in Figure 1. Any solution that falls outside these boundaries is discarded because it does not meet basic requirements.

Step 2: Use the objectives to compare design ideas against one another and select the idea that best fits the objectives. The goal is to find an optimal solution that is closest to ideal.

Step 3: Reflect on the chosen idea and the selection process. Does it make sense? Are you satisfied with the solution? What questions need to be answered about this solution before proceeding to the next stage of the process?

3. Step 1: Evaluating Individual Solutions

In Step 1 each potential solution is assessed against the functions and constraints specified in the requirements. A solution must conform to the required functions and constraints, or must be adjusted so it conforms to these requirements. Otherwise it is discarded.

To accomplish Step 1, each potential design must be well enough described that the design and implementation can be evaluated. Any undetailed parts of the solution must be straightforward, standard practice, or purchasable parts. For example, in the design of a new gasoline engine (e.g., a traditional car engine), a solution may call for an "electronic ignition timing system" to control when the spark plugs will spark, igniting the air/fuel mixture in the cylinder. Any potential solution must include a description of the technology that will be used to control the spark plug timing or specify an existing technology.

Choosing among Potential Solutions for "Electronic Ignition Timing System"

Suppose there are several proposed solutions for the electronic ignition timing system, for example: dedicated discrete electronics, a dedicated microprocessor, a part of the processing time of another microprocessor or programmable hardware cell arrays (called FPGAs), or a mechanical system. The proposed design solutions must be this detailed, because it must be possible to determine if a potential solution meets the requirements (e.g., fast enough, temperature tolerant enough) and is feasible (e.g., are there large enough FPGAs available to perform this job?).

In summary, for each proposed design idea, we determine:

1. Does it meet the functions? (Will it be able to control the spark plug?)

2. Does it meet the constraints? (Will it meet government regulations and safety requirements?)

3. Is it feasible? (Does current available technology exist to implement this solution?)

Any proposed solutions that do not satisfy these basic evaluations are considered for adjustments, as shown in Figure 1. If no adjustments are possible to make the solution work, the solution is discarded.

From this step we could exit with many potential solutions that satisfy both the required functions and constraints. It is a general problem if there is no or too few solutions.

Problem: Too few Potential Solutions Pass Step 1

Sometimes few or none of your proposed solutions will pass the first step. If the design team has spent considerable amounts of time looking at alternative solutions then the project may have to be cancelled, but the following steps should be explored first:

- Revisit idea generation. Now that you have had the experience of evaluating some ideas against the functions and constraints, perhaps you can use this experience to inspire new ideas that avoid the problems encountered with your previous ideas.

- Talk to the client about relaxing or changing the functions or constraints. Possibly bring solutions that almost but don't quite meet the existing requirements to the meeting with the client. Show them that a solution is close, but just outside the existing bounds. Be prepared to back up your claims with a logical, well-documented analysis of the situation.

Example: Modifying a Design Idea to Move It Inside the Constraint Boundary

Suppose you are creating a design for cutting fabric into custom shapes and you propose a cutting machine as a solution. However, the design fails a safety constraint as first proposed, because there is a possibility that the operator could contact the cutting blade and be injured. The proposal could be adjusted by adding an operator guard to the design. In this way the design can be adjusted so that it meets the safety constraint. The increased costs and complexity caused by the guard would be included in the design solution during the next steps in evaluation.

---

4. Step 2: Comparative Evaluation

Proposed solutions that pass Step 1 can now be compared against one another. This is done using the full set of objectives: the solution that best fits the objectives is the best solution. Not all objectives will be met or maximized by any particular solution. Generally no solution will work perfectly; it is up to the engineer to balance the trade-offs. Doing these comparisons often requires using **metrics** to create data that can be used for comparison purposes. A metric is a methodology for measuring, calculating, or estimating the characteristics of a design idea. Metrics can be quantitative (e.g., estimating the speed of a microprocessor) or qualitative (e.g., using a focus group to get people's opinions on taste or visual appeal). There needs to be a metric for each of the important objectives. Without metrics, and resulting data that comes from implementing the metrics, there is no rational way to compare ideas (#metrics).

Once you have data, or estimates of the characteristics of each design idea, you are then in a position to compare the ideas. There are a wide variety of methods that can be used for this comparison process.

At the last pass through the process this stage should result in the one (or a few) proposed solutions to undergo detailed design and implementation.

5. Step 3: Reflection

You exit the evaluation and selection process in one of three ways.

1. One solution. If you feel you have generated an adequate set of solutions, have a comprehensive set of requirements, and have selected the optimal solution for these requirements, then you can proceed to detailed design and implementation of this one solution.

2. No solution. None of the solution ideas you have generated will work or will meet the requirements to an adequate degree. You need to decide if you can generate additional solutions, change the requirements, or if you have exhausted the possibilities and should terminate the project.

3. You leave to revisit the solution generation process. This is actually the most usual result. You may have eliminated some of the inferior, unsalvageable potential solutions. You take with you the information you generated by evaluating some of the potential solutions. This information may be essential for:

   - Adjusting and combining the solution ideas to generate new, better solutions.
   - Adjusting and adding to the requirements based on discoveries during the evaluations. Often you will find that small adjustments and additions are needed in the earlier work on the requirements, and sometimes your discussions of potential solutions with the client will reveal new functions and remove others.

Normally you will exit through result 1 once, with the final design solution, but through result 3 many times before that. When you leave this process for the last time, you have selected the solution you will work with.

The next steps of detailed design and of implementation are very costly and take a considerable amount of time. If you choose a solution that proves overly difficult or does not meet the goal and project requirements, you will have major problems to deal with.

Like most engineering work, there is huge benefit from consideration of the quality and comprehensiveness of the decisions being made. Reflect on the result: Is it reasonable to assume it is the best or close to the best solution that you could have reached under the circumstances of the decision?

6. Leaving This Process Module

A decision about what to do next

- Either one solution (or a couple of variations) that your team will work on further to detail and implement.
- Or a decision to revisit one of the other design process stages before deciding on a solution, and agreement on what to do next.
- Or a decision to cancel the project (unlikely for a design project you are doing for a course in engineering school!).

Additional knowledge that will prepare you for the next stage of the design process

- From the evaluation process you should have a deepened understanding of the design problem, and of the technical issues related to the solutions you have generated.
- This knowledge is useful if you are revisiting the requirements or idea generation phase.
- If you have selected one design solution with which to move forward, then you should have a more solid understanding of the technical implementation requirements for this solution.

Documentation of your decision process

- Documentation should support the decision your team has made.

KEY TERMS

metrics

7. Questions and activities

1. Explain, in your own words, the three steps that are used to evaluate potential solutions.

2. In each step you are comparing the potential solutions to the project requirements. Review the solution list you have generated and the requirements you have developed for the design project you are working on currently.

   a. Are the requirements clearly defined enough to use for this evaluation?

   b. Are the solution ideas detailed enough to use for this evaluation?

2 Selecting a Design Solution

#skill/tool module: #decisionmethods

Learning outcomes

By the end of this module, you should demonstrate the ability to:
- Select an appropriate decision tool
- Understand and guard against the problems with the decision tools adopted

Recommended reading

Before this module:
- **Design Process > Decision-making > 1. Design Evaluation and Selection**

Alongside this module:
- **Design Process > Multi-use Design Tools > 6. Pairwise Comparison**

After this module:
- **Design Process > Decision-making > 3. Decision Methods for Teams**

1. Introduction

This module describes **decision-making methods**. Your team may "feel" that you already know what choice to make. However, this can result in a biased decision and it does not provide a good justification for a choice. The methods described here are used to try to remove bias and ground your decisions firmly in engineering reasoning. They also will assist you in developing the justification for your decisions when you document your design process and reasoning.

The methods presented here can be used for a variety of different tasks both in the engineering design process and in other applications. They can be used to choose which phone to buy or what to do on a Saturday night. However, here the methods will be explained in the context of selecting a design solution to pursue further (i.e., reducing the design space).

The methods described fall roughly into three categories, as shown in Figure 1. Simple models are relatively easy to use and thus appropriate when a quick decision

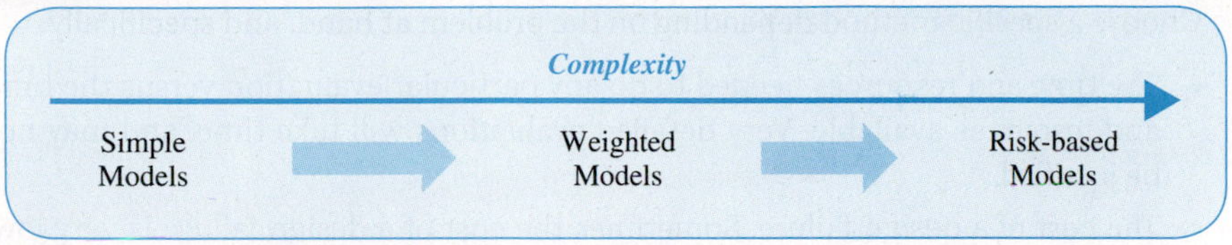

FIGURE 1 **There are a range of decision methods, from simple methods to complex risk-based methods.** Which method you should choose will depend on the type of decision you are making.

must be made, and when a poor decision is not too damaging. Weighted models require more work, but the results are more reliable (see Table 1). Risk-based models add the effects of probability to a decision, and thus are more difficult to use. However the results will give more insight into the decision such as information about the risk a decision incurs. You may decide to use a simple method to eliminate some clearly inferior alternatives, then a more complex method to make a final choice.

Choosing Techniques for Solution Comparison and Selection

At this point you should have enough information about each design alternative to evaluate each idea with respect to the key objectives.

TABLE 1 **Characteristics of the methods.**

METHOD	COMPARES	COMPLEXITY OF USE	CONFIDENCE IN RESULT	USE
Multi-voting	Multiple alternatives	Simple	Low	Rank alternatives. Captures the design teams collective opinion of the alternatives. Prone to bias.
Graphic decision chart	Multiple alternatives	Simple	Moderate	Compare based on objectives to eliminate those alternatives significantly underperforming other alternatives
Pugh Method	Multiple alternatives	Simple	Low	Compare several alternatives using one as the "middle" alternative across multiple objectives. Works best with only a few alternatives
Weighted decision matrix	Multiple alternatives	Difficult	High	Compare alternatives, allowing for the relative importance of each objective and for the scaled performance of the alternatives relative to what would be "perfect" for that objective.
Risk-based models	Multiple alternatives	Very Difficult	High	Evaluate alternatives where risk is involved

Choose a specific method depending on the problem at hand, and specifically:

- The time and resources needed to do any particular evaluation versus the time and resources available. Very detailed evaluations will take time, and may not be justified.
- The cost of a design failure. Sometimes the cost of a design failure is very low. If the design fails, there is little impact on people (welfare, economics, or otherwise). Other times the cost of failure is high: One would not want to find out that a bridge structure was inadequate after a major bridge was built! A high cost of failure justifies spending more time and resources on evaluation.
- The likelihood of design failure. If this is a design problem that requires a risky solution, then more effort should go into the evaluation stage before proceeding to implementation.

2. Simple Models

2.1. Multi-voting

Multi-voting is a good method for reducing a very long list down to a more manageable long list and it is used extensively by professional design teams. If you have had a good idea generation session, for example, you will have many ideas. If the team is good at this process there may be several hundred ideas generated in a 30 to 40 minutes session. A simple way to reduce this list so it is more manageable before you get into more detailed idea selection is to multi-vote. Multi-voting is often used as a ranking method. While multi-voting can be used to rank, there are other idea selection methods that are more powerful, so instead we suggest that multi-voting be used early in the idea selection process to narrow the design space to a long but manageable list of potential solutions for a design problem.

How to use

1. Make a clear list of the alternatives, and make sure there are no duplicate ideas on the list.
2. Each person then receives a number of votes, usually about 25 to 30% of the total number of ideas. So if there are 50 ideas, each person gets 10 to 15 votes.
3. Provide a colored pen, marker, or stickers to each team member.
4. Each person puts a check mark, or sticker next to their top choices.
5. Each person can only vote for each idea once. You can't put lots of your votes next to your favorite idea. And everyone has to use up all of their votes each round.
6. Remove those ideas with no votes or the lowest number of votes.
7. Repeat the process until you have reduced the number of ideas to a short list that will be evaluated in more depth.
8. Then move to another decision-making tool to evaluate the short list more thoroughly.

2.2. Graphical Decision Chart

A **graphical decision chart** (also called a graphical decision matrix or method) is used to compare ideas to two criteria—for example, to compare design solution ideas to the top two objectives on your objectives list. This method works particularly well if

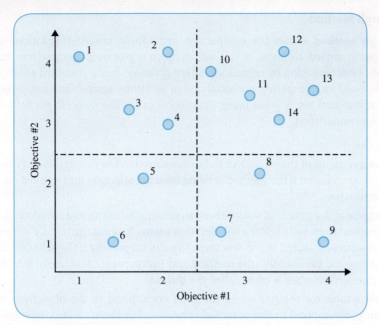

FIGURE 2 **A graphical decision chart is used to visually compare potential solution ideas in relation to one another and in relation to the top two objectives.** The blue dots represent potential solutions and they are placed on the graph using team discussion and comparison with other ideas.

there are two objectives, or criteria, that are clearly more important than the other evaluation criteria for the project.

How to use

1. Put the top criteria (objective #1) on the X axis of a graph and put the second most important criteria (objective #2) on the Y axis (see Figure 2).

2. Through discussion, score each design idea relative to objective #1 and objective #2. This is an approximate method and is only meant to compare the ideas qualitatively, but having a basic scoring system can help with discussion. For example: 1 = poor, 2 = satisfactory, 3 = good, and 4 = excellent.

In the example, design idea #6 is clearly inferior. It scores poorly on both objectives. Design idea #9 is very good with respect to objective #1 but is poor with respect to objective #2. Similarly design idea #1 meets objective #2 well, but fails to meet objective #1. The design ideas in the top right quadrant of the chart (ideas 10, 11, 12, 13, and 14) are clearly superior to the other ideas with respect to the top two objectives. Your team might decide through this process to eliminate ideas 1 to 9 from consideration and just focus on these top five.

Advantages: Quick evaluation of two objectives to eliminate alternatives.

Disadvantages: Ignores objectives that are not charted. Does not indicate the importance of each objective.

Variation: Computerize and put in more objectives (a multi-dimensional space instead of a two-dimensional space) with the disadvantage that additional evaluations must be done.

Selecting a Design Solution

153

2.3. Pugh Method

The **Pugh method** allows the comparison of multiple potential solutions against a selected standard solution. It is a variation on a pairwise comparison that also builds on benchmarking by comparing alternatives against a standard solution. It is generally used for comparing potential design solutions against an existing technology, or a standard set of goals using the objectives for the project (#benchmarking, #pairwisecomparison).

How to Use

1. Select a number of the objectives to be considered. All of the significant objectives should be included if the method is being used for selection and not just comparative evaluation.

2. Select one of the potential solutions, or an existing solution, as a standard. To make the method work well, select a solution that seems "around the middle" of several of the objectives selected in the first step. There is no penalty if the selected solution is not "around the middle"; the method may just generate less useful information. The standard selected is often called the **datum**.

3. Create a table (or matrix) where the rows correspond to the objectives and the columns correspond to the potential solutions. Put the standard, selected in the second step, into the second column.

4. Now you go through the process of evaluating each potential solution against the standard with respect to each objective. If the potential solution is about as good as the standard on a particular objective, put a "0" in the cell. If the potential solution is worse than the standard, put a "–1" in the cell. If the potential solution is better than the standard, put a "1" in the cell. Each cell running down a column should have a –1, 0, or +1. The column for the standard will be filled with S, denoting standard, or with the value 0.

5. Sum up the entries in each column and enter the total in the last row. These sums will give a relative indication of how each proposal compares to the standard, and to some extent to each other. The sum for the standard will be 0.

This example of the Pugh method considers can openers, with the objectives of safety, operation time, cost, and counter space. In this case, solutions that are *safer*, require *less* time to open a can, cost *less*, and use *less* counter space than the standard will receive a "1" in every cell in the column. We would create the Pugh Matrix shown, using a manual two-handled can opener as the datum.

TABLE 2 The Pugh method.

SOLUTIONS OBJECTIVES	STANDARD	ALTERNATE SOLUTION 1	ALTERNATE SOLUTION 2	ALTERNATE SOLUTION 3
	S			
	S			
	S			
	S			
SUM	0			

TABLE 3 Example of Pugh matrix.

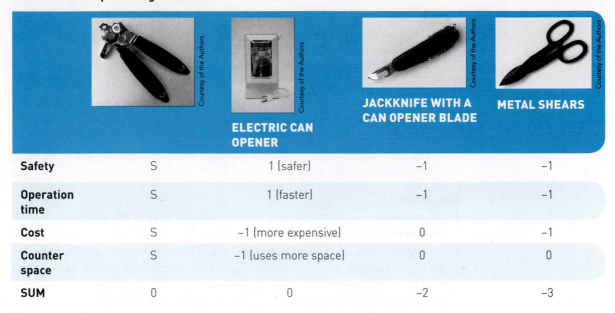

	Two-handled manual can opener	ELECTRIC CAN OPENER	JACKKNIFE WITH A CAN OPENER BLADE	METAL SHEARS
Safety	S	1 (safer)	−1	−1
Operation time	S	1 (faster)	−1	−1
Cost	S	−1 (more expensive)	0	−1
Counter space	S	−1 (uses more space)	0	0
SUM	0	0	−2	−3

By this rating, the two-handled manual can opener and the electric can opener are about the same, the jackknife can opener is third, and the shears are last.

Advantages

- Quick and simple to use. This allows the quick rejection of poor alternatives, such as the shears in the example.

Disadvantages

- Selections are often not differentiated by the sums.
- The significance of each of the objectives is considered the same. In the example safety and counter space are considered equivalent in importance in the sum. In reality this is probably not true; one of these objectives (i.e., safety) would be more important.

Variations

- Instead of just assigning +1, 0, and −1, assign larger ranging numbers (say numbers between +5 and −5) to further differentiate the options.
- Use the method to eliminate only one or two alternatives, then reevaluate.

3. Weighted Models

3.1. Weighted Decision Matrix

The *weighted decision matrix method* (also called a numerical decision matrix method) includes weights for each objective. Weights are numerical values. Important objectives are weighted higher (valued more) and less important objectives are weighted lower (valued less). This affects the decision, giving more priority to solutions that better fit the most important objectives.

This method can be used in any decision-making process where you are trying to make a decision among a set of alternatives, based on a set of objectives. For example:

- Deciding which design solution to pick: which one best fits the requirements, specifically the objectives.
- Deciding which project to fund if you have many projects to choose from and you want to pick the one that best fits your organization's objectives.
- Deciding which homework assignment to work on first if you have many courses and competing objectives (want to do well, homework assignments each carry a different weight and value, etc.).
- Deciding which outfit to wear; if you have many possibilities and competing objectives (comfort, style, etc.).

How to Use

1. Do a pairwise comparison, or use another method, to rank the objectives: O1, O2, O3, and so on, where O1 is most important.

2. Through discussion, choose a weight for each of these objectives. O1 should receive the highest weight. The weights are assigned as a percentage and the percentages of all the objectives should add up to 100%. See the table for an example (note that weights shown are only to illustrate; your weights will be different).

OBJECTIVE	RANK (FROM PAIRWISE COMPARISON)	WEIGHT (DETERMINED THROUGH DISCUSSION)
01	1	35%
02	2	30%
03	3	20%
04	4	10%
05	5	5%
Total		100%

3. Compare your alternatives against the objectives. Through discussion, decide how well each alternative meets each objective. See suggested scale and next table for illustration.

Suggested Scale: any value between 0% and 100% can be used

0%	Does not meet objective at all
20%	Meets objective very weakly
40%	Meets objective somewhat
60%	Mostly meets objective
80%	Meets the objective strongly
100%	Outstanding with respect to the objective

Example of rating design alternatives against objectives. Question asked: How well does this design idea meet this particular objective? Numbers do not need to add up to 100%.

OBJECTIVE	ALTERNATIVE #1	ALTERNATIVE #2	ALTERNATIVE #3
01	10%	50%	5%
02	10%	30%	90%
03	50%	70%	20%
04	90%	80%	5%
05	5%	40%	25%

4. Combine these two tables and add up the score in each column. Example of combined table:

OBJECTIVE	ALTERNATIVE #1	ALTERNATIVE #2	ALTERNATIVE #3
01	$.35 \times .10 = 3.5\%$	$.35 \times .50 = 17.5\%$	$.35 \times .05 = 1.75\%$
02	$.30 \times .10 = 3.0\%$	$.30 \times .30 = 9.0\%$	$.30 \times .90 = 27\%$
03	$.20 \times .50 = 25\%$	$.20 \times .70 = 14\%$	$.20 \times .20 = 4.0\%$
04	$.10 \times .90 = 9.0\%$	$.10 \times .80 = 8.0\%$	$.10 \times .05 = 0.5\%$
05	$.05 \times .05 = 0.25\%$	$.05 \times .40 = 2.0\%$	$.05 \times .25 = 1.25\%$
Totals	40.75%	50.5%	34.5%

Advantages

- This method gives a much stronger indication of what is the "best" solution than the Pugh method
- This method can give you an idea of comparative strengths of each proposal.
- Forces ranking of the objectives. It is important that the team agrees to the ranking, otherwise there will not be consensus on the solution that results from this method.

Disadvantages

- Deciding on the relative weightings of the objectives can take time and be problematic: The weightings may not reflect the wishes or needs of the client.
- There is sensitivity to the "strengths" assigned. If, upon further research, you found that one idea is actually much, much more expensive than you expected, this information could substantially impact your weightings.
- Only the selected objectives are tested. In our example the jackknife will not only open cans, but will also cut other things, tighten screws, and open bottles. These are attributes that may have value and could change the alternative, despite what is indicated in the table. This type of situation is, in fact, quite common at this stage of the design process.

Variations
- Weights can have different scales.

One should never use a tool like this without looking carefully at the results. Do the results make sense? If they don't then perhaps you need to reexamine your objective weightings. But when you start changing the weights it can sometimes feel like you are "fixing" the weights to get the answer you want. This is not the right way to use the method. And you have to be careful that dominant team members are not trying to use the weights to get the solution they prefer. This process requires team discussion (i.e., group processing). However, when used well the weight decision matrix method not only can provide a good justification for selecting a design, but it can also give you insight into why your alternative is good.

4. Risk-Based Models

In our examples thus far we have considered only fully-formed design ideas. Sometimes the development of an idea has risk of non-success, or risk of increased cost or time. Risk-based models can be used to account for these risks in the decision process.

Risk is effectively problem significance, which is a combination of the probability of the problem happening and the costs if it does.

We will not explain these methods here. Many of these involve management decisions and all will involve probability and statistics. You may encounter these types of decision-making tools later in your engineering career, particularly if you take courses in project management or operations research.

5. Strategy and Summary

We suggest that you actually use several of these methods in concert to reduce the design space (i.e., select a design idea for further work). This is because some of these methods (such as multi-voting) work best for choosing a long list of design ideas from a much longer list of design ideas (i.e., a first course selection process), whereas others, such as a Pugh or weighted matrix method, are better suited to working with a shorter list of possibilities that require finer differentiation. So if you are starting with a very good, long list of potential design solutions, then we would suggest:

1. Use multi-voting to get your list down to a long but more manageable number of ideas.
2. Use a graphical decision chart method next, particularly if your requirements have two competing, important objectives that stand out from the other objectives.
3. Use a Pugh or weighted decision matrix to differentiate among the remaining solution ideas.

This may be more methods than are really needed for a design project in engineering school, but it will give you experience working with the different methods. Make sure that you document the decision-making processes you use, in addition to the results, so you can explain to your instructor (or client) how you decided which solution is the best, or optimal, solution.

5.1. Teams

When a decision has to be made, it is important that there is general agreement from your team at least as to the process. This means that your team supports the evaluation process that is being used. It is also important to have general buy-in on the design solution that will be move forward to the detailed design and implementation phase. You may disagree with some parts, or aspects, of the solution, but the whole team has to be willing to put in the work it is going to take to detail out the design. This is difficult if people don't support the decision process or result. Your team should have decided how decisions would be handled during the team formation phase.

KEY TERMS

multi-voting	graphical decision chart	Pugh method
datum	weighted decision matrix method	decision-making methods

6. Questions and activities

1. Research and compare several different types of the same product or service using the methods in this module. For example:

   - Chewing gum
   - Pens
   - Backpacks
   - Calendars, or ways to keep track of due dates
   - Methods for getting from you residence to class (walk, bike, hovercraft, etc.)

   Compare the cost, durability, convenience, and longevity (how long the product will last), or other relevant features of the product or service.

2. Use a Pugh chart to differentiate methods of cutting paper, wood, or bagels.

3. Redo question 2 using a weighted decision matrix. Did the relative order of the potential solutions change?

   a. Check the sensitivity of your results to the parameters you have chosen. Does a small change in weights or evaluations alter the relative order?

   b. What additional characteristics could you evaluate?

4. Apply several of these methods to the potential solution ideas you have generated for the design project you are working on currently.

   a. Do they result in the same ranking of the solutions?

   b. Do some methods work better than others for your team and the problem you are working on?

3

Decision Methods for Teams

#skill/tool module: #teamdecisionmethods

Learning outcomes

By the end of this module, you should demonstrate the ability to:

- Use different strategies for team decision making
- Understand the strengths and weaknesses of each strategy

Recommended reading

Alongside this module:

- **Implementing a Project > Working in Teams > 2. Organizing**

1. Introduction

Presented below are very simple tools for ***team decision making***. These, in combination with the more highly structured tools available to teams, can help you make effective decisions. Often the design team will want to adapt techniques to their own situation to make them more appropriate and more reliable, so you should not feel bound to use these methods exactly as they are described.

Decision making can be a considerable source of frustration for many teams. Decision making is one of the most difficult parts of teamwork because you will need to live with, and support, decisions that you might not agree with. This is part of being a good team player. So what is the best way of deciding? There are many choices: consensus, voting, letting the team leader decide, and so on. We have found that undergraduate design teams typically choose consensus or voting for decision making. Choosing a method is called ***deciding how to decide*** and it is one of the most important team rules you will adopt. There are advantages and disadvantages to each of these strategies. You should also probably have in place a crisis decision-making strategy.

When a decision has to be made, it is important that the team agree on the process that will be used (i.e., that the team has decided how to decide). There should also be agreement that every member of the team will support the final decision. For this reason it is useful for the team to choose a decision-making method before they encounter a difficult decision to make.

1.1. Consensus

Team decisions can be made by *consensus*. Consensus is formed through discussion. Consensus does not mean that everyone agrees with a decision. It means that everyone agrees that they can support the decision. Some members of the team may disagree with a choice, but agree to go along with, and support, the choice. This is consensus.

Advantages

Consensus generally requires discussion, which is an important part of the design process. Teams that are good at consensus will make sure that every member has an opportunity to express their opinion, and be listened to, before a choice is made. This process can build team coherence and lead to decisions that are better thought through.

Disadvantages

Consensus can take longer than a simple voting process. Or teams can get stuck, going around in circles trying to reach consensus and never make difficult decisions. If you choose to use consensus it is important to also have a strategy for moving forward if consensus fails on an important decision that has to be made.

1.2. Voting

Team decisions can be made by a simple majority vote. *Voting* can be done secretly, but more typically is done through a public declaration (e.g., raising hands). If your team has an even number of people then you will need to decide how a tie will be broken.

Advantages

Voting is quick and definitive.

Disadvantages

Voting may cut short discussion (i.e., reduce group processing). In the design process, discussion is one of the most important activities. Listening to everyone's perspective and valuing their ideas is an essential step in moving from a pseudo team or potential team to a real team. If you choose voting, then you should also put in place a strategy to make sure that there is adequate discussion, and that everyone has a chance to express their opinion before a vote takes place.

1.3. Technical Guru Decides

Using this method, a single team member or an outside expert makes the decision. This has the risk that the individual has a bias toward one of the choices, perhaps because of experience with one of the choices. This experience could be positive or negative. There is a natural tendency for a team to let the technically strong team member decide. However, studies show that while individuals create concepts more effectively than a group, group evaluation is superior. So use this strategy carefully.

1.4. Client Decides

Letting the client decide has the advantage that the designer then takes less responsibility for failures to perform as expected. However, an engineer is usually hired as an expert in the technology, and is trusted to guide the design to a good solution. The client may not have the expertise to properly evaluate the choices without assistance. Getting the client involved, however, will often reveal previously unreported objectives, or objectives with more weight than was first indicated. You should also try to keep the client informed of the design solutions under consideration as the design proceeds.

We suggest that you document several "best" choices well, discuss these (and your recommendation) with your client, and let the client make the final choice from these. The client may not like any of your choices, indicating that there may be no acceptable solution or that the project requirements need further work.

1.5. Knockout the Worst Idea

Sometimes a team will spend a lot of time trying to deal with a large number of potential solutions, frequently too much time. Reducing the list by eliminating solutions that are clearly inferior can help. By agreeing to eliminate at least one possibility every time through the list, the list will soon become a more manageable size. Multi-voting is a fast way of doing this.

1.6. Crisis Decision Making

There will be times when you do not have time for consensus and not all of the team members are available for a vote, such as 15 minutes before a deadline when you realize that your report is missing a key figure or table. Your team needs to have a process for making decisions in these tight situations. This may never be needed, but it is useful to have in place just in case. The typical strategy is to empower the person in charge to make an ***executive decision*** on the spot, which basically means one person decides. So if you have tasked that one person to submit the report and they find the error, then they are empowered to decide what to do: include a rough hand-drawn sketch, leave the page out, add in a note, or wait for a printout of the missing material and get a late penalty on the report. Crisis decision making is very stressful and difficult, unless you have a lot of practice at it (e.g., emergency room doctors). Everyone has to agree that whatever the person decides to do, you all will do your best to support the decision.

KEY TERMS

team decision making	**consensus**	**voting**
executive decision	**deciding how to decide**	

2. Questions and activities

1. With your team, discuss which of these methods might work best for you as a group.

    a. Identify how you will address the disadvantages of the method you choose.

    b. Decide on a strategy for making decisions during crisis periods (e.g., close to deadlines).

Stages in Iteration:
Generate, Select, Reflect

1

#process module: #iterating

Learning outcomes

By the end of this module, you should demonstrate the ability to:

- Determine an optimal conceptual solution
- Explain the use of tools and techniques for reaching the optimal solution
- Use a structured approach to reach an optimal solution for a well-defined design problem, including the abilities to:
  - Populate a design space
  - Select the optimal solution
  - Reflect on the solution reached and take appropriate next actions
- Move beyond the first reasonable solution found and spend appropriate time on alternative solutions

Recommended reading

Before this module:
- **Design Process > Decision-making > Modules 1 and 2**

After this module:
- **Design Process > Iterating > 2. Suggested Iteration Process**

1. Introduction

Once the design problem is defined in sufficient detail, the engineer can begin to identify the best conceptual design. A sequence of three key activities may be followed:

1. First, increase the number of ideas from which to choose (Generation).

2. Next, take out those ideas that are inferior to the others (Selection).

3. Finally, carefully consider the next best course of action (Reflection).

To arrive at an ***optimal solution***, this chain of activities will often differ in timing, complexity, and volumes of solutions considered, depending on the project, but will be similar in intent.

1.1. Optimal Solution

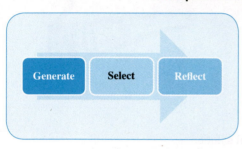

Engineers use the word "optimal" to denote a solution that is closest to ideal, given the circumstances and the technologies available. There is no delusion that an ideal solution will be found, because it is not likely to exist. If you have to solve a problem here and now, you cannot wait around for the best solution that will ever exist. You must use technology that is available or can be invented now. However, you should take the time to select from the options available, rather than just picking whatever is most obvious, or what first comes to mind.

An optimal solution:

- Fits the functions: Does what it is supposed to do. Otherwise, it is not a solution at all.
- Fits the constraints: Meets all of the non-negotiable restrictions. Otherwise it is an unacceptable solution.
- Maximizes the objectives for the project situation: For example, it is as safe as possible, as cheap as possible, as durable as possible, where each objective carries a relative importance or weight.
- Is feasible: It will be possible to implement and does not violate laws of nature or practical limitations on materials or processes.
- Is comprehensive: Deals with all possible DFX considerations. The design does not overlook anything important, even issues that were not part of the original design brief, and properly incorporates the concerns of all stakeholders to the extent possible.
- Is as simple as possible: The design should be simple in implementation and use, not necessarily in concept.

The last point is critical, and more difficult to adopt than the first five, but it is the key to really good design. It is based on the premise that thinking is cheap, while implementation is expensive, so you want to focus your energy and effort as early as possible in the process to creating a simple, elegant solution. In other words, you want the design to be as simple to build and operate as possible, even if it is harder to develop the design in the first place.

2. Generation

Generation involves creating ideas for designs that satisfy the requirements (#idea-generation). It is a ***divergent process***, which means the ***design space*** is deliberately expanded to provide the widest possible spectrum of solutions, or what we would call "solution candidates," from which to choose. Generation requires creativity, aided by technique, to produce variation in solution candidates. The word "creativity" conjures up images of artists and poets, but the invention of new and diverse ideas to solve an engineering problem is very much a creative endeavor.

The process of generating creative design solution candidates is fundamentally different than the familiar process of finding a solution to a closed-ended analytical problem. In a typical engineering exam question in a university course, the goal is to understand the given problem, recall the pertinent equations (which represent physical relationships), and to transform the given information into the required solution using these equations. A problem of this kind almost always has a single correct solution. In contrast, a design problem has many possible solutions, and a talented designer is one who develops the skill of finding the ***optimal solution***. It would be unusual if this were the first and only solution developed, so a key to the process is an ability to generate many different solutions so that the optimal one can be chosen. This is referred to as populating or expanding the ***design space***.

3. Selection

In the selection activity the design engineer will examine solution candidates and determine their suitability and relative merit (#decisionmethods). This is done by evaluating these solutions and comparing the results to the project requirements and to the results of other candidates. The design engineer will eventually use this process to determine which solution to use to move forward to detailed design and implementation, but most of the selection activities will be to reduce the number of candidates under consideration. Many methods are available for selecting between solution candidates.

4. Reflection

It is important for an engineer to reflect on the work they have just done. After every activity a good engineer will consider whether the activity was sufficient or not, and will consider what is indicated as the most-effective next action.

This is more than ensuring that each activity is reasonably executed. Here the team would look at the candidate solutions still remaining after the last selection process. Do the solutions include ideas they feel have high potential to be the optimal idea? Are there too few solutions and general idea classes for this point in the process? Do the solution candidates suggest changes in the requirements and perhaps even the goal of the project? Do they raise questions that the client must answer? How the team proceeded would depend on the answers to these questions.

For instance, consider a team of design engineers generating designs for a jar opener for people with arthritis. They would pause after a first round of solution candidate generation and selection. Imagine the following two situations:

- All of the solutions involved rotating the lid while holding the body of the jar. The team would likely be dissatisfied with this and proceed in a direction that generated other candidates that were more varied in technique.
- One of their candidate solutions involved heating the metal lid so it expanded and loosened. The team would revisit the project requirements, being sure that there were appropriate operator safety constraints in the requirements to prevent operator injury if this solution were used.

It would be incorrect for the team to proceed to further evaluations and selection without at least considering these questions. The *reflection* activity is important to help keep us from following the wrong path in our project. Reflection is what makes a good design engineer.

5. Conclusion

There is no perfect project. Most have some solution that will perform better than others in a weighted comparison, but not be the best for every objective; it will be a compromise.

Some projects will have no acceptable solution. The requirements are beyond the range of available technologies—it cannot be done at all, or is too costly, or will take too long. Or perhaps there is a solution, but it performs very poorly against the objectives. The design engineer must be prepared to terminate or delay the project in cases where there is no workable solution that will satisfy the client or the situation.

On the other hand, most projects will proceed to the next steps. At this point much of the "large picture" creative work is done and mostly standard-practice engineering work remains.

KEY TERMS

optimal solution	**divergent process**	**design space**
reflection		

6. Questions and activities

1. Use the following problems to generate, select and reflect:

   a. A system for keeping track of the number of people on a bus/streetcar/subway train.

   b. A kitchen aid for handling materials in jars for a person with a single, fully functional hand.

   c. A TV remote control for a person with a Parkinson's-like disease where fine motor control is lost (poor spatial orientation, unsteadiness).

   d. A GPS system for a bicycle.

2. For the design project you are working on currently: Use the generate-select-reflect cycle to move closer to an optimal design solution. Observe where your progress from the beginning of the generate-select-reflect cycle to the end. Did you move closer to an optimal design? If so, in what ways are you closer than at the beginning of the iteration?

Suggested Iteration Process

2

#process module: #iterationprocess

Learning outcomes

By the end of this module, you should demonstrate the ability to:
- Use the formalized iterative approach of this unit to reach a single proposed solution to a problem

Recommended reading

Before this module:
- **Design Process > Iterating > 1. Stages in Iteration:** Generate, Select, Reflect

After this module:
- **Design Process > Iterating > 3. Reflection Considerations for Iteration**

1. Introduction

The meaning of *iteration* seems quite straightforward. It is a process we repeat over and over, in a loop, in order to come as close to an ideal solution as we can, given current time, resources, and technology. In practice, however, you may wonder, what exactly am I supposed to do in each iteration? What am I looking for? How am I supposed to "improve" the design? These are good questions. In this unit we introduce a three-round method with specific targets for each of the three iterations. The method is based on the Generate-Select-Reflect stages. The method in this unit will also give you the advantage of knowing in early iterations the targets of later iterations.

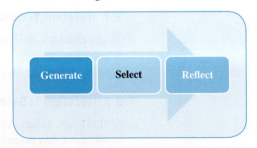

In industry the process will be more variable and unpredictable. The iteration process will be dictated by the flow of the project, changing technologies and circumstances, and input from the client. However, in school projects or when you have more control over the process, we suggest this three-round sequence of iteration steps to reach an optimal solution.

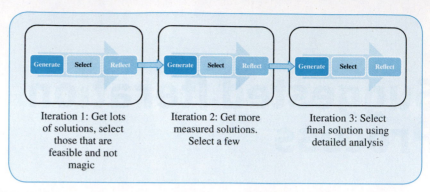

FIGURE 1 **A three-round iteration process.**

The three iterations, as shown in Figure 1, have the following purposes:

- In iteration round 1: the design team will generate the bulk of the solutions. This is the "Fast Generation of Solutions" iteration.
- In iteration round 2: the design team will define the solutions more carefully, produce added solutions using more involved techniques than were used in Iteration 1, and compare solutions to reduce the number of solutions being considered to about a half-dozen. This is the "Evaluation/Reduction" iteration.
- In iteration round 3: the design team will further analyze and compare the solutions. Then, often with client input, they will reduce the number of solutions to one *optimal solution*. This is the "Final Selection" iteration.

2. Iteration 1: Fast Generation of Solutions

In iteration 1 the design team generates a large number of solutions over a wide variation of solution methodologies. The purpose is to expand the design space to the largest extent possible (i.e., "cast a wide net"). The selection process of iteration 1 is important, but it is used to retain a wide range of solution candidates rather than identify one best solution.

2.1. Iteration 1: Generation

To generate the bulk of all the solution candidates, two techniques are used: *benchmarking* existing designs that are solutions or close to solutions, and *brainstorming* or other *creativity methods*.

2.2. Iteration 1: Selection

At this stage, solution candidates are evaluated against the functions, objectives, and constraints. Any completely undetailed parts of the solution must be straightforward, standard practice, or purchasable.

Solutions we want to eliminate fall into one or more of the following categories:

1. They do not meet one or more functions or constraints, and there is no feasible method of modifying the solution to do so.
2. They involve technologies that do not exist and are impossible to develop in the timeframe, budget, and other constraints of the project. This is a very common

situation. Think through a basic cost estimate and estimate an implementation timeline. Do you think the solution candidate will come in on time and on budget? Or is it very unlikely that this idea, or a variation of this idea, will be feasible in terms of economics and implementation time?

3. **They are physically impossible.** For example, they may break one or more laws of physics. They are infeasible, even "magic."

4. **They may be ethically or legally unworkable.** As a person you are required to act within the law, and as an engineer you are bound by a code of ethics. This is true whether you are licensed as a professional engineer or not.

5. **They may be economically unworkable.**

At this point; remove any infeasible design ideas from the list of solution candidates or find a way to modify the idea to make it feasible.

While some ideas retained in this iteration round may seem inferior, they might contribute to better ideas later. You should also keep solutions that, while they don't fully meet functions or constraints, might be modified to do so. In the second iteration round you will consider such modifications; in round 1 you only eliminate unsalvageable solutions.

2.3. Iteration 1: Reflection

The generation activity of this iteration should have yielded many diverse solutions. The team should consider primarily whether they are satisfied that the solutions they have generated cover the design space well. If the team feels the solutions (although feasible and meeting the requirements) seem to fall short of their expectations, then the team should repeat this round, including idea generation.

The team should also revisit the requirements and goal of the project. Each time you evaluate and select you increase the knowledge you have of the design project. Often you will want to use this new knowledge to revise your project requirements.

For example, you might realize that some of the solutions generated could result in hazards for the operator of the design. If this aspect of operator safety is not explicit in your project requirements, you might want to add these to your requirements.

3. Iteration 2: Evaluation/Reduction

In iteration 2 the design team uses comparative techniques and more detailed analyses to expand the number of solutions and then reduce the number severely during the selection activity.

3.1. Generation at Iteration 2

Iteration 2 techniques are not expected to create a large number of solution candidates, but rather to ensure "no stone is left unturned"—that is, that the design team has broadened the search and made sure the design space contains the optimal solution. Techniques used at this stage may include SCAMPER, morphological charts, functions-means trees, or other creativity methods.

3.2. Selection at Iteration 2

At this stage, solution candidates are compared against one another, in relation to functions, objectives, and constraints. The comparisons will often depend on interpretations of client wishes, which still must be considered changeable or even potentially incorrect. Keeping more choices "active" through this iteration will help guarantee that one of the choices will be in line with the true needs of the client.

Selection, at iteration round 2, greatly reduces the number of solutions, but does not identify a single, optimal recommendation yet. That will happen in iteration round 3, but the formal, detailed methods at that stage take time and are costly. Because many of the solutions generated and being considered in round 2 are clearly inferior to others, they can be eliminated so that you don't spend unnecessary time and cost in developing them. Typical processes used in this round are multi-voting and graphical decision chart.

Not all objectives will be met or maximized by any particular solution. There will be trade-offs, and we will introduce methods to compare and balance the choices. Generally no solution will work perfectly; it is up to the engineer to balance the trade-offs. By the end of the round 2 selection activity there should be typically about a half-dozen solution candidates left, with significant differences among them.

Example of Trade-off

Consider the task of choosing a material, such as a particular type of metal, for a project. Which metal is best depends on the situation. Steel is inexpensive and hard but reacts to acid. Gold is expensive, softer, and doesn't react with mild acids. Steel, protected by paint, is a good choice for an automobile body. Gold is a good choice for switch contacts and caps on teeth. However, there are trade-offs. Steel rusts, which is a problem for automobiles particularly in wet, salty conditions. Generally no solution will work perfectly; it is up to the engineer to balance the trade-offs.

3.3. Reflection at Iteration 2

The reflection iteration of iteration 2 is very important, since there will be no new (or few) solution candidates generated in the next and last iteration. For this reason the design team should be sure that the solutions remaining are complete in themselves (that is, they truly solve the problem in acceptable ways) and they have been taken from a well-populated, well-covered design space.

While not exhaustive, the following is a list of some attributes of the design that could greatly differentiate the solution candidates. Many of these are ***design for X*** considerations.

1. Installation and training: Designs often will need to be installed and operators trained. These aspects should be evaluated for each solution candidate.

2. Maintenance: Some designs will have reduced maintenance and, where maintenance is done, some designs will offer easier access to areas requiring maintenance, less frequent maintenance schedules, and/or less costly maintenance cycles.

3. Breakdown/Repair: The damage caused by breakage and the cost of repair should be evaluated whenever the design could break in use.

4. Already available solutions: Are solutions already partly or fully available? There is no point in "reinventing the wheel" when the most cost-effective solutions already exist.

5. Measures against use cases: If you developed use cases or task analyses during the design process it is useful to work through them using each of your solution candidates. This may reveal deficiencies or advantages in each that was not initially apparent.

If any of the issues raised at this point seem to apply to your project, make sure they are reflected in the project requirements. This is also a good time to have your client involved. You have just spent time weighing objectives and looking at solution candidates. Your client may see things differently than you do. Attributes you saw as important could be less so to the client and similarly attributes you do not think are important could be quite important to the client. Now is the time to clarify these issues.

4. Iteration 3: Final Selection

Unless, as often happens, a decision is made to revisit early iterations, you will leave this iteration with a single design solution to move forward for detailed design work and execution. In a typical project the bulk of the cost and time are incurred **after** iteration 3. It is therefore necessary to understand the importance of determining the optimal solution at this point. Often no new idea generation is done in this round; instead the concentration is on careful analysis to choose an optimal solution from the remaining candidates.

4.1. Evaluation and Selection

At this point you should have enough information about each design alternative to evaluate each idea with respect to the key objectives. You will use weighted or risk-based decision-making techniques to compare the various solution candidates against one another. Techniques such as the Pugh method and the Weighted Decision Matrix method are appropriate in this round.

Choosing the depth of evaluation depends on the problem at hand, and specifically:

- The time and resources needed to do any particular evaluation versus the time and resources available. Very detailed evaluations will take time, and may not be justified.

- The cost of a design failure. Sometimes the cost of a design failure is very low. If the design fails, there is little impact on people (welfare, economically, or otherwise). Other times the cost of failure is high: a high cost of failure justifies spending more time and resources on evaluation. Here we are considering only cost, but the actual concern of the design engineer is *risk*, which is the product of probability of failure multiplied by cost of failure. So the design engineer worries about failures that are more likely to happen and more costly when they do happen.

4.2. Reflection at Iteration 3

You exit the iteration 3 selection process in one of three ways.

1. **One solution.** If you feel you have generated an adequate set of solutions, have a comprehensive set of requirements, and have determined the optimal solution for these requirements, then you can proceed to detailed design and implementation of the solution.

2. **No solution.** None of the solution ideas you have generated will work or will meet the requirements to an adequate degree or perhaps the client has rejected all solutions. You need to decide if you can generate additional solutions, change the requirements, or if you have exhausted the possibilities and should terminate the project (unlikely in a project for a school assignment).

3. **You leave to revisit the earlier iterations.** You may have eliminated some of the inferior, unsalvageable solution candidates. You take with you the information you generated by evaluating some of the solution candidates. This information may be essential for:

   - Adjusting and combining the solution ideas to generate new, better solutions.
   - Adjusting and adding to the requirements based on discoveries during the evaluations. Often you will find that small adjustments and additions are needed in the earlier work on the requirements, and sometimes your discussions of solution candidates with the client will reveal new functions and remove others.

Going through the iterations will result in a solution candidate determined to be "best" *among those generated* (i.e., optimal). It is very important that the engineering team pause at this iteration and reflect carefully upon the selected solution.

5. Conclusion

In this unit a design method using three adapted iterations of the basic Generate-Select-Reflect activity set was proposed to help you iterate in order to find an optimal solution. Successfully done—and such success depends on the problem as well as the engineering team—the team will be able to move forward with a single proposed solution to the problem for detailed design and execution.

KEY TERMS

iteration	benchmarking	brainstorming
creativity methods	design for X	risk
optimal solution		

6. Questions and activities

1. Try applying the three-round iteration process to the design project you are working on currently. Were you able to use this process to get closer to an optimal solution?

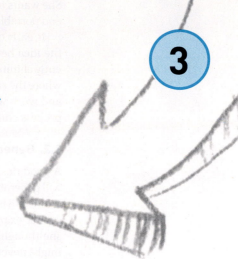

Reflection Considerations for Iteration

3

#review/reflection module: #iterationconsiderations

Learning outcomes

By the end of this module, you should demonstrate the ability to:

- List several keys to success in the design process
- Apply several strategies for success to improve the outcome of the design process
- Identify several problems that can emerge in the design process
- Apply strategies, such as additional idea generation, to overcome these problems

Recommended reading

Before this module:

- **Design Process > Iterating > 2. Suggested Iteration Process**

After this module:

- **Design Process > Post-conceptual Design > 1. Intermediate Design**

1. Keys to Success

Here are four suggestions for successfully managing the Generate-Select-Reflect process.

1.1. Keep Generation and Selection Separate

It is important that the generation of solution candidates be done without evaluation, and not simultaneously as many undisciplined thinkers will do. Often the first thoughts will fall short of a full solution, but adaptations, morphing, additions, and other further work will produce a much better solution. Premature evaluation could prevent these better solutions from being discovered.

To illustrate this, let's look briefly at one of the case studies used in this text. A wedding photographer wants to distinguish herself from her competitors by offering aerial shots of the bridal party and all the guests at a wedding in an open space. To do this, she needs to be able to lift her camera to be able to get a wide-angle shot (50 m² to 500 m²). She wants a solution that is inexpensive, reliable, easy to operate in all sorts of weather, and portable, so that she can carry it with her in her mid-sized car (#aerialphotography).

If, early on, the design engineer decides one answer is to use a long pole, but rejects the idea because it would be difficult to hold in place, the engineer might inadvertently eliminate ideas like tripods or poles with guy wires, and solutions involving poles where the stability issue is addressed. A good design engineer will keep the pole design and work with it to generate other alternatives as the optimal design determination process continues.

1.2. Generate Many Solutions

A good design engineer will not stop at a single design even if it seems optimal, but will generate many designs and not reduce the selection to a single idea until satisfied that there is not a better idea waiting to be uncovered. Consider the wedding photographer example again: Suppose the design engineer stopped generating ideas when she thought of a first good idea, say a pole supported by guy wires. Other solutions might never be considered, such as using a ladder or quadcopter drone.

1.3. Decide How to Decide

Even if there is disagreement about some parts or aspects of the solution, the whole team has to be willing to put in the work it is going to take to get to a detailed design. This is difficult if people don't support the decision process that produced the result. At this stage, your team should have decided how to make decisions. An effective decision-making process not only affects the decisions themselves, but also the productivity that results from those decisions.

1.4. Seek Client Feedback Support

The final selected design must be approved by your client. Using formal idea generation and selection techniques to generate comparative information will strengthen your arguments for particular solution candidates. However, these will be based on the design team's evaluation of the client's needs and wants and so could be misguided. Regularly consulting with your client can be a kind of reality check to keep the process moving toward an acceptable design solution.

2. Sample Problems

Certain problems come up predictably and so it is possible to discuss them in categories. If any of these show up in your design process, here are a few strategies for dealing with them.

2.1. Too Few Solution Candidates Pass Iteration Round 1

When selection is being done, having only a few solution candidates is often an indicator that too little work was done at the requirements or idea generation steps. Ideally you want lots of ideas coming out of round 1. To ensure you stay creative, you should probably reexamine the requirements and idea generation steps. Make sure you have not unnecessarily constrained the design, and have fully populated the design space.

Alternately, this may be a highly constrained problem, possibly a routine design problem. You may want to look for an off-the-shelf solution.

2.2. New Classes of Solutions

Often the solutions generated using the techniques of iteration round 2 will create new classes of solution. The design team should not hesitate to revisit round 1 when new classes of solution are revealed. Such repeats of previous iterations will be faster than the first time through, but have great potential for generating an innovative, optimal solution.

For example, consider again the case of the wedding photographer looking to take aerial photographs of the wedding party. Combining the idea of a balloon support with a toy-airplane-like structure to stabilize and position the camera in the generation part of round 2 might open up a set of other solutions that had not been considered.

2.3. Completeness of the Solution Candidates

Often analysis of solution candidates, brought to the level of analysis of iteration round 2, will suggest hidden requirements that were not considered earlier. They might even suggest that changes in the requirements are necessary.

> **Example:**
> Suppose a solution requires recoating a surface periodically to maintain protection from the environment (e.g., painting a surface with some protective substance). Chemical handling during maintenance might not have been considered as a requirement before, but is important to conform to safety and legislated handling procedures. While some solutions may not have chemical-handling issues, it is important that this aspect be reflected in the requirements so that proper comparisons are made.

2.4. No Solution Candidate Passes Iteration 1

Sometimes none of your solution candidates will pass the first step; that is, none of the ideas are feasible and satisfy the functions and constraints. The estimated cost of the completed project could be excessive, or the computing requirements are beyond present technologies, or the time to finish the project is unacceptable.

If the design team has spent considerable amounts of time looking at alternative solutions then the project may have to be canceled, which is unlikely in a design project for a school assignment, but happens frequently in industry. However, before canceling the project, the following steps should be explored:

- Revisit idea generation. Now that you have had the experience of evaluating some ideas against the functions and constraints, perhaps you can use this experience to inspire new ideas that avoid the problems encountered with your previous ideas.
- Talk to the client about relaxing or changing the functions or constraints. Possibly bring solutions that almost but don't quite meet the existing requirements to a meeting with the client. Show them that a solution is close, but just outside the existing objective goals. Be prepared to back up your claims with a logical, well-documented analysis of the situation.

2.5. No Optimal Solution Candidate at the End of Iteration 3

If, at the end of iteration round 3, the solution you found to be optimal does not seem to be the best you can do, consider the following:

- The comparative evaluation methods have subjective elements in them. Does your team have an unfounded bias toward a particular design?

- Does the chosen solution seem to cost more than it should?

- Does the chosen solution seem to be less safe than it should be?

- Was there enough range in the design space to explore a significant variety of options, or were the differences between design ideas almost too small to matter?

- Was the final choice almost arbitrary? Could more analysis help the team make a better informed and supported choice?

- Does the final choice truly make sense for the situation, or does it seem like the problem is being recast to fit the design solution idea that the team or the client likes best?

- Is the choice being made to be expedient, and is this a good decision?

Often asking questions such as these will prevent a costly mistake. As you think through these reflection questions, document your thoughts in your notebook and discuss these questions with your team. Your team needs to decide whether to go back to add to the requirements or explore more design solutions, or whether you have reached a sufficiently solid solution decision and can move on to detailing and implementing the design.

3. Conclusion: The Value of Experience

In your design process, you may encounter some of the situations described above. The solutions you come up with will then become part of your store of knowledge, to be applied to future situations. Documenting your design process clearly and completely in an engineering notebook is a way of protecting the knowledge that you are gaining. Not only will your engineering notebook remind you of what you have done, but it will also help you to avoid repeating mistakes and enable you to build on what you have accomplished.

4. Questions and activities

1. Review the results for the design project you are working on currently using the criteria in this module. Have you generated enough ideas to find an optimal solution? Does the solution you have identified fit the project requirements effectively? Or is there a need to revisit the iteration rounds?

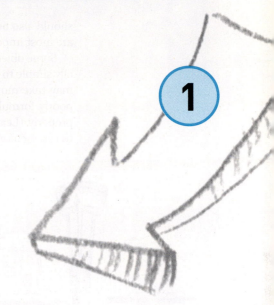

Using Metrics

1

#skill/tool module: #metrics

Learning outcomes

By the end of this module, you should demonstrate the ability to:

- Explain the concept of a "metric"
- Develop metrics for the objectives of a design that can be used to evaluate the characteristics of the design idea and its fit with the requirements

Recommended reading

Alongside this module:

- **Design Process > Requirements > 3. Objectives**
  Design Process > Decision-making > 1. Design Evaluation and Selection

After this module:

- **Design Process > Investigating Ideas > 2. Investigating Ideas through Models and Prototypes**

1. Introduction

A ***metric*** is a method for measuring a characteristic of a piece of technology. Metrics are used to measure quantitative characteristics such as size, weight, and speed. They are used to measure performance characteristics such as responsiveness, portability, and output rate. They can also be used to measure qualitative characteristics such as taste, smell, and visual appeal.

Comparing proposed solutions against each other requires that you have methods for measuring the solutions with respect to each of the objectives (i.e., a metric). It should be clear at this point why it is important to develop good project requirements and good objectives in particular. Poorly formulated requirements are difficult to use for evaluation. It is hard to compare solutions against one another in relation to a vague objective. Well-developed requirements will have clear methods that can be used to generate the comparative measures needed for assessment. Your objectives should also be prioritized; you should have a clear understanding of which objectives are most important. You will use this information in the decision-making process.

Some objectives will be simple to measure. Size and weight, for example, are generally simple to determine (depending on the accuracy required). Others, like reliability, may take more analysis to determine. Any objectives, like "fun to drive," which are poorly formulated as engineering measures, are difficult or impractical to evaluate properly. (Leave this one for the people in the marketing department who know how to run focus groups.)

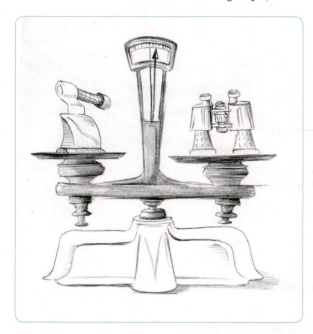

Important here are techniques for estimation. You need not determine exact values, particularly before detailed design is done. And at this stage developing very accurate values is not time efficient. You should determine values to an accuracy and confidence level that allows you to differentiate between various proposed solutions, which is good enough for now. This inexact analysis means any numerical results are also inexact and results should be critically considered.

For example, suppose you find that the estimated weight of one solution is 3.2 kg and the estimated weight of another is 3.5. Is the first solution idea really lighter? Or is the margin of error on your estimates such that effectively these two solutions are basically the same in terms of weight? Probably, given the level of detail of your solutions at this stage and your experience level, you should conclude that these solutions are basically equivalent in terms of weight. A ~10% difference in the results of an estimate is not sufficient to draw a definitive conclusion. There are a number of ways to test alternatives:

1. By design: If the design must be florescent orange, any proposed solution will have this as part of the specification. This is a quick yes/no evaluation, in this case of color. (There are actually industry standards for color, such as the International Color Consortium and ISO 15076-1:2005.)

2. By estimation: The weight of a proposed solution, for example, can be estimated by summing the weights of the heaviest components, and adding some estimate of the rest of the components. You can also estimate other quantities such as volume, distance, cost, or energy requirements this way.

3. By analytic or computer model: Mathematical models or computer analyses can be used to determine measures—lift of an airplane wing for instance, stress on a component, mixing of fluids, or volume of traffic that can be handled on a roadway.

4. By prototyping: Building a small or full-sized model of part or all of the design will help determine some of the measures we are unable to determine reasonably otherwise. This is called a prototype. There are a wide variety of standard tests that can be performed on the prototype to evaluate performance. Every discipline has these and most industries or companies will have a set of test

procedures that they use frequently. If a test procedure is not standard, then you will need to develop your own (i.e., design your own experimental procedure).

In addition to the four methods we have listed here for determining engineering measures, there are also a wide variety of methods that are used by other professionals for evaluating suitability of a design solution. For example, many design companies will employ social scientists (or hire specialized consulting companies) to run focus groups, do user or customer surveys, or do other types of psychological testing or branding work to determine the suitability of a design solution. They may also hire business consultants, legal experts, and other professionals to evaluate the feasibility of a design idea in greater detail before committing to a solution. These other avenues of evaluation are not generally done by engineers, nor are you generally expected to do these types of measures for an undergraduate design project. However, it is useful to be aware that people from other professions will often be working with, and sometimes on, engineering design teams. As an engineer, you need to be able to communicate effectively with these other professionals and factor their opinions and results into the design process.

KEY TERMS

metric

2. Questions and activities

1. Identify two or three standard metrics that are used in your discipline (mechanical engineering, electrical engineering, industrial engineering, chemical engineering, etc.). These can often be found in handbooks or standards documents.

2. For a selected technology, list four or five basic metrics that could be used to test the quality of the technology. For example, a bicycle:

   a. Range of gear ratios: measured by counting the gear teeth and calculating ratio.

   b. Weight it can carry before failure: use destructive testing to weigh down the bicycle until point of failure.

   c. Brake capability: ride at a specified speed and measure distance to stop when breaks are safely applied (i.e., applied in a way that doesn't send the rider over the handlebars).

   d. Weight of the bicycle: weigh the bike.

   e. Size: measure the volume the bike takes up when parked.

3. Identify a metric for every objective in the requirements for the design project you are working on currently. The metric can be taken (with reference) from a source such as an industry handbook, or developed by the design team.

2

Investigating Ideas through Models and Prototypes

#skill/tool module: #models&prototypes

Learning outcomes

By the end of this module, you should demonstrate the ability to:

• Use models and prototypes during the development of your designs

Recommended reading

Before this module:

• **Design Process > Investigating Ideas > 1. Using Metrics**

Alongside this module:

• **Design Process > Decision-making > 1. Design Evaluation and Selection**

After this module:

• **Design Process > Investigating Ideas > 3. Feasibility Checking**

1. Introduction

Prototyping and ***modeling***, with the associated testing, are essential components of the iterative design process. While we present the basic design process in a linear fashion—Problem Definition, Solution Generation, Selection of Optimum Solution, Implementation—what happens in the Sandbox, our fictional design company offices, is closer to reality. In a real design process, engineers are frequently forced to iterate as new information becomes available. The engineering team goes back to revisit the library, back to talk to stakeholders, and back to generate new solutions (or pieces of solutions) frequently.

In an ideal world, perfect designers would work out everything on paper or in a computer in advance of any physical work. However, the real world is not simple or ideal. In fact, it is often necessary to test ideas during the process of conceptual design. And the more innovative the technology, the more testing that is necessary because it is harder to predict how something really new will perform.

For this module we will use the word "model" in its sense of an inexact, perhaps scaled, replica of the final project. It could be in software, as a simulation, or it could be a physical item. As a physical item, it is usually a non-working version of the final project, often scaled, to show relationships of form and function. Prototypes are working versions of the final product, although they generally are missing some requirements, may not be in a final form, and may often only be a piece of the final product. Models and prototypes can be tested to verify that a design is viable—meaning that it will perform in certain necessary or desirable ways. This could relate to the performance of the design itself, or to the ability to use certain manufacturing or assembly methods in its execution.

2. The Steps in Modeling and Prototyping

Imagine that you have been tasked to design a water sterilization system for a Mount Kilimanjaro base camp. The water can be boiled, but this is energy intensive, and the boiling temperature is low at altitude. You know that water can also be sterilized using ultraviolet (UV) light. There are commercial units with UV lamps, but you are not sure if a viable system can be configured to supply the camp with clean water using the power of the sun only. You are not ready to compare UV to boiling; you don't even know if UV can be made to work. What would an experienced design engineer do to take this idea to a point where it could be realistically evaluated against alternatives? She would most likely produce a prototype unit and test the concept using it.

2.1. Estimation

Since models and prototypes take time to construct, the first step in the process is to *estimate* whether the idea to be modeled/prototyped has any merit at all. This ability to make quick estimations of things such as part sizes, cycle times, and energy consumption is a critical skill during the iterative conceptual design phase. These estimates are a type of early modeling, where a very simple system model is conceived to do these early tests.

For your proposed solar UV system, you would look up the UV intensity at the base camp and the dose/sterilization response curve for the bacteria in question. Then you could do a quick estimation of how much time each bit of water must be exposed to the sun and hence the length of plastic pipe needed to expose the water to sufficient UV for the projected water usage.

2.2. Modeling

Even when the final design calculations are complex and time-consuming, it is often possible to use a numerical *model* to answer the feasibility questions. For UV sterilization, you might use rows in a spreadsheet to represent discrete volumes of water as they journey along the pipe, and columns to represent time and UV dose. You could then include the idea of mixing between adjacent cells that would be present in turbulent flow, building in more complexity than could be handled with a pencil and paper.

There are commercial software packages to model things like stress distribution, thermal dynamics, fluid flows, current flows in a circuit, complex scheduling, and chemical process design. These are used to validate the final design, but are also used extensively as part of the idea generation process. You should ask the question: What simulation software could I buy or write to shed light on the design problem?

2.3. Prototyping

A **prototype** is a specially built one-off example of the proposed design or a subcomponent of it. It is typically built with just enough care that it realistically represents the proposed design at minimum cost. For the UV example, a sample section of pipe could be built, dosed with bacteria, and exposed to UV. Experiments would then be done to determine if the estimates and models corresponded with reality.

Many designs are complex enough that calculations and modeling just aren't sufficient. Consider the design of the ICON A5 Light Sport Aircraft. This has an integrated V hull for amphibious operation, and the ICON engineers tested the hull configuration for buoyancy and stability using an inexpensive model built from plywood. It was it was designed and built by one person in two weeks, and was just good enough to last the few hours they needed to confirm their basic design assumptions. This is a very typical prototyping situation: to model the stability of the hull using equations or a numerical model would have been difficult.

Prototyping is used in all branches of engineering. In software design, for example, a prototype is a simplified version of an algorithm used to see if the code behaves as expected, or very commonly to test a graphical user interface (GUI) for suitability, appeal, and human-design principles.

Courtesy of the Authors

FIGURE 1 Early stability prototype for the ICON AS aircraft. (Image Courtesy of ICON Aircraft, Inc.)

3D Printers

The prototyping of physical objects has benefited greatly from the maturing field of *rapid prototyping,* in particular the **3D printer,** also called additive manufacturing. These printers work from designs produced on computer workstations using a *computer aided design* (CAD) program, which are then produced physically by these 3D printers. In a matter of hours, it is possible to "print" a fully functional coffee cup, a set of functioning mechanical parts, or a replica of a bone from MRI data.

These printers are now used for some end-use goods as well, such as joint replacements and some specialized car parts. With the ability to deposit multiple different materials, including metals,

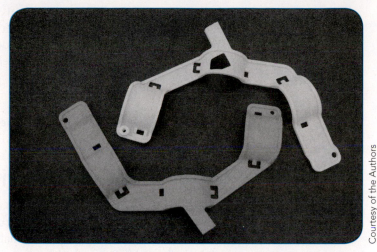

Courtesy of the Authors

3-D printing is becoming a common method for creating parts for prototypes. Objects can be printed out of many different types of substances, creating rigid or flexible components.

in larger pieces, the use of these printers will only expand. For engineers, working physical models is becoming commonplace.

You could use a 3D printer to produce a prototype housing and case for the UV water purifier discussed earlier. This would give you a chance to test its portability and ability to be transported to the site.

3. Models and Prototypes in Late Design Stages

In many projects a full physical design prototype is not possible. Bridges are only built once. Deep-space satellites are never sent on practice runs.

Exactly as estimations can be superficial and quickly produced for early analyses and can be refined for later, more exact calculations, models and prototypes can become more intricate during later stages of design. A simple model of your UV water purifier would be replaced with a more complex model that would give you closer, more reliable determinations of the time to purify the water you need, and thus the size of tubing involved.

For such designs as bridges and deep-space satellites, very intricate simulation and analysis models would be used in the final design work. These models would not only handle the ordinary, but would be able to simulate extraordinary events and problems to aid in proving and improving the design.

4. Conclusion

Creating a simple, effective model or prototype can help you quickly assess whether a design idea will be feasible. Your team will continue to create more detailed models and prototypes as the design process progresses, each one more completely approaching the characteristics of the final design solution. These models are a tool for thinking

Investigating Ideas through Models and Prototypes

through and examining the way the actual design will work in practice. It can be very important where operator interaction with the design is important. As you test your model or prototype, make sure that you document your observations and use this activity as an opportunity to refine your ideas about the design problem.

KEY TERMS

prototype, prototyping   **model, modeling**   **3D printer**
estimation

5. Questions and activities

1. Pick a piece of technology, and develop a metric to measure one characteristic of the design. If you were developing a new design and needed to prototype the design to measure this one characteristic, what would or could it look like? What could it be built out of? Remember that a prototype does not need to resemble the real system to be useful for testing.

   For example: We're developing a new bicycle and we want to measure its size. We could build a prototype out of cardboard or use a computer model to measure its size.

2. Team activity: You are designing a two-story building for use in the far north. What questions might be answered by models and prototypes in the design process?

3. What are the advantages and disadvantages of using a model versus a prototype?

4. Develop a plan to create a model or prototype for the design project you are working on currently.

   a. What will the model or prototype be used for (i.e., what metric or metrics will be applied to measure its characteristics)?

   b. If time and resources are available, build the model or prototype and test it. Ideally, you would have enough time to build several models or prototypes or time to rebuild one several times so you can improve your design.

Feasibility Checking

3

#skill/tool module: #feasibilitycheck

Learning outcomes

By the end of this module, you should demonstrate the ability to:
- Explain how to perform a feasibility check
- Explain why feasibility checks are important to the design process and when they are used

Recommended reading

Before this module:
- **Design Process > Investigating Ideas > 2. Investigating Ideas through Models and Prototypes**

Alongside this module:
- **Design Process > Decision-making > 1. Design Evaluation and Selection**

After this module:
- **Design Process > Investigating Ideas > 4. Routine Design**

1. Feasibility Checks

Throughout a design project you should perform *feasibility checks*, also called *reality checks*. It is easy to say, "Anything is possible," but really it isn't. Serious feasibility issues need to be dealt with. If the requirements are infeasible this needs to be brought to the attention of managers or your client; if design solutions are infeasible they need to be discarded. Many things can make a project or a solution idea unfeasible, but we will just discuss the three most common ones: physical problems, ethical or legal problems, and economic or time problems.

1.1. Physically Unfeasible

The project may be physically impossible. It is surprising how often engineers propose designs that are physically impossible. It is just very easy to imagine systems that

Example

Suppose the client is asking you to design a lifting system that will use a 1 hp motor they have on site to lift weights up to 500 kg, and they want the weight lifted up 10 m in 1 minute or less. This is physically impossible. A 1 hp motor can only produce about 44.7 kJ of energy in a minute and the energy needed to lift a 500 kg weight through 10 m is 49 kJ (assuming standard gravity on earth).

Even without taking into account all of the frictional loses in the system, it is easy to quickly calculate from basic physics that this just won't work.

seem like they should work, but don't. It is surprising how often engineers are asked to design systems, or evaluate design ideas, that are physically impossible. So before you waste a lot of time trying to design something that violates the laws of physics, do a quick estimate to see if the idea or requirements are feasible in this respect.

1.2. Ethically or Legally Unfeasible

The project or solution idea may be ethically or legally unworkable. The requirements may be such that any solution will violate a law, or be unethical. Or a particular solution idea may be illegal or unethical. You have a duty to point this out to your manager and the client, or eliminate solution ideas that have this issue. As a person you are required to act within the law, and as an engineer you are bound by a code of ethics. This is true whether or not you are licensed as a professional engineer. Also, technology designed for legal, ethical purposes may be misused by others, but you should never knowingly engage in a project in which the primary purpose is creating a design for illegal or unethical use.

Example: Removing a Potential Idea from Consideration

Aerial-view camera device: A design team is trying to find a solution for taking aerial pictures at weddings and other events. At the first level of evaluation, the team looks at the solution "bird with camera attached." They quickly estimate the size of bird required to hoist a photographer's camera, the advanced training, the adverse temperature and weather conditions, and the maintenance required for the solution. The conclusion is that the potential solution is not feasible.

1.3. Economically Unfeasible or Unworkable Timeline

The project may be economically unworkable, or may be infeasible in the given timeline. This is a very common situation. The client wants a system that costs less than what is possible, and is finished sooner than is possible. They may specify a budget and timeline that are simply not viable. You may have a really great solution idea, but the client will not be happy with it if the project comes in way over cost or over time.

Opportunity cost means that you have to give up an opportunity for one thing in order to achieve another: for example, deciding between going to a friend's birthday party and going out to dinner with your cousin. There is an opportunity cost to each of the choices. In design, picking one solution may have a substantial opportunity cost that makes it unfeasible.

Assessment of future technologies means making an educated guess about the future technology that will be available in a particular field. You may guess that there will soon be a better technology available that will supersede the choices you have right now. In which case you might put a project on hold, or judge it to be unfeasible if it is likely that the future technology will undercut the design you are working on currently.

Feasibility issues related to physics, ethics and legality, and budget and timeline are three of the most common situations you will come across on projects. However, there are others such as political feasibility, *opportunity cost*, or *assessment of future technologies* that may come into the decision of whether a design problem can be, or should be, solved right now. As you gain more experience as an engineer you will be able to provide expert advice to your company and your clients about the feasibility of a project. As an engineering student you should begin getting into the habit of assessing feasibility at every step in the design process.

KEY TERMS

feasibility check or feasibility analysis   **reality check**   **opportunity cost**
assessment of future technology

2. Questions and activities

1. Describe an unfeasible set of requirements or design solution for each of the following:

   a. Unfeasible for physical reasons

   b. Unfeasible for ethical or legal reasons

   c. Unfeasible for economic or timing reasons

2. Evaluate the requirements for the design project you are working on currently. Are there any feasibility issues related to these requirements? Explain how you have checked your requirements for feasibility in the three most common areas (physical, ethical/legal, and economic/timeline). Support this explanation with calculations, estimations, research results, citations for information sources, and/or references to codes, regulations, and standards, as appropriate.

3. Evaluate a design solution that you are currently considering for the design project you are working on. Are there any feasibility issues related to this solution? If so, explain what they are. Explain how you have checked your design solution for feasibility in the three most common areas (physical, ethical/legal, and economic/timeline). Support this explanation with calculations, estimations, research results, citations for information sources, and/or references to codes, regulations, and standards, as appropriate.

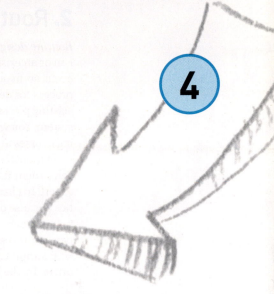

Routine Design

4

#skill/tool module: #routinedesign

Learning outcomes

By the end of this module, you should demonstrate the ability to:
• Select routine design alternatives where appropriate

Recommended reading

Before this module:
• **Design Process > Investigating Ideas > 3. Feasibility Checking**

Alongside this module:
• **Design Process > Decision-making > 1. Design Evaluation and Selection**

1. Invention versus Routine Design

Invention is the process of solving a problem in a new way that would not be obvious to engineers practicing in the field. However, engineering design is often more routine. The design of a ventilation system for a factory floor is the work of an engineer, but this doesn't mean that new types of air-moving equipment are designed as part of this design. Instead, the engineer consults code books to determine what the suitable air flows are, goes to the equipment suppliers to find commercially available fans and ducting, and puts it all together to do the job. It is engineering design because it is an open-ended technical solution, and relies on the engineer's knowledge of math, applied science, and the technologies of heating and ventilation.

Sometimes one of the most difficult things for a creative design engineer is to admit that there is an adequate solution that already exists. Engineers are often looking at how to make something better: more features, smaller, lighter, and better-connected. But going with what has already been invented is often better: less expensive, faster to implement, and more reliable.

2. Routine Design

Routine design means that there is a commonly accepted method for solving the problem or an existing piece of technology that can be used. Some types of design problems occur so frequently in a particular industry that engineers have developed a routine process for developing solutions for this type of problem. And there is a wide array of existing pieces, subsystems, and parts for sale that can be put together to create a new system. You will be trained further in routine problem solutions in your other engineering courses in your discipline because this is a critical skill.

In industry it is crucial that you recognize routine design problems in your field and solve them using standard processes. In many cases you are not allowed, even as an expert, to change the routine design solution process without permission from authorities because of safety issues and regulations. Routine designs are also frequently easier to implement than novel designs. So if there is a really good existing solution or solution type or process then, as engineers sometimes say, "Go with what you know."

Examples of routine design include structural design for routine buildings (heavily driven by the building code and standard practices), minor modifications to an existing design, or algorithmic design (e.g., layout of a printed circuit board; components entered and computer software does the layout for you, which may require minor tweaking after the software completes the job).

2.1. Why Use Routine Engineering

To fully evaluate a solution from scratch is time-consuming and costly. The adoption of a new solution may require training, new tooling, and the design of new methodologies. New solutions may take longer to prove that it meets existing standards.

Let us look at a simple example. Say a new wiring box used for electrical connections is being considered instead of one of the standard wiring boxes, because the new box is less costly, and will take time off the installation. There could be many reasons not to choose this new box.

You might have to train the electricians to use the new box. If one electrician left, you would have to train an incoming electrician before she was able to take on work. There could be limited stock of the new box and few places to get them, whereas you may already have inventory of the older-style box, and it likely is available from many local sources. The electrical inspector might not know about this box, leading to delays in approval, which might delay the whole project.

Courtesy of the Authors

A standard commercial wiring boxes.

2.2. Why Not to Use Routine Engineering

Even though there are advantages to using routine engineering, there are times when routine engineering should not be used. Routine engineering should not be used when it does not satisfy the project requirements well. Any design should meet the functions and constraints, and the best choices will score high in the objectives. If this is not true of a routine engineering method, alternate methods should be strongly considered. To do otherwise is to "fit a square peg in a round hole."

Routine engineering should often be considered against new technologies as they become available, and to changing social requirements. Engineering is about creating technology and the last few decades have shown incredible leaps in the technologies that help us design and produce our own creations. The focus of our work has moved to environmental, power-efficient, cost-efficient, and safety-efficient solutions. Routine solutions may have to give way to new ideas that better match new technologies and revised requirements.

3. Types of Routine Engineering

While routine engineering will be very discipline-specific, here are some common methods. It should be noted that careful selection and testing is still required to show that the design meets the project requirements.

3.1. COTS

COTS means *commercial-off-the-shelf* or *custom off-the-shelf*. It refers to proposed designs that use parts, components, and other systems that are readily commercially available (or open source). The cost of design and production are spread over the number of units sold. This drastically reduces the cost of the technology. In addition to being less expensive, off-the-shelf items have already been tested and certified if this is necessary. Many innovative engineering designs are the result of putting together off-the-shelf technologies in a new and unique way.

Examples of off-the-shelf items include subroutine libraries, operating systems, and standard protocols; standard materials such as alloys, chemicals, gases (e.g., standard purity level) or concrete mixtures; and standard electrical, mechanical, and construction components such as pressure vessels, I-beams, wire gauges, and computer chips.

A version of this is *design by inventory*, which means you try to make your design from parts your company already uses and resources your company already owns or obtains for other projects. The software version of this is to use software libraries and IP (intellectual property) that has already been developed.

3.2. Platform Design

Using an existing *platform* is a type of routine design. A platform is a base on which the custom design is built, such as an operating system, a standard vehicle chasse, or a standard shampoo formulation. For instance, a company may produce a wide range of variations, such as different scents or additives to create a line of shampoos, or a model line of cars that are all designed to use the same chasse. There are many new pieces of technology or customized systems that are built using the same platform.

3.3. Rework an Existing Design

By using existing designs most of the design decisions that go into the new design will have already been made and validated. Also, the company may already own, or have a license for, any intellectual property (IP) that is in the existing design.

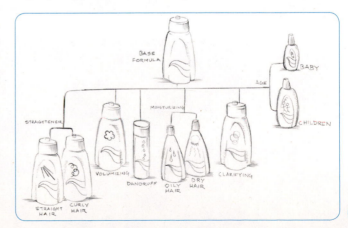

3.4. Use Existing Facilities and Processes

A new design that uses knowledge and equipment that is already familiar to a company will speed implementation and reduce potential problems that come with a learning process. For example, if the new design can make use of existing manufacturing facilities it saves time in redesigning the manufacturing process and retooling the manufacturing line.

3.5. Use a Complete Existing Solution

Sometimes the entire design problem can be solved by purchasing an existing system, technology, or facility. Although engineers like to create their own designs, purchased technologies are more cost-effective, are more time-effective, and sometimes meet objectives that can't reasonably be met with a new solution. Note also that new designs may require new patents or licenses, whereas purchasing an item automatically gives the owner the use of the IP in the technology (i.e., the license is built into the purchase agreement).

4. Routine Processes for Selection

In some instances an engineer needs to follow specified routine selection procedures to pick the right part. Often this type of procedure is used to correctly size an item. In engineering school you may be taught how to size things from first principles (e.g., how thick the wall of a pressure vessel needs to be in order to hold a compressed gas). However, in industry there are codes and standards that will govern the sizing of many things, such as pressure vessels. These codes ensure a *factor of safety* is built into the component. The standard sizing procedures will also ensure that the part that is chosen will operate effectively.

Examples of this type of calculation and routine design practice can be found in manufacturer catalogs, as shown in this example from the Parker O-ring catalogue, and handbooks (e.g., *Mark's Handbook for Mechanical Engineers,* or *Perry's Chemical Engineering Handbook*). They are also often found in handbooks published by professional societies and industry groups (e.g., the ASHRAE handbook series).

eter assemblies of this kind, it is well to use an O-ring one size smaller than indicated, but then the gland depth must be reduced as indicated above because the stretch may approach five percent.

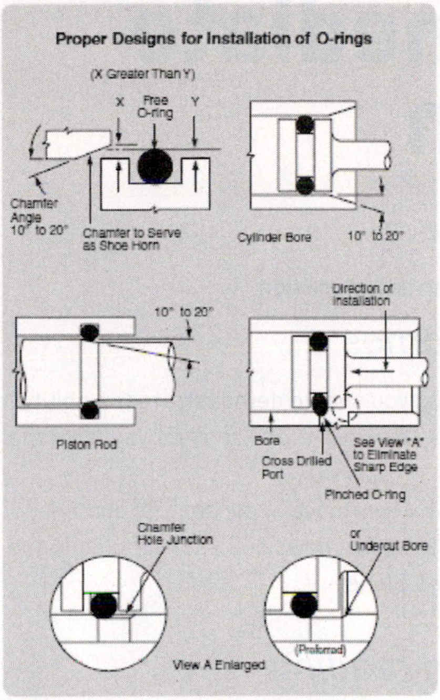

Figure 3-4: Proper Designs for Installation of O-rings

Reproduced by permission of the Parker Hannifin Corporation

KEY TERMS

routine design	invention	custom off-the-shelf (COTS)
commercial-off-the-shelf (COTS)	platform	factor of safety
design by inventory		

5. Questions and activities

1. Discuss the questions that you would ask when you are deciding whether a piece of software should be totally rewritten for a new system, or revised from the existing software. What analyses would go into your decision?

2. Automobile manufacturers have the problem of creating a new model but (usually) not going far from standard manufacturing techniques. Look at two models of automobile from the same manufacturer, and look for:

   a. Common methods of construction and common or very similar parts

   b. What is used to differentiate the two models and the difficulty of the engineering involved to affect these differences

3. Identify a handbook or some industry catalogues relevant in your field of engineering that give information on routine sizing calculations, standard protocols, or other requirements for using commercially available parts in a custom design.

Intermediate Design

#process module: #intermediatedesign

Learning outcomes

By the end of this module, you should demonstrate the ability to:
- Describe the process by which a design is moved from the conceptual to the final stages
- Describe the relative importance of conceptual and final design stages
- Categorize particular design decisions as belonging to the Conceptual, Intermediate, or Final Design Stages

Recommended reading

Before this module:
- **Design Process > Decision-making > 2. Selecting a Design Solution**

After this module:
- **Design Process > Post-Conceptual Design > 2. Final Design**

1. Introduction

What is required to turn a conceptual design into a real product, process, or procedure? You should not assume that the creation of an optimal conceptual design is most of the work. In fact, there is a great deal of work still to be done to "make it real." Like all parts of the design process, these subsequent steps can be done well or poorly. The best conceptual design in the world will still fail if a lack of attention to detail results in poor implementation. In fact, engineering disasters can often be traced to the failure of a single erroneous computation, poor decision, or faulty component during the implementation phase.

Engineers and technicians are usually responsible for most of the effort needed to turn a conceptual technical design into a real product or service. It is an engineer that will choose specific components, provide exact dimensions, and provide details of connectivity between software modules. In the simplest view of the world, the engineering is done when the design plan can be turned over to non-engineers for implementation. A bridge designer provides blueprints to the general contractor, who oversees construction. A mechanical engineer provides detailed shop drawings for the machinist, who then builds the device exactly as specified. In practice, the division of labor is not that simple and the process is often highly interactive. Engineers continue to consult as the implementation of the design progresses, approving the progress along the way and making modifications as necessary.

The process for implementing a design depends very much on the nature of the design itself. However, the creation and choice of the optimum concept clearly depends on some foreknowledge of the implementation process. We will describe the most important general concepts here. We cannot provide all of the details of implementation because there are so many variations across different disciplines; these are more appropriate for discipline-specific upper-year courses.

FIGURE 1 The Hubble Space Telescope mirror polishing unit at Perkin Elmer. The initial images from the telescope were out of focus because of a single misplaced lens in a device used to monitor the curvature of the main mirror [1]. The telescope was repaired, in space, at great expense, by fitting it with corrective lenses. (Image Courtesy of NASA/JPL-Caltech.)

2. Major Steps of Implementation

In other learning modules, we have discussed the process for formalizing requirements, developing creative ideas, and choosing the best option from these. We have talked about the processes used to move through these three critical stages of design in some detail, and have touched on the associated skills and tools employed by engineering designers. The principle creative work is now complete, and the engineer now has produced a conceptual design report, specifying the design without too many details.

If the design were being done for an external client, it would be normal to go back to the client at this stage for approval of the design direction chosen. Assuming that a decision is made to proceed through to a final design and appropriate resources are committed to the project, it is necessary to fill in all the details, and there are many.

Consider the wedding photographer case study (#aerialphotography). The photographer would like to take pictures from a high vantage point to get a wide-angle picture. At this point in the design process, an optimal solution has been established: a pole with the camera mounted at the top. Exhibit 1 summarizes three of the many remaining decisions and actions.

EXHIBIT 1

Part of a decision list for intermediate and final design of an aerial camera setup.

1. Pole
  a. Decide on the material to use for the pole: polymer, metal, or composite.
    o *If metal: aluminum or steel?*
      ▪ If steel: mild or stainless?
        • If stainless: what specific grade?
    o *Given the properties of the chosen material, how thick are the walls?*
    o *What shape? Round is usual, but square is also available.*
    o *Where will the material be purchased?*
  b. Choose a prefabricated extension pole or manufacture from tubing?
    o *What post-purchase processing and treatment will be required? (This could lead to questions of storage of parts and processing/assembly process.)*
    o *If from tubing, how will the nesting mechanism lock: pins, collar locks, other?*
      ▪ If pins, how will they be retained: bolt and nut or indented sprung ball bearing or cotter pin?
        • Once the type of pin is chosen, what material will it be made from?
        • What diameter will the pin be?
        • Who will supply the pin?
2. How will the interface between the camera and the pole work?
    o *How will the interface be attached to the pole?*
    o *Attachment to camera is standard, so we know it will be a ¼ 20 bolt clipped inverted under a base plate.*

      o Hinge mechanism or fixed camera angle?
-    *If hinge: set on ground or adjusted while in the air?*
-    *If while in the air, how will this be done?*

3. What kind of grip material will be for the hand-hold pole support: custom molded parts? Taped gripping area? Dipped in plastic material?

      o Colors?

      o Suppliers: install in-house or use outside service/supplier?

As you can see, there is a great deal of detailed work still to be done and Table 1 is far from complete, excluding the focusing and shutter mechanisms, for example.

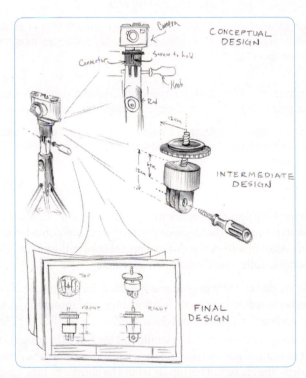

FIGURE 2 **The design evolves from a rough idea to a detailed set of drawings ready to be fabricated.** This involves many decisions that will determine whether a good concept will actually work well in reality.

3. Intermediate Design

For convenience, we can separate the detailed design work into two stages. The first or *intermediate design* stage, is sometimes referred to as *embodiment design* [2,3] (see Table 1). In this stage of the work, the objective is to "rough out" the design. This means different things in different disciplines, but essentially, the engineer is trying to get all of the major implementation decisions made, so that in the last phase of work, he or she can focus only on the details. It is much the same as roughing out a paragraph-by-paragraph outline of an essay, so that you can focus on the sentence structure when doing the final writing.

TABLE 1 Description of typical intermediate design determinations from different engineering fields.

TOPIC	CONCEPTUAL DESIGN	INTERMEDIATE DESIGN
Software	System structure, general communication between parts. General user interactions, overall timing	Synchronization determinations, memory usage estimations, testing requirements, input/output details, security measures, grouping of user control and feedback into logical sections and data structure details
Chemical Plant	Choice of major unit operations and sequence of these: Process Flow Diagram	Piping and Instrumentation Diagram. Specific types of components chosen. Material classes determined
Manufacturing Schedule	Balance between human and robotic elements. Sequence of operations. Input and outputs for each step	Plant layout. Approximate timing of all sequences. Flow of goods, people, and equipment
Electronic Circuit Board	Choice of major components and connections between them	Specific types of components and interconnection schematic. Ratings for critical components
Bridge	Architectural concepts, bridge type, design sketches to check for compatibility with surroundings	Detailed sizing of abutments, piles, spacing, truss configuration, and deck construction

During intermediate design there are many decisions and calculations to be made. In fact, in Exhibit 1, most of the decisions listed would be considered intermediate design decisions. After this stage the design engineers have the basics of configuration finalized and some details would have been specified:

- For software, the design team might have specified the communications sequences, software methods, software objects, relationships, and interactions, but would not have started coding the software program itself.
- In a product design, the design team would not have decided exactly which screws will be used to put the pieces together or what surface finish was required on the machined parts. The classes of materials would be chosen but the design would not yet have reached the level of detail that would be needed in a set of drawings to be sent to a fabrication shop.
- In a chemical plant design, the design team would have chosen the type of pump or valve at each particular location, but not the make and model number of these parts.

The process is best explained with our example. Let's focus on the extension pole that chosen in the conceptual design phase for the wedding photographer. It has to fit in the photographer's car, and extend to ~10 meters to elevate the camera. What decisions are made moving from conceptual to final design? (Note that only some of the calculations and deliberations to justify each of these required decisions are shown here.)

Decision 1:

A typical car would easily hold a 2-meter pole, so it will have to collapse to a minimum of five 2-meter sections.

Decision 2:

A type of material needs to be selected. Experience, confirmed by simple bending calculations, indicate a hollow tube is needed. It cannot be expensive, so a quick reference to a materials database would suggest steel, aluminum, fiberglass composite, or perhaps some type of plastic. Since the pole might be used in all types of weather, including approaching thunderstorms, a nonconductive material would be preferred, reducing the choices to fiberglass or a polymer.

Calculations will determine if a thermoplastic will do the job, or if fiberglass is better, since it is much stiffer. (A drinking straw is commonly made from a thermoplastic; a fishing pole is commonly made from fiberglass.) The calculation would be simple: more than a back-of-the-envelope estimation, but less than an exact calculation. The camera should not swing too much, so the stiffness of the pole will be limited. If the pole is held horizontally, it should deflect by no more than 1 meter. This is a simple bending stiffness computation, which gives the wall thickness and hence mass and cost of a pole needed to meet this criteria. The quantitative comparisons needed to choose materials in cases like this is described in great detail by Ashby [4]. For an inexpensive, nonconductive, lightweight pole of sufficient stiffness, the computations will support the use of fiberglass.

The results are supported by a good technical analogy found in the hardware store: insulating extension ladders often have fibreglass side rails.

Decision 3:

Speed of assembly is not an issue, so special efforts need not be made to accomplish this. A simple solution for making the tube would be to buy a 10-meter tube and cut it in sections, providing simple fasteners to connect the sections. These fasteners could be a larger-diameter pipe that just fits over the main sections, and cut to 20-centimeter lengths. With an indent in the pole sections at 10 centimeters from the one end, the connector will fit snugly over the lower section and provide a receptacle for the upper section. A hole and pin could be used to hold the assembly securely.

With relatively little information, many of the specifics of the telescoping pole component of the overall design have been determined. Approximate computations have been used to determine that the pole cannot be made of a simple thermoplastic and that fiberglass is a better choice. More details have been added to describe how the sections of the pole will be attached together.

4. DFX during Intermediate Design

Design for X (DFX) is the process of considering everything relevant to the manufacture, use, and even decommissioning of a designed product or service. X may represent many of the objectives (not related to basic functionality) in the original design brief. Typically an engineer will want a design to be the best combination of easy to manufacture, clean, and repair; and inexpensive, modular, and scalable. Each of these objectives is a DFX consideration (#DFX).

DFX can be considered during any of the three key design stages. The engineer considers main objectives and functionality as early as possible, and preferably during the conceptual design. It is much easier to build safety in from the beginning by choosing an inherently safe chemical or procedure than it is to guard against the hazard as an afterthought. However, for most designs, it is not possible to consider all possible X in the DFX until reaching the intermediate design stage. At this stage, the design team can and should run through a comprehensive list of all possible X, and crosscheck against the design. It is generally too late to do this when you have entered the final design stage, although we stress again that design is iterative, and there will always be blurry lines between the various stages.

5. Unintended Functions

At this stage of the process you should analyze the design for **unintended functions**. Unintended functions are functions of the design solution that were not deliberately enabled. Some unintended functions are useful and can be marketed as special features of the technology. For example, facial tissue was originally developed for make-up removal, but is now used and marketed for blowing your nose. However, more often unintended functions point to possible safety hazards or failure modes, and can result in legal action if the technology is improperly used.

An experienced designer will start thinking about unintended functions early in the design process. It is difficult, however, to identify unintended functions until you have a clear design solution in mind that is more detailed. So we suggest that this is the stage when you need to carefully analyze your design solution to identify unintended functions. Once identified you can decide how to embrace or counter the function.

For example, in the design solution for the wedding photographer, the engineer should be considering the consequences of the pole blowing over onto the wedding guests. The pole, in this way, could have the unintended function of becoming a projectile that hurts people or lands in the wedding cake. It would therefore be prudent to carefully analyze the design to make sure the guywires or other supports are sufficient to withstand normal conditions and then add in a **factor of safety**. Engineers will also specify the safe usage of the technology in the user's manual or warning labels to instruct the user on safe practices (#handlingrisk, #designforsafety).

6. Conclusion

The intermediate and final design stages are critical steps in the overall process of design and implementation. If the conceptual design is the "fun" part, where creativity is paramount, during the remaining parts of the process attention to detail and a methodical, careful approach can mean the difference between the success and failure of the design in service. Even the best designs can be ruined by poor implementation.

It is impossible to adequately describe the process of moving from a conceptual design through implementation in any detail unless the discussion is restricted to a specific field. In upper-year design projects, you will have an opportunity to explore this process in your discipline. Our purpose in this chapter is simply to highlight the complexity of this task, and to identify some generic features common across disciplines.

KEY TERMS

intermediate design   embodiment design   unintended functions
factor of safety

7. Questions and activities

1. What are the consequences of making a change in the requirements during the post-conceptual design phase as compare to making the same change in the conceptual phase (see "**Scope Creep**")? Consider cost, time, and lost effort.

2. Part of a design is the requirement to cut material. What intermediate design decisions must be made to keep the operator safe if the method selected is:

   a. A rotating blade

   b. A hot wire cutter

   c. A laser cutter

   d. A punch press

3. Maintenance (or lack of maintenance) will be a consequence of design decisions. Where some choices might generate maintenance requirements and others would not, at what stage should these decisions be made? If the answer is "it depends," what does it depend upon?

4. Unintended functions exist for practically every technology people use. Try identifying some unintended functions by creating a unintended function chain like this one:

   a. Nail polish can be used to label tools such as screwdrivers.

   b. Screwdrivers can be used to pry lids off of paint cans.

   c. Empty paint cans can be used as buckets to carry supplies such as brushes and paint scrapers.

   d. Paint scrapers can be used to scrap gum off of sidewalks.

   e. Gum can be used patch a small hole in a bucket or tire.

   f. And so on.

   Please note: NONE of these are approved usages, and all come with risk of injury or harm to self or equipment.

5. Work through the intermediate design steps for the design project you are working on currently. What DFX considerations are important for your project? Have these been taken into account in your design? What are the unintended functions of the design you are creating? How are you taking these into account (designing them in as features or designing them out or guarding against them)?

8. References

[1] Waldrop, M.M. (1990) "Hubble: The Case of the Single-Point Failure," *Science* 249(4970):735–736.

[2] Ashby, M.F., and Jones, D.R.H. (1980). *Engineering Materials: An Introduction to Their Properties and Applications.* Oxford: Pergamon Press.

[3] Dieter, G.E. (1991). *Engineering Design: A Materials and Processing Approach.* New York: McGraw-Hill.

[4] Ashby, M.F. (2005). *Materials Selection in Mechanical Design.* Amsterdam: Butterworth-Heinemann.

2

Final Design

#process module: #finaldesign

Learning outcomes

By the end of this module, you should demonstrate the ability to:

- Describe the level of detail needed for a final design specification
- Describe the tools used during the final design stage
- Explain the concept of scope creep within the iterative design process
- Explain the use of prototyping specific to the final design stage

Recommended reading

Before this module:

- **Design Process > Post-conceptual Design > 1. Intermediate Design**

After this module:

- **Design Process > Post-conceptual Design > 3. Post-final Design Engineering**

1. Final Design Specification

In general terms, a final engineering design specification should yield a plan for implementation that does not require further involvement of the engineer except in an oversight role. That is, it should describe a detailed design.

- In the design of a product, the detailed design is shop drawings of the product with complete parts and materials lists, and all of the details such as tolerances and surface finish specified.
- For software it is the code with commenting and descriptions.
- For a new airport terminal design it would include a detailed schedule of work and materials, of tests and acceptances.

The details of the final design depend very much on the discipline under consideration. The design of a complex manufacturing schedule, a satellite navigation system, and a hydroelectric dam are obviously very different. Table 1 shows some detailed design information for various disciplines. Regardless of discipline, however, the test is whether non-engineers could use the specification to implement the design. Thus there are a lot of details to be finalized after the basic intermediate design is complete.

TABLE 1 Types of design information determined at conceptual, intermediate, and final stages for various disciplines.

TOPIC	CONCEPTUAL DESIGN	INTERMEDIATE DESIGN	FINAL DESIGN
Software	System structure, general communication between parts. General user interactions, overall timing	Synchronization determinations, memory usage estimations, testing requirements, input/output details, security measures, grouping of user control and feedback into logical sections, and data structure details	Specific algorithm selection, language selections, and full modular breakdown. Details of tests to be conducted on code
Chemical Plant	Choice of major unit operations and sequence of these: Process Flow Diagram	Piping and Instrumentation Diagram Specific types of components chosen. Material Classes determined	Plant blueprints, Piping diagrams. Model numbers of all components, details of all piping
Manufacturing Schedule	Balance between human and robotic elements. Sequence of operations. Input and outputs for each step	Plant layout. Approximate timing of all sequences. Flow of goods, people, and equipment	Specific location of each manufacturing cell, and detailed timeline for the movement of the product through the plant
Electronic Circuit Board	Choice of major components and connections between them	Specific types of components and interconnection schematic. Ratings for critical components	Detailed printed circuit board plan. Full specifications of all components
Bridge	Architectural concepts, bridge type, design sketches to check for compatibility with surroundings	Detailed sizing of abutments, piles, spacing, truss configuration, and deck construction	Details of the location and part number of each truss element, each fastener

1.1. Sizing

In the final design, it is a given that all the technical decisions are made, and so the details of part sizes are known with precision. For example, at this stage:

- The dimensions of and properties of the I-Beams in a civil engineering project are known, because the engineer has completed detailed stress analysis.
- A chemical engineer has details of flow rates and acceptable pressure drops in a pipe, and thus specifies an exact pipe diameter and wall thickness.
- An industrial manufacturing engineer has specifics of the assembly times in a multistep assembly process, and can specify exactly how fast the conveyor linking two stations should run.

1.2. Materials Selection

In the intermediate design stage, you would have chosen a specific class of material for each component. For example, you might have made the decision to use plastic pipes rather than copper pipes in the water line feeding the icemaker in a modern refrigerator. In the final design you have to choose:

- What type of polymer: cross-linked polyethyelene (PEX) or polybutylene or something else?
- What diameter: 1/4″, 3/8″, or 1/2″?
- What supplier: Thomasnet.com lists more than 100 polyethylene pipe suppliers [1]?

In the case study we have been using the example of an engineer creating a design for a wedding photographer to take wide-angle photos from a high vantage point (#aerialphotography). A fiberglass pole was chosen for the solution, but at the final design stage the specific grade of fiberglass must be chosen. For commodity items like two-by-fours, ¾″ copper plumbing pipe, or standard fasteners, the implementation team may choose any product that meets the *spec* (*specification*), but for a more specialized material such as a specific type of fiberglass, the supplier will often be chosen as part of the design process.

1.3. Component Selection

Much of engineering design involves the assembly of preexisting components purchased *off-the-shelf*. When designing a ventilation system, chemical plant, or custom electronics board, an engineer typically makes extensive use of components that don't need to be designed from scratch. Even in a novel design, there is great advantage in using existing components unless custom components will perform much better. Existing components are already manufactured and tested in service, so the risk is much lower. Car companies building new vehicles, for example, often specify many components used in existing vehicles, including entire drive trains. In fact, they only create new components when the existing ones will be significantly inferior in the new application (#routinedesign).

Example

Imagine a design calls for a servo-motor-driven linear stage, to serve as the adjustable platform for a moving observation camera in a laboratory experiment. The designer could figure out how to machine and mount a rail on posts, then buy bearings to fit the grooves machined in the rail, and then design rail cars to house the bearings. Or the designer could buy a matched rail and rail car off-the-shelf and design the drive system. Or the designer could simply buy the whole unit, with the rail, bearing block, screw drive, and motor mount already put together. Even though this preassembled piece is expensive, there is a significant savings in design effort, and a significant reduction in risk, since the assembly will be sold as a functioning unit. As a bonus, the manufacturer may supply 3D drawings ready to use in a larger design.

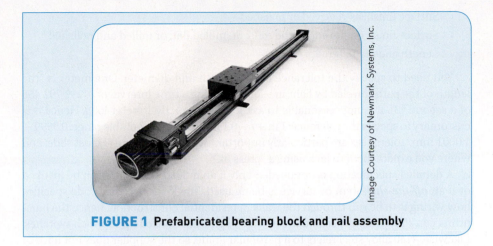

Image Courtesy of Newmark Systems, Inc.

FIGURE 1 Prefabricated bearing block and rail assembly

Where does an engineer find such components? We recommend you use a hierarchical approach, similar to that used for your literature search, trying each of these in turn:

1. Distributors of multiple product lines (e.g., Mcmaster.com for mechanical components or Digikey.com for electrical components).
2. Individual companies selling the products, using an industrial database such as Thomasnet.com or Globalspec.com.
3. Handbooks such as *Mark's Standard Handbook or Mechanical Engineers* or *Perry's Chemical Engineers' Handbook* to identify the specification of the components.

For products in a different discipline—software components, for example, or prefab truss work for house construction—you would modify your search locations but still try to use a hierarchical sequence.

Many of the decisions associated with sizing and sourcing the photographer's pole, for example, could be bypassed by using an existing product. Thomasnet.com actually lists 33 suppliers of telescoping extension poles [2]. Thus it may not be necessary to design and custom fabricate this component at all!

2. Sweating the Small Stuff: Tolerances, Finishes, and Other Important Details

Experienced designers are aware of many small details. It is a great exercise in critical thinking to see if you can imagine what these would be in your discipline. Ideally, the persons implementing the design would need only the instructions and information provided in the final design specification.

For example, a cylindrical tie rod with threaded ends would need the following specifications:

- Material: 4140 alloy steel (the number refers to a specific grade of steel and alloy composition)
- Thread length and type

- Surface finish: as-received or polished
- Surface finish on the end of the rod: cut, milled flat, or milled and polished
- Length and diameter

You need to specify the tolerance on the length and diameter. One meter is "the length of the path traveled by light in vacuum during a time interval of 1/299 792 458 of a second"[3]. It is not reasonable to ask for a bar exactly 1 meter long. Hence it is customary to specify the tolerance: 1 m +/− 0.1 mm, or the acceptable range: 0.9999 − 1.0001 mm. Tolerances are particularly important where mating parts must slide, and where you expect them to lock using a "press fit."

A detailed specification is not needed only if an industry standard can be invoked or if an *off-the-shelf* item or service is being used. The local electrical code specifies how wiring is to be run through the walls in residential construction, hence the blueprints can show a simple schematic of which lights are hooked up to which switches. Likewise, 4140 alloy steel refers to a particular grade, so the supplier does not have to be determined, nor would the specific supplier of a standard fan motor or pump be specified in drawings or plans.

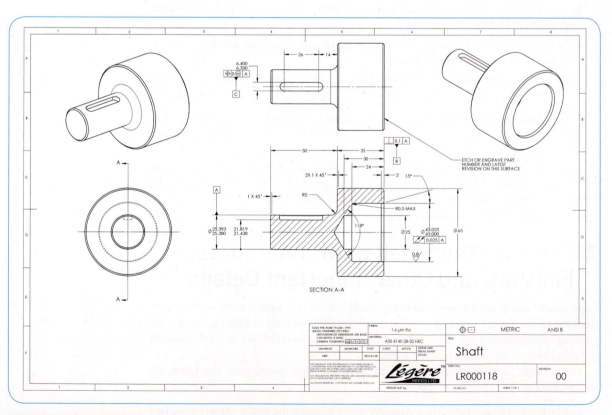

FIGURE 2 **Part of a CAD drawing with all required details.** Tolerances are specified on the shaft dimensions. Other tolerances, materials, and surface finishes are specified in the legend. (Image used courtesy of Legere Reeds Ltd.)

3. The Role of Iteration in Final Design

Iteration is an inherent part of engineering design. A linear or **waterfall design process** is simply not practical in a complex situation, and this is true of the intermediate and final design processes as well. Suppose a conceptual design uses gravity to move fluid through a pipe from the second floor to the first in a chemical plant. During the intermediate design, when more detailed flow calculations are being done, the engineer may discover an auxiliary pump is required to achieve the required flow rates. Perhaps a simple piston pump is added to the system at this stage. However, in trying to find the actual make and model number of a suitable piston pump for the final design, it turns out that a gear pump is the preferred choice for a variety of reasons, and so a particular gear pump is chosen at the very end. These design modifications are quite common, since engineering designers are not omniscient.

At each stage of the process, a willingness to revisit previous decisions is required to achieve the best possible design. Yet this iteration must be finite in scope to complete the project in the allowed time. Hence in a formal process, particularly those used in some disciplines, design decisions will be locked down as of a specific date and not subject to further changes. Once a hardware **platform** is chosen for the server farm supporting a large online retail operation, this decision must be locked down so that the software engineers can get on with their business, and the electrical engineers can set about designing the power supply system. Iteration is no longer allowed in that element of the design. There is a balance between flexibility and the demands of actually executing the project in a timely way.

3.1. Iterative Techniques

In some engineering areas, such as software development, a development method will be used that assumes iteration will be used even into the detailed design stage. Methods such as **spiral development**, **incremental development**, and **agile development** produce working versions often released to a client for use. The expectation is that then another round of development will happen that will add features, including some that are newly added. This is often done in situations where **scope creep** is anticipated in a project.

3.2. Scope Creep

The term **scope creep** describes a gradual change in requirements during the design and implementation process. The engineer is initially tasked with the design of a specified product or service, but with scope creep, the client adds new requirements while the design is under way. For example, a software engineer is demonstrating the user interface for a new program, and this generates a new idea that the client pushes to have added after the fact.

Scope creep becomes a problem for the engineer if the initial contract is vague about exactly what is to be delivered. There are a number of methods to address this problem:

Use a fixed-price contract: A **fixed-price contract** has a clear set of deliverables, backed by a strong set of requirements for a development contract. When those deliverables are delivered, the contract is done. Any changes to the deliverables are negotiated and will usually result in a higher cost to the people hiring the contractor.

Use a cost-plus contract: In contrast, a **cost-plus contract** pays the contracting company for any materials and other costs directly associated with the contract (for example, long distance phone charges, permits, travel time, subcontractors, materials, and supplies) plus compensation at a negotiated rate for the time put into the project by the contracting company employees. In this case, any changes requested by the client are covered by the hourly or daily rate specified in the contract plus the cost of the added materials, supplies, and expenses.

Fixed-rate contracts tend to be very detailed. A contract to refurbish an office space might run to hundreds of pages. The contract will specify the placement and type of every light, the exact type and quantity of carpeting and paint, the height of the vinyl baseboard strip, and so on. Changes to these specifications by the client after the contract is signed are usually charged at a premium rate. A similar level of detail is required in an engineering design or design/implement contract. Fortunately, much of the detail consists of **boilerplate clauses**, generic specifications that can be reused over and over again in new contracts. Cost-plus contracts are used where scope creep is expected, or where the time to develop a fixed-price contract is not available.

4. Tools Used during Final Design

An engineering design is characterized by its technical nature and dependence on knowledge of math and science. Engineering designs typically involve computations in order to predict how the system will perform when it is implemented. Sometimes simple, direct calculations or even **estimations** are sufficient to predict performance. However, for complex systems, simple calculations are not enough, and yet a prediction of performance is always a critical part of the design process. How are such predictions made?

4.1. Calculation

As described, simple equations are sometimes enough to predict performance. For example, suppose you need to design a 2-meter long, 0.4-meter wide wooden gangplank for a ship, and it is to deflect no more than 0.25 meter when end-loaded with a 100-kilogram person. This is an easy calculation for a first-year engineering student:

$$\delta = FL^3 / \left(3E \left(\frac{bd^3}{12} \right) \right)$$

where δ is deflection, F is force, L is length, E is modulus, b is width, and d is the unknown thickness of the plank. Equation X is simply rearranged and solved for d.

Similar computations can be done for many classes of engineering problems.

4.2. Simulation

Complex systems involving interactions of many parts can sometimes be represented by large systems of equations and solved analytically. Often however, a **numerical simulation** is employed for such systems. Furthermore, commercial simulation packages are widely available, and engineers rely on simulation because it is inexpensive when compared to experiments. Automobile car crashes are serious, and the companies work hard to engineer vehicles to protect the occupants in the case of a crash. It is now possible to model the structure of a car so completely that a numerical simulation of an impact can be used to predict its outcome. Some specific examples:

- A mechanical engineer is analyzing the complex main wing spar (beam) in a commercial executive jet. The beam is a sophisticated I-beam, but has many cut-outs and thickened areas to optimize its structure, so Equation 1 is of no use. It is necessary to know the maximum stresses and deflections in the event of a high-g emergency manoeuvre. In this case, *finite element analysis (FEA)* is used for stress analysis. The solid beam is represented by a large number of discrete elements in a computer program that can compute the full three-dimensional stress field resulting from imposed loads or displacements. FEA is typically very accurate.

- A chemical engineer is monitoring flow through a complex system of pipes, pumps, and heated tanks. *Process simulation* software is used to build a system in the computer, where the behavior of each individual unit is used as an input, but the simulation determines the outcome of the complex interactions between units.

- A computer engineer can use simulation software to predict the operating characteristics of a multi-core processor working on a set of problems. The simulation software can show the usage of the parts and the bottlenecks in the connection system of the processing cores.

It is also possible to program simple numerical simulations using code or even standard spreadsheet software. Suppose you are designing a chiller for a chocolate production line. You need to know how long it takes for the chocolate to solidify and hence how long your chilling belt has to be. To do this, you could set up a *finite difference* simulation in a spreadsheet, where each column represents a thin zone of chocolate located x micrometers from the outer edge of the chocolate, and each row represents a tiny increment of time. Using only very simple equations governing the heat flow between two slices of different temperatures, the spreadsheet can record the heat flow in and out of each slice for each increment in time, and the new temperature of each slice at the end of the increment. By marching through time, it can plot the temperature profile inside the chocolate.

Monte Carlo analysis is another common technique used in simulation. Monte Carlo is the site of a famous casino in Monaco, and a Monte Carlo simulation is one in which you use a random number generator to create simulated experimental data. For example, if you are designing a traffic flow system for a new subdivision and you need to predict whether a stoplight is needed at a particular intersection, you could use a random number generator to simulate flows of commuter traffic and children walking to school. As with any "experiment," you would run the simulation multiple times to provide an average result. As you would expect for an important problem such as this, there are a wide variety of sophisticated commercial simulation packages in existence for traffic engineers.

4.3. Prototyping

Prototyping is used throughout the design process (#models&prototypes). A *prototype* is a physical representation of a complete design or a subsystem that can be performance tested. In the conceptual design stage, we use prototypes to check the feasibility of basic design ideas. In the final design stage, however, a prototype is used to validate the design work before full implementation of the design. A beta-test version of computer software is a prototype of the final code that is circulated to a

(a)

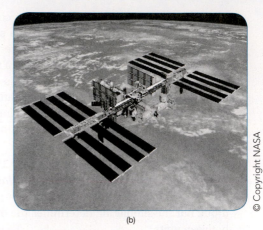
(b)

© Copyright NASA

FIGURE 3 (a) Before the advent of computers and CAD software, people built accurate scale models to check fit and finish of their plans. (b) Modern CAD programs allow NASA engineers to build a fully detailed virtual model of an image of the International Space Station.

limited number of users for real world testing to find undiscovered flaws. Prototypes are usually constructed if it is not cost-prohibitive to do so. In the case of a building or ship design, it is clearly not possible to build a full-sized prototype. Before the advent of sophisticated computer modeling programs, described in the next section, fully detailed scale models were built to validate ship design prior to commencing construction.

4.4. Computer Aided Design/Computer Aided Manufacturing (CAD/CAM)

Computer-aided design (CAD) describes a process in which a sophisticated software package is used to create a detailed two- or three-dimensional representation of a physical structure. In a modern CAD program, each physical part in a full assembly is represented by a detailed description of its mass, volume, and composition in a virtual 3D space. A 3D CAD model serves the role of a virtual prototype, and can be easily modified to meet the design specifications. Once a detailed 3D model is constructed, it serves a multitude of purposes:

1. The mass, center of mass, moment of inertia of each component and the entire assembly are reported by the software.

2. The model can be "meshed" to feed into a finite element analysis to determine stresses and strains under load.

3. The model can be converted into a stereolithography (STL) file suitable for input into a rapid prototyping machine that can build a 3D physical object using a variety of digital printing techniques.

4. The model can be used to send instructions to a computer numerical controlled (CNC) milling machine that will cut parts from solid blocks of material. This is known as computer-sided manufacturing (CAM).

5. The model can be used to produce photorealistic 2D pictures or 3D pictures or videos of the design.

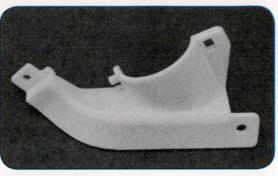

Courtesy of the Authors

Courtesy of the Authors

FIGURE 4 A CAD drawing (left) and the actual component that has been 3D printed based on the computer model. This process is called additive manufacturing.

CAD systems can be used to create a computer model of any physical structure: a plant, a building, infrastructure such as a tunnel or bridge, the design of a golf course, or a product like a toaster or steel mill width gage.

5. Conclusion

Although the conceptual design phase is critical to the success of a design project, the intermediate and final design phases are often more expensive and time consuming, and can determine whether a project will ultimately be a success or a failure. These costs and times are dictated during the conceptual design phase and are controlled by the design decisions there. The details of intermediate and final design are very much discipline-specific, and are normally addressed in the upper years of an engineering undergraduate program.

KEY TERMS

final design	intermediate design	off-the-shelf
waterfall design process	components	spiral development
fixed-price design contracts	computer-aided design (CAD)	agile development
boilerplate clauses	estimation	cost-plus contracts
finite element analysis (FEA)	incremental development	process simulation
scope creep	finite difference	numerical simulation
prototype	Monte Carlo analysis	specification (spec)

6. Questions and activities

1. You are designing a washroom. Which stage (conceptual design, post-conceptual design: intermediate, or post-conceptual design: final design) would the following activities fall into? Give reasons for your answers.

　a. Selection of the specific make of taps to be used for the sink

　b. Placement of the sink, tub, and toilet

　c. Location of the light switch and lights

Final Design

211

d. Method of waterproofing the joint between drain and sink

e. Type of surface to be used in the floor

f. Selection of cement to be used to fasten the floor surface

g. Placement of the mirror

h. Placement of pictures on the wall

i. Selection of the type of tiles to use around the tub

j. Selection as to whether to use an ordinary or whirlpool tub

k. Routing of the water supply pipes

l. Selecting the paint

m. Determining the size of the door opening in the framing

2. You are into the final design stage and discover that a single support beam in the design cannot be specified using standard parts as it must be larger/support more than standard parts allow. What might you do?

3. A design is going to use a battery. What factors are going to determine the type of battery selected? Use this information to answer the following: When would the selection of the battery type be delayed until the final design stage, when at the intermediate stage, and when during the conceptual stage?

4. Most of the tools discussed are for mechanical/physical designs. Software does not have such tools, since the design implementation does not have physical elements. Software also has a very poor track record in terms of prediction of project time and complexity. Discuss how these two statements are related.

5. Work through the detailed design process for the design project you are working on currently.

7. References

[1] ThomasNet (2013) "Engineers' and buyers' choice for finding trusted suppliers." Retrieved October 31, 2011, www.thomasnet.com/nsearch.html?what=Polyethylene+Pipe&heading=58545401&cov=NA&act=M&navsec=modify

[2] ThomasNet (2013) "Engineers' and buyers' choice for finding trusted suppliers." Retrieved October 25, 2011, from www.thomasnet.com/products/poles-telescoping-extension-62088505-1.html

[3] Bureau International des Poids et Mesures. *The International System of Units (SI)*, 8th ed., 2006.

Post-Final Design Engineering

③

#process module: #postfinaldesign

Learning outcomes

By the end of this module, you should demonstrate the ability to:
- Describe some additional services that engineering firms can provide
- Explain the reasons that an owner would include post delivery services in a design contract

Recommended reading

Before this module:
- **Design Process > Post-conceptual Design > 2. Final Design**

1. Introduction

Engineering is often not finished with the last iteration of the final design. A product may go to manufacturing, a building reconstruction will requiring process monitoring and tests, a plant will require maintenance and upgrades, and software will need revisions as bugs are discovered and functionality needs extension. Training in the operation of a piece of equipment, or in the application of a process, or the maintenance of a plant is often required. Engineers are also often used to oversee the decommissioning and disposal phase of a design. Sometimes this is the same engineering team that did the original design work, but often it is a different engineering team with specialized skills.

2. Contracts Past Design, DBOM

DBOM = Design/Build/Operate/Maintain

Large engineering firms can be contracted not only to design a system, but to build operate, maintain, and even decommission the system (see example in Figure 1). In other words, the buyer is purchasing not a design, or even a design that is also executed, but is actually purchasing a complete solution to their underlying problem.

FIGURE 1 On May 31, 2010, Bombardier announced that it had been awarded the contract to Design, Build, Operate, and Maintain (DBOM) a transportation system for the King Abdullah Financial District (KAFD) in Saudi Arabia. Bombardier will be implementing a version of its fully automated BOMBARDIER INNOVIA Monorail 300 system. (© Bombardier Inc. or its subsidiaries. All rights reserved.)

One simple form of this more complete engineering service package is referred to as a ***turnkey solution***. In this case, the owner of the chemical plant, for example, can begin operating by simply "turning the key," since the design firm has handled all the details of both the design and the construction of the plant. In the construction industry, this is referred to as a ***design/build*** project, where the engineering firm is responsible for the civil engineering and the construction of the complete structure or building. Automobile manufacturers provide a literal turnkey solution for the consumers who purchase their vehicles.

Taking this one step further, the design firm can also operate the plant or system, maintain it, and possibly even decommission it at the end of its useful life. A variation of this is a ***design/build/operate/transfer*** contract, where the designers are obligated to operate the system for a period of time before transfer to the owners. This provides the design firm with additional incentive to get it right. If they don't, all of the headaches of getting the system up and running properly are theirs. In these cases, the engineering firms don't own the system, but are doing all this for a fee on behalf of the owners. The owners pay a premium but can focus on their own business, which is the use of the design.

3. Commissioning Engineering

Running a complex building, factory, or other system can be highly technical, and it is possible that the owners, who have paid for the design and construction, are not able to determine if their new purchase is operating as expected. Automobiles come with a warranty to cover this problem, but in a complex structure this might not be enough. One option is for the owners to hire an independent company of *commissioning engineers* to confirm that they are getting what they paid for. For example, imagine that a small city enters into a design/build contract for a new City Hall. The commissioning engineers come in before the final payments are released to make sure that the air flows, fire alarms, heating, and so on, are all performing as specified in the contract. This is clearly a job that requires technical expertise, and hence falls within the scope of engineering practice.

4. End of Life: Decommissioning, Refurbishing, and Recycling

Decommissioning is a term describing the processes needed at the end of the useful life of a design. Simple products like aluminum cans can be discarded or recycled, but more complex things need to be carefully disassembled. For example, old computers may contain toxic heavy metals, but may also contain valuable silver and gold. Services exist to disassemble and safely dispose of dangerous compounds, while recovering valuable materials. If the engineer designing the computer did an appropriate *design for disassembly*, this decommissioning process will have been anticipated and planned for, and will therefore be as straightforward as possible. The parts into which the design can be disassembled will be sortable into recyclable, reusable, or other categories.

Note that *design for disassembly* and *design for assembly* are two entirely different DFX considerations. For example, when we design for rapid and inexpensive assembly, we try to aggregate parts and use snap fittings instead of screws. When we are designing for disassembly, we want to separate materials easily by partitioning them into individual parts, and we use reversible fasteners. It is quicker to use a nail gun to assemble a fence from pressure treated wood. If you need to disassemble the fence in a year, however, you will choose to sacrifice some assembly speed by using screws, which can be more easily removed.

Note that *design for maintenance* usually means that some parts will have to be made for disassembly in order to get access to parts for observation and testing, and to some parts for replacement when worn or otherwise near the end of their life.

One method of promoting design with decommissioning in mind is to build this provision into the contract. For example, most people have experienced problems associated with removing or upgrading a large computer program on their personal computer. An IT professional responsible for the student information system on the servers in a large university would be advised to build decommissioning provisions into the purchase contract, providing financial incentive for the software designers to make removal of the outdated software easy to do.

Legislation is another method of forcing engineers to consider end-of-life. In Europe, for example, legislation dictates end-of-life recycling targets for automobile manufacturers. This becomes a constraint for any new vehicle design process.

> Directive 2000/53/EC of the European Parliament and of the Council [2], dated 18 September 2000 dictates: 1. In order to promote the prevention of waste Member States shall encourage, in particular:
>
> (a) *vehicle manufacturers, in liaison with material and equipment manufacturers, to limit the use of hazardous substances in vehicles and to reduce them as far as possible from the conception of the vehicle onwards, so as in particular to prevent their release into the environment, make recycling easier, and avoid the need to dispose of hazardous waste;*
>
> (b) *the design and production of new vehicles which take into full account and facilitate the dismantling, reuse and recovery, in particular the recycling, of end-of-life vehicles, their components and materials.*

There is a distinction between design with the decommissioning in mind, as dictated by a contract or legislation, and burdening the design firm with the actual task of decommissioning. A company that has to decommission and dispose of all products that it sells has the strongest possible incentive to make sure that this task is easy. For example, commercial photocopiers are often leased rather than purchased. This means that at the end of life, the manufacturers have to collect and dispose of the machines. Since companies have to pay for disposal, especially if there are hazardous materials to be disposed of, they generally want to be able to recycle as much as possible. As a result they have incentive to design copiers with components easy to separate, and to choose nonhazardous components that can easily be recycled.

Construction firms are also faced with decommissioning costs if they erect temporary structures for big events like festivals, trade shows, one-time sporting events, and so on. In these cases, the structures are designed to be assembled and disassembled easily. In some cases, they are designed to be modified for a different use at the end of life.

5. Conclusion

The essential purpose of engineering is turning science into useable systems. Engineers solve routine problems, but more often are called upon to address open-ended, complex problems when there is no preexisting solution that can automatically be employed. The engineering design process begins with the specification of requirements, but where it goes from there depends on the project. Some projects will end with conceptual design, whereas others will progress through the detailed design phase. As described in this module, the engineering involvement may last all the way through the lifetime of the technology. At every stage engineers apply their technical knowledge, ethics, and concern for the public welfare and environment to guide their work.

KEY TERMS

design/build/operate/
 transfer

design for assembly

design for Disassembly

design/build

design for Maintenance

decommissioning

design/build/operate/
 maintain (DBOM)

turnkey solution

commissioning

6. Questions and activities

1. Discuss why a DBOM project could result in different decisions being made for the design than for one where one group or company is in charge of the design/build and a second in charge of maintenance.

2. Explain why using commissioning engineers could save money over situations where the design engineering group also was responsible for the commissioning.

3. Much sensitive information is held in an encrypted pattern by proprietary software. Write an agreement that deals with the decommissioning of the software that includes the encryption/decryption algorithm in such a way that the encrypted data is still recoverable.

4. An engineering firm designs a pipeline with a 50-year estimated lifetime.

   a. How long should the firm be responsible for the pipeline? What are the factors involved?

   b. What other groups should be responsible?

   c. If the pipe breaks after 50 years, should the firm be responsible?

   d. If the pipe is damaged because of deliberate damage by an extreme environmental group, should the firm be responsible?

   e. If the pipe is damaged by an earthquake, should the firm be responsible?

   f. If the pipe is damaged because a tractor accidentally runs into it, should the firm be responsible?

7. References

[1] "BOMBARDIER INNOVIA Monorail – Riyadh, Kingdom of Saudi Arabia" Retrieved October 25, 2011, from www.bombardier.com/en/media-centre/newsList.html?filter-bu=all&f-year=2010&f-month=4&f-type=all&show-by-page=50&page=1

[2] "Directive 2000/53/EC of the European Parliament and of the Council of 18 September 2000 on end-of life vehicles" Official Journal L 269, 21/10/2000 P. 0034 - 0043 Retrieved October 25, 2011, from http://eur-lex.europa.eu/LexUriServ/LexUriServ.do?uri=CELEX:32000L0053:EN:NOT

3

Implementing a Project

Introduction to Teamwork

1

#narrative module: #teamwork

Learning outcomes

By the end of this module, you should demonstrate the ability to:

- Recall the definition of a team
- Describe the difference between a team and a group
- Determine what type of functional team you are on
- Determine the performance level of your team from the five performance categories
- Describe the key aspects of a high performing team model (see Figure 2)
- List and explain the five factors that lead to successful teams
- Identify the five stages in the Tuckman team model and explain each stage
- Given a team scenario (or video of a team working), identify the stage they are in

Recommended reading

After this module:

- **Implementing a Project > Working in Teams > 2. Organizing**

1. Introduction

As an engineer you will find that you work extensively in teams. Effective teams bring diverse perspectives to design and generally provide better solutions than individuals working alone. This is because problems can be solved more quickly, more and better new ideas can be generated, and ideas are implemented more efficiently by teams than by individuals. Understanding how teams work is an important part of any work environment. Understanding teams will help you be successful!

A *team* is a group of people who come together to work in an interrelated manner toward a common goal. The key difference between a group of people and a team is the common purpose or goal and the reliance on the skills of all the members to meet the goal [1]. A group of people who come together to make independent decisions to reach a goal do not form a team. Put another way, team decisions are not simply a sum of independent decisions made by individual people. And the work that a team accomplishes is not a set of individual isolated pieces that are stapled together. Teams operate as an entity and often develop characteristics (almost personalities) apart from their members.

1.1. Types of Teams

Teams are often categorized either by their function or by their performance. There are *cross-functional teams*, across disciplines and skills, and *within-function teams,* within a discipline or skill set.

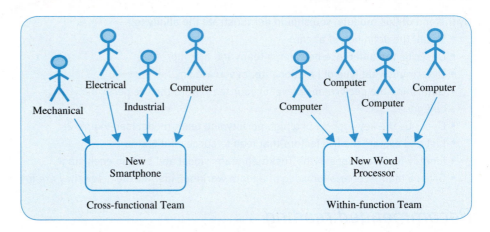

Alternatively, teams can be considered in terms of their performance rather than the function of the individual members [2,3]. Note that all types of experience, knowledge, and skill sets do not have to be represented on a team. Often expertise in some areas related to the project can be hired or contracted and components of the project can be licensed or purchased by the team.

Sometimes a group of people working together performs below what would be expected from the sum of individuals in the group; this is called a *pseudo team*. Pseudo teams work as a set of isolated individuals. This type of "team" is characterized by poor communication and a lack of commitment to a team purpose. *Potential teams* perform at or slightly above the average team member. Potential teams communicate better than pseudo teams. There are some synergistic work habits in this type of team.

Real teams perform well. Real teams communicate actively and have developed cooperative, synergistic work habits. *High-performance teams* go well beyond the capability of the individual members. High-performance teams are highly communicating, cooperative, and synergistic. They fully actualize the potential of every team member. Creating a high-performance team requires ongoing effort from every team member and a commitment to the shared goal.

1.2. Building Successful Teams

Research on effective teams has shown that there are five underlying common factors that make teams highly successful [1–5]:

1. The team shares a common goal or purpose.
2. There is both individual and group accountability.
3. Real work is undertaken: work done is directly relevant to the project and the project is perceived as relevant and valuable by the team members.
4. Processes, skills, and mechanisms are in place to deal with both task and people issues: the team has effective procedures for dealing with team conflict and project difficulties.
5. Group processing occurs: the group reflects on their work, celebrates together, and resolves issues together.

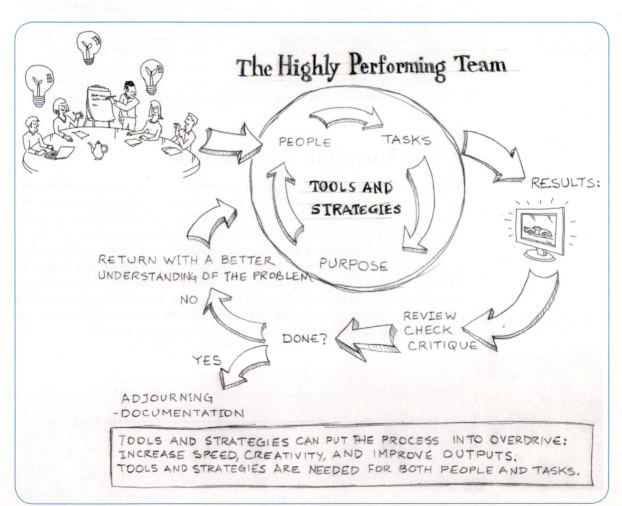

FIGURE 1 High-performance teams successfully balance people issues, tasks, and purpose (goals). They use tools and strategies to keep everyone engaged and working together.

Simply stated, a team needs to have a clear purpose or goal that is understood by everyone and the team itself must have methods it will use to keep itself on track, resolve issues between people, and resolve problems associated with getting the work done. In high-functioning teams, people and their relationships are seen as equally important as the tasks. Your team needs to meet these five themes to be highly successful and this can be challenging.

Part of being a high-performance team is developing the habit of **reflection**. This means reflecting on an experience and intentionally thinking through what you learned from it. In business there are formal methods for reflection of this type; they are called **lessons learned** activities. We suggest you practice reflection and use past experiences to help you learn for your next team assignment.

> ### Reflection (Lessons Learned)
> Think about the best team you have ever been on. It could be a sports team, a band, your theater group, or a club. What were its characteristics? Was it a real team, or was it a pseudo team, or a potential team? Did it meet the criteria for a high-performance team? Was it a functional team or cross-functional? Were there conflicts? How were they solved? Was there a clear goal that everyone was striving for?
>
> Consider the five factors that contribute to a successful team. How did your team rate in these five areas? How could the team have improved in these areas?

2. The Tuckman Team Model

Engineers generally work in teams, so understanding how teams can become successful is important both when you are in school and when you are at work. There are many different models of how teams form and ultimately work effectively together. One of the most commonly used models was developed by Bruce Tuckman in 1965 (#Tuckman). It is a four-stage model that was later expanded to five stages [4,6,7]: **forming**, **storming**, **norming**, **performing**, and **adjourning** (Figure 2). While the model in the figure appears linear, it is really iterative. The addition of a new team member, a major change in the project, or other disruptions and creative conflicts will cause the team to go back to an earlier stage. A team may repeatedly go through forming, storming, norming, and/or performing on their way to successfully completing a project.

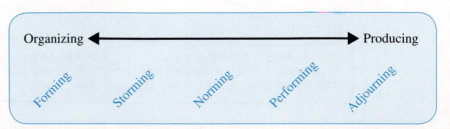

FIGURE 2 Modified version of Tuckman's team development model [6-8].

The first three stages of team development—forming, storming, and norming—can be categorized as *organizing*. During these stages it is particularly important that your team defines the roles and responsibilities of its members, chooses a team leader, sets team rules, and gains a clear understanding of its purpose. If these things do not happen, then the team will stall and will not move into stages four and five, which can be categorized as *producing*. It is while producing that the team will have its greatest productivity in reaching the goal. It is where the team is tuned and cohesive.

Effective teams are not easily formed. Every team member has to understand how a team works, and how to interact with the individuals in the team over the course of a project. The remaining modules in this section describe the stages of team dynamics (the Tuckman model, as described) and the team-oriented project management aspects.

KEY TERMS

team	reflection	cross-functional team
lesson learned	within-function team	forming
pseudo team	storming	potential team
norming	real team	performing
high-performance team	adjourning	Tuckman team model
organizing	producing	

3. Questions and activities

1. Which of the following designs might require a cross-functional team and which might require a within-functional team? Discuss your answers with your colleagues. Are there reasons for both?

   a. Design of a toaster

   b. Design of a calculator app for an iPhone®

   c. Redesign of automobile assembly plant

   d. Design of coffee room in automobile assembly plant

   e. Design of an automobile steering wheel

   f. Design of an entertainment facility for an automobile

2. Think about the best team you have ever been on. It could be a sports team, a band, your theater group, or a club.

   a. What were its characteristics of the group?

   b. Was it a pseudo team, potential team, real team, or high-performance team? Explain your reasoning for identifying the team as one of these.

   c. Was it a functional team or cross-functional?

   d. Were there conflicts? How were they solved?

   e. Was there a clear goal that everyone was striving for?

   f. In the list given in the section "Building Successful Teams," which of these factors did your team have? And which were lacking? Explain your answers.

3. Some things you learn because someone teaches you, but often the best lessons are those where you learn by experience. In your groups, talk about things you have done and the lessons you have learned from doing those things that you can reapply later.

4. You and others have probably encountered many new people when you started your engineering education. Discuss situations where you have seen aspects of the Tuckman model happening in the relationships you have seen being formed.

4. References

[1] Hensey, M. *Collective Excellence: Building Effective Teams,* 2nd ed. Reston, VA: ASCE Press, 2001.

[2] Smith, K.A. *Teamwork and Project Management,* McGraw Hill Higher Education, 2004.

[3] Katzenbach, J.R., and Smith, D.K. *The Wisdom of Teams: Creating the High-Performance Organization,* Boston: Harvard Business School Press, 1993.

[4] Tuckman, B. Developmental sequence in small groups. *Psychological Bulletin* 63 (6):384–399, 1965.

[5] Tuckman, B.W., and Jensen, M.C. Stages of small-group development revisited. *Group & Organization Studies (pre-1986),* December 1977, p. 419.

[6] Hensey, M. *Collective Excellence: Building Effective Teams,* 2nd ed. Reston, VA: ASCE Press, 2001.

[7] Tuckman, B.W., and Jensen, M.C. Stages of small-group development revisited. *Group & Organization Studies (pre-1986),* December 1977, p. 419.

Organizing

2

#process module: #teamorganizing

Learning outcomes

By the end of this module, you should demonstrate the ability to:
- Clearly describe the important characteristics of the forming stage
- Recognize when teams are in the forming stage, storming stage, and norming stage

Recommended reading

Before this module:
- **Implementing a Project > Working in Teams > 1. Introduction to Teamwork**

After this module:
- **Implementing a Project > Working in Teams > 3. Tools for Organizing**

1. Organizing

The ***Tuckman team model*** has three stages that happen when teams are starting up: forming, storming, and norming (see Figure 1). These stages can also recur repeatedly during the project when there is a disruption in the project or team or a conflict. These are typical stages a team will go through as it organizes and coalesces into a working unit.

A companion module to this one, "Tools for Organizing," contains a set of specific tools and techniques that can be used to reduce the length and negative interactions during these stages (#toolsfororganizing).

1.1. The Forming Stage of Team Organizing

In the *forming* stage, the team is coming together but still operating as a set of individuals (#forming). This stage is dominated by individual team members thinking more about themselves than about the team. Think of how you feel each time you start in a new team. Do you ask: What is my role? Why did I end up in a team with that person? How will I fit in to this group? Will I still be able to get the grade I want? Who will be the team leader?

During forming, people are generally more polite and interactions are more formal.

All these feelings are natural in this stage, which is dominated by the team getting to know and trust each other. In this stage, if you are the team leader you may find that you will need to be directive and focus the team on the tasks that need to be accomplished. It is also important to get the team members to define the project they are working on.

Forming is often a comfortable stage because team members are being careful with one another as they get acquainted. It is like being at someone else's family gathering with a lot of people you don't know. People usually actively work to keep conflicts from starting.

If you are observing a team in this stage you might see:

- Polite conversation
- People being quiet or tentative
- Focus on task definition
- Exchange of limited personal information
- Use of the word "me" and "I"
- Very little expression of strong opinions

These behaviors are a natural part of team formation. It is important to let people get used to and gain trust in each other. During this phase you should exchange contact information and define the *purpose* for your team (i.e., the design project).

At the end of this stage, you will need to be ready to start making decisions as a team. Your roles and responsibilities should be clearly defined, the project goal or purpose should be clear to all team members, and you should have selected your team leader.

Organizing ⟵——————————⟶ Producing

Forming   Storming   Norming   Performing   Adjourning

FIGURE 1 Modified version of Tuckman's team development model [1–3].

1.2. The Storming Stage of Team Organizing

In the ***storming*** stage the team is in conflict (#storming). This is the least comfortable stage of team development but is a very important one. Teams can get stuck here all the way through the project if they are not careful. In this stage, different ideas come out as team members get more comfortable with one another. Opinions are expressed, work habits are revealed, and expectations come to the surface. One of the most common points of conflict in storming is related to assumptions. You assume that your teammates will behave the way you want them to, and they don't. Roommates sharing a kitchen or bathroom will be familiar with this stage—as the roommates become familiar with each other, their habits emerge and cause tension and conflict.

During storming, disagreements over work practices and expectations arise.

Resolving these issues requires discussion, agreement, disagreement, and ***compromise***. Roles and responsibilities, if not already clear, should be finalized and assigned to people. The leadership pattern for the team will emerge. Resistance to the leader may start to occur, as well as resentment toward the project itself. Conflict can occur as the design problem is elucidated in more detail and the direction of the team made clearer. If you are the team leader, you will need to be directive and sometimes assertive to keep the team on track and to avoid getting stuck in this stage. Be prepared for team members to resent any exercise of authority in this phase.

If you were observing a team in this stage you might see:

- Conflict and anger over disagreements
- Resistance to decisions
- Intolerance
- Focus on small details
- Frustration with the behavior of teammates

Teams can get stuck in this stage and may revert to individual decision making and the project work returns to a strategy of tacking together individual pieces. If this happens, your team members will do what they individually think needs to get done, rather than what the team has decided needs doing. The team reverts to a ***pseudo team***. The work quality will suffer. To get through this stage it is helpful to have picked an assertive team leader (not aggressive and not timid) who keeps the team focused on the task rather than on personalities. Team leaders who are too aggressive or timid during this phase of team development will lose the support and trust of their team. It will be difficult for them to effectively manage the team moving forward. It is also helpful to start defining how the team will make decisions: deciding how to decide. This process will need to be clear in the norming stage for the team to move forward from organizing phase to the producing phase of the model.

Individuals should try not to become unduly upset at what happens during the storming phase. When upset however, avoid insults or personal attacks. When others are upset, work to moderate the situation and to work out the problems in a logical, fair fashion. The golden rule is important in this stage: treat others as you would have them treat you. A team whose members are willing to forgive behavior of others during times of stress will quickly move past this to a strong working unit where the deficiencies of each individual are more than compensated for by the coordinated strengths of the team.

The module "Managing Teams" considers the type of teammates you might run into, and how to build a strong team despite the individual weaknesses that we all have (#managingteams).

1.3. The Norming Stage

In the **norming** stage of team development, the team has essentially "got its act together" (#norming). It has determined its common goal; in the case of a design project, the project has been defined and planned out. While team members will still have their own ideas, they will be willing to **compromise** in order to make the team work effectively. In a high-performance situation, conflicts are sometimes seen as opportunities for creative development and people adopt the ideas of others and build on them rather than simply compromising. In the norming phase team members are taking responsibility for team decisions. The team has agreed on how to work together and on what the standards for the team are, and roles have been defined.

During norming, team members develop strategies for accommodating differences and actively engaging all team members.

None of the teammates will likely fully agree with each decision made, but they will understand that every decision is not right or wrong, but depends on perspective. Often team members will decide to soften or change their opinions as they come to understand the viewpoints of their teammates, but will in any case support the team in the collective decision.

This phase of team is characterized by:

- Agreement on how the team will behave (i.e., the norms of behavior)
- Agreement on a decision-making process
- A leadership style that is less directive and more supportive
- General **consensus** on the team goals and activities
- Processes and procedures are agreed on and followed not by directive but willingly as an accepted and valued part of team practice
- Members begin to trust each other and appreciate every member's contributions more fully
- Greater focus on tasks rather than resolving people issues
- The emergence of a team personality that is separate from any one of the team members

1.4. Moving to Performing

Once the team has reached the norming stage, they can move into performing. The members know how to work with one another and share a common goal and a pathway to the goal. They are ready to spend less time negotiating the process and more time on working toward the goal.

KEY TERMS

Tuckman team model	purpose	compromise
pseudo team	consensus	forming
storming	norming	organizing

2. Questions and activities

1. Describe the stage your team is currently in. Is it forming, storming, norming, or have you moved on to performing?

2. Using the stage you are in currently, or thinking back to when your team started a project together, can you identify the behaviors of the team and relate them to the behaviors listed in this section? For example, did you observe "agreement on a decision-making process" when you were norming?

3. Give three specific examples of events that occurred in your team and relate them to the relevant stage the team was in at the time. For example, consider an exchange between team members that occurred at a team meeting, or the way the team made a particular decision, such as who would work on which piece of the project.

3. References

[1] M. Hensey. *Collective Excellence: Building Effective Teams*, 2nd ed. Reston, VA: ASCE Press, 2001.

[2] B. Tuckman. Developmental sequence in small groups, *Psychological Bulletin* 63 (6):384-399, 1965.

[3] B.W. Tuckman and M.C. Jensen. *Stages of Small-Group Development Revisited, Group & Organization Studies (pre-1986)*. December 1977, p. 419.

3

Tools for Organizing

#skill/tool module: #toolsfororganzing

Learning outcomes

By the end of this module, you should demonstrate the ability to:

- Apply tools and strategies relevant to the forming, storming, and norming stages to successfully help move your team through these stages
- Determine the appropriateness of each tool and strategy for your team in its current stage, or for application in a given scenario (i.e., case study)

Recommended reading

Before this module:

- **Implementing a Project > Working in Teams > 2. Organizing**

After this module:

- **Implementing a Project > Working in Teams > 4. Producing**

1. Tools and Strategies for Forming

The forming stage is the stage where a team comes together for the first time (#forming). A good practice in the forming stage is to put in place **team rules**, sometimes called **team beliefs** or a **team charter**. This is a set of behaviors that the team members agree will govern their interactions with one another. This may seem silly, but these "rules of the road" will help the team members and team leader define how they will behave. Such rules often include items like: come prepared to meetings, don't be late, no cell phones on during team meetings. The rules can be used to deal with behavior that is getting in the way of achieving the task. Even if you do not reread the rules frequently, the act of negotiating them is an important forming activity. Just like children negotiating the rules of a game on the playground before playing, it helps to lay the groundwork for the activity.

For every team meeting use **agendas** and keep **minutes** (#teamdocuments). Agendas and minutes are used routinely in business for a reason, not merely for the purpose of being more bureaucratic or pretending to be more professional. These documents are going to help keep your team on track and moving forward. It is easy to forget who told what to whom, who is doing which part when, and what your team did just a few weeks ago. Agendas and minutes are part of documenting the process. Assuming that the project is too simple to require these practices invites problems. Getting into the habit of using these practices early on will help to avoid problems later.

Examples of Constructive Team Rules

- Do not answer cell phones, play games, or work off topic (e.g., social media, text) during team meetings.
- Treat teammates with respect.
- Show up on time (the team should decide what "on time" means).
- Give each other the benefit of the doubt, unless proven otherwise (especially in email and other written communication). Always respond constructively to written communication.
- If a member breaks the rules, call them on it. This means immediately bringing it to their attention respectfully and directly.
- Let people know if you are in trouble as soon as you know you're in trouble (e.g., getting overwhelmed, unable to deliver work on time).
- Decide how the team will communicate and how the members will collaborate on documents (e.g., Google docs, Dropbox, or other file sharing systems).
- Answer emails, texts, and phone calls from teammates; decide what is a reasonable response time.
- A problem with a member is a team problem. Everyone needs to take responsibility for doing things differently to make the team work. Don't play the blame game (i.e., it is their problem so they need to fix it). Instead say, "It is our problem we need to fix it."
- All members should ensure that they know what the work expectations are for an assigned task, and the deadline for that task at the time the work is assigned. This rule will ensure that people are clear on what they are being asked to deliver and when. If the team member responsible for section D of the report thinks this means two paragraphs but everyone else thinks it means two pages, there is going to be conflict.

Other questions to answer:

- If work isn't delivered on time, at what point does the rest of the team take on the job of doing the person's work for them (and removing their name from the author list)?
- How much warning needs to be given before a team member who is not responding is cut out of the process?
- What are the team's expectations for quality of work? Is everyone striving for the highest grade possible, or would people be happy just to pass?
- What happens if the team rules are broken? What are the penalties? Consider every problem, such as late for or missing a task, late for or missing a meeting, not responding to emails, very substandard work, among other issues.

(continued)

Examples of Destructive Team Rules*

- If you are late for a team meeting you have to buy everyone a treat.
- You can play on your laptop during team meetings if you agree to do more of the work.
- People who don't get their work done have to wear a hat that says, "I'm a loser who let my team down."
- If you are 5 minutes late even once for a meeting you have to leave; you are not allowed to attend the meeting.
- Team members should tell each other exactly what they think of them at the end of each meeting (or publically online).
- If a team member delivers inadequate work for one report, they have to do more of the work on the next report.
- The team leader is responsible for cleaning up any team messes and rewriting anything that is poorly written.
- If a member doesn't deliver their work at least 2 hours before a deadline, then the rest of the team is responsible for writing the missing sections. (This is a bad rule because 2 hours is not nearly enough time for the team to remediate the situation. It isn't fair to the team.)
- Team meetings are optional for people who finish their part of the work early.

*Adapted from actual team rules that we have seen undergraduate design teams try to use.

A note on rules: Consequences should be part of the rules. These consequences should be generally strong; the team can decide to reduce a penalty if they think it is warranted, but it is not fair nor is it easy to increase a penalty. Consider as well repeat offences: Should being late a second time be dealt with more harshly than the first? When do you invoke harsher consequences? Ideally consequences should improve the overall team performance, not be used to punish individual team members.

2. Tools and Strategies for Dealing with Storming

Storming is when your team rules become really important (#storming). During this stage review your rules frequently, and address issues as soon as they arise. Don't wait until things get terrible. If the rules are not working for the team, then revise them to be more effective.

A few brief tips to use in the storming stage:

- During the storming stage a lot of time will need to be spent on resolving people issues, leaving less time for tasks. Make sure you account for this in your project planning; it is an important necessity.

Working through conflicts requires negotiation, decision making, and commitment to shared goals.

- Recognize that you have flaws too, and there will be times you won't be able to deliver on promises also.

- Set early deadlines for the first deliverables (i.e., work that needs to be delivered to the team to be incorporated into a team document such as a report). This will allow you all to see how the team operates, and have time for recovery from misconceptions and performance issues.

- Remind yourself of what you value about each team member. Even though you see the weaknesses at this phase, every person has strengths too.

- Discuss the issues with a positive sense of humor and work toward positive goals.

- Do not wait until people are really angry to resolve a conflict. If you see a problem looming, call it out and discuss it before it becomes critical.

- If your team has to resolve a major conflict close to a deadline, be especially calm, responsive, and professional (pick your words and actions carefully) to combat the tension. Sometimes the team can agree to postpone dealing with the underlying issues until immediately after the deadline (and be sure to do it, or it will again be an issue at the next deadline).

- Reaffirm your commitment to making this team work.

- Recognize that even the most wonderful teams will go through storming periods.

- Keep communicating.

2.1. Negotiating Conflict

In *Getting to Yes*, Roger Fisher and William Ury define a system for mediation that is highly applicable to the storming and norming stages of team development [1]. They claim that there is little to be gained by arguing over "positions." A "position" or a "stand" is a statement of a belief that something is absolutely true or right. It might be your position, for example, that the people who work harder should get the greater reward and that people who do not work hard should be punished. The problem is that you can waste a lot of time arguing about whether one person's "position" is more valid than another's. Fisher and Ury's four-step process is adapted here: for engineering design teams:

1. **Separate the "people" from the "problem."** It is easy, when you are having a conflict, to get emotional and blame a person for the problem. But if you can back away from that, perhaps by allowing time to "cool off" a bit, you can start to see the problem as separate from the person or people involved and turn your attention away from "blaming" and onto "solving." In other words, it is not really going to help to blame someone and then punish or humiliate the person. While it might seem like some kind of justice, the problem will still be there and have to be solved anyway. Try to define the problem itself and move away from "blaming" the people responsible. Then focus the team on addressing the problem and finding strategies for better team performance.

2. **Focus on "interests," not "positions."** Instead of arguing about who is right and who is wrong, ask one another, What do you actually *want to accomplish?* The best solution to a problem is not a matter of rewarding the right people and punishing the wrong people. The best solution gets people most of what they want and helps the team accomplish its goals.

Sergei Bachlakov/shutterstock

FIGURE 1 Even experienced groups will go through every stage of team development; however, people who have spent their careers successfully working on many teams are able to more quickly deal with issues and move into productive collaborative work.

3. **Develop solutions that will benefit everyone.** Once you have achieved step 2, start to generate ideas that will achieve your team goals. This is much like generating alternative solutions to a design problem and you can use many of the same tools, such as brainstorming.

4. **Define effective performance measures.** After you have agreement on beneficial solutions, you have to come up with a systematic way to put them into practice and monitor whether the solution works. Some project management tools would be effective here. Defining some simple performance measures to monitor the strategy you are using to improve team performance enable you to move onto the next stage, Performing.

Note that there was a cause and an effect. Deal first with the effect. Next, adapt your team processes and rules to deal with the cause so it will not happen again. This second step should be done without personal attack if at all possible: The revised rules and processes should apply to all team members.

3. Tools and Strategies for Norming

3.1. Making Decisions

The team must decide how to make decisions and needs to put in place a clear decision-making process (#teamdecisionmethods). This needs to start in the forming stage and continue to evolve in the norming stage or the team will never move to the producing stages of the Tuckman model (e.g., performing) (#norming). Decision making can be a considerable source of frustration for many teams. It is one of the most difficult parts of team work because you will need to live with, and support, decisions that you don't fully agree with.

Two of the more common decision making methods are **consensus** and **voting**. For design process decisions there are many tools that can be used. However, most simple team decisions do not require anything more complex than voting or consensus. There are pros and cons to each of these strategies. You should also probably have in place a "crisis" decision-making strategy for when there is not the time to come to consensus or the team is deadlocked and voting will not work (see next section).

Making Decisions in a Crisis

There will be occasions when you do not have time for consensus and not all of the team members are available for a vote–for example, 15 minutes before a deadline when you realize that your report is missing a key figure. Your team needs to have a process for making decisions in these tight situations. This may never be needed, but it is useful to have in place just in case. The typical strategy is to empower the person in charge to make an **executive decision** on the spot. An executive decision means the person makes an individual decision on behalf of the team. So if you have tasked one person to submit the report and they find the error, then they are empowered to decide what to do (e.g., include a rough hand-drawn sketch, leave the page out, add in a note, or wait for a printout of the figure and get a late penalty on the report). Crisis decision making is very stressful and difficult, unless you have a lot of practice at it (e.g., emergency room doctors). Everyone has to agree that whatever the person decides to do, you all will do your best to support the decision.

Performance Measures

By the norming phase your team should have developed effective process strategies, such as the use of agendas, minutes, and status reports. These, used in conjunction with your project plan, will help you formulate performance measures, which indicate how well your team is performing. To norm effectively, make use of these performance measures to monitor the team activities, and use this information to continue to improve team performance.

KEY TERMS

team rules	**team beliefs**	**team charter**
agenda	**minutes**	**consensus**
voting	**executive decision**	

4. Questions and activities

1. Pick a few of the destructive team rules. Explain why each rule is destructive, or give an example of where the rule would fail to improve the performance of the team.

2. A team member has failed to deliver a section of a report at the designated time for the second time in a row. What should the team do?

3. A team member becomes unavailable by email, is not appearing in class, and nobody seems to know where the person is. What does the team do?

4. Discuss why each of the stages of organizing (forming, storming, and norming) are not as productive as the performing stage, which comes next.

5. A productive team has reached the performing stage. What might happen in the following instances (relate your answer to the Tuckman model):

a. A vital piece of design equipment breaks.

b. A team member leaves.

c. A team member, working under a misconception, spends a significant amount of time producing part of the design, which has to be thrown out.

d. There is suddenly a need for a significant change in the design, which will change the structure of the design and also the amount of time and cost of the technology.

5. References

[1] R.F. Fisher and W. Ury. *Getting to Yes: Negotiating Agreement without Giving In*. New York: Penguin Books, 1991.

Producing

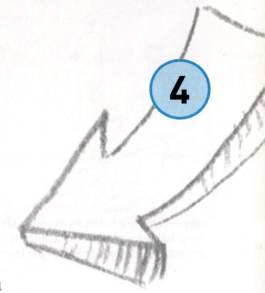

4

#process module: #teamsproducing

Learning outcomes

By the end of this module, you should demonstrate the ability to:
- Identify the two stages in the producing phase of the Tuckman model
- Clearly describe the important characteristics of the performing stage
- Recognize when teams are in the performing stage
- Clearly describe the important characteristics of the adjourning stage
- Recognize when teams are in the adjourning stage

Recommended reading

Before this module:
- **Implementing a Project > Working in Teams > 3. Tools for Organizing**

After this module:
- **Implementing a Project > Working in Teams > 5. Managing Teams**

1. Producing

Most of a team's time will hopefully be spent in the Producing stages of the Tuckman model: Performing and Adjourning (#Tuckman). In these stages the team is cohesive and dedicated to working on the project. In the Performing stage many of the team issues have been settled and the team is functioning effectively. However, as new issues arise the team may revisit the Storming and Norming stages and then return Performing. Even high-performing team will go through these iterative processes, but they will do this quickly and efficiently and with a focus on returning to Performing.

2. Performing

When a team enters this stage it has achieved the balance between focus on tasks and focus on people (#performing). The team is ***strategic*** and clearly understands what it is doing. Strategic means the team focuses on developing efficient and effective

When teams are in the Performing stage, the focus is on the tasks at hand and achieving the goals of the project. Less time is spent on the team relationships.

processes and procedures for accomplishing shared goals. The team members will have a shared understanding of the problem, project, or design and their focus is generally on the goals. The members also look out for one another and have a shared sense of responsibility and ownership of the project.

This stage of team development is characterized by:

- Roles are clearly defined.
- The team organizes itself; every member takes responsibility for involvement.
- There is an understanding of and respect for each team member's strengths and weaknesses.
- Strengths are celebrated.
- Weaknesses are supported.
- The members are interdependent and individually accountable and responsible.
- The team leader is only required to provide limited direction; the team is self-directive.
- Team language is characterized by the use of "we" and "us" as opposed to "me" and "I."

2.1. Tools and Strategies for Performing

To perform well and continue performing well throughout the remainder of the project, the team needs to:

- Continue monitoring performance and feed this information back to the team members.
- Continue using this information to find new ways of improving performance.
- Not become complacent; continue communicating actively.
- Not assume that you know what the other team members are thinking, even though you now know them well enough to guess.
- Recognize that the team may occasionally Storm and Norm again, and build time for this into team meetings.
- Regularly reassert expectations, and discuss process.
- Make sure everyone is listened to and everyone is engaged.
- Enjoy the project; not every team you work on will get to the performing stage, so enjoy working on a well-performing team, and reflect on how you got here.

3. Adjourning

Adjourning is when the team breaks up but may also occur when in long-lasting teams if a member leaves or there is significant change in the team structure or purpose. In the case of an engineering design project it is usually marked by the submission of the finished design project, and sometimes occurs when an intermediate goal is reached. The team will feel good about what it has accomplished, but team members who have become closely bonded may also feel some sense of loss.

This stage of team development is characterized by:

- A lack of energy as members resist the change.
- Evaluation of the team's efforts.

3.1. Tools and Strategies for Adjourning

- Celebrate your achievements.
- Thank your teammates.
- Reflect on what you have learned from working with this team; some time teams will use a formal *lessons learned* activity to improve their chances of success on the next project.

Not every group will become a high-performing team. Understanding this team model, and the tools and strategies for the different phases, can help you achieve a better level of performance. However, much of what differentiates a poor team from an excellent team is attitude. There is a saying in industry that "companies hire for attitude, and train for skills." This may not be 100% true (your grades do matter), but it makes a point. Your attitude matters and it will affect what you get out of this experience. You may not be fully satisfied with the results when the team adjourns. However, this experience will be an important part of your engineering education.

KEY TERMS

producing	**performing**	**adjourning**
strategic	**lessons learned**	

4. Questions and activities

1. What differences might you see in the performance of a team member that is part of a Norming team and one that is part of a Performing team?
2. List two things you could do to help your team reach the Performing stage and remain in that stage. Explain why these strategies would be effective.
3. Explain two strategies you could use for quickly and efficiently Storming and Norming again if a disruption occurs in your project.
4. What are two ways of celebrating a successful project and appropriately (and professionally) thanking your colleagues for their efforts?
5. Many organizations will talk about the importance of "networking."
   a. What online facilities are available to you for networking?
   b. Does it make sense to establish networking connections with your teammates?

5

Managing Teams

#skill/tool module: #managingteams

Learning outcomes

By the end of this module, you should demonstrate the ability to:

- Pick a team leader
- Apply strategies to manage both people and tasks
- Analyze some standard team problems and suggest remediation strategies for them

Recommended reading

Before this module:

- **Implementing a Project > Working in Teams > 4. Producing**

After this module:

- **Implementing a Project > Working in Teams > 6. Management Strategies**

1. Introduction

Every team is as different as the members in the team and the project they are working on. To make a team perform effectively is a continuous, dynamically changing process as unique as the team itself. Despite this uniqueness you will find that certain strategies and the understanding of certain situations and people will help you and your teammates work effectively together.

2. Team Leader

One of the most important decisions that you will make as a team will be the selection of the team leader. In undergraduate engineering design teams, your leader should have several duties. He or she will be the member of the team whose role is the closest to that of a ***project manager*** found in engineering companies. They will often assign or oversee the assignment of tasks to team members, keep track of where the team is in the project, deal with issues that arise relating to people, and will likely bear the responsibility of communicating with the instructor if issues become critical and cannot be resolved by the team without help.

Typically, one of the team leader's primary duties is managing team communication and activities. Team leaders certainly can and should do technical work and writing but will be overloaded if they try to do too much and also lead the team. Being a team leader, as distinct from a project manager, means they are also full members of the team. It would be a mistake for a team leader to take on too much or too little work on the project.

There is generally one key difference between the teams that you are part of when you are in school and those when you are working. The team leader in the work environment is often, but not always, someone who has the responsibility and the authority to assign tasks and to discipline team members. In university or college design teams, you are all peers. Being on a team of peers adds complexity to the role of team leader. The leader must provide direction and guidance to the team when all the members are equal. They must be highly persuasive and respected by the team, and able to help the team reach consensus while maintaining mutual respect. Select your team leader with care!

> **Reflection**
>
> Take a moment to think about what you want a team leader to do. Discuss what you want your leader to do as a group before the selection of that leader. Then take some time to think about the characteristics that would make a good leader based on what your group has decided the team leader is to do. Do you want the person who is the most technically able, the person with the best marks, the person with the best organizational skills, the person people like the most, the person who wants to control the situation, the relaxed person on the team, or the person who is not afraid of confrontation?

3. Managing Tasks

Things can go wrong with tasks within a team. Here are some of the issues that occur most commonly in student engineering design teams. Your team rules should specify who has the lead responsibility for managing these issues. Usually it will be the team leader, or an assigned team member, who will have the first level of responsibility for helping to prevent and in dealing with these problems.

> **Managing ≠ Micro-managing**
>
> *Managing* a team or its activities means guiding the work and collaboration between team members to make sure everyone is working effectively together. This is different from *micro-managing,* or being highly and specifically directive. Being very controlling or directive rarely works in a team where everyone is equal and the team leader or other members have no more authority than anyone else on the team (see *hijacking* below).

3.1. Meeting Focus

One of the most common challenges that teams face is having difficulty conducting focused meetings. Too often meetings are used just to exchange information, not to get work done. The solution is to set an *agenda* and keep *minutes* so you do not repeat tasks and you can get work done in your team meetings (#teamdocuments).

Design teams frequently find that they do not have enough time in team meetings to get everything done. Regular team meetings should be no more than an hour or two. This is assuming that these meetings are primarily work sessions with decisions being made and work being completed. Make sure the agenda is not too long; a good agenda should be short enough to complete in one meeting, and long enough to put the time to good use. Assign a person to chair the meeting and keep it on track. The chair position should rotate so that all team members have an opportunity to manage a meeting, and provide the agenda. Someone other than the chair should keep the minutes. Often a status report will provide the key elements to an agenda.

There is a simple technique for saving time by linking your agenda with your minutes. The minutes should be a list of tasks assigned, including deadlines and decisions made at the meeting. It should have a brief review of previously assigned tasks, including progress against deadlines. The minutes from the last meeting will become the agenda for the next meeting. Other information that needs to be recorded from the meeting can be put in a separate area of the agenda document for the next meeting and should be kept to a minimum. When done this way, your agenda and minutes become very closely linked to your project plan.

3.2. Task Timing

Another common issue that teams face is underestimating how long it takes to do things. This results in an inability to complete tasks on time and reflects a lack of effective planning (#tasks). You may find that actual times are as much as twice your initial estimates or more. Experienced designers will often add half or more to their first task time estimates because they know they tend to be optimistic.

Sometimes the problem is that tasks cannot be done in parallel; they must be completed serially. In order to make a realistic project schedule, engineering design teams use project planning tools. Two of the simplest and most frequently used are **PERT charts** or **Gantt charts** (#Ganttchart). A Gantt chart is a list of tasks illustrated as a bar graph on a calendar (see Figure 1) to record the tasks and task timings, and to order the tasks in a project.

3	Rough in electrical circuits	1 day	2	electrician
4	Rough in plumbing	2 days	2	plumber
5	Plumbing inspection	2 days	4	
6	Finish walls	3 days	5,3	carpenter
7	Finish electrics	1 day	6	electrician
8	Build cabinetry	4 days	1	carpenter
9	Finish cabinetry	3 days	6,8	carpenter
10	Install sink, tub, toilet	1 day	6	plumber
11	Install tub	1 day	6	plumber
12	Install final lighting	1 day	6,7	electrician
13	Final inspection	1.5 days	10,11	
14	Electrical inspection	2 days	12	
15	Done	0 days	11,13,14,9	

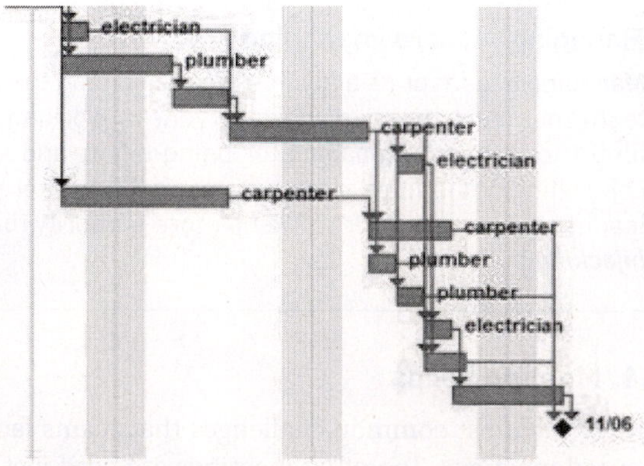

FIGURE 1 Part of a Gantt chart for a bathroom renovation project.

To use a scheduling tool effectively, be sure that you have listed all the tasks that are needed to complete the project. Then start from the project completion date and work backward to figure out how long you have to complete each task and when you need to start. This tool will not only help you keep on schedule but will provide you with the agenda items for your meetings and will show when certain tasks have to be completed before others can be started.

4. Managing People

In an undergraduate design team, there is often an early expectation that everyone is going to contribute equally. Unfortunately, things do not always work out this way. Because you are all peers, dealing with people issues can be particularly problematic. It helps to first identify the type of problem behavior you are dealing with. We have found that there are generally four major "people" issues in undergraduate engineering design teams. They can be grouped as *hitchhikers*, *hijackers*, *isolationists*, and *enablers*. Hitchhikers are by far the most common.

4.1. Hitchhikers

Hitchhikers are people who contribute significantly less to a project than everyone else [1]. They seem to always have an excuse ("too busy," "test tomorrow so I couldn't make the meeting," "I forgot," etc.). Or they just go silent: they don't return emails, calls, or texts, and they often miss meetings without communicating. However, they expect to get the same credit for the work as everyone else on the team.

If you have a hitchhiker on the team and there seems to be a legitimate reason for their non-performance (e.g., they have to work a lot of hours outside school, or they have health issues) then encourage the person to seek help. There may be a part-time option, or other types of assistance from the university or college that could help to meet the person's needs. Shirking work on the design project is not an acceptable way to deal with the underlying problem.

Regardless of the reason, continuing to shift the work to their teammates is not an acceptable option. This overloads the other team members and is unfair. Warnings, early deadlines, and prescribed actions for repeated infractions of team rules may solve the problem. However, if these actions are ineffective the behavior should be brought to the attention of your instructor in a professional and polite manner.

4.2. Hijackers

Hijackers are people who tend to be very anxious about their own grade in the course. They want to make sure everyone on the team does work that is up to their standards. When they disagree with the standard of work, they will often redo it themselves; "If you want something done right, do it yourself" is their motto. Hijackers will use a team leadership position to gain control of the team. While guidance from a team leader can be very important, particularly in the early stages of team development, too much control exerted by a team leader is abusive and can destroy team motivation.

Hijackers tend to have very little trust in other people's abilities. Hijacked teams often do well initially when the hijacker has enough time to redo work but as the pressure builds these teams tend to break down. The hijacker is overwhelmed with work, or has some personal crisis, and no one else in the team is in a position to step in to help because they have been so left out of the decisions and the tasks that they do not know what is going on or what to do. They have little motivation to fix the problem because they no longer feel any sense of ownership of the project. In the end, work will not be completed and the team generally does poorly. Interestingly, team members of hijacked teams frequently "sit back and enjoy the ride" if the hijacker appears to be able to do most of the project themselves. Those team members will focus on other courses, overloading the hijacker further until the situation implodes and everyone ends up hurt and angry.

Teams should be very careful that they do not place a person with a high need for this kind of control in the position of team leader. The members should work with the person to build trust, and continue to insist on doing their shares of the work.

4.3. Isolationists

Isolationists are willing to get the work done, usually competently, but do not want to interact with the team more than minimally necessary. They are the "just tell me what to do, and I'll do it" people. The work they deliver, while generally acceptable, may not fit very well with the rest of the project because they have not adequately communicated with the rest of the team. They are not involved in team decisions and do not want to take the time to listen to other teammates. Generally they do not value input from others. Isolationists tend to cause the formation of subgroups within a team; their presence causes team fractures. They cause the other team members to spend time and energy trying to make the isolationist a happy team member by accommodating the person's behavior. The balance between tasks and people will be moved too far toward the people side as the other team members try to draw this person in.

Strategies that you can use to manage isolationists include having them be the chair of team meetings and a reporting and updating requirement on their tasks. Basically this mechanizes their team involvement and requires them to take part. However, the most important strategy is to hold the rest of the team together and try not to spend inordinate energy and time accommodating this behavior.

4.4. Enablers

Enablers help everyone out. They want to be good citizens and can end up getting too much work handed to them because they do not say "no." If everyone else is "too busy" they will volunteer for tasks, and more tasks. They do not mean to take over the project, they are often not the team leader, and they are not behaving this way because they want to control things; they are just trying to make sure everyone else on the team is happy. As a result, they become overwhelmed and are unable to deliver on their promises. The likely consequence of this behavior is that the team does not meet its goals.

The team and team leader need to be very careful that tasks are assigned evenly and that they do not overload enablers simply because the person keeps agreeing to take on more work. When tests and other course assignments loom they will let the team down. While it may be tempting to take advantage of an enabler, in the end the project generally suffers.

5. Using the Information: Dealing with Team Issues

Having a good decision-making strategy and a clear task list helps avoid a number of the people issues just described. A task list will help to define roles and responsibilities and the team and team leader can make sure the work is distributed evenly. Well-defined, appropriate team rules will deal with minor problems.

If you have a team problem, then a team member (usually the team leader) must speak to the person *in private* and clearly set out what needs to change and why. This is not easy. If this is your responsibility, you must at all times speak objectively and calmly to the behavior, not to the person's character. Do not wait too long. If not dealt with, both the team and the person involved are going to struggle as the workload increases and all team members become more stressed as they try to meet deadlines and commitments. If the behavior continues, then move to other resolution strategies.

For school projects, you may want to involve the instructor in helping you deal with the issue. Often it is a good strategy for the team leader to inform the instructor on an "information only" basis. It means that the instructor will understand the nature, severity, and duration of the problem if ever called on for action if the problem becomes severe.

Challenges with people will happen throughout your professional career. This means that the strategies that you learn and the experience you gain in your student design teams will serve you well in your career. If you, yourself, have some anti-team behaviors—that is, if you are in some respects a hitchhiker, hijacker, isolationist, or enabler—now is the time to recognize and to stop those behaviors. Finding ways to cope with your need to control (if you are a hijacker) or your need to make people happy (if you are an enabler) in a manner that is effective for you and the people you work with is essential to make you a better team member.

Remember, a team is not simply the sum of its parts. Highly successful teams are able to leverage the differences in people's abilities and styles to be productive. Learning team skills will provide you with valuable knowledge for success in university or college and in your career.

KEY TERMS

project manager	**managing**	**micro-managing**
hijacking	**agenda**	**minutes**
PERT chart	**Gantt chart**	**hitchhikers**
isolationists	**enablers**	

6. Questions and activities

1. Practice writing up an agenda for a team meeting for your project. Do you have a project plan to help you construct the agenda?

2. Develop a template for your meeting minutes that you can use for every meeting. Make sure it has a place to put the project title, date, and people in attendance at

the meeting. Then add in a structure for the other elements that should go into the minutes. This template can be shared with your team and used for every team meeting.

3. At a team meeting a new task is identified as needing to be done by the next meeting. How might the response differ among the following types of team members:

    a. A team member if the team is in the Forming stage?

    b. A team member if the team is in the Storming stage?

    c. A team member if the team is in the Norming stage?

    d. An ideal team member if the team is in the Performing stage?

    e. A team member if the team is in the Performing stage, and the member is a hijacker?

    f. A team member if the team is in the Performing stage, and the member is a hitchhiker?

    g. A team member if the team is in the Performing stage, and the member is an isolationist?

    h. A team member if the team is in the Performing stage, and the member is an enabler?

4. When should a team member who is not performing properly and is not changing their behavior have to leave the team?

    a. Is it all right to have a team member who does not pull their weight, because they can't?

    b. Is it all right to have a team member who does not pull their weight, because they won't?

    c. What if the team member is an isolationist?

    d. What if the team member is an enabler?

    e. What if the problem team member is the leader?

7. References

[1] B. Oakley. It Takes Two to Tango: How "Good" Students Enable Problematic Behavior in Teams. *Journal of Student Centered Learning* 1(1), Fall 2002, pp. 19–27.

Management Strategies

6

#resource module: #teammanagementstrategies

Learning outcomes

By the end of this module, you should demonstrate the ability to:

- Describe several strategies that can be used to address common problems that arise in undergraduate (and professional) design teams
- Apply strategies to address some common situations that occur in teams

Recommended reading

Before this module:

- **Implementing a Project > Working in Teams > 5. Managing Teams**

After this module:

- **Implementing a Project > Working in Teams > 7. Sample Team Documents**

1. Strategies for Dealing with Task and People Problems

This module will describe strategies for dealing with task problems and people problems on undergraduate design teams. In industry similar strategies are used to deal with task and people problems. Here we have adapted industry practices to address some of the most common issues that arise in undergraduate design project teams.

Problem: Everyone is responsible for everything, and nothing is getting done.

Strategy: Avoid the "6-year-olds-playing-soccer" problem (i.e., everyone running for the ball). Instead, play positions. Make sure you have a clear task list and each task has a lead person who is responsible. Others may help with a task, but you should have one person who is assigned to deliver on each task. When you develop

your action items for agendas think of them not just as assignment of tasks, but assignment of responsibility. Avoid task lists that read:

1. Researching the problem: everyone

2. Writing the report : everyone

Divide up these large tasks and make sure everyone has a defined piece to contribute. In industry there are many variations on project management methods related to this strategy. One example is a **RACI matrix,** which designates different levels of responsibility for each task: Responsible, Accountable, Consulted, and Informed.

Problem: Work delivered late.

Strategy: What to do if one (or more) of the team is delivering their work so close to a deadline that the team leader (or whoever is assigned to compile the document for submission) doesn't have time to do a good job? The most effective strategy is to move up internal deadlines. Everyone on the team needs to deliver their work even earlier. This puts **slack** in the schedule, which allows delays without catastrophic consequences. Consider these two scenarios:

Scenario #1: The team agrees to get their document together at 5:00pm the day before the deadline so they can edit and proofread the report before submission. However, one person is running late with their work. Their part took longer to write than expected and when their piece shows up 5 hours late it is not well written. The team is now in crisis. Arguments erupt over whether to rewrite this section, or just leave it as is. Everyone is tired, angry, and resentful. The problem was not with this one team member, the problem was the plan. The team's plan was overly optimistic. It left no room for delays or problems.

Scenario #2: The team agrees to get their document together at 5:00pm two days before the deadline so they can edit and proofread the report before submission. However, one person is running late with their work. The team rules say that the team will give the person a "final notice" (i.e., a warning that the work must come in now or the rest of the team will begin writing up the missing sections). The final notice is sent at 11:00pm. The person responds by sending a piece of work that is not well written. Now the team has at least 30 hours to decide what to do and remediate the situation. There is time to ask the person to rewrite their section, working with them to support the effort, or implement another plan to address the problem.

It is important that the whole team take responsibility for putting together their work earlier, not just the person who is "always late." Everyone has things that go wrong in their lives from time to time. There will be a time when you are running late with your work too. It is important that you build in some buffering to your team schedule to account for the realities of life. In project management in industry this is called building in slack time and it basically removes from the **critical path** tasks that are likely to experience delays.

Problem: Poor quality work is delivered by a team member.

Strategy: We have two suggested strategies for coping with poor-quality work. Both revolve around the philosophy that the person who delivered the poor-quality work should not be rewarded. In engineering school, one of your most valuable resources is time; you never have enough to do everything. A person who delivers poor-quality work is hoping to save some of their time, at the cost of your time (if you are the person who is expected to "fix" their work).

Strategy #1: *Time blocking*. Arrange team work sessions that are 3 or 4 hours in duration. Hold them in a location where people can have access to computers, or can plug in their laptops. Agree on a realistic set of fixed goals for the session (e.g., everyone needs to finish a first draft of their section). If the team all meets the goal early, trade around work, start editing, and then reward yourselves by ending the work session early. Ideally, everyone arrives with their work already done, or partially done, and you can all go home early. But if people are not done, then everyone stays until the work gets done. This type of work arrangement mimics a common strategy used in engineering design companies: the *war room*. A war room is a common team workspace used to motivate creativity and productivity. It is based on evidence *team collocation* is an effective development strategy [1].

> Sometimes when you use time blocking for the first time the team uses the time to build social relationships instead of working. Social relationships are important to team cohesion and shouldn't be underestimated. But it is important to have clear goals to accomplish during the time block, and hold the team to substantial progress on those goals. By practicing this technique repeatedly you will find that the social side becomes secondary to the work productivity as the team moves into norming and performing.

Strategy #2: *Out-loud editing*. This is how out-loud editing works: Team member Alice gives her work to team member Bob for editing. Both sit together and Bob starts going through the work explaining, out loud, what his impression of the work is and what needs to be fixed. Bob must do this in a respectful and constructive way. Bob makes changes to the first few paragraphs to demonstrate the process for Alice. After a few paragraphs Bob hands the work back to Alice. Alice now knows what Bob "sees" as he is reading her work. She is given some time to fix it up before submitting it. Bob should also have his work out-loud edited by someone else. Overall this process improves both the writing and the editing skills of everyone on the team. This strategy is based on best practices in feedback techniques. It utilizes specific, constructive feedback to help the writer improve. Ideally (time permitting), each person on the team would go through several out-loud editing sessions during the development of the project.

Notice that in both of these strategies people who produce poor-quality work are neither rewarded (with more free time) nor punished. They are given support to help them accomplish the team goal, and treated with respect. Everyone needs support of this kind at one time or another in their career. Engineers are constantly teaching one another and learning from one another; it is an important part of the profession.

If you have a teammate who is not the best colleague or writer but who is really trying, give them credit for their effort. Make sure you praise your teammates for what

Management Strategies

they are doing, and for the energy they are putting into the project. You will work with plenty of people who have a poor attitude and don't make an effort, so make sure you show your appreciation when you work with good people. All of us have areas of strength and weakness, but bringing a good attitude and a willingness to work to a project should be supported.

2. Warning and/or Firing a Team Member

Not all instructors will allow you to fire a team member. However, if they do then firing a team member should be a strategy of last resort. All fair and reasonable attempts should be made to keep the team functioning and keep all of the members actively participating. However, if remediation fails and a team concludes that a member is causing substantial difficulty or repeatedly not contributing, that member can be fired. Typically firing can only occur with the approval of the instructor and must follow a specific procedure. This is the procedure we recommend:

1. The team will issue a warning letter to the person informing him or her of the situation (i.e., that the person will be fired from the team unless change occurs). This letter should be reviewed by the instructor before being sent to the team member. The letter must use a professional tone and cite specific behaviors that the person must change in order to remain part of the team.

2. After the warning letter is sent, the instructor will meet with the team together and, if appropriate, separately with the person who is under warning. The goal of these meetings is to:

   a. Determine what, at a minimum, the team member must change to be retained. The instructor will act as a mediator in this discussion, but will also be allowed to veto requested behavior changes if he or she feels the request by the team is excessive or inappropriate (i.e., vindictive).

   b. Work with the person under warning to develop strategies to allow him or her to meet this minimum (and preferably exceed the minimum) requirements necessary to remain on the team. Identify if there are underlying problems (personal crisis, health problems) that are affecting the person, and help them to connect with appropriate services on campus.

   c. Determine if the person does not intend to change the behavior that is causing the problem.

3. If the person refuses to meet with the instructor, or ignores requests for a meeting, the team may proceed with the firing procedure subject to approval by the instructor.

4. If the situation persists, or if the person has refused (or ignored) the request to change the behavior, the instructor may approve the firing. This approval shall be in writing and shall be copied to other administrators if necessary.

5. The team may then inform the person that he or she is fired. This is done in a letter copied to the instructor. The letter shall be reviewed with the instructor prior to sending it. The letter must have a professional tone, and should be short and to the point. The letter shall state the specific date on which the team member is officially fired. The letter shall briefly cite the reasons for the dismissal.

Note: For the purposes of this procedure, a "letter" may be a properly formatted email.

We recommend: Allowing a student who is fired from a team to be be hired by another team in order to finish out the course. However, it is the responsibility of the student to find a team willing to hire him or her. The student will receive a **zero** for any assignments that were not accomplished because the student was not part of a design team after being fired. Work submitted for a team assignment that has been developed solely by the fired student, in the absence of the team's contribution, will not be acceptable.

We have found that a clear, well-written warning letter is often sufficient to get a person back on track and contributing, at least minimally, to the project.

3. Quitting a Team

Not all instructors will allow you to quit your team. However, if they do then quitting a team is a strategy of last resort. All fair and reasonable attempts should be made to keep the team functioning and keep all of the members actively participating. However, if remediation fails and you have substantial cause, you can quit your team. You can only quit with the approval of the instructor and must follow a specific procedure. Here is the procedure we recommend:

1. You will issue a warning letter to your team informing them of the situation (i.e., that you are ready to quit the team unless change occurs). This letter shall be reviewed with the instructor before it is sent. The letter must use a professional tone and cite specific behaviors that the team must change in order for you to stay.

2. After the warning letter is sent the instructor will meet with the team together and, if appropriate, separately with the person who is considering quitting. The goal of these meetings is to:

   a. Determine what, at a minimum, the team must change to retain the team member. The instructor will act as a mediator in this discussion but will also be allowed to veto requested behavior changes if he or she feels the request is excessive or inappropriate (i.e., the instructor will act as the judge of what is reasonable).

   b. Work with the person who is considering quitting to develop strategies that allow him or her to have needs met within the team.

   c. Determine if the team does not intend to change the behavior that is causing the problem.

   d. If the team is unwilling to change to retain the team member, the team member may quit, subject to approval by the instructor.

3. The approval from the instructor shall be in writing and shall be copied to other administrators if necessary.

4. The team member may then inform the team that he or she is quitting and give a specific date when this will occur. This is done in a letter copied to the instructor. The letter shall be reviewed with the instructor prior to sending it. The letter must have a professional tone, and should be short and to the point. The letter shall briefly cite the reasons for the departure.

Note: For the purposes of this procedure, a "letter" may be a properly formatted email.
We recommend: Allowing a student who has quit a team to be hired by another team and finish out the course with the new team. However, it is the responsibility of the

student to find a team willing to hire him or her. The student will receive a **zero** for any assignments that were not accomplished because the student was not part of a design team at the time of the assignment due date. Also, the quitting student must leave copies of any work developed for the project up until their departure date with their team to use.

Work submitted for a team assignment that has been developed solely by an individual student, in the absence of the team's contribution, will not be acceptable.

4. Conclusion

Remember, a team is not simply the sum of its parts. Highly successful teams are able to leverage the differences in people's abilities and styles to be productive. Working in teams can be a challenge but they can also be a fun and effective way to produce high quality work. The strategies provided in this module should help you and your teammates work together effectively. Learning team skills will provide you with valuable knowledge for success in university or college and in your career.

KEY TERMS

RACI matrix	**slack**	**critical path**
time blocking	**war room**	**team collocation**
out-loud editing		

5. Questions and activities

1. Try using time blocking with your team. How does working together for a few hours change the team dynamic? Did the focused activity allow you to get more done?

2. Analyze your work breakdown schedule (WBS), (i.e., your list of tasks to complete a project):

   a. Are the tasks small enough? Is there one person who is the lead, responsible for making sure that task is completed?

   b. If you are working with project management tools, such as a Gantt chart, check the resource allocation to make sure work is evenly distributed across the team.

   c. Check the Gantt chart or your schedule; have you built slack into the schedule to allow for late work delivery? Does the team have a strategy (i.e., team rules) for dealing this possibility?

3. Use out-loud editing with a teammate and then listen while they out-loud edit your work. When listening, don't argue about the points they are making, just listen. When necessary, ask clarifying questions to make sure you understand clearly their perspective.

   a. What did you learn about your writing?

   b. List one or two things you learned that you can use in your writing in the future.

5. References

[1] There are many articles on the effect of team collocation on productivity. One example is S.D. Teasley, L. Covi, M.S. Krishnan, and J.S. Olson, "How Does Radical Collocation Help a Team Succeed?" *Proc. ACM Conf. Computer Supported Cooperative Work (CSCW `00)*, December 2000, pp. 339–346.

Sample Team Documents

#resource module: #teamdocuments

Learning outcomes

By the end of this module, you should demonstrate the ability to:
- Develop an agenda for a team meeting
- Develop a set of minutes from a team meeting
- Create a simple, effective status report

Recommended reading

Before this module:
- **Implementing a Project > Working in Teams > 6. Management Strategies**

After this module:
- **Implementing a Project > Project Management > 1. Introduction to Project Management**

1. Introduction

This module has templates and sample documents that demonstrate the use of an agenda, minutes, and a status report. This type of documentation can help your team stay focused and efficient when working on a design project. Tools like agendas, minutes, and status reports are communication tools that serve to keep the whole team informed about the process, and focused on tasks. The templates are given at the end of the module and can be adapted to your project needs.

2. Agenda

An agenda is used to keep a meeting on-track. It should help your team complete all of the necessary tasks that need to be done during the meeting. Some important tips for a good agenda:

- Don't make it too long. Make sure you are able to complete the agenda.
- Have people circulate items prior to the meeting, such as updates, and then very briefly review them at the meeting.

- Make sure people leave the meeting with clear tasks (i.e., action items). In the sample below the "who" and "when" columns should be filled in during the meeting and then circulated to the team after the meeting.
- It is useful to include a snapshot of the Gantt chart or task list in the agenda so the team can see where they stand in the project execution.
- Some people like to include a time next to each agenda item to keep the meeting on track. For example, you could allocate 15 minutes for updates, and 30 minutes for the discussion of the client meeting.

The easiest way to construct an agenda is to start with the action items from the last meeting, and then fill in upcoming tasks from the project plan.

Sample Agenda

Agenda: ShareCare Produce Warehouse Design Team

Thursday March 22, 2013, 10:00–11:00am

People attending: Harry, Sue, Dev, Sam, Shilpa

Regrets: Matt (report from Matt was circulated prior to the meeting)

1. Opening remarks by Sue. Selection of person to take minutes.
2. Review individual progress (brief reports circulated prior to the meeting):
   a. Harry: Analysis of warehouse construction
   b. Sue: Delivery information—goods, times
   c. Dev: Visit to similar warehouse—Chang Bakery
   d. Sam: Visit to similar warehouse—Leon's Dry Goods
   e. Shilpa: Requirements development
3. Discussion of next client meeting (March 27, 9am)
   a. What key points should be presented?
   b. What questions do we need to ask the client?
4. Tasks for this coming week: who will be doing what this week
5. Closing remarks: Team members involved in client meeting will continue with a working session to prepare for the meeting

Action items:	Who	When
- Complete the presentation preparation:		
- Attend the client meeting:		
- Compile benchmarking report:		
- Complete warehouse analysis:		
- Continue requirements development: (by next team meeting, how far along should this be?)		

3. Meeting Minutes

Meeting minutes should record the decisions and main points made during a team meeting. They should not be a complete transcript of what was said. Good minutes will keep the team moving forward and prevent miscommunication. They can also keep the team from revisiting the same decision multiple times. Good minutes:

- Should be clear and to the point
- Should record team decisions
- Should track action items
- Should identify who was at the meeting, and who contributed key points in the discussion

The easiest way to construct minutes is to start with the agenda, and fill in key points.

Sample Minutes

Meeting minutes: ShareCare Produce Warehouse Design Team

Team meeting: Thursday March 22, 2013, 10:00–11:00am

People attending: Harry, Sue, Dev, Sam, Shilpa

Regrets: Matt (report from Matt was circulated prior to the meeting)

1. Opening remarks by Sue. Selection of person to take minutes.
   a. Dev took minutes
   b. Sue updated group on feedback received on the last report
   c. Key point; more coherence needed in the final version
   d. Decision: Shilpa will review the next report for cohesion before submitting
   e. Decision: Material for the next report must be given to Shilpa at least 24 hours before the deadline.
2. Review individual progress (brief reports circulated prior to the meeting):
   See brief circulated reports for information
   a. Harry: Analysis of warehouse construction
   b. Sue: Delivery information—goods, times etc.

Action items: Who When
- Complete the presentation preparation: Sue and Matt by March 25 (and practice on 26th)
- Attend the client meeting: Sue, Matt, Sam, Shilpa and Harry on March 27. Sue will email out travel arrangements.
- Compile benchmarking report: Dev and Sam by March 27
- Complete warehouse analysis: Harry with the help of Sue by March 29
- Continue requirements development: Shilpa with the help of Matt and the team, draft by March 29

* Next meeting: March 29. Add to the agenda: review of the draft requirements and review outcomes of the client meeting.

Action items added at the meeting:
- Shilpa will edit the next report before submission, which is on April 2.
- All materials for the next report will go to her at least 24 hours before the deadline.

4. Status Report

A brief status report produced weekly during the project will help your team track the progress on the project. This is particularly useful if you have a supervisor, advisor, or project manager with whom you meet regularly. The status report can be used to show your supervisor where you are on the project, and serve as a focus for discussion of progress and obstacles. We recommend that the status report include:

- A snapshot from your project Gantt chart showing one or two weeks back and two weeks forward from the date of the report. This shows what should have been completed and the tasks coming up.
- Work done during the last period (e.g., over the last week or two). Who did what, and when they completed it.
- Work planned during the next period. Again this should include who will be doing what, and when the task is scheduled to be completed.
- Challenges that your team has dealt with during the last work period.
- Strategies you have been using, or steps you have taken to deal with these challenges.

Sample Status Report

Date of Report: March 29
Report Prepared by: Harry Colman
Project: ShareCare Produce Warehouse Design
Project Manager: Prof. Pearl
Team Leader: Sue Liang
Team Members: Sue Liang, Harry Colman, Shilpa Montgomery, Sam Ng, Matt Serrano, and Dev Prasad
Activities since the last report (i.e., progress to date):

	SPECIFIC TASK	PERSON ASSIGNED	DATE DUE	DATE COMPLETED
1	Prepare status report	Harry	March 29	March 28
2	Presentation slides for client meeting	Sue and Matt	March 25	March 26
3	Client meeting	Sue, Matt, Sam, Shilpa, and Harry	March 27	March 27
4	Benchmarking report	Dev and Sam	March 27	March 25
5	Warehouse analysis	Harry (helped by Sue)	March 29	Not completed
6	Requirements development	Shilpa with the help of Matt and team	March 29	Continuing; 90% complete

Problems Encountered: The warehouse analysis is taking longer to complete than anticipated. However, we are just finishing it up and it should be done later today. The requirements development depends on completing the warehouse analysis. We anticipate the requirements can be finished tomorrow after the warehouse analysis is done.

Lessons Learned and Strategies to Address Problems: We put the deadline for the requirements draft the same day as the completion date for the warehouse analysis, but the requirements depend on the analysis so this was not a good plan. Also, the warehouse analysis took longer than expected, pushing both tasks late. We should have given Harry and Sue more time or help working on the analysis, realizing that it would impact other tasks.

Team Decisions: The team will be getting drafts of all report sections to Shilpa by April 1. She will be combining these with the draft requirements to create a cohesive report for submission on April 2. We have also decided to begin idea generation earlier than originally planned.

ADDITIONAL COMMENTS (items to discuss: documents, client meetings, upcoming due dates and activities, etc.): The client meeting on the 27th went very well. They are pleased with our progress to date. Also, we were able to get a number of questions answered that will help us with our report due on the 2nd. We completed most of our requirements a week ago, so we believe we are in good shape for the upcoming report submission.

ATTACH GANTT CHART (one week past, two weeks forward only)
Plans for the next work period (1–2 weeks)

	TASK	PERSON ASSIGNED	DATE DUE	COMMENT
1	Complete warehouse analysis	Harry and Sue	Today	Running behind schedule
2	Complete requirements draft	Shilpa and Matt	New due date: March 30	
3	Drafts of all sections to Shilpa	Harry: W. analysis Dev: Benchmarking Shilpa: requirements	April 1	
4	Complete editing	Shilpa	April 2	
5	Submit report	Shilpa	April 2	
6	Begin idea generation	Sue to lead the team	April 3	
7	Interview of warehouse staff	Matt	April 2	
8	Team meeting	All	April 5	

KEY TERMS

agenda   minutes   status report
action items

5. Templates

Agenda Template

Project title: _____

Meeting date: _____

Meeting time and location: _____

People attending: _____

Regrets: _____

Agenda items:

1.
2.
3.
4.
5.

Action items

TASK	WHO	WHEN

Status Report Template

Date of Report _____

Report Prepared by* _____

Project Title _____

Project Manager _____

Team Leader _____

Team Members _____

This report may be written by one team member (as delegated by the team), but all members of the team take full responsibility for the accuracy of the final version.

Activities since the last report (i.e., progress to date):

	SPECIFIC TASK	PERSON ASSIGNED	DATE DUE	DATE COMPLETED
1	Prepare status report			
2				
3				
4				
5				

Problems encountered _____

_____

_____

_____

_____

Lessons learned and strategies to address problems _____

_____

_____

_____

_____

Team decisions _____

_____

_____

_____

_____

ADDITIONAL COMMENTS (items to discuss during meeting with project manager: documents, client meetings, upcoming due dates and activities, etc.):

_____

_____

_____

_____

_____

_____

_____

_____

_____

ATTACH GANTT CHART (one week past, two weeks forward only)

Plan for the next work period to be approved by your Project Manager before leaving the meeting.

PM Initials: _____

Plans for the next work period (1–2 weeks)

	TASK	PERSON ASSIGNED	DATE DUE	COMMENT
1	Prepare Status Report			
2				
3				
4				
5				
6				
7				

Introduction to Project Management

#process module: #projectmanagement

Learning outcomes

By the end of this module, you should demonstrate the ability to:

- Develop a plan for a project, including work schedule, cost schedule, and resource schedule
- Break larger activities into tasks for execution of a project

Recommended reading

After this module:

- **Implementing a Project > Project Management > 2. Project Management Concepts**

1. Introduction

By the time you reach university or college you have undoubtedly seen some major engineering projects: apartments or homes being built, roads being constructed, or sewers being laid. If you have watched any of these projects, you will have noticed various stacks of materials appearing on the job site from time to time, then slowly disappearing as the materials went into the construction, or noticed holes being dug, then shored up by putting up retaining walls to prevent collapse, and eventually filled. **Project management** is the ordering of the activities and the resources in a project and it is about directing that order to make the project run smoothly.

Project management can seem quite simple: Just put the activities in order and bring in the materials. The problem is that small mistakes can cause major cost and time overruns. If a few special screws are missing, many thousands of dollars of custom equipment will sit idle waiting for the delivery of these screws. Not planning for the time required to properly test and debug coded routines has put many software

projects well behind schedule. The basics of project management are simple; the execution often is not.

What constitutes "the project" will be different in various industries, and in various aspects of those industries. A project might include a feasibility study, building a prototype, creating a response to an RFP (Request for Proposal), designing a device for production, creating the design for a device (but not going forward into production), designing and building a single piece of technology (like a bridge), or designing and creating a design for volume sales (like a commercial software package). Each of these will have variations in what constitutes the technical efforts, and will have variations in the project management efforts and techniques used. Here we describe some of the typical project management methods often used in industry.

2. Overview of Project Management

Consider the installation of a new bathroom. The installation involves dealing with the water supply, the sewage connections, the facilities (toilet, sink, tub/shower), the electrical (lights, power outlets), and the walls. A simplified plan for doing this is shown in Figure 1. This type of diagram, called a ***Gantt chart***, is a typical representation of a plan for project management (#Ganttchart).

On the left side of the figure is a list of the tasks that must be completed to finish the bathroom. Each unit of work, called a ***task***, takes a certain amount of time, called a ***duration***, which is shown in the "Duration" column. Each task usually requires people and/or equipment, known as ***resources***, as assigned in the "Resource Names" column. The ordering is established by indicating what tasks must precede other tasks. For example, the walls must be built (task 2) before the electrical can be roughed in (task 3). Immediate ***predecessors*** of each task are shown in the "predecessors" column. The duration, resources, and ordering of the tasks is shown visually in the connected bars

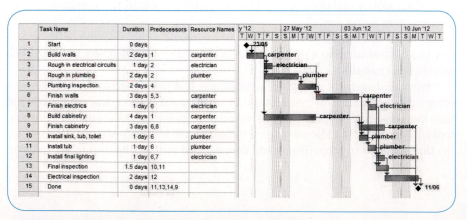

FIGURE 1 **Gantt chart for a bathroom renovation project.**

IMPLEMENTING A PROJECT

at the right side of the figure. Visualization is the essence of the Gantt chart. You can compose similar plans on paper, using a spreadsheet program or project management software packages, and then use these plans to execute the project.

Any moderate or large project requires many people, often many teams of people, and many hundreds or thousands of individual tasks. To make sure that efforts are coordinated efficiently there must be a well-organized plan in place.

2.1. Tasks

Engineering project managers organize projects in the same way engineers often handle the technical issues of the project, by breaking the larger problem into smaller problems that are easier to solve (#tasks). In fact, the project manager and the design team must coordinate closely and for some projects, and in some situations, the project manager will actually be a member of the design team.

To be most useful, "the *task*"—the smallest activity managed—should meet the following criteria:

1. The work done for a task is closely interrelated.
2. The task has a testable outcome.
3. A task requires one interconnected set of resources (people and resources).
4. The duration of the task is known (to reasonable accuracy) and is typically completed in one *monitoring cycle* at most; from a day or two to a week or two.
5. The people and equipment involved in the task will usually be required for the entire execution time of the task.
6. The task will typically utilize the special competence of the person or group executing the task.

Table 1 shows some examples of tasks. Breaking a large project down into manageable tasks allows for efficient management and monitoring of the project. You can schedule people with specialized abilities to be on the job at specific points in the project so people are not waiting around for their part to start. And you can make sure the resources and supplies are available when needed.

On a design project, a list of tasks helps your team make sure that every task will be completed before a deadline. It also helps you divide up the work evenly so everyone is working on their own part of the project, but is also aware of what other people in the team are doing. A list like this prevents overlap between work (e.g., two people writing up the same section of a report) and gaps (e.g., one section that no one wrote up because they thought someone else was doing it).

2.2. Milestones

Some points in the work flow are of particular note because they indicate a well-defined checkpoint in the execution of the project. These points are called *milestones*. A milestone is a zero-duration task that marks the end of a well-defined stage in the project, for example: "First five floors completed" or "prototype available for delivery to customer" or "test procedure ready." Milestones are also useful to denote progress through a project, particularly if a payment is associated with this progress or where a critical project decision (such as go/no-go) is to be made when the project reaches

TABLE 1 Examples of well-formed and poorly formed tasks.

TASK EXAMPLE	WELL-FORMED?	WHY?
Install stadium seats and install toilets in washrooms	No	Two parts are not closely interrelated
Install plumbing and toilets in washrooms	No	First step can be done before toilets are available for installation. Resources are not interconnected
Install the plumbing on third floor	Yes	This could be broken down into more specific tasks, but is likely a reasonable task for a plumbing team, and will then make the floor available for wall finishing and for the installation of toilets and other washroom fixtures
Research how to plumb a bathroom	No	There is no test built into this task to indicate when this task is completed (i.e., when an appropriate amount of research has been done)
Cap the ends of the water supply inputs and turn on the water to test the joints	Maybe	The instruction given is really just part of a task. This instruction might be given to a novice plumber who needs more detailed guidance. An accomplished plumber or plumbing team would only need instructions to "install the water supply to the third-floor washroom," which is a full task
Build the forms then fill them with concrete	No	The concrete supply is not needed until partway through the task. This would be better as two tasks
Install the plumbing in a condominium complex	No	Too much work involved. Execution time is too long. This would be further broken down further into shorter duration tasks

that stage. In a university design project for a course, a milestone could indicate the due date for a project report or presentation, for instance.

In the example of the bathroom the milestones are represented as black diamonds and zero-duration tasks numbers 1 and 15, as shown in Figure 2. Here they are used to start and to end the project so every task is on a predecessor chain going between these two milestones.

2.3. Resources

The example shown in Figure 1 has a list of the **resources** assigned to each task. In this case the resources are people, such as plumbers to do the plumbing in the bathroom installation, and electricians to do the wiring. The resources can also be equipment,

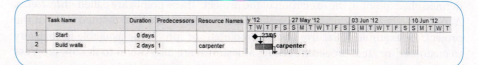

FIGURE 2 Part of Figure 1 showing a milestone.

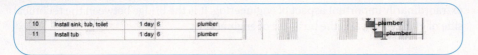

| 10 | Install sink, tub, toilet | 1 day | 6 | plumber |
| 11 | Install tub | 1 day | 6 | plumber |

FIGURE 3 **Close up of tasks 10 and 11 from Figure 1.**

such as cranes, compressors, workstations, and software simulators. They include the **consumables** and the project materials: fuel, cleaning supplies, the electrical power, the drywall, the beams, and so on. Resources can also be financial: the money available to fund the project at any particular time.

Availability of resources can affect the project timing. Consider tasks 10 and 11 shown in Figure 3. These tasks could be done at the same time, except that they are both being done by the same plumber. Since the plumber cannot work on both at once, they have to be staggered to make the schedule practical. Any wait for a resource to become available for a task will affect the timing of that task.

3. Conclusion

As with many engineering endeavors, setting a proper direction at the beginning will save time and cost and be more satisfying in the execution. At the heart of the project management process is the **project plan**. The project plan describes what will be done in the project, the order of the tasks, the time scheduling of the tasks, and the costs and the resources involved with the tasks. During the planning stage the project plan is created and during the control stage the project plan is executed.

A plan consists of a number of what are called schedules, since they have a time basis. Typically these are:

1. A work schedule. This is a set of tasks frequently described in an ordered graph (like a Gantt chart).

2. A cost schedule. Costs are associated with each task and thus the financial needs of the project at any point in time can be determined from the work schedule.

3. A resource schedule. Equipment, materials, and workers need to be at the right place at the right time. These are associated with the tasks and thus can be scheduled using the work schedule.

Project management is the glue that holds the project together and helps everyone on the team stay coordinated. It requires knowledge of the technical processes and operations involved and of the project requirements. Project management courses abound to address the problem of poor project management, which leads to cost overruns, schedule delays, and safety issues. A globally recognized organization, the Project Management Institute [1], certifies project management professionals. Groups such as INCOSE (The International Council on Systems Engineering) [2] promote the process of engineering development. Models such as CMM (Capability Maturity Model) from Carnegie Mellon University are widely adapted and used as a model to control projects, particularly software projects, through careful management [3]. These advanced professional organizations and models are beyond what you need for an introductory design project. However, you may want to consider furthering your education in project management during your career because this is an extraordinarily important aspect of engineering. It is the bridge between the technical side of engineering and the business of engineering.

KEY TERMS

project management	Gantt chart	task
milestone	resources	consumables
monitoring cycle	project plan	duration

4. Questions and activities

1. Consider a lab for a physics course or chemistry course. Are the following well-formed tasks? Why or why not? Under what circumstances would the activity need to be broken down into smaller tasks? And when would the activity need to be added to other activities to make it a complete task?

   a. Perform the lab with your partner or team.

   b. Perform one step in the lab.

   c. Record the results.

   d. Write up the results section for the lab report.

   e. Write up the entire lab report.

   f. Edit the report.

   g. Submit the lab report.

2. What advantages are gained by being able to take a chain of sequential tasks and make them into two or more parallel chains? In particular:

   a. How might the cost be reduced overall?

   b. What sorts of additional resources might be added to make this split possible? Give example situations.

   c. In what situations would this not be of any advantage?

3. Consider starting a month before to prepare for three final examinations. Assume the first exam is 9:00 on Tuesday morning and you have two other exams after that in the same week. Set up a Gantt chart to plan out your study and physical activity schedule. (Physical activity is included because it is important to balance studying with exercise.)

4. Create a plan for building a bird house, baking a cake, or making some other simple physical item. Ideally, pick something that you have made before so you are familiar with the procedure and resources needed. Set up a simple work schedule (Gantt chart), a cost schedule, and a resource schedule to plan out the activity.

5. References

[1] Project Management Institute. Retrieved from www.pmi.org/
[2] International Council on Systems Engineering (INCOSE). Retrieved from www.incose.org/
[3] CMM/CMMI. Retrieved from http://cmmiinstitute.com/

Project Management Concepts

2

#process module: #PMconcepts

Learning outcomes

By the end of this module, you should demonstrate the ability to:

- Develop a plan for a project, including work schedule, cost schedule, and resource schedule
- Break larger activities into tasks for execution of the design
- Determine the trade-offs between the elements found on the project triangle

Recommended reading

Before this module:

- **Implementing a Project > Project Management > 1. Introduction to Project Management**

After this module:

- **Implementing a Project > Project Management > 3. Creating a Project Plan**

1. Concepts in Project Management

Project management is about the ordering of the activities and the resources in a project and about directing that order. While it is associated with the design and design decisions, project management is about how those decisions are executed.

Consider the installation of a new bathroom. The installation involves dealing with the water supply, the sewage connections, the facilities (toilet, sink, tub/shower), the electrical (lights, power outlets), and the walls. A simplified **Gantt chart** for this project is shown in Figure 1 (#Ganttchart).

	Task Name	Duration	Predecessors	Resource Names
1	Start	0 days		
2	Build walls	2 days	1	carpenter
3	Rough in electrical circuits	1 day	2	electrician
4	Rough in plumbing	2 days	2	plumber
5	Plumbing inspection	2 days	4	
6	Finish walls	3 days	5,3	carpenter
7	Finish electrics	1 day	6	electrician
8	Build cabinetry	4 days	1	carpenter
9	Finish cabinetry	3 days	6,8	carpenter
10	Install sink, tub, toilet	1 day	6	plumber
11	Install tub	1 day	6	plumber
12	Install final lighting	1 day	6,7	electrician
13	Final inspection	1.5 days	10,11	
14	Electrical inspection	2 days	12	
15	Done	0 days	11,13,14,9	

FIGURE 1 Gantt chart for bathroom renovation project.

1.1. Task Order

When a project is reduced to tasks, the tasks are unordered (#tasks). However, as in our bathroom example, the tasks will have an order in which they must be performed. In the Gantt chart of Figure 1 this ordering is done by defining precedence tasks. In other techniques, such as the **_PERT_** chart, arrows show the precedence for a simple set of tasks. (PERT stands for Program Evaluation and Review Technique.)

Figure 2 shows a simple example of a PERT chart. This chart indicates that you can put on either your left sock and shoe first or your right, but you cannot put on your shoes before your socks. Predecessors in the Gantt chart show this type of precedence (i.e., in what order the tasks must be performed). Tasks not linked by predecessors or chains of predecessors can be done in any order or at the same time if you have the resources to do so.

Besides the Gantt chart, there are variations of the PERT chart and the Network Diagram that are also commonly used to represent tasks and ordering.

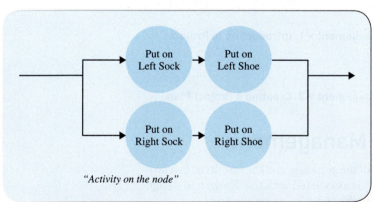

"Activity on the node"

FIGURE 2 An example of an "Activity on the node" PERT chart for putting on your socks and shoes.

1.2. Gateways: A Special Kind of Milestone

Sometimes in projects there are points where evaluations are made to see if the project is proceeding well enough or if it is costing too much or taking too much time or resources to continue.

These are decision points, where a decision is made to proceed with the project, or to modify it or terminate it, based on the evaluations done at that point in the project. The milestones where these types of decisions are made are called *gateways*. Many projects will have gateways required by company executives and by financing institutions as ways of bringing accountability to the projects. These are often referred to as *gated projects*. Viable projects will continue after these gateway decisions, or there may be modifications to the project direction based on the gateway evaluation. In some cases the decision will be made to terminate a project.

2. Trade-offs in a Plan

When a project plan is being formed there are trade-offs, meaning that one factor can be improved at the expense of another. For example, you could take a taxi to class and get there more quickly, but at greater cost than taking public transit, walking, or biking. The trade-offs are easily seen in the Project Triangle, which shows the basic elements of a project (see Figure 3).

The top vertex of the triangle is the *scope* of the project. This includes the project requirements: the features or functions of the design; the objective goals; and the constraints that the design must meet. Further, the scope also includes client expectations, quality, and ethical concerns that may not be explicit in the project requirements. The left vertex is the *cost* of the project. It will be a result of design and execution decisions and of the costs of unplanned events during execution. Finally, on the right vertex, is the *time* that the project will take, which will also depend on the design and execution decisions and on the effects of unplanned events. This is the time it will take to get a product to market (ready to be ordered or bought), the time to complete and commission a construction project, or the time to install or implement a service or system and test it.

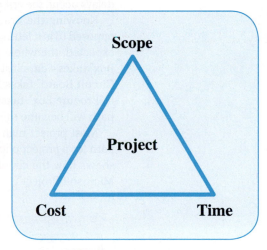

FIGURE 3 **The project triangle.**

The elements of the triangle come into play in every project. When Apple® designs a new version of the iPhone®, they consider carefully what functionality to have in that version (the scope) because this is key to keeping sales high. But they must also release that new phone before the competition starts selling a phone with the targeted technology (the time) and they are accountable to the shareholders for the money spent on this development (the cost). Every project has a mix of these three elements: scope-time-cost, although one or two of them may be dominant in a particular project.

2.1. More about the Time Trade-off: Critical Path

When project timing is being considered as a trade-off, one aspect becomes apparent: There is at least one path from the start to the end of the project that will take the same or more time than any other path. This is called the *critical path*.

Consider Figure 4, which is a *PERT* chart representation of a simple circuit board project. There is a path shown as a solid line and a path shown as a dotted line. The solid path, which goes from "start" to "done," takes 5 days, but the parallel path shown as a dotted line takes 6 days. This means the project will fall behind if there is any delay

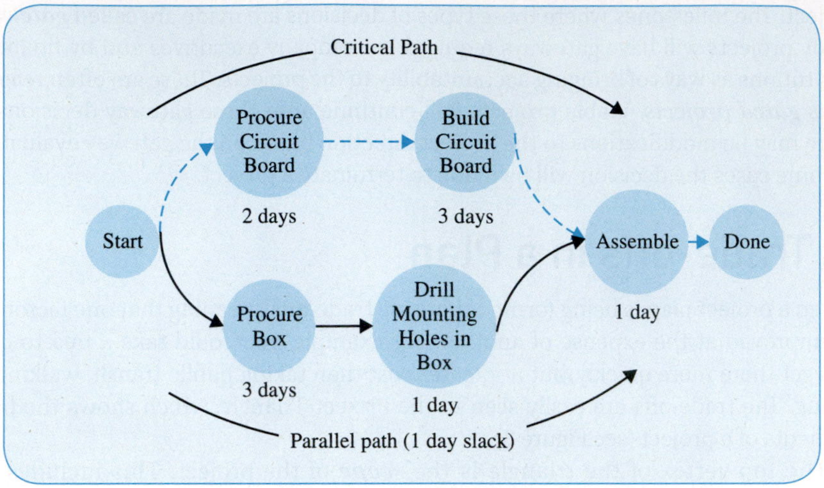

FIGURE 4 Illustration of an "activity on node" PERT chart showing the critical path.

in the tasks on the dotted path. The dotted path is the ***critical path***. More complex projects could have several paths that are critical paths. And as a project proceeds and delays occur, the critical path may change.

Knowing the critical path is one of the first steps in managing a project, as opposed to just letting things happen. If any task on the critical path takes longer than expected, the whole project will take longer than expected. In Figure 4 if "Procure Box" takes 4 days instead of 3, the project will still finish on time. However, if "Procure Circuit Board" takes 3 days instead of 2, the project will take an additional day. But if "Procure Box" takes 5 days instead of 3 the project will take longer and the solid path will become the critical path.

Most project management software will automatically find and mark the critical path for a project once the tasks, durations, and predecessors are identified. Figure 5 shows how the circuit board project, illustrated in Figure 4, looks when entered in Microsoft Project®.

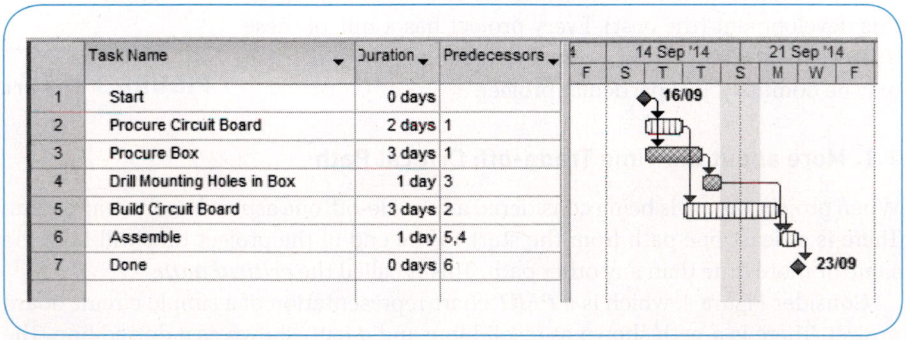

	Task Name	Duration	Predecessors
1	Start	0 days	
2	Procure Circuit Board	2 days	1
3	Procure Box	3 days	1
4	Drill Mounting Holes in Box	1 day	3
5	Build Circuit Board	3 days	2
6	Assemble	1 day	5,4
7	Done	0 days	6

FIGURE 5 A Gantt chart showing the critical path for a circuit board assembly project.

3. Conclusion

Forming a plan is necessary to prevent inefficient use of resources and time. It often includes key evaluation points such as gateways. It also allows the determination of the critical path, which can be important in optimization of the project execution and managing the project when actual task timing of tasks differs from the projections in the plan. Plans consist of trade-offs in the three vertices of the project triangle—scope, time, and cost—and the project manager has the responsibility of managing the balance among these three elements.

KEY TERMS

Gantt chart	**PERT chart**	**gateway**
gated projects	**scope**	**critical path**

4. Questions and activities

1. Critical path can also be shown on a Gantt chart.

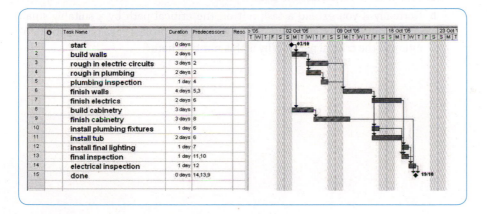

a. What is the critical path in this project?

b. How long will the project take (in elapsed days) if Task 6 takes two extra days?

c. How long will the project take (in elapsed days) if Task 10 takes two extra days? (Note that Task 6 is *not* extended as in part b.)

d. Task 9 takes 3 days, yet goes from Thursday to Monday. Why?

e. What advantages are gained by being able to take a chain of sequential tasks and make them into two or more parallel chains? In particular:

- How might the cost be reduced overall?

- What sorts of additional resources might be added to make this split possible? Give example situations.

- In what situations would this not be of any advantage?

2. Compare designs of three barbeques or other consumer product that you find in online advertising from a single store that are in high-, medium-, and low-cost categories. Look at various advertised features. Argue how these features (the "scope") trade off in terms of cost.

3. Consider starting a week before to prepare for a final examination. You will have to use time to prepare, you may have costs of materials and help to prepare, and you may choose to work on selections of the total material only. These are the time-cost-scope values of this project. Indicate how you might trade off cost-time, time-scope, and cost-scope in planning your preparations.

4. Consider the plan below for building a bird house.

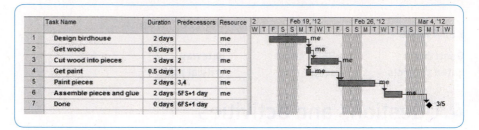

	Task Name	Duration	Predecessors	Resource
1	Design birdhouse	2 days		me
2	Get wood	0.5 days	1	me
3	Cut wood into pieces	3 days	2	me
4	Get paint	0.5 days	1	me
5	Paint pieces	2 days	3,4	me
6	Assemble pieces and glue	2 days	5FS+1 day	me
7	Done	0 days	6FS+1 day	

a. Which tasks constitute the critical path?

b. Normally if you extend a task not on the critical path by a suitably small amount it does not affect the completion of the project (because of the definition of the critical path). Why is this not true here?

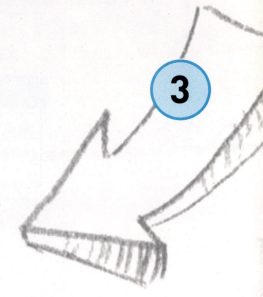

Creating a Project Plan

3

#process module: #creatingaprojectplan

Learning outcomes

By the end of this module, you should demonstrate the ability to:
- Decompose a project into tasks using the *work breakdown structure*
- Show task ordering using PERT charts in both *activity on the node* and *activity on the arrow* forms
- Organize a project using *network blocks*
- Use advanced task relationships in the building of a plan

Recommended reading

Before this module:
- **Implementing a Project > Project Management > 2. Project Management Concepts**

After this module:
- **Implementing a Project > Project Management > 4. Estimating Cost and Time**

1. Introduction

There are many different project management tools and methods used in project planning. This module will expose you to some common tools that are used in industry. As you progress through your career you will learn about tools that are specific to your industry or company.

2. Decomposition of a Project: Work Breakdown Structure (WBS)

A **work breakdown structure** (WBS) is an organized list of activities, called **tasks**, that is used to plan out a project (#tasks). To deconstruct the work into manageable pieces, a project is organized into subprojects and tasks in a tree structure. Consider the example of building a bathroom. The WBS for a bathroom construction project is shown in Figure 1.

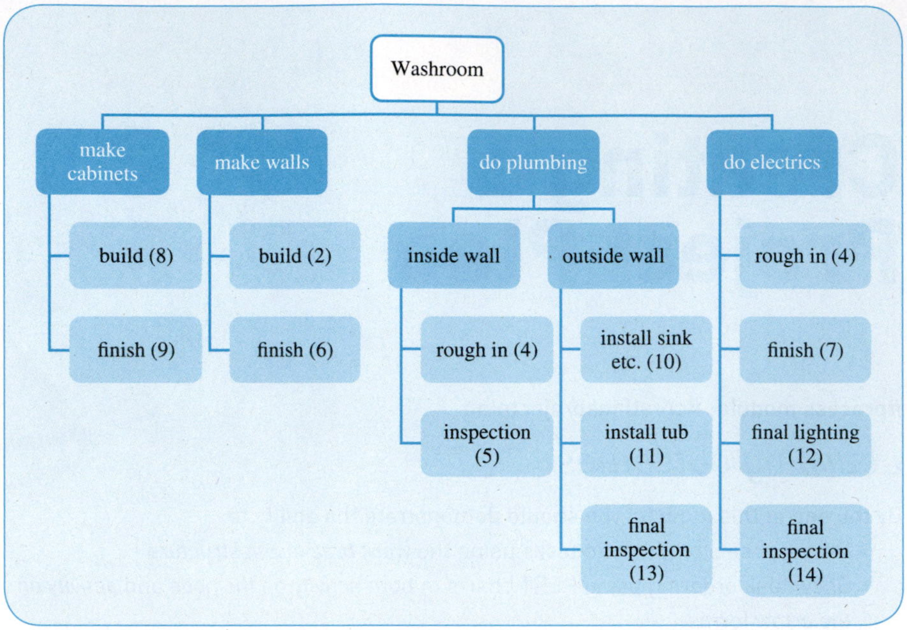

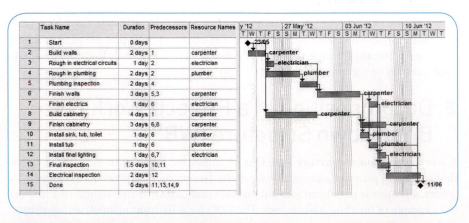

FIGURE 1 **Work breakdown structure (WBS) for a bathroom construction project.**

Each color is a different level; the lightest entries are the tasks and the task numbers for identification are given in parentheses. The work breakdown structure is created as a precursor to the creation of a ***Gantt chart*** (#Ganttchart). The darkest blocks indicate breakdowns of the project into parts that are considered separately for project management, because they use separate sets of resources (by different people with different expertise or with different sets of equipment). These darkest blocks might be called subprojects, activities, or have some other name (e.g., subassembly in product manufacturing).

	Task Name	Duration	Predecessors	Resource Names
1	Start	0 days		
2	Build walls	2 days	1	carpenter
3	Rough in electrical circuits	1 day	2	electrician
4	Rough in plumbing	2 days	2	plumber
5	Plumbing inspection	2 days	4	
6	Finish walls	3 days	5,3	carpenter
7	Finish electrics	1 day	6	electrician
8	Build cabinetry	4 days	1	carpenter
9	Finish cabinetry	3 days	6,8	carpenter
10	Install sink, tub, toilet	1 day	6	plumber
11	Install tub	1 day	6	plumber
12	Install final lighting	1 day	6,7	electrician
13	Final inspection	1.5 days	10,11	
14	Electrical inspection	2 days	12	
15	Done	0 days	11,13,14,9	

FIGURE 2 **A Gantt chart showing the tasks and scheduling for the construction of a bathroom.**

The middle hue blocks show a further breakdown of an earlier division, here a breakdown of the plumbing activity. A very complex project might have many, many layers in the WBS, and be distributed over many pages. The responsibility for a branch of a complex structure could be a project in its own right, where the heads of that project are responsible to the head managing group. The tasks (lightest blocks) along with project **milestones** can be entered into a list to create a **Gantt chart** (see Figure 2), which is a common method for structuring a project and showing the structure visually.

3. Other Forms of Plan Representation

3.1. PERT Charts

PERT stands for **program evaluation and review technique**. This representation of a project plan is often used for small projects and in small subprojects that are part of larger projects because it is easy to read and understand. There are two common representations of this graphic illustration of the scheduling. The first is **activity on the node,** as in the simple example shown on the right in Figure 3. In this representation the task is shown in the node bubbles at the end (or beginning) of each arrow. The second representation, **activity on the arrow**, is shown at the left in Figure 3. In this method the nodes are numbered and the tasks, A to F in the figure, are shown on the arrows. Different industries favor one of these representations over the other.

Note that there is sometimes a need for a **dummy task** to show the dependencies completely when doing activity-on-the-arrow. The reason for this can be seen if you closely examine the activity-on-the-node version. Task E cannot be started until task B is complete. This is explicitly shown in the on-node chart, but not in the on-arrow chart without a dummy task. The dependency of task E on task B is shown by putting a dummy task into the on-arrow chart.

Timing information can be added to the plan so tasks can be scheduled. For example, in Figure 3, if task A takes exactly a week, then there are two implications.

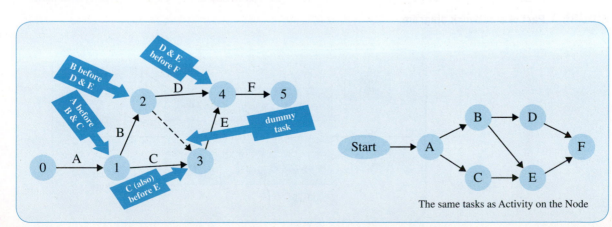

FIGURE 3 Diagrams showing activity on the arrow and activity on the node.

First, whatever resources (people, equipment, and available finances) that are required for task A must be available before task A can start and be available for a week. Second, tasks B and C can be scheduled no earlier than a week after the start of task A, because they are dependent on A. Task E will have to wait until A, B, and C are complete. Task F, because it falls last in the order, cannot be started until all the other tasks are finished.

3.2. Network Diagram

For larger projects computer packages are often used to plan the work flow. Two types of computer-generated representations that are widely used in industry are Gantt Charts and network diagrams.

In a ***network diagram*** the start, finish, and duration of a task is indicated at the node. A network diagram for a small portion of a project is shown in Figure 4. Depending on the software used and the configuration of that software, various other pieces of information can go into each network block. Figure 5 shows the typical information that is entered. In this block there is an identification number, task description, time the task will take, and when the task will take place (relative to the start of the project).

The ***slack*** time (also called ***float*** time) is also typically shown (see Figure 5). Slack time indicates how late a task can start before it will affect the end date for the whole

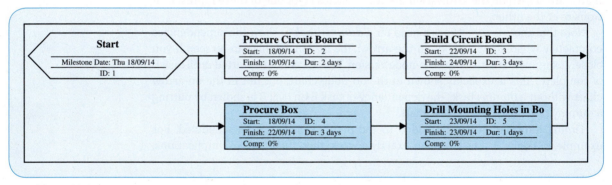

FIGURE 4 Part of a network diagram.

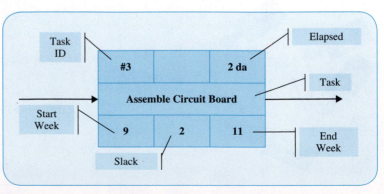

FIGURE 5 A typical network block.

project. For the simple example shown in Figure 4, ideally task 3 in Figure 4 will start on 11/05/10 as shown. However, if it starts 1 day late it does not matter because the middle hue path only takes 4 days to complete and the darkest path will take 5 days. So task 3, "Procure Box," and task 4, "Drill Mounting Holes in Box," together have 1 day of slack. (The "Comp" value indicates how much of that task is completed to date.)

The key to all of these representations of work plan information is communication. Managers, supervisors, workers, clients, government officials, and others will all have an interest in the elements of the plan and will look to the planning documents to understand what is going on in the project. The representation of the project plan must deliver accurate, current information that allows members of these groups to communicate with one another, and to avoid costly and potentially dangerous mistakes and/or miscommunications.

4. Planning: Other Task Timing and Relationships

4.1. Task Start Time

Besides "as early as possible," which is generally the default expectation, a task can be scheduled to start "as late as possible," or by a certain date, or not before a certain date (#tasks). You can also schedule it to finish before or not before a given date. In a more complex plan you may need to use these alternative ways to indicate timing of a task. For example, certain pieces of equipment may not be available until a certain date, and speedup of predecessor tasks will not change this. It is important that the schedule reflect this information so everyone working on the project is aware of this constraint.

4.2. Task-to-Task Connections

A task that begins "as early as possible" means that a task will begin as soon as its predecessors have completed. This is also known as a "*finish-to-start*" relationship between the tasks, since the predecessor must finish before the dependent tasks start. Finish-to-start relationships are the most common, but there are three other possibilities that could be used:

"Start-to-start" is used where a second task must wait on the start of the first, but can begin as soon as the predecessor task begins. For example, installation of the toilets in a new condominium building can be scheduled to start right after the start of "Uncrate the shipment of toilets" because the expectation is that the toilets will all be unpacked much more quickly than the installations will take place, and the installations can start as soon as the first toilet is uncrated (but not before).

"Finish-to-finish" is used where engineers want to get two tasks to finish at the same time. Cooking parts of a meal has this relationship; you want the turkey and the vegetables to be finished at the same time for a traditional Thanksgiving dinner.

The last option, "start-to-finish," says the second task will start before the first finishes. This could be used where engineers need continuous coverage of some aspect of the project, such as making sure there is constant drainage of a basement area by not discontinuing the use of an auxiliary pump until the permanent sump pump is installed and working.

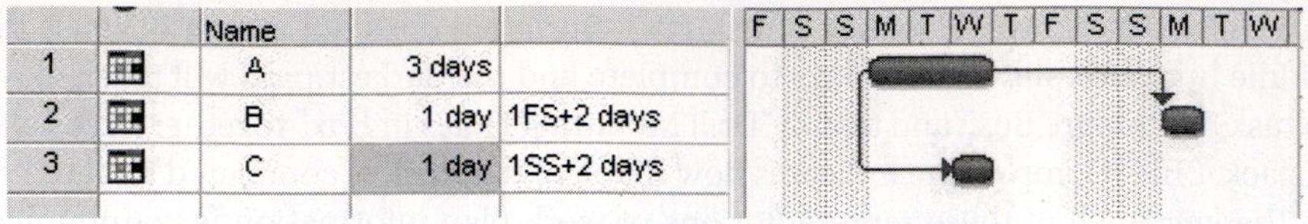

		Name			F	S	S	M	T	W	T	F	S	S	M	T	W
1		A	3 days														
2		B	1 day	1FS+2 days													
3		C	1 day	1SS+2 days													

FIGURE 6 Portion of a Gantt chart showing lag between tasks A and B.

4.3. Lag

In many instances a **lag** is indicated between tasks. This is a delay of the start of a task after its predecessor. For example, in Figure 6, task B has a two-day lag (plus a weekend) after the end of task A in a finish-to-start relationship. Task C has a start-to-start relationship, so the 2-day lag means it starts 2 days after task A starts. A lag can be put into a project plan for several reasons. For example, sometimes there is a need to wait for resources; people and specialized equipment are busy with other projects. Or there is a need to wait for materials to dry (e.g., paint) or cure (e.g., concrete).

5. Conclusion

Work breakdown structures, Gantt charts, PERT charts, and network diagrams are all ways of documenting the tasks needed to complete a project. The charts and diagrams help the project team visualize the schedule and have a shared understanding of the work needed to complete the project. When you are working on a project, especially with a team, it is useful to prepare a project plan to keep everyone organized. For engineering design projects the task list will ensure that every member of your team knows what they are supposed to be doing and when their work should be delivered to keep the project on time. Having clear expectations and communication can help the team focus on tasks and avoid miscommunications and wasted effort.

KEY TERMS

work breakdown
  structure (WBS)
milestones
activity on the node
slack or float

tasks
program evaluation and review
  technique (PERT Chart)
dummy task
finish-to-start

Gantt chart
activity on the arrow
network diagram
lag

6. Questions and activities

1. Create a Gantt chart from the tasks for the project "Building a Bright Green Birdhouse."
   a. Are there tasks that must be done before other tasks? Why?
   b. Are there circumstances where there is lag? If so, what is the cause?
   c. Assume that you and a friend are building the birdhouse. Are there tasks that then can be done in parallel? What tasks must wait on resources if there is only one saw?

Estimating Cost and Time

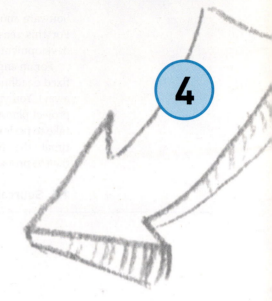

4

#resouce module: #estimatingcost&time

Learning outcomes

By the end of this module, you should demonstrate the ability to:

- Complete a basic time and cost estimate for a plan
- Recognize sources of error and risk in established work models when determining cost and time information for a specific project
- Work back from a deadline to plan the timing for a project using advanced timing relationships between tasks

Recommended reading

Before this module:

- **Implementing a Project > Project Management > 3. Creating a Project Plan**

After this module:

- **Implementing a Project > Project Management > 5. Project Cycles**

1. Creating the Plan: Finding the Time/Cost Information

At the heart of successful project management is correct estimation of the time and costs to do the various tasks. Getting those numbers together is difficult and error-prone. As a result, most industries rely heavily on previous experience. Some of this experience has been compiled and published. Construction, in particular, has very detailed information available for estimation. Available books and digital information will tell you the per-square-foot (or meter) cost estimates for installing dry wall, insulation, flooring, concrete, and so on, for various types of structures like residences, apartments, and airport terminals. Some other industries, such as software, have huge problems with time estimation and the estimation methods are not as good.

Software models can produce errors of 50–100% even when well calibrated [1]. For this reason, among others, the software industry has generated progressive development-management techniques to help minimize risk.

For an engineering design project you are working on in school you probably have fixed deadlines that are relatively close (i.e., a few weeks or months rather than years away). You will also have a fairly well defined set of deliverables. This helps with the project planning, but you still need to be careful not to underestimate the time it will take to perform each step of the design process. In particular, students often underestimate the amount of time it will take to build a prototype, test it, and the time it will take to properly write up project reports.

1.1. Sources of Error in Time-Cost Determinations

In industry project planners must think carefully when using published information, and consider issues such as *scaling*. For example, you may find information on how long it takes to install a tile floor. However, tiling a long, narrow hall will take longer than the same area in a large square room. A thousand lines of computer code will take considerably longer than 10 times 100 lines of code and even longer for safety-critical and sensitive financial applications. Mechanical designs with experimental elements to them will have wider variations in time and cost than straightforward duplications of previously successful work.

Here are some typical sources of major error in time and cost estimation:

1. Not allowing sufficient time for preparation tasks and disruptions in the workplace. Sometimes there are delays (e.g., due to weather or the need to obtain a permit or inspection). Tasks often require site preparation, movement of materials, instruction, and training and equipment setup that are not adequately taken into account. Breakdowns, strikes, and mistakes can also disrupt work flow.

2. Not sufficiently allowing for the conditions of equipment and systems. These are design deficiencies that become scheduling problems during execution. These can include:
   - Setup and initializations of systems and equipment. You may have planned for a steady-state condition, but getting started can often take a huge amount of time.
   - Erroneous inputs and misuse. Equipment or systems may be slow because an operator is new to the technology.
   - If the equipment or systems are old, not well maintained, or infrequently used they may be in poor condition and not perform at the speed anticipated or be unreliable.

3. Testing. This can be testing and qualifications on a job site, or software testing to prove reliability (i.e., prove the product works reliably without error).

The amount of time and number of resources to do this work can be significant, and in the case of software, electronics, and critical devices, can surpass the initial effort to produce the technology.

4. Informing and communicating. Typical new engineers will spend upward of 40% of their time in communication-related activities [2]. This percentage increases with responsibility. Communication activities will take the majority of a project manager's time. Not allowing for this in time estimates can result in estimates that are easily wrong by orders of magnitude. If your design team leader is also serving as your project manager, you need to make sure they are given time in the schedule to do the communication work necessary to keep the team on track.

5. Using ideal estimation. People, including engineers, do not like to admit that every project includes mistakes and time spent doing things that turn out to be unnecessary or need to be reworked. Many people do not realize they are clouding their judgments in this way; in hindsight it seems so simple to do things right the first time through. Any planner who relies on optimistic, unrealistic estimates that do not take into account these types of detours in work is in trouble from the start.

The Pi Rule for Novice Estimators

If you have a piece of work to complete, estimate the amount of time it will take. Be conservative (what do you think is really the maximum time it will take). Then multiply by π (approx. 3) to get a more accurate estimate.

Scaling your homework: If you can write 2 pages of a report in an hour then you should be able to write 16 pages in 8 hours, right? This is very difficult to do, and almost impossible to do well. You have to build into your plan time between work periods to regain your energy and mental ability. Even professional writers have difficulty working for more than a few hours straight. Be careful about this type of idealistic scaling.

1.2. Planning: Task Time versus Actual Time

Effort driven versus *fixed duration*. Most tasks are considered to be *effort driven*. This means that the time can be reduced by increasing the effort or, in other words, by increasing the people (or pieces of equipment) working on the task (often referred to as *man-hours* or resource-hours). However, doubling resources will not usually halve the time for the task exactly, although there will be a reduction in the time to completion. Tasks that cannot be done more quickly by adding available resources are known as *fixed-duration* tasks. For example, this could be because of limits of available resources, or a type of operation that only allows limited access to the work area.

Work time versus elapsed time. Usually a person is needed to work on a task to complete it. However some tasks, such as waiting for paint to dry, do not require effort, just time. The time assigned to these tasks is called *elapsed time*. Elapsed time can happen at night or on weekends or other times when you usually would not schedule a task for workers.

2. Trading Cost for Time

The blue solid line in Figure 1 shows a typical project cost curve. Project cost will tend to decrease if more time is allowed for finishing the project up to a point because more time allows resources to be used more efficiently. However, the cost of the project will increase if the project takes too long because idle time in the project costs money and this will start to outweigh efficiencies. This would suggest an optimal target *time to completion* (*duration* for the whole project) shown by the vertical line on the right.

However, the stakeholders, such as the owner, will be considering more than just the project cost. They will also be considering revenue they would have received if the project was finished and in service or on the market. These costs are shown in Figure 1 as the dotted curve. For a volume product, it is important to introduce new features well before the competition, or to quickly match the features of the competition; delays will result in lost sales and a quicker product introduction will result in increased sales. For something like a rental unit there is rental income not being realized while the construction takes place [3]. A city could lose revenue if road construction slows transport of goods and people. Other projects will have other unrealized revenues. As shown, if the project took no time there would be none of these revenue losses. These lost opportunities for revenue increase as the project time increases.

The total cost will be the sum of these costs; the black curve. This suggests that there are trade-offs between the time and cost of the project. As a result engineers plan projects to balance these trade-offs and try to plan in a way that optimizes the system, such as by planning to finish the project at the time shown by the vertical line to the left in Figure 1.

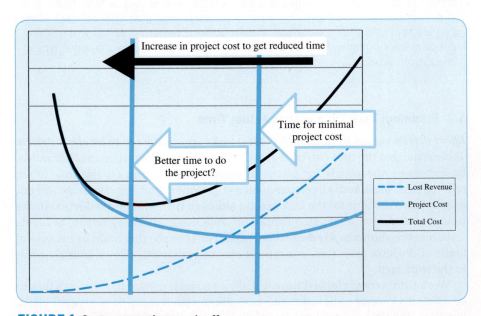

FIGURE 1 **Cost versus time tradeoff.**

3. Conclusion

It should be clear by now that the methods and tables used to create the task times and costs are estimates. The key to controlling a project is to take account of the risk in your estimates. For some areas of work—software development, for example—the estimates have been notoriously bad [4]. For a project manager it would be quite reasonable to use task times in planning two or three times the estimates given by programmers. In other areas with better-known characteristics, such as construction, it may be easier to reach a good estimate.

A task will generally require resources in order to be executed. If the resources are not available, the time for the task could be lengthened beyond budget. Worker illness or other unavailability, breakage or non-delivery of tools, supplies not arriving on time or not meeting specification (i.e., not to spec), and predecessor tasks not being done are examples of possible problems that could result in these resources being unavailable at the time they are needed. All of these issues represent risks to your project plan. In good project management, one needs to continually plan for things to go right and take into account that things will not always go right.

KEY TERMS

scaling	**effort driven**	**man-hours**
fixed duration	**elapsed time**	**time to completion**
duration		

4. Questions and activities

1. Explain why:

   a. Tiling a long, narrow hall will take longer than the same area in a large square room.

   b. A thousand lines of computer code will take considerably longer than 10 times 100 lines of code and even longer for safety-critical and sensitive financial applications.

   c. Mechanical designs with experimental elements to them will have wider variations in time and cost than straightforward duplications of previously-successful work.

2. You allocate a half hour to do a lab preparation. What problems in the work could extend this time? What distractions in your workplace could extend this time?

5. References

[1] Kemerer, C.F. An empirical validation of software cost estimation models, *CACM,* 36(2), 1993.

[2] Ganssle, J. *The Art of Designing Embedded Systems,* 2nd ed. Elsevier, 2008, p. 33.

[3] Pasternak, J. Example: Nest egg of red brick. *Financial Post,* Sunday, April 12, 2009.

[4] Standish Group Report 1995. Retrieved July 20, 2012, from http://spinroot.com/spin/Doc/course/Standish_Survey.htm.

Estimating Cost and Time

10

MS Project Instructions

#resource module: #MSProjectinstructions

Learning outcomes

By the end of this module, you should demonstrate the ability to:

- Use MS Project software to plan out a project

Recommended reading

Before this module:

- **Implementing a Project > Project Management > 2. Project Management Concepts**

1. Microsoft Project®

This is a brief overview to allow you to get started with and to use Microsoft Project® for projects. The version used in this example is Microsoft Project 2007®.

We use an example of the design/build of a microcontroller-directed switching power supply. In this power supply the microcontroller reads the output voltage and adjusts the rate of energy transfer from the main power input to a lower-voltage, regulated output. We won't worry about the technical details, just the project management.

For this project, there are three parts that have to come together: electronics, software, and packaging. Some simplifications have been made and the numbers used are for convenience of demonstration. You will also find alternate methods of doing the work here as you become familiar with the program.

1.1. Starting the Project

A project is started either using a "wizard" to guide you through a detailed project setup, or by bringing up the Project Information window as shown.

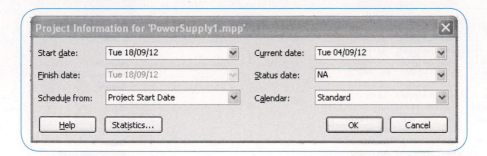

If you want the end of the project to be determined by the task lengths and the start date, set Schedule from to Project Start Date as shown (#tasks). You can also give a finish date, such as a project due date, and the program will calculate a last possible start date using the task times.

In this window you can also set up the default units and scaling to suit your project.

1.2. Listing the Tasks

The next step is to list the tasks, shown in the next figure. You can see that the program assumes default times of 1 day, and also assumes that every task can be done simultaneously.

If you are not seeing the lists shown, select View > Gantt Chart.

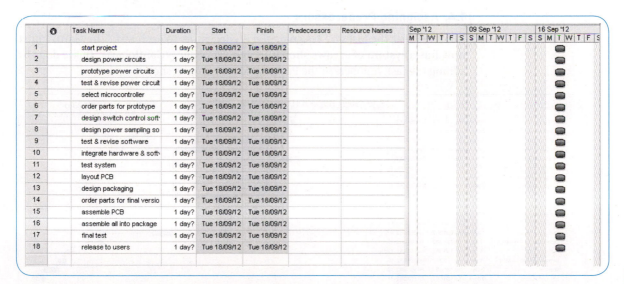

1.3. Assigning Task Durations

Now working down the Duration column, we add the times for each task. Note that milestones, like task 1, are given zero duration. Milestones are shown as diamonds on the Gantt chart.

	ⓘ	Task Name	Duration	Start	Finish	Predecessors	Resource Names	09 Sep '12	16 Sep '12	23 Sep '12
								F S S M T W T F	S S M T W T F S	S M T W T
1		start project	0 days	Tue 18/09/12	Tue 18/09/12			◆ 18/09		
2		design power circuits	2 days	Tue 18/09/12	Wed 19/09/12					
3		prototype power circuits	3 days	Tue 18/09/12	Thu 20/09/12					
4		test & revise power circuit	5 days	Tue 18/09/12	Mon 24/09/12					
5		select microcontroller	3 days	Tue 18/09/12	Thu 20/09/12					
6		order parts for prototype	1 day	Tue 18/09/12	Tue 18/09/12					
7		design switch control soft	2 days	Tue 18/09/12	Wed 19/09/12					
8		design power sampling so	2 days	Tue 18/09/12	Wed 19/09/12					
9		test & revise software	5 days	Tue 18/09/12	Mon 24/09/12					
10		integrate hardware & soft	2 days	Tue 18/09/12	Wed 19/09/12					
11		test system	5 days	Tue 18/09/12	Mon 24/09/12					
12		layout PCB	2 days	Tue 18/09/12	Wed 19/09/12					
13		design packaging	3 days	Tue 18/09/12	Thu 20/09/12					
14		order parts for final versio	1 day	Tue 18/09/12	Tue 18/09/12					
15		assemble PCB	1 day	Tue 18/09/12	Tue 18/09/12					
16		assemble all into package	1 day	Tue 18/09/12	Tue 18/09/12					
17		final test	1 day?	Tue 18/09/12	Tue 18/09/12					
18		release to users	1 day?	Tue 18/09/12	Tue 18/09/12					

Note that the program default is no work on the weekends (an unlikely situation for your student design team). See task 9, for example, where 5 days of duration take a week of calendar time. We can change this by using elapsed days in our duration (for example, "3 edays", meaning three elapsed days), or by changing the underlying scheduling calendar in use.

1.4. Assigning Predecessors

Now we order the tasks by assigning the Predecessors to each task. **The Gantt chart now has a system of arrows showing the task order** (#Ganttchart). This is done by adding the task numbers in the Predecessor column, as shown.

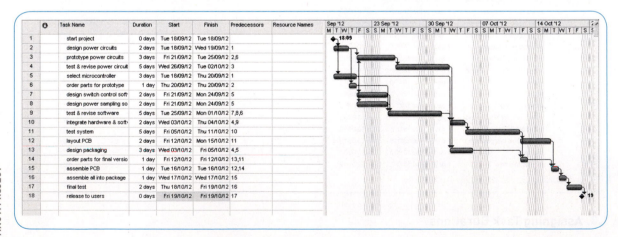

	ⓘ	Task Name	Duration	Start	Finish	Predecessors	Resource Names	Sep '12	23 Sep '12	30 Sep '12	07 Oct '12	14 Oct '12
1		start project	0 days	Tue 18/09/12	Tue 18/09/12			◆ 18/09				
2		design power circuits	2 days	Tue 18/09/12	Wed 19/09/12	1						
3		prototype power circuits	3 days	Fri 21/09/12	Tue 25/09/12	2,6						
4		test & revise power circuit	5 days	Wed 26/09/12	Tue 02/10/12	3						
5		select microcontroller	3 days	Tue 18/09/12	Thu 20/09/12	1						
6		order parts for prototype	1 day	Thu 20/09/12	Thu 20/09/12	2						
7		design switch control soft	2 days	Fri 21/09/12	Mon 24/09/12	5						
8		design power sampling so	2 days	Fri 21/09/12	Mon 24/09/12	5						
9		test & revise software	5 days	Tue 25/09/12	Mon 01/10/12	7,8,6						
10		integrate hardware & soft	2 days	Wed 03/10/12	Thu 04/10/12	4,9						
11		test system	5 days	Fri 05/10/12	Thu 11/10/12	10						
12		layout PCB	2 days	Fri 12/10/12	Mon 15/10/12	11						
13		design packaging	3 days	Wed 03/10/12	Fri 05/10/12	4,5						
14		order parts for final versio	1 day	Fri 12/10/12	Fri 12/10/12	13,11						
15		assemble PCB	1 day	Tue 16/10/12	Tue 16/10/12	12,14						
16		assemble all into package	1 day	Wed 17/10/12	Wed 17/10/12	15						
17		final test	2 days	Thu 18/10/12	Fri 19/10/12	16						
18		release to users	0 days	Fri 19/10/12	Fri 19/10/12	17					◆ 19	

1.5. Assigning Resources

For our power supply we will assume that we will use an electrical engineer, a computer engineer, and a technician to do the various parts of the work. We use the short forms EE, CE, and Tech for these people, and allot them to the appropriate tasks in the Resource Names column. The program uses the names to create a set of resource information for each unique name.

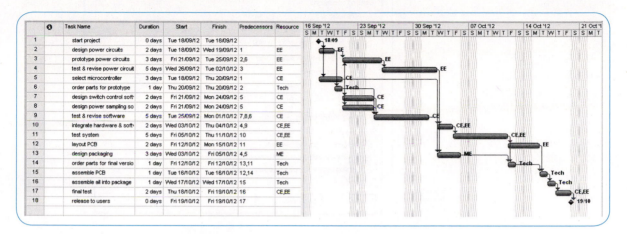

	Task Name	Duration	Start	Finish	Predecessors	Resource
1	start project	0 days	Tue 18/09/12	Tue 18/09/12		
2	design power circuits	2 days	Tue 18/09/12	Wed 19/09/12	1	EE
3	prototype power circuits	3 days	Fri 21/09/12	Tue 25/09/12	2,6	EE
4	test & revise power circuit	5 days	Wed 26/09/12	Tue 02/10/12	3	EE
5	select microcontroller	3 days	Tue 18/09/12	Thu 20/09/12	1	CE
6	order parts for prototype	1 day	Thu 20/09/12	Thu 20/09/12	2	Tech
7	design switch control soft	2 days	Fri 21/09/12	Mon 24/09/12	5	CE
8	design power sampling so	2 days	Fri 21/09/12	Mon 24/09/12	5	CE
9	test & revise software	5 days	Tue 25/09/12	Mon 01/10/12	7,8,8	CE
10	integrate hardware & soft	2 days	Wed 03/10/12	Thu 04/10/12	4,9	CE,EE
11	test system	5 days	Fri 05/10/12	Thu 11/10/12	10	CE,EE
12	layout PCB	2 days	Fri 12/10/12	Mon 15/10/12	11	EE
13	design packaging	3 days	Wed 03/10/12	Fri 05/10/12	4,5	ME
14	order parts for final versio	1 day	Fri 12/10/12	Fri 12/10/12	13,11	Tech
15	assemble PCB	1 day	Tue 16/10/12	Tue 16/10/12	12,14	Tech
16	assemble all into package	1 day	Wed 17/10/12	Wed 17/10/12	15	Tech
17	final test	2 days	Thu 18/10/12	Fri 19/10/12	16	CE,EE
18	release to users	0 days	Fri 19/10/12	Fri 19/10/12	17	

2. Advanced Project Planning and Tracking

2.1. Leveling Resources

When you assigned the resources you may have seen changes happening in the task start and end times. In our example this did not happen. The reason it may have happened was that the program recognized that the single resources could not do two jobs simultaneously. In our case we are going to have to do this adjustment "manually."

Note that CE, for instance, is designing two pieces of software at the same time in tasks 7 and 8. If we look at the Resource Graph (View > Resource Graph) and go to the CE for this time period, we can see that the allocation bars exceed 100% as shown below.

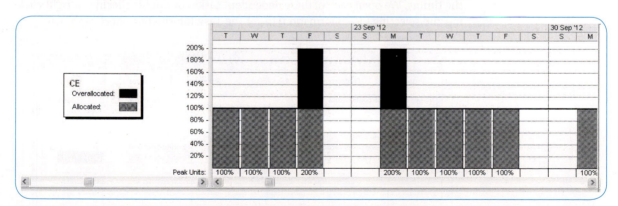

To solve this we do resource "leveling." Select Tools > Level Resources and click on "Level Now" in the window shown below. You will see the Gantt chart change so the CE is no longer working at 200%. You can recheck the Resource Graph and will see the black shaded areas are gone.

Leveling may cause the project end to move to a later date or the start date to move to an earlier date, as might be expected.

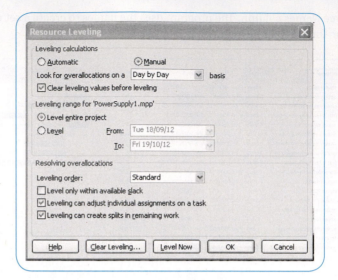

Note in the window the radio buttons for Leveling calculations. If Automatic had been selected we would have seen the leveling changes as we assigned the resources to the tasks.

2.2. Lag

Two of our tasks involve parts ordering, and each is given a duration of 1 day of Tech time. However, the parts will not arrive instantaneously if they are ordered from external suppliers; we can use "lag" to account for this delivery time in the tasks that must wait for the parts.

Each task that has these tasks as a predecessor will need this delivery lag added to the timing. We open each of these dependent tasks (by double clicking, or right clicking and selecting task information) to get the Task Information sheet as shown.

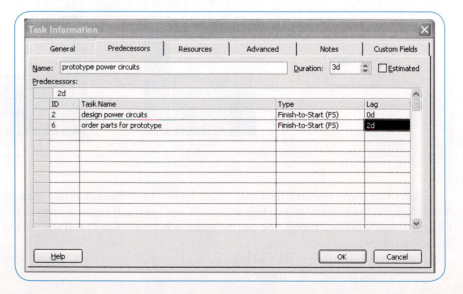

At the right side we can set the lag for the predecessor task of ordering the parts, as shown. The change will be reflected in the Gantt chart and in the task list predecessor information. (See the figure in the next section for Gantt chart changes. Note task 15 in particular.)

2.3. Critical Path

By choosing View > Tracking Gantt we can see the Critical Path, as shown.

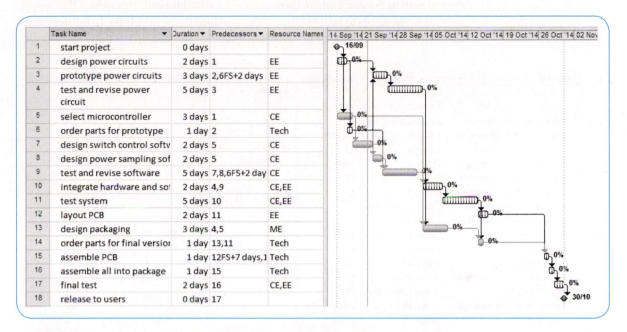

	Task Name	Duration	Predecessors	Resource Names
1	start project	0 days		
2	design power circuits	2 days	1	EE
3	prototype power circuits	3 days	2,6FS+2 days	EE
4	test and revise power circuit	5 days	3	EE
5	select microcontroller	3 days	1	CE
6	order parts for prototype	1 day	2	Tech
7	design switch control softw	2 days	5	CE
8	design power sampling sof	2 days	5	CE
9	test and revise software	5 days	7,8,6FS+2 day	CE
10	integrate hardware and so	2 days	4,9	CE,EE
11	test system	5 days	10	CE,EE
12	layout PCB	2 days	11	EE
13	design packaging	3 days	4,5	ME
14	order parts for final version	1 day	13,11	Tech
15	assemble PCB	1 day	12FS+7 days,1	Tech
16	assemble all into package	1 day	15	Tech
17	final test	2 days	16	CE,EE
18	release to users	0 days	17	

The tasks shown hatched are the tasks on the critical path.

This same view shows the percentage completion of each task, shown to the right of each bar.

Note that many projects have fixed start and end times. If the amount of work does not take 100% of the time between start and end times, there will be no critical path according to the program. This is because every path has "slack" in it.

2.4. Adding Resources

If our power supply will not be done on time, one option is to add resources. We can do so through the Resource Sheet (View > Resource Sheet) as shown below. Changing the Max Units for EE to 200% allows us to add EE resources to tasks without over-allocation.

	ⓘ	Resource Name	Type	Material Label	Initials	Group	Max. Units	Std. Rate	Ovt. Rate	Cost/Use	Accrue At	Base Calendar	Code
1		EE	Work		E		200%	$0.00/hr	$0.00/hr	$0.00	Prorated	Standard	
2		CE	Work		C		100%	$0.00/hr	$0.00/hr	$0.00	Prorated	Standard	
3		Tech	Work		T		100%	$0.00/hr	$0.00/hr	$0.00	Prorated	Standard	
4		ME	Work		M		100%	$0.00/hr	$0.00/hr	$0.00	Prorated	Standard	

Note that we can also add costing information and calendar (availability) information about the resource using the Resource Sheet.

To assign the extra capability to a task, go to the Task Information sheet (double click on the task, or right click and select Task Information). Pick the resources tab and change the Units to 200%. The change will be reflected in the Gantt chart, shown in the Tracking Gantt chart for task 3.

Note that the critical path has changed since we reduced the duration of a task on the critical path to the point where there is now slack in this path. This means that assigning double resources to task 4 will not further reduce the project time. To further reduce the project time we will have to shorten the duration of tasks on the new critical path.

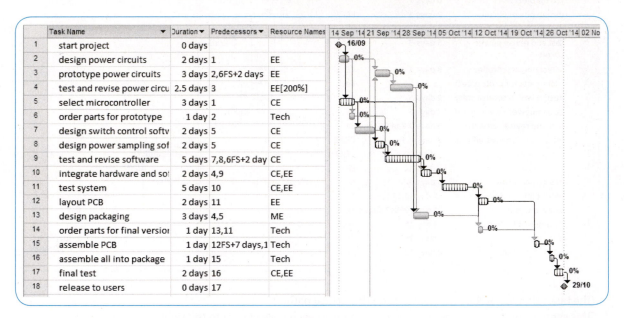

	Task Name	Duration	Predecessors	Resource Names
1	start project	0 days		
2	design power circuits	2 days	1	EE
3	prototype power circuits	3 days	2,6FS+2 days	EE
4	test and revise power circu	2.5 days	3	EE[200%]
5	select microcontroller	3 days	1	CE
6	order parts for prototype	1 day	2	Tech
7	design switch control softw	2 days	5	CE
8	design power sampling sof	2 days	5	CE
9	test and revise software	5 days	7,8,6FS+2 day	CE
10	integrate hardware and sof	2 days	4,9	CE,EE
11	test system	5 days	10	CE,EE
12	layout PCB	2 days	11	EE
13	design packaging	3 days	4,5	ME
14	order parts for final version	1 day	13,11	Tech
15	assemble PCB	1 day	12FS+7 days,1	Tech
16	assemble all into package	1 day	15	Tech
17	final test	2 days	16	CE,EE
18	release to users	0 days	17	

A caution here: two people do not usually finish the work twice as fast as one! However, you can't have one person working 200% of their time.

2.5. Tracking the Project

In the General tab of the Task Information sheet (double click on the task, or right click and select task information) you can indicate how much of a task is done. This will result in black bars appearing in your basic Gantt chart that allow you to quickly see how the project is progressing.

2.6. Task Grouping

For larger projects it is useful to group projects together. To create a group:

1. Move the various tasks to be grouped together. To move a task: click on a task, then click and hold on the top or bottom border. Move the task to the position desired and release the mouse button.

2. Create a Summary Task: click on the top task in the group and press the Insert key. Name the task appropriately.

3. At the left, select the tasks in the group, not including the Summary Task as shown.

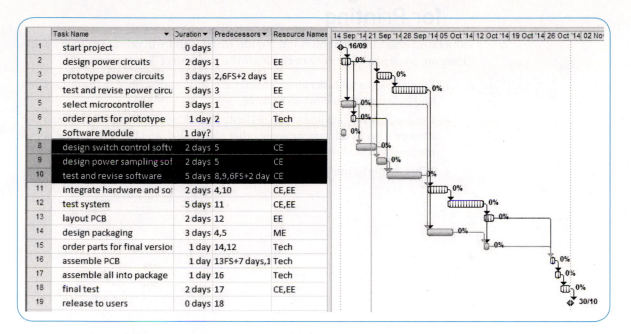

	Task Name	Duration	Predecessors	Resource Names
1	start project	0 days		
2	design power circuits	2 days	1	EE
3	prototype power circuits	3 days	2,6FS+2 days	EE
4	test and revise power circu	5 days	3	EE
5	select microcontroller	3 days	1	CE
6	order parts for prototype	1 day	2	Tech
7	Software Module	1 day?		
8	design switch control softv	2 days	5	CE
9	design power sampling sof	2 days	5	CE
10	test and revise software	5 days	8,9,6FS+2 day	CE
11	integrate hardware and so	2 days	4,10	CE,EE
12	test system	5 days	11	CE,EE
13	layout PCB	2 days	12	EE
14	design packaging	3 days	4,5	ME
15	order parts for final versior	1 day	14,12	Tech
16	assemble PCB	1 day	13FS+7 days,1	Tech
17	assemble all into package	1 day	16	Tech
18	final test	2 days	17	CE,EE
19	release to users	0 days	18	

4. Press the arrow (➜) to "indent" the tasks. This will result in a grouping such as the following. The group can be collapsed and expanded using the + and – buttons shown at the top of the figure, or using the box to the left of the Summary Task.

A hierarchical structure is possible with multiple levels of indentation.

	Task Name	Duration	Predecessors	Resource Names
1	start project	0 days		
2	design power circuits	2 days	1	EE
3	prototype power circuits	3 days	2,6FS+2 days	EE
4	test and revise power circu	5 days	3	EE
5	select microcontroller	3 days	1	CE
6	order parts for prototype	1 day	2	Tech
7	⊟ **Software Module**	**9 days**		
8	design switch control so	2 days	5	CE
9	design power sampling :	2 days	5	CE
10	test and revise software	5 days	8,9,6FS+2 day	CE
11	integrate hardware and so	2 days	4,10	CE,EE
12	test system	5 days	11	CE,EE
13	layout PCB	2 days	12	EE
14	design packaging	3 days	4,5	ME
15	order parts for final versior	1 day	14,12	Tech
16	assemble PCB	1 day	13FS+7 days,1	Tech
17	assemble all into package	1 day	16	Tech
18	final test	2 days	17	CE,EE
19	release to users	0 days	18	

3. Reducing the Size of the Gantt Chart for Printing

The Gantt chart information is often quite lengthy and will print over many pages. Here are some ways to reduce or eliminate the problem of "too many pages":

- Lump tasks as subtasks under a Summary Task, as described in the last section, and hide the subtasks when possible.
- Double click on date line at top of Gantt chart. Under timescale, go to "one tier." Change label to dd/mm, or some other short label, and the units to the largest reasonable scale for your project.
- When printing: use the "Dates from/to" option to print limited information.

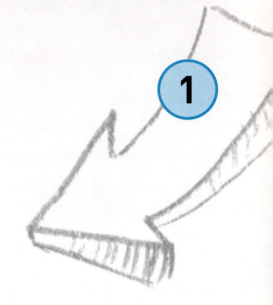

Basic Concepts

1

#skill/tool module: #criticalthinking

Learning outcomes

By the end of this module, you should demonstrate the ability to:
- Define Critical Thinking and key related concepts
- Identify sources of your own bias and obstacles to objectivity

Recommended reading

After this module:
- **Implementing a Project > Critical Thinking > 2. Critical Thinking in Design Documents**

1. Introduction

There are many ways of defining *critical thinking (CT)*. What the definitions have in common is that they focus on the ability to analyze a situation, idea, or problem *for yourself*: to come up with *your own unique ideas* about it. They also emphasize objectivity, the importance of recognizing bias, and other factors that affect the *validity* of information. People need to develop critical thinking in order to deal with situations in which there is no single right answer, even though there are, in fact, wrong answers. Critical thinking also helps in situations in which there is no set procedure and you cannot just "follow the rules."

In critical thinking, you personally consider a problem from several perspectives (see Figure 1), develop a number of alternative solutions, and come up with an "evaluative judgment." An "evaluative judgment" is not a "right answer." Rather, it is judgment based on criteria that you develop and that you can persuasively and logically explain to somebody else. The final step in CT is looking back on the solution you have come up with, as well as the process you used to come up with it, and considering their strengths and weaknesses, as well as possible ways to improve for the future.

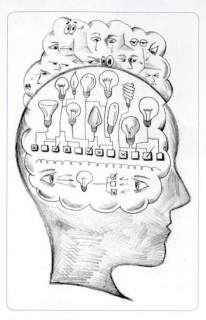

FIGURE 1 In the engineering design process, four aspects of critical thinking are used: multiple perspectives going into the problem, multiple solutions going out, methodical selection of a solution, and then, finally, reflecting on the overall process.

The engineering design process is a parallel to CT. When approaching a design problem, engineers consider it from the multiple perspectives of stakeholders, the environment, human factors, and economics. They generate alternative solutions and criteria to use when selecting one to recommend. They use documentation to logically, objectively, and persuasively explain their design choices. Finally, they reflect, using reality checks and other methods to evaluate and improve their processes. Critical thinking and complex problem solving are the hallmarks of engineering practice that distinguishes the engineering profession from the work done by technicians and technologists [1].

2. Useful Critical Thinking Concepts

Critical thinking is a form of independent judgment that incorporates fair-mindedness, awareness, *objectivity*, and freedom from prejudice. You might agree that these seem to be very desirable qualities, but how can you be sure that you are using them when making design decisions? The principles of critical thinking include:

- Becoming aware of how personal values affect decisions by understanding frame of reference, bias, and purpose
- Understanding the obstacles to objectivity that people face
- Evaluating sources of information
- Using skeptical thinking tools to question and analyze information

2.1. Understanding Frame of Reference, Bias, and Purpose

Frame of reference, *bias,* and *purpose* are related concepts because they all affect the objectivity of information in similar ways. You may know the term "frame of reference" from physics: the perspective of a viewer will affect the viewer's observations. It is also used, however, in psychology to identify the coordinates of the perspective with which each of us views, and judges, the world. Your frame of reference is how you, consciously and unconsciously, put together everything you have observed, felt, and learned from others. This frame affects how you perceive the world, how you connect new ideas and experiences to what you have previously seen, felt, and known, in order to help you understand and respond appropriately in your present life. Although a person's frame of reference is connected to her or his sense of community, family, country, and profession, each frame of reference is unique. Because of this uniqueness, the same fact may be interpreted differently by different people. Understanding your own frame of reference, and getting a sense of those of others, will allow you to examine facts and ideas separately from their sources and evaluate them more objectively.

Your frame of reference, where you are and what your perspective is on the world, is more than just a physical location. It is psychological, intellectual, and emotional. It is made up of values, experiences, beliefs, and ideas you have learned from others. It is your *subjectivity*. To get a sense of your own frame of reference, see if you can identify:

1. A key value that characterizes how you make decisions
2. A memory that you feel has changed the way you think or behave—what were you like before this event? What changed?
3. One or two beliefs you gained from your parents, religion, or community—what are some you continue to adhere to, even if they are inconvenient or cause discomfort? What are some you feel are important even if you do not always adhere to them?
4. An idea that is important to you that you learned from someone else

Now, ask yourself how any of the values identified above influenced your choice of school or program.

The exercise might have shown you how your frame of reference affected the objectivity of your choice. That is, the exercise might have led you to see how *bias* operated in your decision making. Bias specifically refers to the tendency to judge things in a certain way, due to the values a person brings to the situation rather than due to the details of the situation itself. At its worst, bias is recognizable as *prejudice*: it is the judgment of a person or their idea based on factors such as gender, age, race, or religion. No one is entirely free of bias, however, since everyone has a frame of reference. Therefore, to make objective design decisions engineers have to work to minimize the interference of bias on objectivity. Being open-minded, or reserving judgment, is a way of preventing bias from interfering with an objective evaluation. Your ultimate decision may agree with your bias, but good judgment comes from examining the situation carefully and as objectively as possible before coming to a conclusion. Bias is an impediment to that process.

Finally, in addition to frame of reference and bias, a person's *purpose* will affect their objectivity. Purpose can be defined as what you want to happen as a result of some action. Purpose can be conscious but it can also be unconscious. However, in as much as you can, being aware of what you want is a good thing because it allows you to be in control of what you do. Being aware of what others want is a key to understanding why they behave the way they do and why they give the information that they give.

A good example of purpose and information can be found in a website for a car manufacturer. The purpose of the website is most likely to sell cars. Therefore, much of the information given will have a *bias.* It will tend to emphasize the positive aspects of the car and minimize—or ignore—the negative. However, some of the information will be useful, because it will be less likely to be affected by bias. For example, the actual specifications are likely to be accurate (e.g., the size of the engine). They are measurements and are easily tested for validity. Other kinds of information—exciting descriptions of performance—are much more difficult to verify and much less likely to provide objective information. If a carmaker says an engine is a 4-cylinder, 16-valve design with 148 horsepower, then that information is far more trustworthy than the statement that the car represents a new dimension in driving pleasure. Therefore, if you are using the website for information, you have to determine which information is more affected by the purpose and which information will be less likely to be affected and, therefore, possibly more reliable.

At least the purpose of an advertising site is obvious: to sell you something. Other sites may not be so obvious but their purposes will still affect the validity and usefulness of their information. So the question "what is the purpose of this piece of information" becomes more important the more difficult the purpose is to determine. Critical thinking helps you overcome the problems of frame of reference, bias, and purpose in that it promotes taking many different perspectives into account. This is precisely what you do, in fact, in considering stakeholders' concerns in a design problem. Thus, critical thinking and the design process work together to help you develop objective solutions to design problems.

2.2. Obstacles to Objectivity

Hard as you might try, you cannot be completely objective. Some attitudes that can get in the way of objectivity include:

1. The belief that there is a right answer and that someone knows it
2. The feeling that it is rude or disrespectful to question authority
3. The feeling that you will betray your own values if you consider an idea that is contrary to deeply held beliefs
4. Dislike of the person or source of the information
5. The dislike of bad news or information that is hard to accept
6. Fear of asking a "silly question," or one that is perceived by others as being silly

1. **There is no right answer.** Your task, in design and critical thinking, is to examine a problem in depth, generate alternative solutions, and consider them according to criteria. You must make a choice and then justify it. There is no single correct answer in engineering design. Developing critical thinking is not like learning a formula and applying it to a given problem. Open-ended problems have no such certainty and looking to the professor, supervisor, teaching assistant, or client, to give or confirm a solution is simply asking for their judgment. Even clients who seem to have a solution in mind will benefit if you are able to increase their understanding of their problem and appreciation of

alternative solutions. While there is no right answer in design, there are better and worse solutions. The quality of your solution will be judged on its fit to the requirements, how well it meets the stakeholder needs, and the quality of the documented thinking process that went into developing the solution.

2. **If you feel it is disrespectful to contradict the client or question authority,** then consider the following:

- Simply repeating what the client says because you feel it is the right answer is not very helpful. It is not giving the client anything new.

- Not asking questions may lead you to making mistakes because you do not truly understand something. So, while you might be trying to be respectful, you might also be underestimating an authority who wants to help you to understand and welcomes questions that will increase your knowledge.

So, ask yourself, when dealing with information from authority:

- Do I believe this just because it comes from an authority (or the client) or can I find some evidence to independently back it up?

- When I repeat something I have been told, do I really understand it? Can I put it in my own words, relate it to my own sense of the world? Am I adding something to it by considering it critically?

- Can I find a respectful way to approach this subject and question authority such that everyone comes away better informed?

It is possible to have a discussion with an older person, or a person in authority, and ask probing questions without being disrespectful.

Courtesy of the Authors

Of course, it is possible to ask a question in a disrespectful way and that is why it is important to develop professionalism. Approaching a subject respectfully allows people to accept new information that may change their perspective.

3. While it is important to have **personal values**, you have to understand that not everyone's values agree and seeing a design problem only through the lens of your own values may prevent you from understanding the way others see it. This, in turn, may result in you coming up with a solution that does not work for all who are affected by it, and may even be damaging for some people.
So, when dealing with ideas that you either agree with or disagree with, due to your own values, ask yourself:

- What objective evidence is there to support or contradict this idea?

- How can I see this from another person's point of view? What values of theirs would lead to an idea such as this? Is there a context where the different ideas make sense, and is that context part of the environment for which the ideas were intended?

- If, for the purpose of argument, I accepted the other person's values, how would this idea be useful to me and to the world? What advantages are gained? What advantages are lost?

A perpetual motion machine is an example of an idea that does not stand up to scientific investigation. It violates the laws of thermodynamics.

4. Disliking another person is a common reason for not accepting their ideas. But finding a way to get around this tendency will not only help you become more open-minded and respectful, it will also help you see more ideas than you could have come to on your own.

5. There may be **an idea that would make you feel badly** if it were true and so you may choose to deny it. For example, the Laws of Thermodynamics can be difficult to accept. It is hard to accept that the efficiency of an internal combustion engine is as low as it is. There are many examples of ideas that are unaccepted because the consequences are difficult to accept. Some ideas take some getting used to, because they are unintuitive, or feel wrong or unfair. However, some ideas that you don't like can be a source of motivation. Historically, engineers have worked hard to improve efficiency. Separating an idea from the feelings it causes will enable you to more objectively evaluate its truth.

> **From a professor in thermodynamics:**
>
> "As a professor in the field of thermodynamics I routinely have inventors who come to talk to me about a new idea they have for power generation. Some of these inventions are valid. However, most will never work in reality and invariably it is because the idea violates the second law of thermodynamics. Also, invariably, the inventor will refuse to believe that the invention is impossible. It is difficult to comprehend how people refuse to accept the reality of a physical law that has been known for over 100 years, but perhaps it is because the second law can feel very unfair."

6. You can combat the **fear of asking a "silly" question** by considering the consequences of not asking the question and not knowing the answer. Going ahead when you do not understand something can cost you marks on an assignment, and can cause real damage in professional life.

3. Leaving This Skill Module

After working through this module you should be able to

- Define critical thinking and key terms related to it
- Identify your own frame of reference
- Identify barriers to objectivity that you might be prone to

KEY TERMS

critical thinking (CT)	frame of reference	multiple perspectives
bias	stakeholder	purpose
prejudice	objectivity	subjectivity
validity		

4. Questions and activities

1. A large city has an area that was built up over and around the mouth of a river. There is now a movement to restore this area to a more natural form, or wetland, which will be home to a wide variety of birds, fish, mammals, and insects. However, a countermovement wants to develop the area as a tourist attraction with a Ferris wheel.

   a. Identify a number of perspectives with which you can view the issue: describe the wetland from each of these perspectives.

   b. Brainstorm 20 other uses for this area, besides tourist attraction and wetland. How would each of the perspectives view these uses?

   c. Define at least three criteria for choosing a solution.

   d. How could accepting both perspectives as valid motivate creative design solutions?

2. Consider a recent election in your home town or country. Looking at an analysis in the news and, choosing one of the candidates who did not win, ask yourself:

   a. What strategies did the candidate use that you agree with?

   b. What strategies do you disagree with?

   c. Why do you think the candidate chose these strategies?

   d. What lesson do you take from the analysis that you think you might apply to your own life?

5. References

[1] International Engineering Alliance. The Washington Accord, "Graduate/professional competence profiles." Retrieved August 27, 2013, from www.washingtonaccord.org/GradProfiles.cfm.

[2] Bassham, G., Irwin, W., Nardone, H., and Wallace, J.M. *Critical Thinking: A Student's Introduction.* 2nd ed. Boston: McGraw-Hill, 2005.

[3] Chaffee, J, McMahon, C., and Stout, B. *Critical Thinking, Thoughtful Writing: A Rhetoric with Readings,* 5th ed. Boston: Wadsworth, 2012.

[4] Kelly, S. *Thinking Well: An Introduction to Critical Thinking.* Boston: McGraw-Hill, 2000.

[5] Moore, B.N., and Parker, R. *Critical Thinking,* 7th ed. Boston: McGraw-Hill, 2003.

[6] Van Gyn, G., Ford, C., et al. *Teaching for Critical Thinking.* London, Canada: Society for Teaching and Learning in Higher Education, 2006.

[7] Westenholz, A. "Paradoxical thinking and change in the frames of reference." *Organizational Studies* 14(1, 1993):37–58.

2

Critical Thinking in Design Documents

#skill/tool module: #CTindesigndocuments

Learning outcomes

By the end of this module, you should demonstrate the ability to:

- Analyze a problem from multiple perspectives
- Identify and represent multiple perspectives in an engineering design problem
- Generate a number of equally valid solutions to a design problem and assess them according to criteria you define in relation to design processes
- Commit to a solution while identifying its benefits and challenges
- Evaluate your own process and predict strategies for improvement

Recommended reading

Before this module:

- **Implementing a Project > Critical Thinking > 1. Basic Concepts**

After this module:

- **Implementing a Project > Critical Thinking > 3. Making and Supporting Statements Effectively**

1. Introduction

Critical thinking and the engineering design process have notable similarities. The critical thinker considers a problem from multiple perspectives (see Figure 1). Likewise, when approaching a design problem, the engineer considers it from the multiple perspectives of stakeholders, the environment, human factors, and economics. Critical thinkers generate alternative solutions and criteria to use when choosing between ideas. Design engineers use creativity and selection methods to achieve the same results in design problems. Critical thinking cannot happen without reflection, looking back to determine the quality of the decision-making process, and design engineers also improve by reflecting, using reality checks and other methods to evaluate and

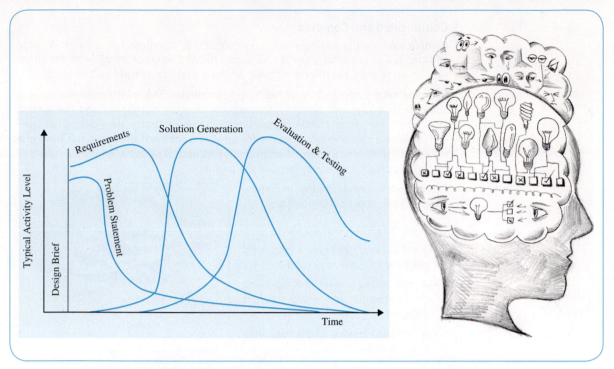

FIGURE 1 In the engineering design process, four aspects of critical thinking are used: multiple perspectives going into the problem, multiple solutions going out, methodical selection of a solution, and finally, reflecting on the overall process.

optimize their processes. Critical thinking is a process, as well, that examines the validity of arguments that are used to prove claims that are being made.

Understanding these concepts is important because engineers have to deal with problems that are both *complicated* and *complex*. When a complicated system has to operate in a complex environment, it is difficult to entirely predict what will happen. Engineering design tools, combined with critical thinking, equip engineers to resolve these kinds of problems more successfully than simplistic or solution driven processes would.

Engineers use documentation to logically, objectively, and persuasively explain their design choices. The ability to look at a problem from multiple perspectives is explored in *project requirements*. The ability to generate a number of solutions and criteria for choosing among them is part of the *conceptual design specification*. The ability to logically justify a design decision, showing its benefits and challenges, is an aspect of the *final design specification*. (See the following sections for explanations of these concepts.) Analytical reflection is shown throughout the process, from reality checks to *lessons learned documents* that accompany the end of a project and adjournment of a team.

Complicated and Complex

Complicated means having many interconnected, interdependent parts. A large commercial aircraft, for example, is a complicated engineered system. All of the parts have to work together effectively for the system to operate as designed.

Complex means non-routine or not fully determinate. The weather, for example, is complex. In a complex system interdependencies are non-linear and the system as a whole may have chaotic properties or behavior that is very difficult to predict. Solving a complex problem means solving a problem that does not have one accepted solution procedure; it often means having to select and apply a variety of methods including creativity to solve the problem.

Examples of a complicated system operating in a complex environment are:

- A jumbo jet landing in a storm
- A vehicle that has to work effectively with any driver on any road
- A cell phone model that is being sold to people from many different cultures
- A medical device implanted in a human body

2. Project Requirements Show a Clear and Detailed Understanding of the Problem Itself

Project requirements present the problem itself from multiple perspectives, not only that of the initiator of the design problem, whether that is a client, in-house team, or the inventor-engineer herself or himself (#requirements). The engineer has to analyze the problem and do research to evolve a more accurate and complete understanding of all of its aspects. The engineer uses critical thinking to ask, and answer, a series of questions, beginning with, "If this is what I know about the situation, what else do I need to know?" Engineering design has a structured approach to answering this question. Engineers ask a series of related questions:

1. Has the client expressed the wants and needs of the project completely and with technical accuracy? Clients often are so familiar with the situation that they do not fully express the problem details. Sometimes they are unfamiliar with the possible solutions and may have already decided, based on inadequate knowledge, what the solution should be. It is up to the design engineers to determine how well the problem is posed.

2. What "gap" is the client trying to fill? What exactly is missing in the world or, if there are existing solutions to the client's problem, what is wrong with them? What is the client trying to achieve that existing solutions do not do?

3. The client has one perspective on the problem; what are some other perspectives? People or organizations that will be affected by a design, either positively or negatively, are considered **stakeholders**. Understanding stakeholders and their **interests** helps the designer to enhance benefits of a design and prevent or reduce possible problems.

4. What has to happen, in engineering terms, for the client to get what they want? This question has many related questions that should be answered. What exactly is the engineering in this design? What will it do? What are the qualities it will have to have in order to be working effectively? How will we test it to know that it works? Might the design have problematic side effects?

5. What kinds of environments will the design operate in and how will those environments affect it?

6. What rules and regulations apply in this kind of situation?

7. Finally, how can I ensure that this design works well for people, society, and the environment, increasing benefits and limiting damage?

These questions have to be answered before the design team can even start considering solutions. These answers enhance the understanding of the problem and add value to the overall process.

3. Multiple Solutions and Criteria for Evaluating Them in the Conceptual Design Specification

Generating a number of valid solutions begins with creative thinking (#ideageneration). Creative thinking is sometimes considered an aspect of critical thinking and other times it is considered a natural partner. In engineering design, it is one component of the overall process of coming up with the best possible solution by first utilizing idea generation tools to create a number of possibilities from which to choose.

Once the ideas have been generated, they have to be organized into coherent design solutions. Decision-making tools are used at this point. What makes this a thoughtful process of value to both the client and the design team is that it creates alternatives of equal quality. If all of your designs are, in fact, equally valid, then the decision about which to implement is not going to depend on an absolute value—one being "best" in all ways—but rather on agreement between the design team and client about which trade-off between objectives works best for the client or situation at this particular time. The success of your decision-making process will depend on how well you defined functions and objectives (designs that do not meet constraints are automatically eliminated) and how well you measured the way each design solution meets those criteria.

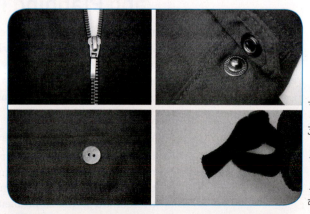

Photos courtesy of the authors

FIGURE 2 A critical thinker will generate multiple solutions to solve a particular problem and then, in addition, will formulate specific criteria in order to choose one solution to recommend.

Critical Thinking in Design Documents

To ensure that your solutions give the client good information, ask yourself:

1. Do these solutions give value to the client? What *unique* information does each provide?

2. What are the *unique* trade-offs of each solution? What are its benefits? What are its challenges?

4. Justifying a Solution: The Final Design Specification (FDS)

When you justify a solution, you show that it best balances the needs of the client, society, and environment (#designevaluation). You have reached this solution through thorough understanding of the problem, the development of appropriate criteria for judgment, examination of a number of alternatives in terms of these criteria, and consultation with the client as well as your supervisors. Your discussion of that process must be logical, objective, and persuasive.

Beyond justifying the solution itself, you owe it to the client to detail as objectively as possible the requirements for implementation as well as consequences that implementation will have, economically, socially, politically, and environmentally. You should address economic, environmental, and social consequences of the design.

- Economics: What are the costs, who will pay them, and when?
- Environment: Over the lifetime of the solution, when and how does it damage the environment and when and how does it benefit the environment? Can the negative effects be lessened and the positive effects increased? How?
- Social impacts: How does this solution change its stakeholders' lives or business practices? What kind of change would occur if this solution was adapted to a different situation?

5. Reflective Thinking: Assessing Yourself and Planning Improvement

Critical thinkers question themselves as much as they question others. They ask themselves questions about what they have done and create personal documents, which they might call something like ***lessons learned***. Lessons learned documents are used in business at the close of team projects, to evaluate a project process and plan for the future.

A *lessons learned* document asks the following kinds of questions:

1. What was the situation?

2. What were we trying to achieve?

3. How well did we do? With this question, you have to determine the measure (metrics) you are using. If your goal was to do as well as possible on an assignment, then a reasonable measure might be the grade you received. But if your goal is to overcome a particular problem from a past assignment, then the measure might be the comments of the marker.

4. What did we do that was effective in achieving our goal?

5. What problems did we have? These may have to do with not achieving the goal or with side-effects. For example, you might have achieved your goal of getting a high grade in an assignment in one course, but at the expense of studying for a mid-term in another course. When describing problems, it is important to use objective, value-free language so that you can identify the problem in a way that is both accurate and allows you to develop a strategy.

6. What could we have done better? This question does not simply address problems, but rather it also addresses aspects of the situation that worked pretty well, with the aim of constant improvement. This kind of question is important for program-mers or manufacturers responsible for developing new versions or models. The old version may work well. How could it work better?

6. Leaving This Process Module

After working through this module you should be able to

- Develop a thoughtful project requirements
- Explain your use of idea-generation and decision-making tools in terms of critical thinking
- Reflect on your process and strategize improvements

Reflection can lead to action and improve-ments in the team and design process.

KEY TERMS

complicated	complex	bias
conceptual design specification	critical thinking	idea generation
stakeholders	decision-making tools	interest
final design specification	project requirements	lessons learned

7. Questions and activities

1. One example of a complicated system in a complex environment is an aircraft en-countering unpredictable weather. In designing a solution to problems that emerge from such a situation:

   a. What stakeholders would you consider?

   b. What would their interests be?

2. Choose a typical everyday activity that currently has some technologies associated with it (drying hair, for example) and brainstorm 40 ideas for a new, different tech-nological approach.

3. Choose an item that you have recently purchased or would like to purchase, one that you spent some time thinking about and some research on, such as a computer or an automobile. List the criteria you think are important for making a choice. Prioritize the list. Have a friend do the same and compare results. Were the criteria the same? The priorities? Can you explain logically to your friend why you made the choices you did?

8. References

[1] Sagan, C. *The Demon-Haunted World: Science as a Candle in the Dark*. New York: Ballantine Books, 1996, p. 8.

[2] Van Gyn, G., Ford, C., et al. *Teaching for Critical Thinking*. London, Canada: Society for Teaching and Learning in Higher Education, 2006.

Critical Thinking in Design Documents

3

Making and Supporting Statements Effectively

#skill/tool module: #makingstatementseffectively

Learning outcomes

By the end of this module, you should demonstrate the ability to:
- Identify the difference between statements of opinion, fact, and claim
- Make statements that are supported with explanations and evidence

Recommended reading

Before this module:
- **Implementing a Project > Critical Thinking > 2. Critical Thinking in Design Documents**

After this module:
- **Implementing a Project > Critical Thinking > 4. Skeptical Thinking**

1. Introduction

There are situations in which engineers must communicate persuasively, with logical argument to support a point. Logical argument is particularly important in showing that a solution, design, or recommendation will not only meet its testable requirements (functions, objectives, and constraints), but *also* yield the *results* desired by the client or situation. These kinds of discussions occur in the motivation sections of proposals, or in discussing a recommended design either in the conceptual design phase or in the final design specification. These are more than explanations; they are explanations that *prove* something.

Engineers are constrained by the communication practices of science—in persuading their audiences, they must be logical, objective, and unemotional. They must support statements or claims with both evidence and logic. Together, evidence and logic help to avoid arguments based merely on opinion. While debate can be quite spirited, discussions based on evidence, science, math, and logic move away from the personal opinions, toward analysis and proofs.

Being persuasive begins with making a credible statement—one that has evidence and explanation. It requires an ability to utilize logic and avoid fallacies, as well as the ability to use objective, non-emotive language.

2. Three Categories of Statement

Three categories of statement relevant to engineering writing are:

1. *Opinion*
2. *Fact*
3. *Claim*

Opinions are of two sorts: 1) a statement of evaluation made by an expert and 2) a statement of a personal feeling, value, or judgment made by anybody. Even though opinions are unique to individuals—or evaluative judgments specific to design teams—their subjectivity does not release them from the requirement of explanation and evidence. If you state your opinion or your design team makes a judgment, you must be prepared to explain it and support it with some form of evidence and logic. Even expert opinions are bound by this requirement in order to be valid and persuasive.

Facts are statements of truth. Like opinions, they have two basic varieties. Scientific facts, laws, axioms, and principles are generally accepted and are supported by formulas or other well-known identifiers. For example, if you are explaining the force required to accelerate an object, you may refer to Newton's Second Law of Motion, or F=ma. It is expected that, at this point, you are well enough acquainted with Newtonian physics that you do not need to look up this formula.

The second kind of fact is one that is supported by information sources. That is, it is a "fact," something that happened in history (such as the names of the winners of the 1923 Nobel Prize in Medicine) but is not part of your general knowledge. You find this information from websites, books, journals, magazines, newspapers, unpublished theses, or interviews with people. Though you may make use of this kind of information, it does not actually belong to you and you must clearly identify its source, using a standard *referencing* system.

Claims are any statements that are not proven or historical facts and are not, like opinions, based on personal belief or subjective judgment. Claims use science, mathematics, and other evidence to come to a conclusion or supposition. The better they are explained and supported, the more useful they become.

Most of the statements that people make are "claims." Engineers generally use facts to support their claims and even if called upon to give an opinion, they usually treat those opinions like claims, explaining them and supporting them with evidence. Therefore, learning how to make a credible claim—or a "complete statement"—is central to becoming an effective communicator and engineer.

3. Making a Complete Claim

A complete claim is a statement that gives a reader or listener enough information both to understand and to accept the ideas being presented. Making a complete statement has three steps:

1. Stating the idea
2. Explaining it
3. Supporting it with evidence

These fundamental three steps form the foundation of effective communication. Once you have established them, you can build complex understanding. However, if these three steps are not taken—if the idea is not stated clearly, explained, and supported—then no matter how interesting or worthwhile the rest of your message is, your reader or listener will find it difficult to follow and appreciate.

3.1. State Your Idea First

State your idea right away, first thing, even if it is a technically complex idea. Writing or speaking this way may be different than what you have learned or practiced; it is absolutely concise and gets right to the point. Many writers or speakers don't want to make their point right away. They believe that the audience (the listener or reader) won't understand it without background.

People will listen or read for a short time to find the main point, and then listen or read to build on that point. However, if the main idea doesn't appear quickly you lose your audience. If you launch into a long, detailed background, your audience is likely asking themselves, either out loud or in their head, "Why are you telling me all this?" Your main idea will get lost at the end.

3.2. Explain Your Idea

Once you have stated your idea, you have to explain it. You must define terms and fill out the details of the ideas. Doing this has a double benefit: it reduces misunderstanding

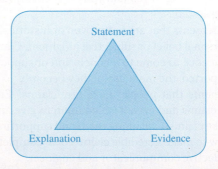

FIGURE 1 **A complete claim has three necessary parts: the claim itself, an explanation of the claim so that someone who is not familiar with the problem can understand its dimensions, and evidence to show that the claim and its explanation have independent, objective support.**

and allows you to identify—and fill in—any gaps in your own understanding. If the first statement of the idea takes one sentence, then you will likely take two or three to explain it in more detail. If the first statement of the idea takes several sentences, then you may need a few paragraphs to explain it.

Graphs, diagrams, charts, and other graphical objects can be used to help explain your idea. Often these will be clearer than textual explanations alone. However, graphics should always be accompanied by text explaining to the audience what they should see in the picture.

3.3. Support your idea with evidence

There are three kinds of objective data that you will likely use, as an engineering student or engineering professional:

1. Scientific method or principle
2. Your own data
3. Other people's data or information gathered from research

The first category of information needs little explanation. As a student in engineering, you are gaining an enormous amount of technical, scientific, and mathematical knowledge. This knowledge is characterized by well-known laws, axioms, and formulas. As you progress, you will be expected to know these and apply them naturally.

When you do your own research and develop your own data, you will be able to use this data to back up claims. This type of evidence is often the foundation for lab reports in engineering. And then there will be times when you use data or information other people or organizations have generated to support your ideas. You must always properly acknowledge the source of ideas, images, or words expressing ideas by using a standard *referencing* system. Also, your idea will be considered better supported if you are able to cite more evidence, and particularly evidence from multiple credible sources.

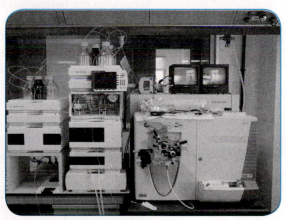

© MARKA/Alamy

The evidence for engineering claims comes from data, often produced through experiment or measurement.

Making a complete claim, from the statement of the idea through to properly supporting the idea with evidence, is fundamental to engineering communication. This approach will be used in every lab report, design project, and research paper you write. It is also used in professional presentations and other modes of communication. It forms the foundation of logical, credible communication.

4. Logic

Logic is a structured means for developing new, valid ideas from preexisting facts. It is a principled approach to connecting the facts so that they reasonably support the new idea being proposed. Even the term "reasonable" can be defined and explained in this context. It means that the structure of the facts, the argument, will stand up to intellectual analysis. It will prove to be both true and valid. An idea is "true" if evidence exists that corroborates the fact being stated. For example, we may say a scientific experiment has taken place to prove a hypothesis, if we have read a journal article about the experiment. What is true, in this case, is the fact that a journal article has been written. It is likely and is almost always the case that the experiment described actually also took place. However, we may still have some doubt that the experiment proved the hypothesis. That remains a claim with the data presented in the journal article as evidence.

Validity is quite different. An idea is "valid" if the idea is supported by evidence or reasoning. It may be true that an experiment took place, but if the data produced points to an entirely different conclusion than the one the experimenters claimed, then they did not prove their hypothesis in a valid manner. A good logical support structure, or argument, must therefore be both true *and* valid.

In many ways, a logical argument is more a matter of what you *don't* do, than what you do. The study of invalid arguments goes back thousands of years and many of these have been categorized; they are known as **fallacies**. Fallacies basically *distract* from the facts, rather than support or connect them. For example, one form of fallacy is to attack the person rather than the person's ideas. Trying to discredit an idea because of the person proposing it creates a distraction from the main point—the idea itself, its merits and its faults. An idea may be good, no matter what problems you have with the person presenting it, or it may be bad, even if you like the person a great deal. Moreover, it makes the entire discussion of an idea personal rather than impersonal and that is the opposite of what you want to do as a critical thinker and engineering communicator.

Other frequent fallacies to beware of and avoid:

- Common sense. Common sense is just not all that "common" and it is not considered an acceptable way to validate a claim. It presupposes that people share common ideas, experiences, understandings, and values. Working in a global context, it is illogical to make the assumption that anything is "common" sense outside of your own, immediate environment.

- Taking lack of evidence for evidence. Not having evidence to prove one side of a question does not automatically prove the other side. For example, saying that "because there have been no accidents, the vehicle is safe" is a fallacious argument.

- Mistaking sequence for cause. Just because one event precedes another does not mean that the first *caused* the second. There could be a connection but not necessarily. Assuming cause due to sequence alone will likely yield a fallacious argument.

- Making an **emotive** or figurative argument. Appeals to emotion or heavily using metaphors and similes equally distract from the purpose of a dispassionate, reasonable, and scientific argument. These can be considered forms of fallacy and so you should consistently attempt to use value-free language.

5. Value-Free Language

In engineering communication we strive to use **value-free language**. This is language that is not intended to invoke emotional responses such as happiness, sadness, or anger, but instead it meant to convey ideas in a clear, concise, and objective manner. The emotional responses good engineering communication should invoke are feelings of trust: confidence, credibility, and professional competency.

Language has two kinds of meaning: explicit and implied. The **explicit meaning** of a word is its dictionary definition, but the **implied meaning** is one that exists for certain people in certain situations. Take the word "blind," for instance. It literally refers to nearly complete vision impairment, but it has also been used to refer to ignorance or neglect, as in "the politician was blind to the needs of his constituents." In such uses, it became derogatory and that negative feeling then transferred back to people with impaired vision. As a result the usage of the word changed, and it is no longer used to describe people with low visual acuity. Becoming aware of the implied value judgments in language allows you to understand information in all its dimensions.

Implied meanings can actually directly contradict explicit meanings. A good example can be found when a tone of voice is used to change the intention of the words, as when someone says, "That's a great idea!" in a way that indicates the person really feels the idea is worthless. This extreme example shows how important it is to ask, "What are the implied meanings?" If they contradict the explicit meaning, and you are not aware of it, you may end up using information that says something other than, or even the exact opposite of, what you intend.

Words that carry a lot of implied meanings based on particular social, political, or cultural values can be referred to as **value-laden language**. This is particularly important if you are dealing with people from a different cultural or religious background than your own or if the design problem is in a social situation with which you have no familiarity—a shelter for the homeless, for example, if you have no experience of the homeless. An acceptable phrase or mannerism in one situation can be derogatory or insulting in another.

Writing formally, not conversationally, using scientific language and avoiding any figurative language helps to prevent value-laden terms from damaging your arguments. Figurative language includes:

- Similes: a form of comparison of unrelated phenomena to create an image. A famous example is from Robert Burns, "My love is like a red, red rose."
- Metaphors: These are also comparisons but they do not use the word "like." They create a direct picture; "The tree of life" is an example of a metaphor.
- Adjectives or adverbs that are not directly measurable: "The recommended design is *superb*."

Figurative language is much like "flowery language" (another simile). If you are already aware of the need to avoid figurative language, then you are already on the way to clear, objective, logical communication.

6. Leaving This Process Module

After working through this module you should be able to

- Develop a logical argument in support of a design decision

KEY TERMS

statement	opinion	fact
claim	referencing	explanation
evidence	logic	fallacies
value-free language	value-laden language	explicit meaning
implied meaning	emotive language	

7. Questions and activities

1. Recently, a man was arrested for a crime in which a truck was stolen and the owner of the truck was killed. The suspect's lawyer argued that he has to be innocent because he was a wealthy man and had no need to steal a truck when he could have easily bought one. Do you think this is a valid argument? If not, what fallacy would it represent?

2. Write a statement you believe to be true. Demonstrate that you understand and can apply the concepts in this module by explaining the statement and backing it up using evidence.

3. Choose a design that you have created and write a paragraph arguing that the design will have the results intended, as opposed to satisfying the functions, objectives, and constraints. You may have to do some research in order to make your argument.

4. If you do not have a design for question 3, try this one: you have designed a better desktop for left-handed people: your design increases the area of a typical lecture hall writing surface for a left-handed user. You have proven that it can support the load specified. Create an argument that shows it will also decrease injuries associated with regular desks and left-handedness.

8. References

[1] Ennis, R.H., and Weir, E. *The Ennis Weir Critical Thinking Essay Test*. Pacific Grove, CA: Midwest Publications, 1985.

[2] Ennis, R.H., and Millman, J. *The Cornell Critical Thinking Test Level Z*, 5th ed. (2005) Retrieved from www.CriticalThinking.com

[3] Irish, R., and Weiss, P.E. *Engineering Communication: From Principles to Practice*. 2nd ed. Toronto: Oxford University Press, 2013.

Skeptical Thinking

4

#skill/tool module: #skepticalthinking

Learning outcomes

By the end of this module, you should demonstrate the ability to:
- Define skepticism or skeptical thinking
- Apply two skeptical thinking techniques to your design decision making

Recommended reading

Before this module:
- **Implementing a Project > Critical Thinking > 3. Making and Supporting Statements Effectively**

After this module:
- **Implementing a Project > Communication > 1. Engineering Communication**

1. Introduction

Briefly, **skeptical thinking** is a process of questioning information. It is not a matter of automatically accepting or rejecting information. Rather, skeptical thinking, or skepticism, means challenging and testing any and all ideas. As a scientific thinker, you are expected to be skeptical. Dr. Carl Sagan, a respected astrophysicist and science writer, considered skepticism to be "central to the scientific method"[1]. In science, he wrote, ideas are accepted with the understanding of their limits, margins for error, and the possibility that they may one day be disproved.

You have probably heard the expressions "Don't believe everything you hear," or "Don't believe everything you read." At a time when we have an unprecedented ability to access more information than ever before—information provided by an enormous variety of sources—this is

Skepticism often addresses non-scientific claims, not with denial, but with questions about evidence that might support the truth of such claims.

probably better advice than ever before. Skeptical thinking is not only a valuable tool when doing research; it applies in other areas as well. It is often important in generating and evaluating ideas, where it prevents the premature rejection of a good idea or the acceptance of a bad one.

Skeptical thinking is a term that goes back to the mathematician and philosopher René Descartes. He formulated the idea of *systematic doubt*. Systematic doubt means that you do not accept an idea as valid without testing and retesting it—a process familiar to anyone in science. A skeptical thinker accepts that an argument or hypothesis is true when there is sufficient evidence to support it. However, if new and contradictory evidence is found, true skeptics are prepared to change their minds. Astronomer Carl Sagan stated that he would welcome evidence of life on other planets, but until he gets that evidence, he cannot state categorically that he believes life exists elsewhere in the universe.

2. What Skepticism Is Not

Skepticism is not about negativity or being disrespectful. It is not about using questioning to disprove ideas just because you may not like them—or the person who suggests them. It is also not about rejecting authority. In fact, questioning an idea and coming to a deeper understanding of it may help you develop a deeper appreciation of the decisions of people you respect.

Skepticism is not about being cynical. A cynic is someone who has a generally negative view of the world and is unwilling to accept new ideas. In the absence of good evidence, however, a skeptic merely withholds judgment. Carl Sagan reported that when pressured to give his "gut feeling" about the possibility of life on other planets, he responded, "But I try not to think with my gut. . . . It's really okay to withhold judgment until the evidence is in" [2].

Finally, skepticism is not about just being argumentative. It is about having—or finding—evidence to support an idea. So, even if an idea sounds reasonable—or like common sense—a skeptic will look for evidence to support it before accepting it. For example, if someone says that you have to sleep 10 hours a night, every night, in order to stay healthy, you might mistake the plausibility of the argument for evidence that it is true. But an argument is merely a hypothesis, a statement of belief that must be supported with testing. If you really want to know how many hours of sleep a person needs, you have to test the hypothesis scientifically, with a statistically worthwhile number of people in a controlled environment. This should be completely obvious to anyone with even a hint of scientific training, but it is amazing that well-educated people often fail to make the critical distinction between a plausible hypothesis and the actual evidence needed to support it, especially when they *feel strongly* about something. Proof requires evidence, preferably from a number of sources or through strong mathematical or experimental work.

Triangulation: finding two or more sources that support the original source makes information more credible.

3. Two Techniques for Skeptical Thinking

Skeptical thinking is important because it represents the questioning skill associated with science and scientists. More important for the team design process, skeptical thinking techniques will save you hours of arguing on the basis of people's opinions. As a process, questioning has a purpose: to find evidence to prove or disprove an idea. Here are two techniques you can use to organize your questioning process.

1. *Triangulate.* Find two or more sources that support the original source. Findings that are confirmed by three or more sources are more credible than findings in only one source.

2. Look for a counter-example. Finding a contradiction to the information or idea you want to use might seem counter-productive, but in fact it isn't. A counter-example can show flaws or weaknesses in ideas and lead you to much more productive research. Skeptics value this sort of exploration since they are interested in testing all ideas as rigorously as possible.

4. Three Obstacles to Skeptical Thinking

Skeptics complain that people who passionately believe in ideas (such as life on the moon) will use any shred of evidence to support their claims. A typical example is the theory that governments are covering up evidence of alien life forms from other planets visiting earth. Lack of evidence does not prove such a conspiracy is going on; lack of evidence neither proves nor disproves such visits. Using lack of evidence as "proof" is one form of *fallacy*—invalid argument. Here are three other fallacies skeptics keep in mind when examining or making arguments:

1. Confusing *correlation* with "cause and effect." Correlation refers to two variables that appear to change together. For example, there appears to be a correlation between power outages and pregnancies. The number of pregnancies in a population tends to increase during long power failure events. However, correlation is different than cause and effect. Power failures do not cause pregnancies.

2. Confusing examples for evidence. A single example does not represent a proof. Moreover, even a number of examples may not really prove something to be true. A scientifically minded person should not be satisfied with any kinds of proofs that are not objective, repeatable, and statistically significant. They will also look for other evidence that substantiates the claim.

3. Creating false dichotomies. A *dichotomy* is a two-sided idea. You create a dichotomy when you organize a concept into two sides—right or wrong, for example. Dichotomies are often expressed with the words "either-or." Dividing a situation into two choices simplifies matters considerably, but may oversimplify to the point that truth is compromised. Open-ended problems generally do not lend themselves to this way of thinking. There isn't one right and one wrong answer and thinking in those terms will lead you away from—not toward—sound thinking.

Example

Two hundred years ago people thought hot, humid air caused diseases such as malaria. This was because the malaria rate was higher in the summer than in the winter. And it was also noted that people living near swamps where the air was warm and moist had higher rates of malaria. This argument would appear to be very logical. Using the concepts in this section, explain why it is flawed and how a skeptical thinker might approach questioning this logic.

The mosquito that carries malaria
CDC—Global Health—Division of Parasitic diseases and Malaria.
Photograph by Dr. Richard Darsie. Public domain. Available at http://phil. http://www.cdc.gov/

Anopheles sundaicus

5. Leaving This Skill Module

After working through this module you should be able to

- Investigate claims for evidence correlated in several sources
- Investigate counter-claims in order to identify weaknesses of ideas
- Identify invalid arguments based on the following three fallacies
  Confusing "correlation" with "cause and effect"
  Confusing "examples" for "evidence"
  Creating false dichotomies

KEY TERMS

skepticism   skeptical thinking   systematic doubt
correlation   fallacy   triangulate
dichotomy

6. Questions and activities

1. Find a science article in your local newspaper. While reporters try to keep science reporting accurate, they also need to make it newsworthy. Can you spot any inaccuracies? Try composing a letter to the editor discussing the article from an objective, scientific point of view.

2. Check out the Skeptics Society at www.skeptic.com. They are an organization dedicated to promoting skeptical and critical thinking, especially in relation to claims that deal with paranormal or extraordinary experiences.

7. References

[1] Sagan, C. *The Demon-haunted World: Science as a Candle in the Dark*. New York: Ballantine Books, 1996.

[2] Sagan, C. The Dragon in My Garage. In *The Demon-haunted World*, chapter 11.

[3] Bassham, G., Irwin, W., Nardone, H., and Wallace, J.M. *Critical Thinking: A Student's Introduction*. 2nd ed. Boston: McGraw-Hill, 2005.

[4] Chaffee, J., McMahon, C., and Stout, B. *Critical Thinking, Thoughtful Writing: A Rhetoric with Readings*, 5th ed. Boston: Wadsworth, 2012.

[5] Kelly, S. *Thinking Well: An Introduction to Critical Thinking*. Boston: McGraw-Hill, 2000.

[6] Moore, B.N., and Parker, R. *Critical Thinking*, 7th ed. Boston: McGraw-Hill, 2003.

[7] Skeptics Society. (1992–2013) *Skeptic*. Retrieved from www.skeptic.com/.

[8] Van Gyn, G., Ford, C. et al. *Teaching for Critical Thinking*. London, Canada: Society for Teaching and Learning in Higher Education, 2006.

1

Engineering Communication

#skill/tool module: #engineeringcommunication

Learning outcomes

By the end of this module, you should demonstrate the ability to:

- Use language and conventions appropriate to professional engineering
- Avoid figurative language and other everyday communication habits that are not appropriate in professional engineering situations

Recommended reading

Before this module:

- **Implementing a Project > Critical Thinking > 3. Making and Supporting Statements Effectively**

After this module:

- **Implementing a Project > Communication > 2. Organizing Communication**

1. Introduction

Engineering communication, like any professional language, is a subset of communication in general (see Figure 1). It uses many, but not all, of the words in everyday language. It uses many, but not all, familiar document forms. It also uses words differently than they are used in everyday language and adds words that are specific to engineering.

Engineering communication is characterized by brevity, clarity, and the use of objective scientific and mathematical data to support claims. Engineering communication is as unambiguous as possible and does not attempt to persuade the reader or listener by appealing to the emotions through dramatic or figurative language, such as the language found in the metaphors and similes of poetry. Such "flowery language" is highly discouraged in engineering.

The emphasis on clarity and the attempt to eliminate ambiguity reflect the enormous responsibility that engineers have and the trust that people put in them. Engineers' communication will affect human lives, property, economics, and the environment. Misunderstandings can be costly and even tragic.

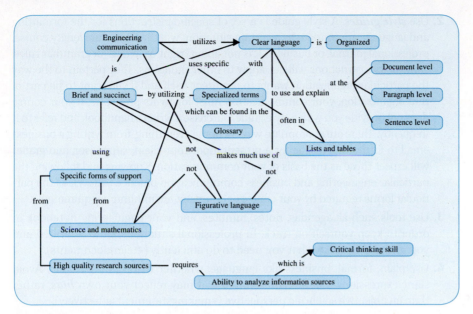

FIGURE 1 **Engineering communication is different than other forms of communication.** It draws on science and mathematics to support clear, concise, and objective expression of information.

2. Developing a Professional Voice and Creating Trust

Developing a *professional voice* means learning how to communicate appropriately in the working world—with business people, engineers, and other professionals. A "voice" is both the way you give presentations and also the tone and style of your writing. Taking part in a design project during your university education is a way of practicing this. It is a preparation, or a rehearsal, for your professional work as an engineer.

For engineering design projects, you will have to draw on your businesslike personas. You will have to dress and speak appropriately for different distinct situations: client interactions, supervisor or project manager meetings, instructional situations with instructors, and, finally, meetings with team members. In the future this will include people you manage. The number one rule in all these situations is respect. You show respect by consistently being polite, ensuring that your language and your tone are not insulting.

In order to develop your professional voice:

1. Make yourself aware of the principles and rules around each particular kind of communication task. Make sure you understand the rules and follow them. A business email, for example, is not the same as a personal email and if you do not appreciate the difference and you use language that is too informal, you may offend your client or the person you are writing to.

Engineering Communication

319

2. Use *style guides*. A style guide is a set of specific instructions for the formatting and language appropriate to a certain situation, whether it is a university course, professional journal, or business. It is not about "right or wrong" or grammar rules; some of its instructions will not apply to other situations. They pertain to the way the particular organization wants to represent itself and if you want to be part of that organization, your communication must follow its principles. If your course does not provide you with a style guide, purchase a writer's handbook and refer to it frequently. There are appropriate ways of doing everything, from writing a business email to formatting a slide for a presentation. Your client, supervisor, and grader will expect these as the basis of your communication. Knowing and following the particular engineering and business communication practices, as well as the particular forms required by your course or company, are minimum requirements.

3. Use tools such as agendas, notes, minutes, and your engineering notebook in order to keep yourself on track. In professional situations, time is limited and you must accomplish what you need to do efficiently (#teamdocuments).

4. Use plain, formal, business-like language, whether writing or speaking. Avoid slang expressions that are *value-laden* and may reflect your own *bias*, rather than an objective scientific perspective (#makingstatementseffectively).

5. Always ensure that your writing is grammatically correct, your sentences are logically structured, and your spelling is correct. These details are almost never taught but almost always expected. Poor grammar, sentence structure, and spelling have a surprisingly negative effect on your credibility. Errors in grammar and spelling undermine the trust of your audience. They give the reader the impression that you are careless, you do not feel that the task at hand was worth the time to do properly or that you are unintelligent because you do not know the basics of the language. Since none of these things are likely to be true, you do not want to give the reader this impression.

2.1. Appropriate Conventions

Comprehensive lists of appropriate conventions can be found in technical writing handbooks, but even the most comprehensive will still be subject to the requirements of a particular organization, laboratory, journal, or course. Some key conventions that are likely to turn up include:

- Engineering writing is formal. Use whole words only, no contractions. That is, always write out "it is" fully; never contract it as "it's." The same goes for "cannot" or "will not."
- Representing numbers verbally or numerically. Numbers are best expressed numerically (i.e., 132 rather than one hundred and thirty two). However, when there are two numbers in a row, one is usually numeric and one is written as words. Which is numeric is going to depend on the situation. Some examples include:
  o Two 5-gram containers
  o The shipment included 1,438 five-meter bars

  Exceptions to the rule occur when both numbers are large. For example, "We ordered 390 550 cm long pipe sections last month." To avoid confusion this might be rewritten as "We ordered 390 pipe sections (550 cm length) last month" [1].
- When a number starts a sentence, it is always written as a word, such as "Five grams were used."

2.2. Figurative or "Advertising Language" and How to Avoid It

If plain language is language in which the **explicit meaning** has far more effect than the **implied meaning**, then **figurative language** is the opposite. What it is *implying* is more important than what it is saying directly. Poetry is a good example of figurative language. When the poet Robert Burns writes, "My love is like a red, red rose," he does not mean that he is in love with a flower. He intends readers to summon up all of their personal, positive associations with roses—their beauty, fragrance, elegance, delicate textures—and imagine a person who embodies these characteristics. In this example, Burns is using a **simile**. It is a figure of speech that compares one thing with another and signals the comparison with a word such as "like." In many cases the two things are completely unalike—a human and a flower (#makingstatementseffectively).

Another figure of speech is a **metaphor.** This compares two things more directly, by describing one *as* the other. A common example of a metaphor is "user friendly" or "environmentally friendly" (see Figure 2). In fact, since friendliness is a quality only found in living creatures—a friendly person, a friendly dog, a friendly dolphin—this figure of speech cannot possibly be literally true. "User friendly" implies that something is easy to use, perhaps intuitive or familiar in some way. "Environmentally friendly" implies that something will not harm the environment or perhaps will help it. Because these terms are vague and imprecise, they are of no use to the design engineer. Thus similes and metaphors are discouraged in engineering writing.

What is more, the purpose of similes, metaphors, or other figures of speech is to create a feeling. "My love is like a red, red rose" or "environmentally friendly" are ideas that make the audience *feel* good. That is why figurative language is so useful in advertising. It appeals to the emotions and facilitates emotional decision making. Making decisions based on emotion, rather than intellect and reason, is the opposite of good critical thinking and sound practice in engineering. So language that appeals to the emotion is characterized as "advertising language" and, like figurative language, is emphatically discouraged. Instead, the engineer *defines* the characteristics other people use metaphors and similes to refer to; for example, instead of "user-friendly" interface, the engineer will define ergonomic characteristics that make the technology more accessible or flexible in measureable ways—font sizes and types that have been tested for clarity on different platforms, or navigation tools such as menus and sidebars.

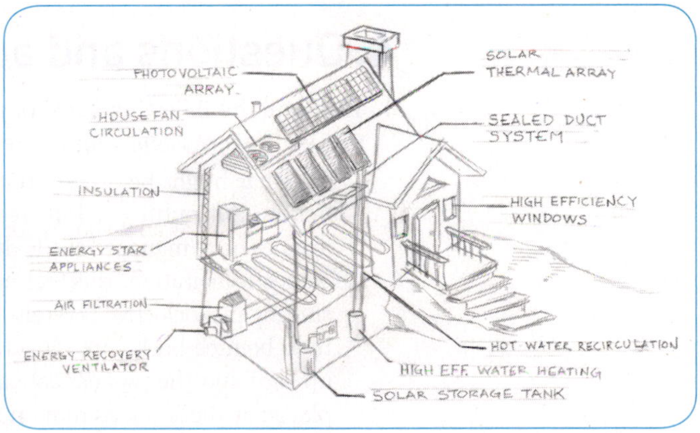

FIGURE 2 "Green" or "environmentally friendly" are metaphors that imply specific specifications to an engineer.

2.3. Bias-Free Language

Avoiding figurative language—metaphors and similes—and being aware of the implicit meanings of words will do a great deal toward making your language objective and bias free. You will also increase the acceptability of your language if you pay attention to, and use, the words that your client or supervisor uses in relation to social or political conditions with which you do not have a personal familiarity. Finally, become aware of your own values and how they affect the way you talk or write about topics such as religion, politics, and economic conditions. Ask yourself if the words you use are clear to others who have other beliefs or political positions or who do not agree with your beliefs.

3. Leaving This Skill Module

After working through this module you should be able to

- Write an brief engineering document using professionally appropriate language and appropriate conventions
- Identify and avoid figurative language and other everyday communication habits that are not appropriate in professional engineering situations

KEY TERMS

style guide	bias	figurative language
explicit meaning	implied meaning	simile
metaphor	value-laden	

4. Questions and activities

1. Translate the following piece of text into appropriate engineering language. The text is 105 words long. Can you revise it so that it is half that length?

    "In what might be a first step toward taking anti-gravity out of pure science fiction and putting it into reality, a Swiss mechanical engineer and his colleagues have managed the levitation of small objects using sound waves. Called "acoustic levitation," this technique makes use of sound waves emitted by shape-shifting piezoelectric crystals. The waves move up until they hit an object and then bounce back. The reflected waves crash into waves that are still moving upward and the two cancel each other out at a point called a "node." Objects placed at these nodes remain suspended, held in place by waves coming from both directions" [2].

2. Create a sample agenda for your next out-of-class team meeting.

3. Take notes at your next team meeting, using the meeting agenda as an outline. Rather than trying to take down what people are saying, organize it in terms of Ideas, actions, and deadlines.

4. Come up with four meaningful objectives you could use for a building design that is supposed to be "environmentally friendly."

5. Take an email, text, or other communication you recently wrote to a friend. Try rewriting it as if you were sending the message to a business colleague.

5. References

[1] Eisenberg, A. *A Beginner's Guide to Technical Communication*. Toronto: McGraw Hill, 1998.

[2] Howgego, J. (2013, July 15) Sound Waves Levitate and Move Objects. *Scientific American*. Retrieved from www.scientificamerican.com/article.cfm?id=sounds-waves-levitate-and-move-objects

Organizing Communication

2

#skill/tool module: #organizingcommunication

Learning outcomes

By the end of this module, you should demonstrate the ability to:

- Organize a document with numbered headings
- Organize a paragraph with a focused topic sentence that establishes relationships between the sentences in the paragraph
- Organize sentences so that ideas can easily be followed, even in longer more complex structures

Recommended reading

Before this module:

- **Implementing a Project > Communication > 1. Engineering Communication**

After this module:

- **Implementing a Project > Communication > 3. Putting Together an Engineering Report**

1. Introduction

In addition to clear, objective language and well explained and supported statements, your writing has to be organized so that your readers can follow your ideas easily. Your documents must make sense as a whole; they must contain all the relevant and expected sections. Within the sections, the paragraphs must be written in a way that allows a reader to find information quickly and easily. Sentences must have a logical structure so that, no matter how long and complex, the reader is always clear on what is happening in the sentence and who, or what, is causing the action to occur.

2. Organizing a Document

In many situations you will be given a template, structure, form, or outline to organize your documents for you. However, there may be times when you have to work without a preset structure, the template provided may be partial or require modification or you may have to develop an outline yourself for your organization. Documents made up of several sections benefit from not only having clear, descriptive section **headings**, but also having **numbered headings**. Headings break up a document and signal readers about the kind of information they are about to read. This makes reading faster and information easier to understand. Numbering sections imposes a hierarchy on the information, identifying main ideas and the supporting ideas that belong to them (see Figure 1) (#writingreports).

Generally, no more than three levels of subsection are recommended. This avoids situations where you get subsections with little information and little connection to the other material. The relationships between paragraphs are the "thinking glue" that allows discreet pieces of data to come together to become coherent, effective ideas. Your ability to identify these relationships and show them clearly is an indication of another dimension of your intelligence, extending beyond your ability to solve a problem. They show that you are not only able to solve the problem, but that you have a good understanding of why the solution works and, by implication, how other, varied problems might be solved. They show that you can put together ideas in new and original ways.

To use numbered headings as an outline, begin by organizing the headings of your document. Make sure that the headings and subheadings contain full ideas so that your outline not only tells you what the sections of your document are, but also what the content of that section is going to be. So, do not simply write "1.0 Introduction." Write something that has real information in it, such as "1.0 Introduction: Increasing Security of Financial Data at Cyberbank Washington."

Many engineering students begin with an outline and find that it helps to keep them organized. Others prefer to start at the beginning and write to the end, in a more intuitive process. The problem with the second, intuitive process, especially in group work, is that it makes planning more difficult and if you do not build in significant amounts of time (and, at some point, an outline to organize what you

1. Topic AA

    1.1. AA – Support point 1

    1.2. AA – Support point 2

        1.2.1. AA – Support point 2-sub-point 1

        1.2.2. AA – Support point 2-sub-point 2

FIGURE 1 **Example of _numbered headings_. Note that all ideas in section 1 must relate to topic AA. If there is a paragraph or sentence that deals with a different topic, it will belong in a different section.**

have written) the coherence of your document and intelligibility of your ideas will be compromised.

3. Basics of Organized Paragraphs

Your goal, when writing, should be to produce documents that can be read once, quickly, and understood immediately. This means that, although the complex technical concepts may need some explanation, the paragraphs and sentences should be straightforward. When you are writing, imagine that your reader is in a hurry, is only looking for specific pieces of information and does not want to read the whole document from beginning to end.

Paragraph size and structure help. It is more difficult to find specific ideas in long paragraphs that have many ideas mixed up together. Shorter paragraphs that deal with one idea at a time and announce the idea at the start of the paragraph are much easier to search through for desired information.

You may have been taught two pieces of contradictory advice. One is that every paragraph must identify and explore a single idea and the other is that essays should have five paragraphs. The five paragraph essay is used to ensure that short high school essays have more than one idea—hopefully, three ideas: one per paragraph, plus an introduction and a conclusion. Applying the five-paragraph essay form to engineering report writing is like trying to pick up water with a fork. It is the wrong tool for the job. Writing a five-paragraph essay is a practice you should now abandon unless *specifically* instructed to use this form by your instructor.

An organized, logical paragraph for a reader in the 21st century should be short, have only one idea, and a full development of that single idea. An essay—or any other document—should not be limited by numbers of paragraphs. It should have as many paragraphs as are required to fully identify every claim that support your main ideas.

You probably know about **topic sentences**, but here is another way of looking at this device (see Figure 2). If you adopt the three-step approach to writing (claim-explanation-evidence), you will always put your main statement in your first sentence. That will clearly identify to the reader what idea you are developing in the paragraph. The reader looking for that idea will be able to find it easily (#makingstatementseffectively).

After identifying your main idea, explain it. That will take at least a couple of sentences. Finally, include the support— the objective evidence that gives credibility to your idea

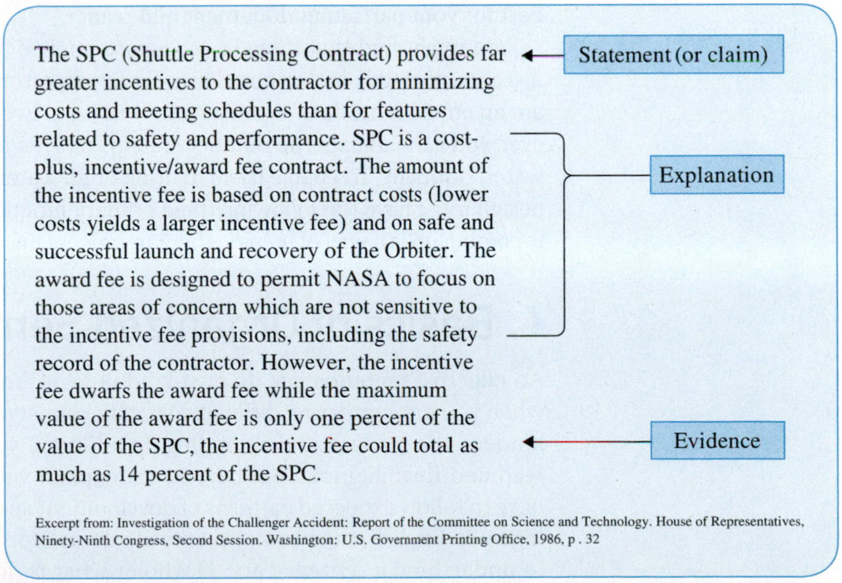

The SPC (Shuttle Processing Contract) provides far greater incentives to the contractor for minimizing costs and meeting schedules than for features related to safety and performance. SPC is a cost-plus, incentive/award fee contract. The amount of the incentive fee is based on contract costs (lower costs yields a larger incentive fee) and on safe and successful launch and recovery of the Orbiter. The award fee is designed to permit NASA to focus on those areas of concern which are not sensitive to the incentive fee provisions, including the safety record of the contractor. However, the incentive fee dwarfs the award fee while the maximum value of the award fee is only one percent of the value of the SPC, the incentive fee could total as much as 14 percent of the SPC.

Statement (or claim)

Explanation

Evidence

Excerpt from: Investigation of the Challenger Accident: Report of the Committee on Science and Technology. House of Representatives, Ninety-Ninth Congress, Second Session. Washington: U.S. Government Printing Office, 1986, p . 32

FIGURE 2 A typical paragraph from an engineering report organized using the claim-explanation-evidence structure.

Organizing Communication

325

and its explanation. That can take one or more sentences. This means that no paragraph can have only a single sentence because it will not have enough information in it to be complete. It also means that if you have a sentence that does not state the main idea, explain it, or support it, but instead talks about something else, then you will know that this sentence is not in the right paragraph. Remove it or move it elsewhere.

3.1. Using Lists

Some people may consider paragraphs to be "better writing" than lists, but in engineering writing, the goal is to provide information as concisely and clearly as possible. Sometimes lists are the best way to do this. The choice of whether to use paragraphs or lists will depend on the kind of information being given. If you are expressing an idea that requires explanation and support, then generally a paragraph is the best choice. The sentences, in that case, will build on one another to develop the particular topic identified in the first line of the paragraph.

But if you are giving a number of different, individual facts—functions, objectives, or constraints, for example—then a list is best. Again, always keep in mind the reader is looking for particular information and introduce your list with a clear sentence or two identifying what the list is about. Make the sentence as specific to the document as possible. "The design has the following functions" is only minimally informative and not at all unique, whereas "To enable users to drink or fill water bottles at the same fountains, the design must have the following three primary functions."

Lists should be revised. You may draft a list as items come to mind, but you should go back to it and give some thought to the best order for those items before you submit the document. Should they go from least to most important, or the reverse? Should they go from most to least expensive? It does not actually matter what kind of organizational idea you use; there is not one that is better than another, but making a decision creates better writing than not making a decision. Determine what would be best for your particular document and reader.

You might find that there is a sequence implied in the list, in which case you might use a numbered list. Numbered lists are familiar from instructions or lab reports. They are an efficient mode of organizing information. Even if the list is not numbered, however, you can indicate the number of items in it, as in the example given above, for the water fountain: "To enable users to drink or fill water bottles at the same fountains, the design must have the following three primary functions." Alerting your readers to what is coming up is a way of helping them to process information sooner, and to read faster.

4. Basics of Organized Sentences

An effective sentence has its own kind of logic. You must master that logic so that when you are expressing difficult concepts your sentence structure helps, rather than hinders, the reader's understanding. In addition, your sentences should be quick to read and unambiguous. So, beyond using plain value-free language, your sentences have to follow expected patterns of development specific to the language you are writing in. In English, the two key pieces of information the reader needs to have in order to understand a sentence are: 1) who or what is doing something and 2) the action that the person or system or entity is performing. The first is often called the "subject" and can be represented as a name, a noun, or a pronoun. The second is often called the "action" and is represented by a verb. Many sentences have a third component, an object or complement. This is the part of the sentence that receives or finishes the action.

In the English language, there are two basic sentence patterns: *active* and *passive*. The *active voice* is the more logical and easy to follow; it puts the subject first and the action second. That is, it identifies who is doing the action and what the action is: this is the easiest order possible for a reader to take in meaning in the English language. A *simple sentence* is one that has only a single action or idea. The following sentence is a simple, active sentence: "A design project often starts with a statement from a client."

But not all sentences can be active and short. In scientific writing, in order to maintain the idea of objectivity, the subject of the action cannot always be easily identified. Consider the case of a lab report. In an experiment, it is the experimenter who is doing the action, but if you are the experimenter, you do not write up your experiment with a bunch of sentences starting with the word "I": "I put the liquid in the beaker," or "I mixed in some sodium chloride." So, you often end up using the passive voice: "The liquid was placed in the beaker," or "Sodium chloride was added to the beaker."

The passive voice creates writing challenges, especially when you are writing *compound sentences* or *complex sentences*. Often the subject is hidden. In the sentence "The hypothesis was not verified by the results," the action is being performed by the researcher, but the researcher has no place in the idea. The idea should exist, according to scientific practice, separately from any individual involved. However, you can make objective active sentences if you work at it. For example, the sentence "The hypothesis was not verified by the results" can be rewritten as "The results did not verify the hypothesis" without losing any objectivity but gaining the value of an active sentence. That value becomes important when combining simple sentences to create compound or complex sentences.

So, the sentence "The results did not verify the hypothesis" is a simple, active sentence. It has one idea and has a subject, action and object in a conventional, easy to understand order. If you add another sentence to it—say the sentence "There was a high degree of error"—then you can have a *compound sentence*: "There was a high degree of error and the results did not verify the hypothesis." A compound sentence is made up of two simple sentences joined with the word "and." It is perhaps the easiest long sentence form to use, but it has two conditions. The units joined with a simple conjunction such as "and" have to be of fairly equal importance and they have to no special relation to one another.

A special relationship between ideas creates dependencies in sentences. This kind of sentence is called a *complex sentence*. One group of words, or clause, is "dependent" on another in order to be understood fully. You can turn the compound sentence above into a complex sentence by turning it around: "The results did not verify the hypothesis but there was a high degree of error." Now, the word that is combining the two ideas is "but," which indicates a contradiction or change in direction. The second way of writing the sentence brings out the idea that the error might have been the reason the results did not verify the hypothesis. This idea is implied in the compound sentence, but made more explicit in a complex sentence.

Because you will need to use compound, complex, and even compound/complex sentences, which combine the two forms, and these sentences can go on for quite a while, you have to have ways of ensuring that the reader does not get lost. The best way to ensure this is to consider the sentence like a road and identify signposts.

Remember, the reader is looking for the subject and the action. So track those first. Let's take the compound/complex sentence that begins this paragraph. First we'll track it for the <u>subject</u>.

Because <u>you</u> will need to use compound, complex, and even compound/complex sentences, which combine the two forms, and <u>these sentences</u> can go on for quite a while, you <u>have</u> to have ways of ensuring that the reader does not get lost.

Now, we will italicize the *actions*.

Because <u>you</u> *will need to use* compound, complex, and even compound/complex sentences, which combine the two forms, and <u>these sentences</u> *can go on* for quite a while, <u>you</u> *have to have ways of ensuring* that the reader does not get lost.

Notice you can break this long sentence into four short sentences:

- <u>You</u> *will need to use* compound, complex, and even compound/complex sentences.
- <u>Compound/complex sentences</u> *combine* the two forms.
- <u>These sentences</u> *can go on* for quite a while.
- <u>You</u> *have to have ways of ensuring* that the reader does not get lost.

These are simple sentences on their own and, if we have to, we can locate their subjects and actions. The phrase "compound, complex, and even compound/complex sentences" is the ***object*** of the first clause and "compound/complex sentences" the subject of the second, but the word "which" allows the writer to combine these two ideas and eliminate the repetition of the words "compound/complex sentences". After the words "two forms," the word "and" allows the writer to include another simple sentence: "These sentences can go on for a while."

However, the word "because" begins the long sentence. "Because" turns any clause that follows it into what is known as a ***dependent clause***. That is a clause that cannot be understood on its own, but requires another clause to complete the idea. The reason for this is that "because" means that there is a cause and effect. Strangely, people usually put the effect first and the cause second, but in this case, the writer didn't. She or he put the effect clause at the end of the sentence: "you have to have ways of ensuring that the reader does not get lost." Once again, this is basically a simple sentence and it could stand on its own. Moreover, you usually see the clause beginning with "because" after the one on which it depends, mostly because we are told in grade school "never begin a sentence with 'because'!" This is not a grammatical rule, but it is easier than trying to teach children the difference between a dependent and an ***independent clause***. The real rule is "Any sentence with 'because' in it must have two clauses, one that sets up the effect and the other that sets up the cause."

The central complex sentence here is: "Because you can have compound, complex, and even compound/complex sentences, you have to have ways of ensuring that the reader does not get lost." The matter in the middle is all just further explanation. But the writer ensures that the reader does not get lost by making the signposts of the sentence, the subject and action, clear and easy to find. Also, the action is right next to the subject, making it easier for the reader to connect them. When you are writing long, complex sentences, the way to ensure that they are clear is to locate their subjects and actions. See if you can get the actions as close to the subjects as possible. This will help ensure the logic of your sentences and really help your reader.

5. Leaving This Skill Module

After working through this module you should be able to

- Create a document with numbered headings that enable a coherent outline of information
- Write unified paragraphs with focused topic sentences and relationships between the sentences
- Write a variety of sentences in which ideas can easily be followed

KEY TERMS

headings	numbered headings	topic sentence
simple sentence	compound sentence	complex sentences
compound-complex sentence	object	dependent clause
independent clause	active voice	passive voice

6. Questions and activities

Take the following paragraph and:

1. Organize it so that the main idea is stated first as a topic sentence and the explanation and evidence follow in a logical order.

2. Revise the sentences so that long sentences make sense and can be quickly read and understood immediately.

Ideas, sentences, paragraphs, and images are all readily available on the Internet and people can cut and paste them into all sorts of documents for all sorts of uses. Over the last few years, high-profile cases worldwide showed how plagiarism can cost people their careers [1,2]. Plagiarism has become a difficult and complex concept for people to understand and respect with the advent of the information revolution. There is no control or doorway and so the concept that these ideas, words, or images are owned by someone who must at least be acknowledged but who must sometimes be paid or has to give permission is not easy to see or accept.

7. References

[1] Schuetze, C.F. (2013, March 11) New Plagiarism Cases Cause Second Thoughts in Germany. *New York Times.* Retrieved from www.nytimes.com/2013/03/11/world/europe/11iht-educside11.html?pagewanted=all&_r=0
[2] BBC (2013, February 9) German Minister Annette Scharar quits over "plagiarism." *BBC Online. Retrieved from* www.bbc.co.uk/news/world-europe-21395102

Diagrammatic Elements

#resource module: #diagrammaticelements

Learning outcomes

By the end of this module, you should demonstrate the ability to:

- Explain the types of diagrams typically used in engineering communication and their purpose
- Explain the essential characteristics of a diagram, table, or graph
- Use basic diagrams, tables, and graphs to enhance understanding in documents and presentations
- Use diagrams to enhance your own problem solving processes

Recommended reading

Before this module:

- **Implementing a Project > Communication > 3. Putting Together an Engineering Report**

After this module:

- **Implementing a Project > Communication > 5. Using Pictures and Photographs**

1. Introduction

Thinking like an engineer means thinking verbally, visually, and mathematically. Engineers take a problem from the real world or text form, such as a client statement, and transform it so that they can solve the problem with science and mathematics. To achieve that transformation, they use models whether it is simply to show inputs and outputs or more complicated phenomena. Models are often represented in ***diagrams***, which are essential to engineering thinking and communicating. They are used in design and problem solving and they are used in documenting the design or recommendation that is being made to a client. The results from problem solving are often depicted visually as graphs or charts before the solution is implemented back into the real world.

For many engineers, using diagrams is natural and intuitive. They draw quick, sometimes messy sketches, put boxes around important ideas, and draw arrows to show connections. Such sketches may seem ordinary, but they are essential to thinking,

Client statement (words) or problem
↓
scientific/mathematical construct (model) and diagram
↓
solution
↓
design or solution interpretation (charts and graphs)
↓
implementation back into the real world

FIGURE 1 Photograph of the existing streetlight that a client wants to change.

communicating, and documenting the problem solving process. Engineering reports may include both formal polished drawings and also sketches. Ultimately, engineering drawings are used to help others execute (i.e., make or build) the technology you have designed. Creating diagrams and sketching out problems may not be a natural habit for you now, but it is an important skill. Diagramming a problem may show you the way to transform your perception of the problem leading to an ingenious engineering solution.

For example, you are a part of a team that has been asked to design a new street light (see Figure 1). New light fixtures are more energy efficient and so the client wants to replace the old version. However, the old versions are mounted on concrete poles that still have a good usable life. You have to figure out the maximum weight of the new fixture so it can be safely installed on the existing poles. You might start by sketching a free-body diagram (see Figure 2) to help you clarify your understanding of the physical system.

2. Five Functions of Visual Forms

Using graphics to explain concepts in an engineering fashion expands on the familiar use of illustrations in documents. To do this effectively, it helps to understand the categories of visual expression. Visual forms fulfill five basic functions:

1. To execute
2. To identify
3. To enhance understanding
4. To instruct
5. To enhance visual appeal

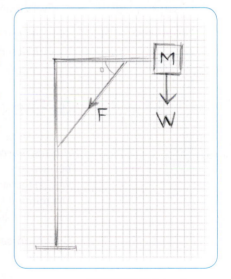

FIGURE 2 A basic free-body diagram sketch used to solve the problem of estimating the safe maximum weight for the new light fixture.

The first purpose, to execute, refers to formal engineering drawings that are used to execute a project. Formal engineering drawings will often start with a sketch of a part or system that is handed off to a technician or draftsman to render. Or you may learn in your upper year courses how to use *computer-aided design (CAD)* packages to create formal engineering drawings. How different types of systems are visualized in these drawings will depend on the discipline. Civil engineering drawings for a structure will use different diagrammatic standards to indicate types of walls and beams than electrical engineering drawings for printed circuit boards showing connections and components. The detailed standards for formal engineering drawings are beyond the scope of this text, but as you progress through your discipline your will learn to create and read formal engineering drawings in your field. Here we will discuss visual diagrams of the type that appear in reports, presentations, and in general instructions. These are also very necessary drawings in engineering communication and share common characteristics across disciplinary fields.

Diagrammatic communication refers to any kind of graphic representation of information: ***charts***, ***graphs***, ***diagrams***, ***flowcharts***, ***schematics***, ***tables***, ***pictures***, or ***photographs***. You are probably familiar with using these to illustrate your documents, but the ways that engineers use visual documentation to understand a problem is a skill you are currently developing. Graphics form a bridge between the problem in words, as it is presented to you by your client and discussed in your team, and the scientific and mathematical formulas and models that you, as an engineer, are going to use to solve the problem.

Information graphics, such as tables, diagrams, and graphs, are communication tools, but they are incomplete without text. The meaning of an information graphic is never self-evident. It is always necessary to explain to the reader what they should be seeing when they look at the graphic. The graphic should also serve a clear purpose within the text, not simply be there for show.

Every graphic you include should have

- A clear ***caption*** or ***title***
- A description in the text of the report or document in which they are imbedded
- A purpose in the context of the document

3. Diagrams and Schematics

The visualization of models and ideas in engineering go by many names: ***diagrams***, ***schematics***, ***flowcharts***, and many others. These drawings are typically a combination of visual elements (shapes and arrows) and textual elements (letters, numbers, symbols, and words) that are meant to assist both the reader and the writer (see example in Figure 3). Diagrams are used by engineers to help them solve a problem and help them communicate their solution to others. Diagrams are used to describe problems or systems.

Communicate a model. Diagrams communicate a model of the system. Diagrams typically only include elements essential for analyzing the system and leave out other aspects of the real system. For example, a cat sitting on a wall is modeled as a mass being operated on by the force of gravity. No information about the color of the cat would be included in the diagram.

Communicate characteristics. Diagrams communicate the characteristics of the system. To do this diagrams often use a set of standard ***visual vocabulary*** to convey ideas. For example, in mechanical engineering a zigzag line is a spring, in electrical engineering a zigzag line is a resistor. The type of diagram and shape indicate the properties or characteristics of the component.

Communicate states. Diagrams communicate the ***state*** of mass, energy and information at particular points in time or space. For example, a flowchart will indicate points when decisions are made, or a phase diagram will indicate the state of matter (liquid, solid, gas).

Communicate relationships. Diagrams communicate the relationship between parts of a system. This may be a physical relationship between parts, or a conceptual relationship. For example, a diagram of a jet engine will show that the compressor stage comes before the combustor, a molecular diagram shows the arrangement of atoms in a compound, or a Venn diagram will show the conceptual relationship between sets.

Labeled to enhance understanding. Good diagrams are clearly labeled. *Labels* allow the writer to easily draw the reader's attention to a particular point in the diagram. Labeling components, states, steps, and other key points in the diagram also makes it easier to connect the diagram to both the text and equations. So the initial position of a mass can be labeled "1" in the diagram, and called x_1 in the related equation. This helps you track key aspects of the problem you are solving from reality, to the diagram, to the equation, and finally back to reality when you have solved the problem.

Communicate direction and motion. Diagrams often communicate the direction mass, energy, or information are moving in time, or space, or conceptually. Vectors and arrows are common elements in diagrams and are used to indicate direction and motion (see example in Figure 4).

Used in problem solving. Diagrams help you debug your work. If you are checking a solution, trying to debug a program, or figuring out an error in a process, then you can use the diagram to visualize and track the key quantities in the system, helping you to find errors or inconsistencies.

In a written report diagrams and schematics are treated as *figures*. This means that the *caption* for the diagram goes below the diagram.

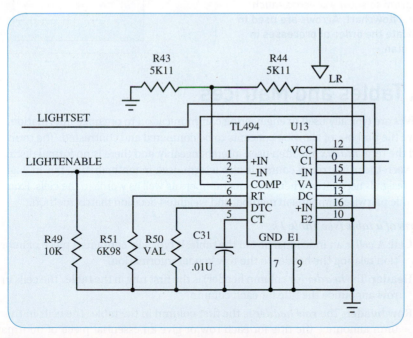

FIGURE 3 **An electrical schematic showing part of a circuit. The components labeled "R" are resistors, C31 is a capacitor, and U13 is a chip. This illustrates the use of standardized visual vocabulary (shapes) and labels to communicate the characteristics of the components.**

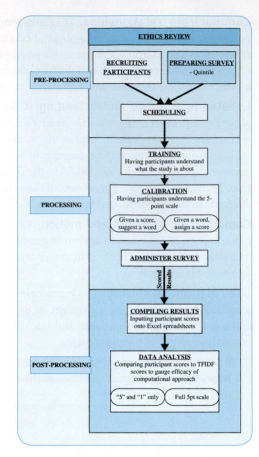

FIGURE 4 **Diagram of the methodology for a research study.** This figure illustrates the use of a diagram to show a process much like a flowchart. Arrows are used to indicate the order of processes in the plan.

4. Tables and Matrices

Tables are typically used in engineering communication to organize information. They serve the purpose of allowing concepts to be compared and contrasted. The term table and the term *matrix* are often used interchangeably, and there is no formal distinction between them. Matrix, of course, is also a term used in mathematics, but in engineering design is used often when there are numbers as well as words in the cells. Examples include pairwise comparison method, and weighted decision matrix method.

Parts of a table (see Table 1):

   Cell: A *cell* is an individual box in the table. It typically contains a piece of information relating the header to the row header information.

   Header: The *header* (or column header) is the first row in the table. The cells in this row announce the title for each column.

   Row header: The *row header* is the first column in the table. The cells in this column announce the title for each row or give an essential piece of information that connects all of the information in the row.

   Row: A *row* is a set of cells along a horizontal path in the table.

   Columns: A *column* is a set of cells along a vertical path in the table.

TABLE 1 Comparison of components found in heat engine systems. This table illustrates the parts of a table. The table title goes here, at the top of the table, and can be more than one sentence.

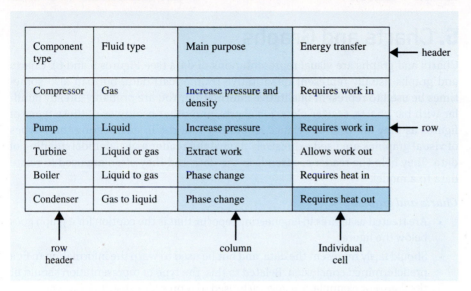

Component type	Fluid type	Main purpose	Energy transfer	← header
Compressor	Gas	Increase pressure and density	Requires work in	
Pump	Liquid	Increase pressure	Requires work in	← row
Turbine	Liquid or gas	Extract work	Allows work out	
Boiler	Liquid to gas	Phase change	Requires heat in	
Condenser	Gas to liquid	Phase change	Requires heat out	

row header · column · Individual cell

TABLE 2 Temperature, pressure, and specific volume at points along the liquid and vapor saturation boundaries for water. This example illustrates a table used to communicate numerical data. Each row gives the data for a point on the saturation line.

T (C)	P (KPA)	V_F (M³/KG)	V_G (M³/KG)
0.01	0.61	1.000E−03	2.060E+02
10	1.23	1.000E−03	1.064E+02
20	2.34	1.002E−03	5.779E+01
30	4.25	1.004E−03	3.289E+01
40	7.38	1.008E−03	1.952E+01
50	12.35	1.012E−03	1.203E+01
60	19.94	1.017E−03	7.671E+00
70	31.19	1.023E−03	5.042E+00
80	47.39	1.029E−03	3.407E+00
90	70.14	1.036E−03	2.361E+00
100	101.35	1.044E−03	1.673E+00

Very often tables are used to characterize items such as components of a system, solution approaches to a problem, stakeholders, or other sets of key considerations. The items are identified in the first column, and their characteristics are described across a row. The header is used to identify the categories of characteristics that are being examined. This gives you and the reader an easy way to visually take in a large amount of information and see how the information is connected and related (see example in Table 2).

5. Charts and Graphs

Charts and graphs are visual representations of data (see Figures 5 and 6). Charts and graphs usually represent quantitative (numerical) data, but can also sometimes be used to represent qualitative information. You are probably already familiar with bar charts, scatter plots, pie charts, and other examples of these types of figures as they are used in science laboratory reports. In engineering these types of visual graphics are used extensively to give the reader a visual understanding of data. They help the reader see trends in the data and they often are used to relate data to a model.

Charts and graphs:

- Are treated as figures in engineering reports; that is the caption for a graph goes below the figure.
- Should fairly represent the data, and not be used to warp the information to fit a predetermined conclusion. Related to this, the type of representation should fit the data. For example, a *histogram* is used to represent a distribution, a pie chart to represent percentage data.
- Should be clearly labeled. In particular the **axes** (if there are axes) must be labeled. For example, it is standard in a pie chart to include both a label for each slice and the numerical data associated with the slice.
- Are often used to relate observed behavior to a model. When using a graph or chart this way it is typical to plot the results predicted by the model on the same graph or chart with the experimental or observed data. This allows the reader to visually compare the observations and the results predicted by the model.
- Are always accompanied by a reference to the figure in the body text of the report. The text in the report must reference the figure (e.g., see Figure 5) and must explain to the reader what can be observed in the figure: "Figure 5 shows a plot of the liquid saturation boundary for water on the temperature-entropy plane." Note that the information in the text may repeat some information from the figure caption, but this is not considered redundant.

6. Conclusion

Concepts in engineering are often best communicated using a combination of text (verbal representation) and diagrammatic elements (visual representation). Diagrams, tables, graphs, and other visual elements give the reader another route to understanding data and verbal descriptions. To make diagrams effective they need to be clear, well labeled, and connected to the body text. While diagrams allow the reader to follow your writing more effectively; equally as important, developing diagrams and other visual representations allows you, as a problem solver, to clarify your understanding of a problem and work through problems more effectively.

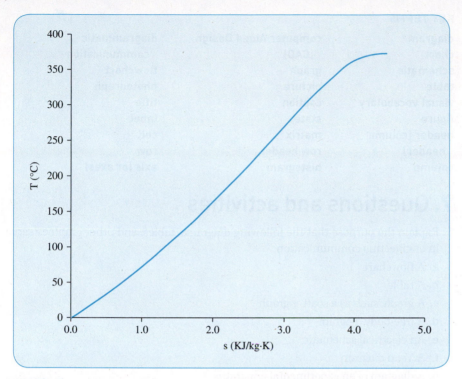

FIGURE 5 Shows temperature versus entropy along the liquid saturation line for water. This is an example of a graph demonstrating the use of labels and a descriptive caption.

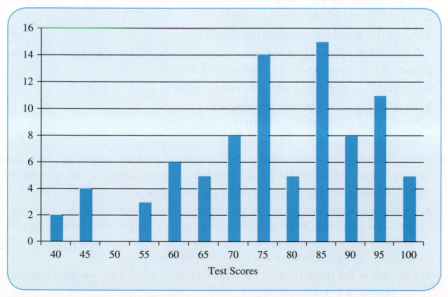

FIGURE 6 Histogram showing test score distribution. This is an example of a chart that allows the reader to visualize numeric data.

KEY TERMS

diagram

chart

schematic

table

visual vocabulary

figure

header (column
  header)

column

computer Aided Design
  (CAD)

graph

picture

caption

state

matrix

row header

histogram

diagrammatic
  communication

flowchart

photograph

title

label

cell

row

axis (or axes)

7. Questions and activities

1. Explain the purpose that the following diagrams, tables, and other graphics serve in engineering communication.

   a. A flowchart

   b. A table

   c. A graph, such as a scatter graph

   d. A free-body diagram

   e. An electrical schematic

   f. A Venn diagram

   g. A diagram of an experimental apparatus

2. In a design report you might include a weighted decision matrix in an appendix to justify the recommendation of a design decision.

   a. Why would you show the matrix?

   b. What do you want the reader to perceive from this visual element?

   c. What title would you give the matrix?

   d. What text should accompany the diagram?

3. Create a diagram (sketch or computer-generated graphic) of a design solution for a project you are working on, or of an existing design.

   a. Add labels to the diagram to enhance its value.

   b. Create a caption to go with the diagram.

   c. Write a paragraph that refers to the diagram and explains to the reader what they are seeing.

4. Select three diagrammatic elements (diagrams, schematics, tables, graphs, charts, or other visual elements) that you included in a recent document. For example, elements from one of your design reports, or a lab report.

   a. Identify key components in the graphics you have chosen, such as labels, axes, rows, headers, identified states, titles, and captions, use of visual vocabulary (i.e., standard shapes) for particular components, use of arrows, and other essential diagrammatic components you have incorporated in your work.

   b. For each of the three elements you have chosen; what purpose does it serve in the context of the document?

   c. Give one example of text in the document that is used to draw the reader's attention to key points in the associated graphic element (i.e., diagram, table, or graph).

Using Pictures and Photographs

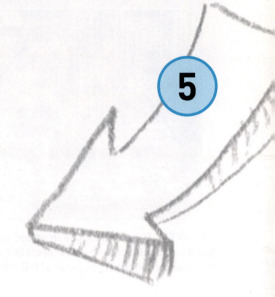

5

#resource module: #pictures&photographs

Learning outcomes

By the end of this module, you should demonstrate the ability to:

- Utilize graphics in both the process and documentation of design

Recommended reading

Before this module:

- **Implementing a Project > Communication > 4. Diagrammatic Elements**

After this module:

- **Implementing a Project > Communication > 6. Influencers of Communication**

1. Introduction

Diagrams such as flowcharts and graphs are used to transmit information and formal engineering drawings are used to execute a project. However, there are instances when a ***photograph*** or ***drawing*** (picture) best serves the purpose of communicating an idea. In engineering reports and presentations photographs and drawings can serve the purpose of enhancing the visual appeal of a document, but they are also used to identify phenomena. Photographs can present information to a reader (or audience) that a simple diagram or graph does not capture, and pictures are often used to instruct people on proper usage of technology.

2. Enhancing Visual Appeal

While formal engineering drawings are meant to be purely informational, there are times in reports and presentations when it is appropriate to use a visual to enhance the aesthetics of the communication. Photographs and drawings are visually interesting and can add to the appeal of a document, presentation, or poster. They tend to contain a lot of information—that is, many shapes, colors, objects, and activities can

Courtesy of the Authors

Courtesy of the Authors

FIGURE 1 **Bascule bridges in raised position, Chicago 2009.** Photographs are packed with information, but it is all subject to the interpretation of the viewer, unlike diagrams, which often contain only essential information. (Photos courtesy of P.E. Weiss.)

be included and these have different meanings for different people. The photographs also cause viewers to have feelings and associations and these too are different from one person to the next. This can be a strength or a weakness.

Take the two views of the Chicago River (Figure 1). They each give a very different sense of the same body of water. The daytime photograph on the left might be useful to show how a bascule bridge operates. But look at all the other activity in the scene: sailboats, tour-boats, pedestrians, not to mention the architecture all around, with unique architectural elements that catch the viewer's attention. There is a lot happening. One hardly knows where to look or what to focus on.

The nighttime photograph, on the right, has a completely different feeling. Looking at it, one gets a different sense of the location. Unless you know otherwise, you would likely not recognize the bascule bridge in the night photograph.

These photographs also have emotional information—the amount of activity, the kinds of objects, the colors and shapes may create all sorts of different feelings in the viewer. In fact, one of the values of photographs is their ability to evoke feelings. When used appropriately, photographs and pictures can make a reader or audience member respond more positively to your ideas. When used inappropriately, they distract the reader and might make your document or presentation seem less valuable, perhaps even silly.

Drawings or paintings have similar issues to photographs. They contain a lot of information and give the viewer feelings, but it is difficult to define and/or control exactly what the viewer is supposed to get from the image. Symmetry, shapes, colors, content and detail all contribute to the feelings that a photograph or drawing can result in. Further, these feelings change from person to person. Drawings, paintings, and photographs are effective ways to create emotional response but must be used thoughtfully. You must consider the emotion you are trying to create and choose a photograph or drawing carefully. In engineering reports pictures are infrequently used for visual appeal only, except perhaps on a report cover. They are more frequently used in presentations, specifically to illustrate the message of the slide or text.

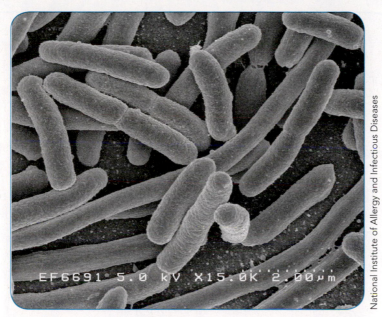

National Institute of Allergy and Infectious Diseases

FIGURE 2 **Micrograph of Escherichia Coli.** Credit: Rocky Mountain Laboratories, NIAID, NIH. Public domain. www.niaid.nih.gov/SiteCollectionImages/topics/ biodefenserelated/e_coli.jpg

3. Identifying Phenomena

In terms of identification, a photograph or picture can show only what something looks like because it only identifies something in a general sense; it is not very precise. Take, for example, a photograph of a hotel in a place you have never been to before. While it will help you recognize the hotel when you are finally right in front of it, it will be no good at helping you get to the hotel from the airport or train station. For that, you need a map, preferably one with the route highlighted.

So, if you wanted to answer the question, "What does E. Coli look like?" you might use a scanning electron micrograph. It gives a very good impression of the tubular shape of the bacteria (see Figure 2).

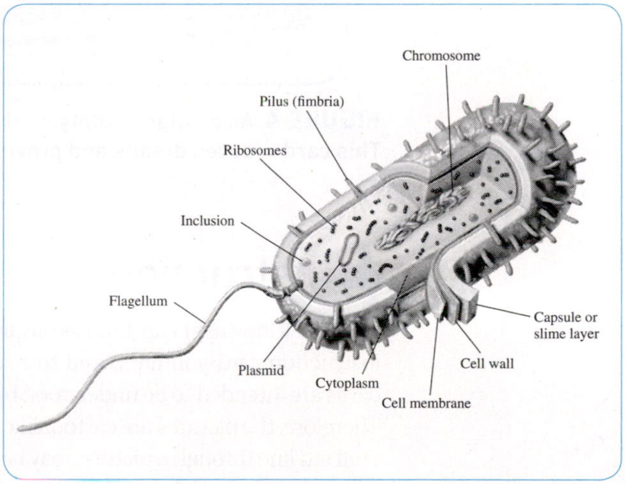

Chromosome

Pilus (fimbria)

Ribosomes

Inclusion

Flagellum

Capsule or slime layer

Cell wall

Plasmid

Cytoplasm

Cell membrane

Jacquelyn and Laura Black, *Microbiology: Principles and Explorations*, 9th edition, John Wiley and Sons, Inc.

FIGURE 3 **A diagram of E. Coli.** Labels draw attention to details that are clarified in the diagram, versus the micrograph.

Photographs and drawings are useful in identifying the way something looks in a general sense, but to enhance understanding diagrams must be used. A diagram is a depiction of an object or process and it includes the most important details in a way that makes them stand out. It should also be well labeled so that all of its important components are brought to the attention of the viewer. In addition, dimensions must be shown when appropriate. These enable the viewer to appreciate the scale of the diagram (see Figure 3).

Using Pictures and Photographs

FIGURE 4 An airplane safety card illustrates the use of pictorial instructions. This card reduces details and provides step by step instructions in pictures only.

4. Instruction

Pictures (drawings) can be used to instruct; one situation that might be familiar is the instruction card you are asked to read on an airplane (see Figure 4). These instruction cards are intended to be understood by people no matter their language or education level. Therefore, the pictures are cartoonlike and simplified. Certain conventions, such as a diagonal red line through a picture, may be used to express abstract ideas. For these to be effective, people have to know what they mean, but many of these, such as the red circle with a diagonal line, meaning "No" as in "No smoking," are commonly seen all over the world.

5. Conclusion

Pictures and photographs are used to illustrate ideas in engineering reports and other documents. They complement the diagrams, graphs, and drawings that are used to convey ideas. Together with text, visual elements such as these help the reader get a more complete and multi-faceted understanding of concepts. Visuals, such as pictures, can also cut across language barriers. This can be advantageous when developing instructions, interfaces, and signage.

KEY TERMS

photograph   **drawing**   **diagram**

6. Questions and activities

Figure 5 shows a photograph of the Canadarm stowed aboard the Space Shuttle Explorer.

1. In words, describe the Canadarm and its function. You will have to do a little research. Try to keep the descriptions to only two paragraphs, no more than 400 words in total.

2. Draw a quick sketch of the Canadarm in action, adding dimensions. How does even a quick sketch enhance understanding?

Courtesy of the Authors

FIGURE 5 **Canadarm in its folded configuration, stowed aboard the Space Shuttle Explorer.** The robotic arm was used for servicing items in space. (Photo courtesy of P.E. Weiss.)

7. References

Irish, R., and Weiss, P.E. *Engineering Communication: From Principles to Practice.* 2nd ed. Toronto: Oxford University Press, 2013.

Manning, A., and Amare, N. "Using Visual Rhetoric to Avoid *PowerPoint* Pitfalls." *2005 IEEE International Professional Communication Conference Proceedings.*

Using Pictures and Photographs

343

6

Influencers of Communication

#narrative module: #communicationinfluencers

Learning outcomes

By the end of this module, you should demonstrate the ability to:

- Describe the three different influencers of communication

Recommended reading

Before this module:

- **Implementing a Project > Communication > 5. Using Pictures and Photographs**

After this module:

- **Implementing a Project > Communication > 7. Organizing Presentations**

1. Introduction

The ways in which we communicate are changing as methods of communication through the Internet and social media accelerate and evolve. However, people have not changed that much in some important aspects of communication.

Research has shown that there are generally three major influencers of communication: what is said (the words), how it is said (the ***tone***), and what the communication looks like (the term ***body language*** is often used to describe this factor). Interestingly, the last has been shown to have the most impact on how a communication is received, particularly if that communication is face to face. Equally interesting, this factor is the one that is perceived as being missing from many forms of communication such as texting and email. Its absence is often blamed for miscommunication. People also frequently associate the importance of being able to "see" someone when communicating with how well they know the other person. If you know them well, trust is understood (or not) in the relationship. This is why it is rare for teams to meet effectively entirely through remote means if the team members have never met one another. The impact of these three factors applies to both written and oral communication.

Depending on the research, what you say or the words you use impacts 7 to 15% of the communication and the message that it is received by the person or group being communicated to (Figure 1). Greater than 50% of the message is through the "body language" component. The rest is received through the tone used in communication.

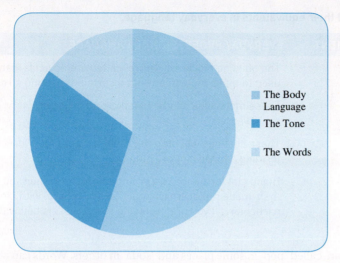

FIGURE 1 Body language impacts the perceived communication more than tone and words together.

2. What Is Said

While the words that are used in a communication are generally felt to have the least impact on the message received, they can still result in misunderstanding. Engineers and scientists have frequently been criticized for "speaking over peoples' heads." Learning to use the language appropriate for the person or group that you are communicating with is critical.

For example, read the following passage and without looking up the answers, see if you can figure out what this means: "*Amino acid based block co-polymers can be used to create non-thrombogenic coatings for application in the cardiovascular system, particularly for vascular grafts below the knee.*" This passage is written at university postgraduate level, and in a specialized field of study, biomedical engineering. As a student in engineering, you are likely familiar with a number of the words and there is a chance that you will get the idea but may be missing some key words that are important. If this example is rewritten and rephrased, it can be understood by a much broader audience (see Table 1).

The result is much easier to follow and the importance of the research can be appreciated by a non-specialist:

> *The building blocks of nature, called amino acids, can be used to make molecules with a certain repeating structure. These molecules can be used to coat artificial blood vessels. These coatings will stop the replacement blood vessels from forming a blood clot or "plug." Blood clot formation is a common problem when you put man-made replacement materials in contact with blood, particularly if you are replacing blood vessels below the knee.*

The words can also be an issue in another way. Some common items can have very different names, particularly in English. For example, the article of clothing called a "sweater" in North America is called a "jumper" in many other countries. Carbonated

TABLE 1 Phrases of the passage and their equivalents in everyday language.

PASSAGE FROM SCIENTIFIC DOCUMENT	EVERYDAY PROFESSIONAL LANGUAGE
Amino acid based block co-polymers	The building blocks of nature, called amino acids, can be used to make molecules with a certain repeating structure
Create coatings for application in the cardiovascular system	These molecules can be used to coat artificial blood vessels
Non-thrombogenic	These coatings will stop the replacement blood vessels from forming a blood clot or "plug"
Vascular grafts below the knee	Blood clot formation is a common problem when you put man-made replacement materials in contact with blood, particularly if you are replacing blood vessels below the knee

beverages are called "pop" in some places and "soda" in others. Words can also change their meaning from one generation to another. The word "jazz," for example, has had a negative meaning and then a positive one (it did not originally refer to music but now does). Something to be particularly careful of is slang. It can be misunderstood if someone is from outside the region or country where the slang is common and it can be offensive. Innuendoes or off-color language should be totally avoided, even if you are quite familiar with the client. In some cultures such language is never used; at best it will brand you as unprofessional.

When communicating using technologies including twitter, texting, and whatever comes next, people often create new words and abbreviations. This can lead to much confusion! As the use of social media and other rapid communications increases, it is important to remember that not everyone with whom you communicate, particularly if there is even a small difference in age, is likely to understand the "shorthand" associated with newer communication modes. Typical examples might be LOL or BTW. It is generally a good idea to stay away from acronyms and short forms in professional written communications, and to stick to traditional salutations and composition. Acronyms, when they are used, must be specific for the field or project, and spelled out when first used in the document.

3. What the Communication Sounds Like

The tone of a communication has the second highest impact on how that communication is received. Think about how writers will put tone into their writing. They use phrases like "she exclaimed" or "he responded softly" to indicate tone. Think of a close family member saying your name and the variations in tone that are used to impact the communication when only using that one word. Generally the message is very clear!

Tone can also be misunderstood in communications like email. Even when communicating in writing, people will attempt to build in tone. If you have a particularly direct way of writing (as many engineers do, "give me the facts and let's move on"), many people may feel you are annoyed. It is important to proofread your communications from the reader's perspective and adjust the tone if necessary to convey the message you are trying to convey.

4. What the Communication Looks Like

When you are delivering the message orally, you also impact the message through the way you look, move, and react to the communication. If you lean forward while speaking or listening, the person to whom you are speaking is likely to interpret that as interest. If you have very little expression and you neither lean forward nor backward, you will likely confuse the person and make them uncomfortable because they will not be able to tell how you are receiving the communication. Your expression and motions are called **body language**. It is your physical demeanor while communicating and incorporates your physical stance as well as the tone of your voice.

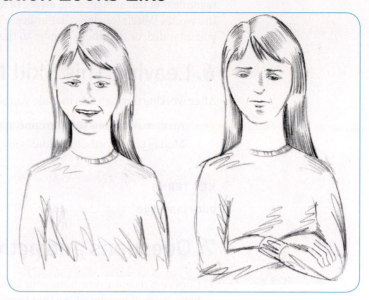

It is important not to forget about "body language" in a written report. It exists. Professional formatting of your material, neatly written, proofread, and using proper referencing provides a significant positive first impression. The reverse is true of sloppy work. The reader will form an opinion about you and your recommendations very quickly from the how the documentation "looks."

Body language can also be culturally influenced and therefore it is important not to misinterpret or "read" too much into body language and to take the whole communication, words, tone and the body language into account. One of the most difficult cues to interpret is eye contact. In some cultures, indirect eye contact is considered a sign of dishonesty and in other cultures direct eye contact may be considered a sign of disrespect. It is important to be aware that you will react to the messages being sent to you through body language and that the person, group, or audience with whom you are communicating will also react to you. It is a significant influence on the communication.

5. Verbal Communication through Third Parties

Sometimes you will give messages through a third party. For example, this could be a client's associate, or administrative help, or subordinate team member.

You should note the following:

- The third party will often interpret your message, so you must take extra care to make sure your message is clear and unambiguous.

- The body language may be stripped out of your message. Make sure that the words contain all of the meaning you wish to convey.

- The third party may also use non-verbal information. If you are disrespectful, or annoyed, or show other messages through body language, these may also be communicated to the client. This may not be what you intended.

Influencers of Communication

347

When communicating, it is important to be prepared, practiced, and professional regardless of the audience. Be aware of all the ways you communicate, not just with the words. Watch for clues that indicate that the client has not received the message you intended, or that there is more to the subject than you thought.

6. Leaving This Skill Module

After working through this module you should be able to

- Analyse situations to determine appropriate modes of communication
- Modify your words, tone, and body language to suit professional situations

KEY TERMS

body language   **tone**

7. Questions and activities

1. Choose a recent event from your life, one that you feel shows an ability that you have. It could be something that happened at work or volunteering or at school.

   a. Write a paragraph describing the event to a friend.

   b. Write a paragraph describing the same event but as you would discuss it in a job interview.

   c. Analyse the two paragraphs. What influenced your language choices and what were the results?

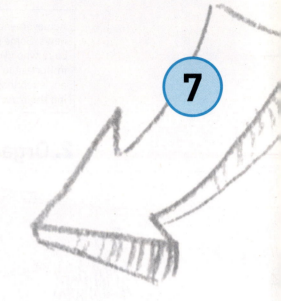

Organizing Presentations

7

#skill/tool module: #organizingpresentations

Learning outcomes

By the end of this module, you should demonstrate the ability to:
- Prepare an organized presentation

Recommended reading

Before this module:
- **Implementing a Project > Communication > 6. Influencers of Communication**

After this module:
- **Implementing a Project > Communication > 8. Effective Slides**

1. Introduction

Person-to-person interactions are key to developing trust. While the written word, especially when carefully done, can provide extensive detail and explanation, in-person communication enables you to establish a human relationship with your audience. In-person communication is flexible, allowing you not only to present the details you think are important, but also to find out what your audience wants to know and to answer their particular questions. Ideally this type of communication actually takes place in person, but it may be necessary to conduct the meeting virtually (e.g., a **webinar**).

Presentations are about taking person-to-person interactions, the development of trust between two parties, to a second level, in which the interaction is one-to-many and the development of trust is general. A successful presentation is one that ends with listeners feeling confident that the speakers have in fact achieved what they said they were going to achieve or are likely to achieve it in the future, as planned. Good questions to ask at the end of any presentation, therefore, are:

- Did we clearly identify ourselves and what we intend (or intended) to achieve?
- Did we adequately prove that we achieved it or will achieve it?

Achievement does not necessarily mean a successful outcome to a project, or good news. Some presentations have the goal of successfully delivering bad news. In the case when you are developing a presentation to deliver bad news it is particularly important to identify the purpose of the presentation, and carefully plan out a strategy such that your audience understands what you are saying without compromising the trust in the relationship.

2. Organizing a Presentation

Successful presentations are the result of planning and organization. In the planning phase, you have to establish a sense of who your audience is and what you are trying to achieve with the presentation: both your goal and your main point. Organization requires, at the very least, a clear, coherent outline of the points you are going to cover. You must not only know *what* you are going to present, but also *why* you are presenting each detail so that the details all come together in the mind of your audience to form a unified idea. This will give them both the sense that they understand and/or appreciate your technical work and also give them a sense of confidence that you will be able to complete your work, solving any problems that might come up along the way.

2.1. Understanding Your Audience

Because a presentation cannot have as much detail as a written document, you must select particular points to make and the credibility of your presentation will be determined by the logic of your selection. Your audience has to understand why you are telling them what you are telling them (and therefore why you are leaving out what you are leaving out).

In order to help your audience understand you, you have to begin by trying to understand your audience, what parts of your project they will be familiar with, what parts will be easy for them to comprehend and what parts will be difficult to appreciate or accept. Every audience is different and audiences are made up of individuals who have different levels of technical skill and different interests in your project.

So, a first step in planning is to ask these questions:

1. Who is my audience?
2. What is their level of involvement with the project? How much do they know at a high level? How much do they know about current details?
3. What does my audience need to know NOW?
4. What is the level or range of technical knowledge in my audience? What is their technical interest or enthusiasm? How much time do I need to set aside for particular, technical explanations?

FIGURE 1 An audience can seem intimidating, but figuring out the audience's need in advance and tailoring your presentation to meet that need will help you develop confidence in your presentation.

Technical explanations are valuable both for audience members who do not have a high technical knowledge and for those that do. The first, less technical group, will be eager to understand what they do not already understand; the second, highly technical group, are enthusiasts. They will want a lot of details because they enjoy the subject area. The nature of the details may be different for each group, or you may have an audience comprising both groups and have to carefully determine the best selection of details to satisfy all.

2.2. Determining Purpose

Once you understand of the needs of your audience, you should determine how the presentation serves both those needs and the needs of your design team and the organization for which the team works. These needs are *both* individual and situational. So, while no two audiences are alike, a single audience will have very different needs at different stages of the design process. Your supervisor, for example, might initially be most concerned with your preparation, vision, and planning. When hearing a progress report, however, she or he might be much more interested in how you have resolved a problem than what the next stage of the process is. Indeed, the supervisor might be unwilling to listen to plans for the future until she or he is satisfied that issues from the past have been adequately dealt with. At the end of a project, your supervisor will want to know how well you have tested and proven your design and how feasible the next steps are going to be.

So, you can see that the purpose will change for each presentation. Early on, the purpose might be to build the audience's confidence in your team's plan and ability to carry out the plan. During a project, the purpose might be to reassure the audience that problems are being resolved and that you are taking all possible steps to keep the project on schedule and on budget. At the end of the project, the purpose is closure, but you may want to reinforce your relationship with the client so that you and/or your team are considered for future projects.

2.3. Defining a Main Message

A *main message* is related to the purpose of your presentation. It encompasses both the topic of the presentation and a sense of what you want the audience to think at the end of the presentation. For example, for a progress report, the topic is the progress to date. But the message "We are now going to discuss our progress from April 2012 to September 2014" is neither memorable nor helpful, since it is likely the audience already knows that part of the purpose. Far more helpful would be something like, "In our progress from April 2012 to September 2014, we completed the design stage and resolved the mass problem by changing materials." A main message that is specific and informative gives the audience something to look forward to and creates a framework so that the audience can process the details that you present as they come up.

2.4. Organize Your Presentation so that You Make Your Point Effectively

Once you have determined audience, purpose, and main message, you can proceed to outline your presentation. It should have an opening that introduces the team, project, and specific presentation (e.g., "Proposal" or "Progress report"). Then, there will be the main body of the presentation and finally a conclusion that consolidates the information and provides a *take-away* message that will help the audience remember the presentation favourably.

So, first of all, identify yourself and your project completely in two ways:

1. Give the title of the project and list all participant names, both first and last names, on the title slide. When presenting, you should clearly and carefully, announce the name of your project and introduce yourselves at the beginning of the presentation. Give your names slowly and clearly; since engineering teams and audiences may come from all over the world, names can be unfamiliar. Hearing them and seeing them in writing at the same time is a great way for the audience to get comfortable with them.

2. Show a slide that gives the main message of the presentation and an outline of its parts (#effectiveslides). The main message is a full sentence that goes at the top of what is often known as the "outline" slide. It predicts what you are going to prove, pulls the presentation together at the start, and makes each of the points in the presentation mean something unique.

While certainly it is not "wrong" to create a slide that slide that says "Outline" or "Agenda" and then gives a list like "Introduction, Design, Conclusion," it also does not take much thought to create such a slide and it most certainly does not give a unique or lasting impression. On the contrary, a slide that has a title such as "Improving Flow of Movement in Occupational Therapy Clinic" is going to catch the audience's attention and get them to start thinking of questions: How are they doing that? How is it possible? What is the cost? How complicated is the refit? Then, such conventional sections as "Introduction, Design, Conclusion" can become

- Current situation: room dimensions, required equipment per bed, number of beds
- Desired increase in numbers of beds: space challenges associated with goal
- Models of layout and workflow
- Optimal balance of space and workflow

Making your points specific and unique to your project makes your presentation far more interesting. There is a hidden challenge to this, however. Choosing conventional headings and an old "tried-and-true" recipe for a presentation might feel more comfortable because you or your team members will feel that you are not making any mistakes, that such an approach is not "wrong." However, just because it is not wrong does not mean it is right. It certainly does not guarantee it is good, effective, or memorable.

The final part of the presentation has two components:

1. Signal the audience that the presentation is coming to a close with a "conclusion." This is a brief summary of the conclusions of each point and it consolidates the information you want the audience to remember. Thus, you would not say, "We had a problem with mass" but rather, "So, titanium-aluminum proved to be the best solution for the mass problem." Summarize each point in this way, but briefly.

2. The final step is the *take-away*. This is the last impression or what people will remember. Reiterating your main message is a good way to consolidate the presentation. Not having a main message and just presenting details from your report will have the opposite effect. What people will remember is that they had nothing to remember, that the presentation was dull and not informative, lots of details but nothing to grasp.

3. Leaving This Skill Module

After working through this module you should be able to

- Determine a main idea that takes into account the situation of the project, needs of the audience and purpose of the presentation
- Figure out the points that best explain and support the idea
- Develop each point
- Reiterate the main point to sum up

KEY TERMS

purpose   **main message**   **take-away message**

4. Questions and activities

1. Try developing a short presentation about your design project
   a. Select an audience and decide on the purpose for your presentation.
   b. Write up an outline.
   c. Develop the outline into a set of slides and a point form script, or set of notes. Engineering presentations are almost never read, but they are also not spontaneous. Cues and notes are used to remind the speaker of main points and facts while they are speaking.
   d. Try delivering your presentation to your audience and get their feedback on how well it fulfilled the intended purpose.

Effective Slides

#skill/tool module: #effectiveslides

Learning outcomes

By the end of this module, you should demonstrate the ability to:

- Prepare effective slides
- Use slides effectively in a presentation

Recommended reading

Before this module:

- **Implementing a Project > Communication > 7. Organizing Presentations**

1. Introduction

Slides, or visual support for oral presentations, have evolved tremendously over the past two decades due to the development of programs and devices for designing and projecting visual material. Various programs allow wide choices of layout, background, font, color, and the inclusion of dynamic material: animations or videos. As tablets are becoming prevalent, we are starting to see the introduction of other approaches to presentation, such as the use of writing during the presentation. With so much choice, however, it can be difficult to make design decisions, to know when more words or images are necessary or when you have too much.

Slides are part of a coordinated set of instruments which engineers use to communicate their ideas. Emails, memos, reports, meetings, and presentations each play a role. Reports are the most complete and detailed. Presentations are more personal and interactive. They can bridge to the report, for the audience. Accessories to the presentation can highlight key points: handouts (on paper or electronic) are excellent for detail that might require careful reading, samples that can be handed around give a kinesthetic understanding, slides can reinforce key ideas.

Slides cannot be as detailed as a report or a hand-out; that would be too much reading, imposing on the audience and distracting from the speaker. They cannot be as experiential as a sample you can touch and examine in different perspectives, but they can augment a presentation with added visual elements, sound, movement, and ideas.

2. Preparing Effective Slides

When slide software first developed, it was used much in the same way as the previous, less-sophisticated technology of overhead projections. Colors were limited. Professional advice was to keep headings and bullet points short and simple. As a general rule, slides were supposed to have no more than 5 equations or 50 words.

Problems developed as these highly condensed summaries of complex projects began to be distributed and read in place of the more detailed reports. Lack of complexity and relationships between ideas were an obstacle to an in-depth understanding necessary for success. Michael Alley, a leader in slide design thinking, proposed a way of creating slides that overcome some of the weaknesses of previous designs [1–3]. Referring to the example in Figure 1, the slide has basically three elements:

1. A clear title in the form of a brief sentence giving a complete idea
2. A set of bullet points that use enough words to give the idea in some depth
3. A visual element that helps to exemplify the idea

This slide design maintains simplicity, but gives enough detail so that the ideas of the presentation do not become trivial. The main point here, though, is that the slides contain whole, unique ideas as opposed to topics or general titles, such as "Introduction." Also, the visual element is used to support and build understanding. It is not just decorative. In fact, you should avoid decorative images, clip-art, and photographs that are not making a direct point related to the presentation.

So, good slides, very briefly, should be clean and uncluttered. They may use whole sentences and full ideas, but are still only a selection of the details given in a full report. The selection should be made carefully, according to the main message you and your team want to present—the message that will express the best of what you can do. If this is supported by objective, tangible evidence (tests, measures, principles) then you will have achieved important goals of effective engineering communication.

The Standard for this presentation is the slide fromat proposed by Michael Alley (*The Craft of Scientific Presentations*):

What: an informative slide

Why: standard PowePoint practices can simplify ideas too much

How:
- The header contains an entire idea, not just a few words signifying the topic
- Whole sentences are used in bullets
- A graphic element is incorporated in the lower right side of the slide

Courtesy of the Authors

space capsule interior

FIGURE 1 **Example of slide design as described by Michael Alley [1-3]. This slide exceeds the "5 equations or 50 words" rule by a few words, but retains full ideas without being cluttered.**

3. Using Slides Effectively

To use slides effectively in presentations

1. Use slides as a backdrop; your oral presentation should fill in most of the details.
2. Do NOT read your slides to your audience.
3. Make slides as graphic as possible.
4. Move forward through your slides.
5. Keep your eyes on the audience.
6. Use guiding language as well as (or instead of) a laser pointer.

3.1. Use Slides as a Backdrop

The audience is expecting you to explain your ideas. They are not expecting to have to read them from a slide. In an engineering context, the audience may already have a report and if they want to read, they can read that. What they want from a presentation is personal contact and verbal explanation. The slides merely reinforce the main ideas. Therefore, you have to be very selective about the details you put on the slides; make sure only the most important ideas are included.

3.2. Do NOT Read Your Slides to Your Audience

Very often in industry presentations you will supply your audience with a copy of the slides either on paper or electronically. You can also assume your audience is literate. Therefore, it is not necessary to read the slide out loud to them. You should, instead, speak to the ideas that are presented on the slide.

3.3. Make Slides as Graphic as Possible

Finding the most graphical way to present your ideas is one strategy to ensure that your slides support rather than replace the oral presentation and enable the audience to use different ways of processing information to more fully understand and appreciate the ideas. Diagrams and appropriate graphs can be very effective, but make sure the labels are large enough to be readable (and are readable on handouts also). Graphics also help you to ensure you are talking through the material and not just repeating a series of points already written on the slides.

3.4. Move Forward through Your Slides

Changing from slide to slide should be as smooth as possible and should not be distracting. If you find it difficult to speak and change slides, you can have a team member do it for you; just make sure you practice in advance so that you are well coordinated. Also, if you have a slide that has information you are likely to want to return to at different points in the presentation, then copy it and place it in those points rather than try to return to it when you need it. We recommend avoiding animated slide transitions (elements bouncing around the screen or dissolving), which do not add value to the presentation.

3.5. Keep Your Eyes on the Audience

It is tempting, particularly if you are shy, to turn away from the audience and read from your slides. This problem is intensified when you are using the slides as an outline to help you remember what you planned to say in the presentation. However, if you turn away from the audience, you lose their interest and with that, you lose credibility. That means that you lose marks in a class context and you lose trust in a professional context. So keep your eyes front. If you have to, you can glance at the monitor in front of you or notes. Never read your entire presentation from a prepared script. When pointing out items on a slide, try to keep as much of your body facing the audience as possible and keep the activity brief so that you can return to eye contact quickly.

3.6. Use Guiding Language as Well as (or Instead of) a Laser Pointer

One of the things that really helps the audience locate information on a slide is if you use guiding or directional language in your oral presentation as well as (or instead of) a pointing device. Phrases such as "in the upper right corner" or "just to the left of center" reinforce the pointer and help verbal as well as visual audience members find information faster. Such terms are also more precise and careful than the vague "here you see . . ." Precision and care build confidence.

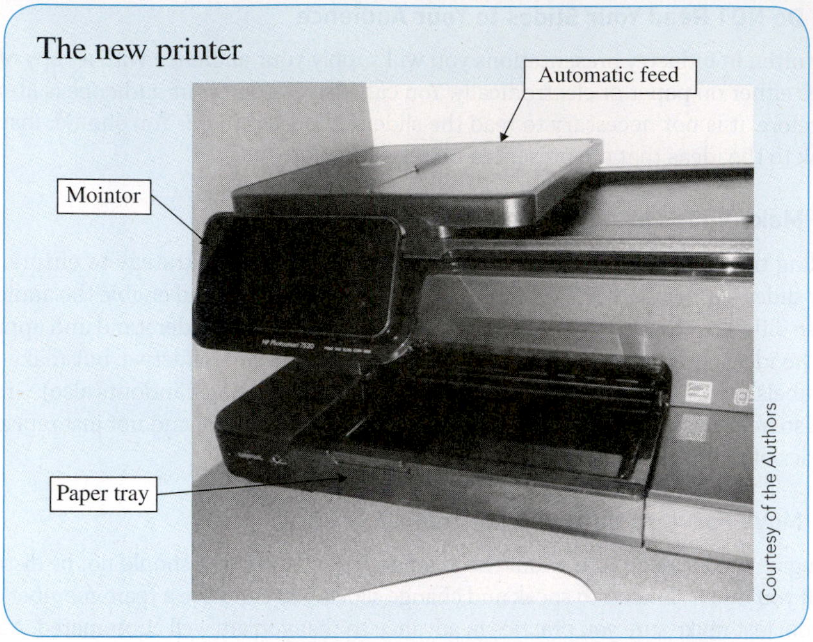

The new printer

Automatic feed

Mointor

Paper tray

Courtesy of the Authors

FIGURE 2 **Guiding or directional language can help the audience follow your explanation of a slide.** In presenting this slide you might say, "Moving from left to right, this slide shows . . ." or "In the lower left we see the paper tray that holds . . . ." Matching the terms you use with the labels on the slide also help your audience follow your presentation.

4. Leaving This Skill Module

After working through this module you should be able to:

- Design slides that are organized, informational, and rely on visual elements as well words to express ideas
- Utilize slides in a presentation so that the slides support your ideas rather than distract from them

5. Questions and activities

1. Use the ideas in this module to develop a short presentation.
   a. Try delivering the presentation to a friend and have them practice delivering a presentation for you.
   b. Critique your friend's presentation and ask them for feedback on your slides and performance.

2. Watch a few professional presentations and identify how the speaker's slides work effectively or could be improved. A good source of professional presentations related to design can be found on TED talks: www.ted.com/talks

6. References

[1] Alley, M. *Rethinking the Design of Presentation Slides: The Assertion-Evidence Structure*. Retrieved from www.writing.engr.psu.edu/slides.html.

[2] Alley, M. *The Craft of Scientific Presentations: Critical Steps to Succeed and Critical Errors to Avoid*. New York: Springer, 2003.

[3] Alley, M., and Neeley, K.A. "Rethinking the Design of Presentation Slides: A Case for Sentence Headlines and Visual Evidence." *Technical Communication* 52 (4 November 2005):417–426.

[4] Tufte, E.R. *Cognitive Style of PowerPoint*. Cheshire, CT: Graphics Press, 2003.

4

Design for X

Design for Durability

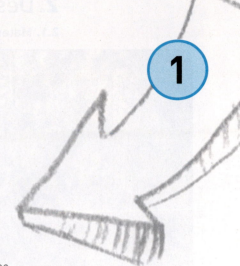

1

#process module: #designfordurability

Learning outcomes

By the end of this module, you should demonstrate the ability to:

- Recognize design decisions that will result in increased design lifetime
- Balance objectives pertaining to durability with other design goals

1. Introduction

When you purchase a piece of technology, you do not expect it to break immediately. You expect it to be useful for a while, but not forever. A student backpack that is designed to last 15 hours is useless. However, a student backpack does not need to be designed to last 100 years either. This section is about engineering design decisions related to the operating life of technology. The goal is usually to make technology that lasts longer before it fails while meeting the other objectives of the design. These decisions constitute design for durability.

Formally, we are talking about extending the ***mean time to failure*** (MTTF), or alternatively the ***mean time between failures*** (MTBFs). This is the average time that can be expected between when the design is first used (or fixed after a failure) to the time it cannot be used because of a subsequent failure. These failures can result from many different causes such as wear, corrosion, or misuse. The primary goal of design for durability is to design first for the prevention of failure, and second to design in a way that takes into account probable failure modes.

2. Design for Durability and Longevity

2.1. Material/Component Reliability

Courtesy of the Authors

By careful choice of material and of the amount of material used, one can extend lifetime. Some materials, such as stainless steel, will last much longer than ordinary steel in most situations. Plastics may be better or may be worse than metal, depending on the circumstance. Where excessive stress or wear is expected, components can be made larger or duplicated. Where corrosion is expected, alternate materials can be found, or components can be coated with resistant materials.

When a design team calls for painting things to avoid rust, or to use thicker materials that will not flex, or when they overspecify parts or when they use redundant parts to share the loading, they are designing for durability.

2.2. Standards and Codes

Often government regulations are linked to durability and following them will produce more durable designs. For example, building **codes** dictate the types of building materials to be used for flooring and walls, as these are materials known to have suitable lifetimes for buildings.

Standards, codes, and regulations can come from three sources:

- Governments, which may legislate regulations and codes
- Industry, where all companies in an industry agree to certain minimum levels of durability to maintain consumer confidence in the industry
- Within a company, where a company requires that all company designs meet minimum standards for longevity. This often is because the company warranties their products to last a certain length of time.

When we design using industry standards and practices, we design using the industry knowledge of durability for the products involved. This is usually published as an ***industry standard***.

2.3. Testing (Initial; Periodic)

Testing protocols or maintenance protocols are an approach to design longevity. Technology is often tested for correct operation when it is first built. However, many types of technology will require periodic testing throughout their lifetime to maintain proper performance. Aircraft, for example, and motor vehicles are often subject to periodic testing, which adds to their durability.

3. Limits on Design for Durability/Longevity

You should also be aware at this point that, when designing for a longer lifetime, one can expect trade-offs in design objectives, such as cost, weight, or function.

> To make a streetcar track assembly last longer requires extensive preparation of the roadbed, specialized track mountings, concrete foundations, and more. The cost is significantly higher than the least expensive options for laying track, but will save money in the long run by lasting significantly longer.

Design for durability, although closely related to design for safety, is not a substitute. Even a product with an expected long lifetime could fail prematurely, and if safety is a concern, safety should be dealt with separately. Similarly, a very safe design does not necessarily have a long lifetime.

It should also be noted that durability can usually be extended significantly through some initial measures, such as application of coatings or moving to another material. However, once these relatively simple techniques have been used, further lifetime extension can become increasingly and prohibitively costly.

The start of design for durability is to know the likely causes of the design failure and to do risk-based analysis. There is no point in spending time making a building's structure stronger if the foundation remains poor and the likely cause of problems. Colloquially, a chain is only as strong as its weakest link.

4. Conclusion

For any technology to be used for a significant time, it is important to design for durability. In some cases it will not be a significant design concern, as the other functional requirements will dictate a design with an appropriate durability. Sometimes, however, it will be a factor, and design efforts will have to be directed toward extending the expected lifetime. Not to do so will result in consequences of failure running from warranty and recall costs, through lawsuits and loss of end-user confidence.

KEY TERMS

mean time to failure (MTTF)   **mean time between failures (MTBFs)**   **codes**
standards (industry standards)

5. Questions and activities

1. Examine a new automobile. What design for durability decisions can you see? Examine:

  a. Material selections

  b. Material treatments

  c. Suggested maintenance procedures

2. For older automobiles, what are the usual failure mechanisms? (You may have to contact a repair service for information.)

3. For the following consumer items, discuss the design for durability decisions that might be involved:

a. clothes iron

b. toothbrush

c. lawn mower

4. What would be the differences in the durability design decisions for a consumer lawn mower and for a commercial mower?

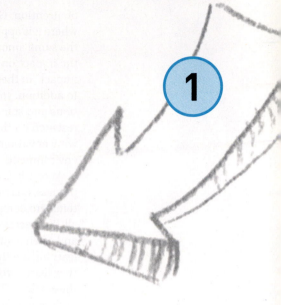

Design for the Environment:
Introduction

#process module: #designforenvironment

Learning outcomes

By the end of this module, you should demonstrate the ability to:

- List and explain the 3 Rs: reduce, reuse, and recycle
- Apply a simple 3 Rs approach to explain how environmental impact was considered during the design process of a technology.

Recommended reading

After this module:

- **Design for X > The Environment > 2. Life Cycle Assessment**

1. Introduction

Assessing the potential impact of a technology on the environment is complicated because the stakeholders may be distant from the technology and because environmental systems are complex and interconnected. The impact of a design on the environment might only be felt over a long period of time, or through a complex chain of cause and effect. However, it is often unanticipated effects on these hidden stakeholders that cause an otherwise excellent design to be considered a failure in the long run. Hence a good designer will investigate the potential impact of the design on the environment, and if possible involve representatives of these stakeholders in the process (government representatives or environmental groups, for example). Disregarding this important consideration can lead to rejection of a project, poor performance of a technology once it is in use, or environmental disaster. However, listening to issues that arise from communities and taking environmental concerns seriously can result not just in good solutions, but in fact can be the motivation for truly innovative design ideas.

There are, unfortunately, numerous examples of destructive environmental impacts that have resulted from the implementation of technology. These often receive a lot of attention. Good design engineers try to take the environment into consideration where it is appropriate. There are plenty of good examples, but they often don't receive the same amount of attention as disasters. Engineers have the opportunity to lessen the impact on the environment considerably by including "minimize environmental impact" in their list of objectives when they are developing the project requirements. In addition, you may have to consider constraints imposed by environmental regulations and standards that govern the project. It is important not only to do your own research on the environmental issues that pertain to the project, but also to consult with environmental experts and organizations that represent the interests of the environment.

While it is easy to say that the natural environment is important to the design process, it is much more difficult to actually implement this idea. Environmental systems are complex and far reaching. It is difficult to see how an engineer can grasp all the necessary information and put it to use when there is so much complexity and often conflicting opinions about what is best for the environment. Listening and learning are the first steps, but engineers need methods to help put the information they learn into a usable framework. This section will discuss strategies for addressing these issues. The methods are taught here at a basic level that is appropriate for an undergraduate design project, but these methods form the foundation for more comprehensive techniques that are used in industry and government.

To understand these methods you need to know a few basic principles that govern the operation of environmental systems. These principles include **conservation of mass** and **conservation of energy**. If you are not yet familiar with these concepts it would be useful to review them because they will be utilized in substantially in techniques for minimizing environmental impact through considerations in the engineering design process.

2. The Three Rs

Designing for the environment involves intentionally thinking through both the mass cycle and the energy cascade in the system you are designing. Virtually all methods for reducing the impact of an engineering design are predicated on intentionally designing to **reduce**, **reuse**, and **recycle** mass and energy. These are called the three Rs and the order indicates the order of preference in terms of design.

- **Reduce**: First, try to design a system that requires a reduced amount of mass and energy. Reduce the amount of mass and energy used in the production of the technology; reduce the amount of mass and energy needed to operate the system; reduce the energy needed to dispose of the system at the end of its life and the mass that is disposed of. For example, reduce packaging of compact disks (CDs) by packaging them in a paper sleeve instead of a plastic jewel case.

- **Reuse**: Second, try to design a system that can be reused as many times as possible. The material and equipment required for production (manufacturing or installation) of the technology should be reusable; the system should be reusable, or components of the system should be reusable; at the end of its life the system should be designed to be easy to disassemble so that components can be easily harvested for reuse. Examples include refillable printer ink cartridges, reusable water bottles, and reusing water (e.g., a grey water system).

- **Recycle**: Third, plan for the system to be recycled. Recycling means using the material (mass) in a different form. Design the production of the technology such that the waste products produced during manufacturing (or construction) can be recycled; design the system so that the waste products produced during operation of the system can be recycled; at the end of its life the system should be easy to disassemble so that the components materials can be easily recycled using a minimum of energy. Examples include recycling plastic water bottles into artificial wood products that can be used to build decks or park benches or recycling car tires into a soft mat product that can be used on running tracks and playgrounds.

The three Rs approach is a basic thinking practice that can be applied during the design process. The goal is to have these ideas in mind at each stage of the decision-making process that goes into the design of a system. Using this approach can result not only in an improved design for the environment, but also a better economic return on the technology. Mass and energy cost money. By reducing, reusing, and recycling these valuable commodities you may be able to decrease costs and improve revenues generated by the design. There are some excellent examples of this principle in action.

> **Other Rs**
>
> In addition to the traditional three Rs, many waste management organizations now include other Rs in their waste planning processes. These include such terms as recover, resell, restore, and rethink.

KEY TERMS

conservation of mass	**conservation of energy**	**reduce**
reuse	**recycle**	

3. Questions and activities

1. In your own words explain the three Rs. Also explain why reduce comes first, reuse comes second, and recycle comes third.

2. Select a common product that you use frequently. Apply the three Rs concept to the design of this product. Can you suggest ways of reducing, reusing, and/or recycling the mass and energy used in operating (using) this product?

3. Apply the three Rs concept in your team design project. Suppose "reduce environmental impact" is one of the objectives for your design.

a. Try creating measureable secondary objectives that would enable the primary objective of reducing environmental impact using the three Rs. For example, if you are designing a product that requires energy to operate you might add the objective "uses the least amount of energy to operate" (i.e., reduces energy usage). Two design alternatives can easily be compared based on this objective. Try creating other objectives that pertain to your design project based on the reduce, reuse, and recycle strategies.

b. Create metrics for your objectives. How will you measure whether one potential design idea is better than another with respect to a particular objective? For example, what method will you use to measure the recycling potential of your design alternatives?

Life Cycle Assessment (LCA)

2

#process module: #lifecycleassessment

Learning outcomes

By the end of this module, you should demonstrate the ability to:
- Describe a life cycle diagram
- Create and explain a life cycle diagram for a product
- List the four steps in a life cycle assessment
- Explain why it is easier to do a comparison of the life cycles of two alternative designs than it is to assess the environmental impact of a single design

Recommended reading

Before this module:
- **Design for X > The Environment > 1. Design for the Environment:** Introduction

After this module:
- **Design for X > The Environment > 3. LCA Goal Definition and Scoping**

1. Introduction to Life Cycle Assessment (LCA)

Design for the environment is difficult to do well because it involves taking into account very complex systems. Using a basic three Rs approach is a first step, but genuinely considering the environmental impact of technology usually requires deep thinking that goes beyond the three Rs. For example, you might assume that it is obvious that ceramic plates are better for the environment than disposable paper plates because ceramic plates are reusable. You might be right, but you cannot be sure until you have carefully accounted for the impact of each stage of the production and use of these products on the environment. For example, to make ceramic dishes requires substantial energy input to a kiln and heating the water to wash them every time they are used. How does this energy compare to that used in the production and use of paper plates? And what is the difference in the environmental cost of disposal of these two alternatives? To answer these types of questions, engineers employ *life cycle assessment*.

Life cycle assessment (LCA), also called *life cycle analysis*, is a rigorous method for analyzing the production, operating, and end of life environmental costs for a technology in terms of mass and energy. In a full LCA all of the mass and energy that goes into, and results from, a system are quantified. It includes all of the resources that go into producing the technology, operating the technology, and disposing of the technology. This type of analysis allows engineers to investigate more thoroughly opportunities to reduce the impact of a design on the environment. It also allows you to think through the trade-offs inherent in design decisions and document the critical thinking process that goes into these decisions. This work is essential for uncovering potential opportunities to reduce environmental impact. It is also a means for demonstrating to project stakeholders the environmental costs that will be incurred during the lifetime of the technology. Once these costs are exposed they can be discussed among the stakeholders in the planning and design process.

The work of developing an LCA is worthwhile when:

- The product or system will be produced in volume (i.e., in large enough numbers to have a meaningful environmental impact).
- A single unit or just a few units can have a meaningful environmental impact (e.g., a power plant or an oil rig).
- You are learning how to do this type of analysis. In professional practice it may not make sense to do a full LCA for a "one-off" item that has only a small impact (e.g., the design of an assistive device, such as a walker, for one person). However, when you are first learning how to do this type of complex analysis it is easier to start with a simple, unique system to hone your skills.

2. A Life Cycle Diagram

An LCA starts by constructing a *life cycle diagram*. The life cycle for a technology includes all of the mass and energy that go into, and results from, producing (constructing or installing) the technology. It also includes all of the mass and energy that go into the technology during its operating life, and that results from (comes out of) operating the technology. And finally it includes all of the mass and energy that go into, and that result from, decommissioning, disassembling, and/or disposing of the system at the end of its life.

Figure 1 shows a very simple generalized graphic of a life cycle for a product or system. The life cycle starts with *raw materials* (RMs) as they exist in or on the earth (ore, lumber, water, etc.) or as recycled materials available for use. In this simple example raw materials undergo processes (Process 1 and Process 2), which result in intermediate products (P1 and P2). To accomplish these processes energy is required ($E1_{in}$ and $E2_{in}$). During the processes some of the mass may be discarded as *waste* into the air or water or as solid waste. This is called a *residual* (Res 1 and Res 2). There will also

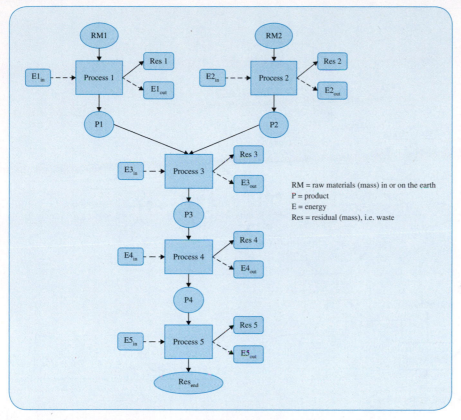

FIGURE 1 **Generic schematic of a life cycle diagram.** At each step there is mass and energy entering the process and mass and energy leaving. The mass leaves each step as a product or as a residual (i.e., waste).

probably be some energy that is discarded as well because (recalling conservation of energy) all of the energy entering with the raw materials or added during the process must either exit with the products or leave as discarded energy (E1$_{out}$ and E2$_{out}$). The intermediate product may undergo further processing (Process 3) before it becomes a usable system or product. The result (P3) is then put into operation. As the technology is operated (used) it continues to undergo a process of use (Process 4) until it is retired from service. The result, at the end of the life cycle, is a product (P4) that needs to be disposed. The last process (Process 5) is the disposal process and results in a final waste product (Res$_{end}$).

This diagram is very basic. Most life cycles will involve many more processes (e.g., the transportation of materials and intermediate products). An example of a simple life cycle diagram for a product is shown in Figure 2. An LCA diagram represents a method for tracking the energy and mass that flow through the life cycle of an engineered product. This method can be applied to any product or system.

There are software packages that can be used to construct a life cycle diagram for more complex systems and analyze it. Examples of LCA software include GaBi

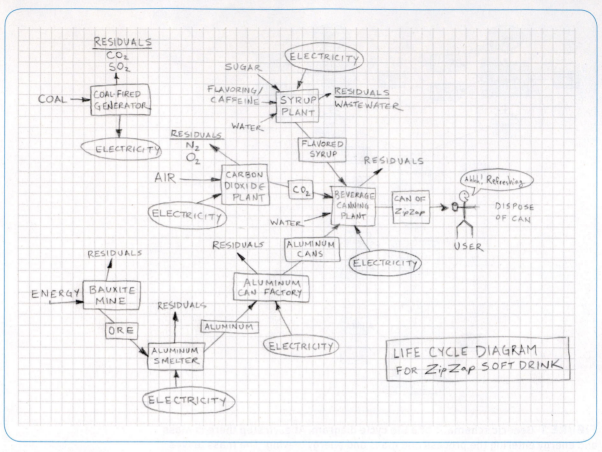

FIGURE 2 Life cycle diagram for a carbonated drink product. Note that the LCA diagram for this product includes both the drink and the packaging. Courtesy of Prof. David Bagley, University of Wyoming, Laramie.

Software and SimaPro. For a simple system a sketch of the life cycle diagram and the use of a spreadsheet to manage the data will suffice.

Drawing the life cycle diagram is a necessary precursor to a life cycle assessment. A standard LCA consists of 4 stages:

1. *Goal definition and scoping*: where the engineer decides what processes will be included in the assessment, and what processes are beyond the scope of the assessment.

2. *Inventory analysis*, data gathering: where the engineer gathers information about each process to inventory the mass and energy inputs and outputs throughout the life cycle.

3. *Impact analysis* of the life cycle: where the engineer assesses the environmental impact of every process, and all of the processes together in the life cycle.

4. *Improvement assessment*: where the engineer uses the impact analysis to identify key points in the life cycle where improvements can be made that reduce the environmental impact.

3. Easy LCA

Working out a full LCA can be a daunting task. Two approaches that can make this process easier are:

1. Perform an LCA comparison between two design alternatives

2. Shorten the process by performing a 3 Rs analysis after defining the goal and scope

The easiest type of LCA to perform is a comparison between two or more alternative designs (i.e., two different designs that are intended to serve the same purpose). An example is shown in Figure 3. We recommend that the LCA be used as a comparison tool at this stage in your career. There are several reasons why using LCA as a comparison tool is easier than performing an LCA for a single proposed design. First, an analysis of this type is often very difficult to interpret. If you determine that the proposed design will produce 500 kg of CO_2 over its lifetime, is this a lot, or a little? It is very difficult to form a judgment in the absence of experience or comparators. Using a comparison you could assess which of the two designs produces less CO_2 and which produces more as one measure of environmental impact.

Second, comparing systems allows you to focus on the specific processes in the life cycle that differ between the two systems. This allows you to neglect, or at least simplify, some of the other parts of the life cycle, which can make the information gathering you will need to do much more manageable. For example, suppose you are comparing the impact of bottling ZipZap (see Figure 2) in plastic bottles instead of aluminum cans.

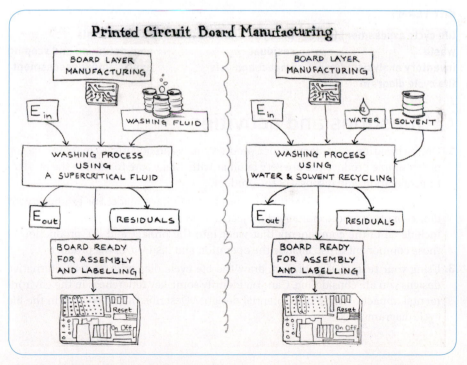

FIGURE 3 The LCA process is simplified by comparing two life cycle alternatives for the same product or system.

In these two alternatives the production of the ZipZap drink is the same so you can focus the LCA just on the packaging process to assess the environmental difference between the two alternatives. Also, related to both of these reasons, a comparison of two LCAs makes it much easier for you to write about the results because it provides a context (i.e., compare/contrast, similarities/differences) for you to discuss.

Another way to do an analysis that is better than a simple 3 Rs analysis but much less work than a full LCA is to combine the two methods. Use the LCA method to draw a life cycle diagram and define the goal and scope of the life cycle. Then use a 3 Rs analysis for each process in the life cycle. This is not actually an LCA, but it is a strategy for examining each stage of the life cycle for possible environmental improvements. This approach is more comprehensive than a simple 3 Rs analysis, but more manageable than a full LCA.

Thoroughly documenting your work throughout a life cycle assessment is very important. The credibility of your conclusions at the end of the analysis depends on how well you have explained and described your decision processes at each step as you perform the analysis. A typical LCA requires that you make assumptions, mix data of different types and qualities, and make significant decisions about what to include or exclude from the analysis. In these respects it is very typical of many engineering analyses you will have to perform in your work as an engineer. A thorough documentation of the process explains this to your reader as they work their way through your analysis; the documentation should "teach" your reader through the LCA so by the end they understand what you have learned by doing the analysis.

KEY TERMS

life cycle assessment (LCA)	life cycle analysis (LCA)	raw materials
waste	residual	goal definition and scoping
inventory analysis	impact analysis	improvement assessment
life cycle diagram		

4. Questions and activities

1. Draw a basic life cycle diagram for a product you use frequently. To make this easier, pick a simple product you are very familiar with. Make sure to include the production, operation, and disposal of the product.

2. Draw a basic life cycle diagram for a product you have made. For example: your dinner or other food, a craft or art project, a building project, or a project for school. Include all of the components that went into the project and the production of those components. Also include the operation and disposal of the product.

3. Using your team design project, draw the life cycle diagrams for two alternative designs you are considering. Can you identify some key differences in the environmental impact of these two potential designs? Describe the differences in the life cycle diagrams.

LCA Goal Definition and Scoping

3

#process module: #LCAscoping

Learning outcomes

By the end of this module, you should demonstrate the ability to:

- Define the goal of a life cycle assessment (LCA) analysis
- Identify the scope of an LCA analysis: select which process to include in the assessment
- Select which processes in an LCA to put in the foreground and which ones to put in the background, and explain your choices
- Identify a functional unit for comparison between two LCA analyses
- Given a case study, identify the goal and scope used in the case and discuss the choices made

Recommended reading

Before this module:
- **Design for X > The Environment > 2. Life Cycle Assessment (LCA)**

After this module:
- **Design for X > The Environment > 4. LCA Inventory Analysis**

1. Goal Definition and Scoping

Goal definition and scoping is the first step in a life cycle assessment (LCA). The approach to LCA discussed in this module is adapted from the process described in *Life Cycle Assessment: Principles and Practice* [1].

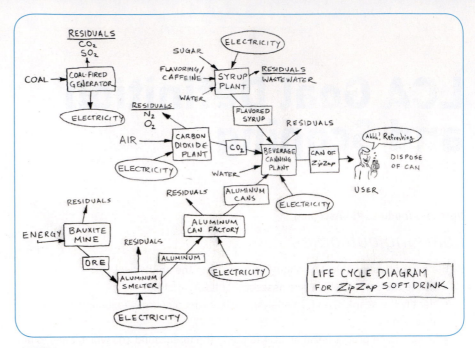

FIGURE 1 Life cycle diagram for a carbonated drink product. Note that the LCA diagram for this product includes both the drink and the packaging. Courtesy of Prof. David Bagley, University of Wyoming, Laramie.

Scoping the life cycle means deciding which parts of the life cycle to include in the analysis. Because of the interconnection between technologies it is virtually impossible to construct a comprehensive life cycle for any system. For example, in the relatively simple life cycle example shown in Figure 1, should you also include the trucks that transport the bauxite ore to the smelting plant? You could try to take into account all of the energy and materials needed to construct and operate these trucks. And what about the ink used to label the cans of ZipZap? The labeling ink is also a product that must be manufactured and supplied to the beverage canning plant. It quickly becomes obvious that it is impossible to trace every aspect of the life cycle for this one simple product. In reality to make a life cycle analysis useful an engineer must choose to focus on the important aspects of the life cycle, and ignore the relatively trivial components. This is called scoping, or specifying a scope. For example, if one truckload of ink satisfies printing needs for the factory for the whole year and there are 50 trucks full of product leaving the factory each day, it is probably reasonable to ignore the ink transportation in the LCA. Another way to reduce the complexity in a comparative LCA is to omit steps common to both alternatives (shipping the final product, for example).

1.1. Step 1. Goal and Scope

This step requires that the engineer make decisions about what parts of the life cycle they will focus on, and which to ignore. Scoping the life cycle is one of the most difficult parts of the LCA process, and the choice of *scope* will depend on the goal of the analysis. An LCA can have as a goal [1]:

- To help guide design decisions. The goal is to use the LCA to compare and contrast alternatives in the design process or production of the design on the basis of environmental impact.
- To identify the steps or processes in the life cycle that contribute most significantly to environmental impact. The goal is to identify these steps so that alternatives can be found or researched.
- To support product or process certification. The goal is to produce a credible LCA that will convince a certifying organization that the process or product is deserving of a special designation for environmental friendliness (e.g., a "green" certification).

There are other possible goals, but these three are the most common. Start the documentation of your LCA by stating your goal.

Next, choose what processes to include in the life cycle. Some guidelines for selecting processes to include when drawing the life cycle diagram:

- Start with the operation of the system, structure, or product. This process is the foundation of the life cycle (in the example, this process is the user purchasing and drinking the ZipZap). Identify all mass and energy inputs that are needed for this operation, and all of the outputs from this process.
- Add the "downstream" processes. These are the processes that must occur to dispose of or decommission the system when it has reached the end of its life. Add the energy and mass flows that occur during this phase.
- Now choose the "upstream" processes to include. What you choose will depend on the goal of the LCA. However the upstream processes should go far enough back in the life cycle to include the extraction or recovery of the primary raw materials needed for the production of the system or product. The raw material could be a substance in its natural form in the environment, or a recycled substance.

At this point you also need to decide what to leave out of the life cycle. The processes in a life cycle can be categorized as primary, secondary, tertiary, and so on. The primary processes are those that are necessary for the primary mass and energy streams needed to produce the technology. For example, in the case of ZipZap, the syrup, the water, the aluminum for the cans, and the carbon dioxide are all primary mass flows required for the production of this product and should be included in the life cycle. The secondary processes would include the ink for labeling the cans, and the production of the sugar or flavoring for the syrup. These are processes that are secondary contributors to the life cycle. You have to decide whether these secondary processes, such as the production of the flavoring, should be included in the analysis or left out. An example of a tertiary process in this example would be the production of packaging for the flavoring (i.e., the production of the bags or barrels used to transport the flavoring ingredient to the syrup plant).

If possible, use the purpose or goal of the LCA to make these decisions. In addition to going back to the LCA goal for guidance on what to include and what to leave out, you can also run a brief *sensitivity analysis* to estimate the contribution of the secondary processes. If you find that a secondary process contributes only a small percentage to the primary mass or energy flow, then it is probably reasonable to leave it out. All decisions about what to include and what processes to exclude should be carefully documented (i.e., explained). You may also want to come back and review your decisions about scope once you have worked through the other stages of the LCA.

Example

If the goal of the LCA is to compare two alternative drink ideas for ZipZap, one that involves using brown sugar and the other refined white sugar, then the sugar production process is an important point of comparison and should be included. However, if the LCA is being used to compare two drink ideas for ZipZap that both use the same amount and type of sugar, then the sugar production process can be left out because it will not affect the conclusions of the LCA in terms of contrasting the environmental impact of the two design ideas. It should be noted that leaving out the sugar production process will remove it from consideration. Thus improvements to the environmental impact of the design based on a better sugar production process will also be removed from consideration with this decision. This should be documented so you and your client are aware that through this decision, improvements in this area of sugar production are not being considered in your analysis.

One class of contributions that is typically excluded from an LCA is the *capital equipment* required for the production of the design. For example, in the life cycle of a building the impact of manufacturing the dump truck or excavator used during the construction process would be excluded from the environmental impact analysis. The analysis would typically include the fuel the truck uses, but not the manufacture of the truck itself. The impact of constructing the equipment that will be used again for other projects is really tertiary to the project's life cycle and therefore can be excluded. In the ZipZap example you would exclude the impact of constructing or manufacturing the aluminum smelting equipment, the equipment in the canning plant, the equipment in the carbon dioxide plant, and other capital equipment. However, you would include the energy used by this equipment during the production process because the energy used in the manufacturing process is primary to the product.

1.2. Step 2. Specifying Foreground and Background

Once you have drawn a first draft of your life cycle diagram and decided which processes to include or exclude from consideration, you need to decide which processes to place in the *foreground* and which to treat as *background* [1]. The processes in the foreground are those that are unique to the life cycle being analyzed. The data for these processes will be specific to the technology you are analyzing. The processes in the background are common, generic processes for which you will gather average data. For example, in the production of ZipZap the electricity production is in the background. The power plants are there already and not specifically dedicated to the production of ZipZap. You would gather data on electricity production for the analysis that is typical (average) for the region in which the plant is located. The same would be true for the aluminum smelter, assuming that ZipZap is getting its aluminum from general sources and not developing a unique manufacturing facility just for the production of this beverage. The foreground processes for the ZipZap life cycle are the syrup plant, the canning plant, the distribution of ZipZap to stores, the using of Zip-Zap (i.e., the user drinking the product), and the disposal process. The aluminum can plant may be foreground if its only function is the production of ZipZap cans, or in the background if the company are acquiring the cans from a general source (i.e., getting cans from several can factories that make cans for lots of drink producers).

- Foreground processes → individual, unique data for the process or plant
- Background processes → average data (e.g., regional data or values that are typical for a particular industry)

1.3. Step 3. Choose a Functional Unit

When comparing two or more possible design alternatives it is important to choose a comparable *functional unit*. A functional unit is an amount that has an equivalent *function* to another system. For example, you could compare the environmental impact of 2000 250 L cans of ZipZap to 1000 500 L bottles of another drink. These units are functionally equivalent because they carry the same amount of fluid. A more complex comparison would be the comparison of disposal plastic water bottles that

are used once to reusable metal water bottles. To carry out this comparison you need to estimate how many times the metal bottle could be reused (maybe 500 times) and compare this to the equivalent number of disposable plastic bottles of the same size. Whatever functional unit is chosen, the amount should be large enough so the data represent the typical average energy and mass requirements for the processes. So you could assess the environmental impact of manufacturing, using and disposing of 1000 metal bottles, and compare this to the environmental impact of manufacturing, using, and disposing of 500,000 plastic bottles, the functional unit equivalent.

LCA Goal Definition and Scoping

Estimating a Functional Equivalent

It would be reasonable to guess that a typical person might buy a plastic bottle of water and refill it a few times during a day before disposing of it. So to determine a functionally equivalent usage for a metal water bottle you need to estimate how many days a metal water might be in use during its lifetime.

(*continued*)

You could get this information from:

- An independent research study by a university, government, or NGO, which is considered to be credible.
- A study carried out by industry (i.e., the plastics industry or the metallurgical industry, which is considered to be less credible because of the inherent conflict of interest).
- Other sources of information.
- An educated guess; "educated" means that we have some experience or knowledge that helps to inform the guess.

If you use an educated guess, based on your own experience with water bottles, you might guess that a metal water bottle would typically be in use for 1 to 3 years, or about 365 to 1095 days. So you might pick 500 days initially to get started on your analysis, and then refine this number later through additional research if necessary. You would document your reasoning to show how you selected 500 plastic water bottles to be the functional equivalent of 1 metal bottle.

Summary of goal definition and scoping steps:

- Select a goal.
- Identify all primary processes in the life cycle from materials extraction through to disposal. Make a list of these processes.
- Decide on secondary processes to include. Use your stated goal, and a brief sensitivity analysis to make your decisions. Make a list of these processes.
- Recall that capital equipment is generally left out of the life cycle assessment.
- Draw out a clear life cycle diagram for your assessment.
- Identify which processes are foreground and which are background, and document these choices (i.e., write this down with an explanation).
- Choose the functional unit, and document your choice.

Applying a three Rs Analysis

At this point you could continue with the LCA methodology and proceed to inventory analysis. Alternatively you can exit the LCA methodology and use a three Rs approach to analyze each of the processes in the life cycle. This will not produce an LCA, but is more comprehensive than a simplistic three Rs approach that only considers a small part of the life cycle. For example, in the ZipZap life cycle you could identify that the environmental impact of the product could be reduced by using recycled aluminum for the cans instead of freshly mined material. You could support it by finding a source that compares the environmental costs of recycled aluminum to the use of freshly extracted metal. This is a qualitative statement that you have not supported with data, so it is less credible than a full analysis. However, if environmental analysis is only a small part of your design project, then you may want to take this short cut. If your instructor is expecting a more complete and in-depth LCA then continue to inventory analysis.

KEY TERMS

scoping

foreground process

goal definition and scoping

sensitivity analysis

background process

capital equipment

functional unit

2. Questions and activities

1. Explain why it is important to define a goal for a life cycle assessment.

2. Select a relatively simple product that you are familiar with, such as potato chips, apple juice, paper clips, a wool sweater, and so on. Ideally choose something that contains only a few components and packaging.

a. Sketch a life cycle diagram for the item including the packaging (if there is any).

b. Write a brief goal and scope report that goes with the sketch to explain which process you included and why. Explain which processes you have neglected in the diagram and why.

c. In the report, identify which processes should be in the foreground and which in the background. Explain your choices.

3. Select a product that can be delivered in two or more formats. For example:

- Music: CD or MP3 file format
- Soap: liquid hand soap or bar of soap
- Plate: paper plate or ceramic plate
- Juice: frozen concentrate or bottle of juice
- Beans: dried bag of beans or canned beans

If you were going to perform a comparative life cycle assessment you would first need to determine a functional unit for comparison.

- Using the product you have selected, identify an appropriate functional unit. For example, how many cans of beans and what size cans are functionally equivalent to how many bags of dried beans and what size bags.
- Explain how you determined the functional unit including any calculations.
- Include in your explanation any assumptions you are making or research you did to inform your answer (e.g., did you go to the store to look at bean cans? Or on-line to get information about dried beans?)

3. References

[1] Scientific Applications International Corporation. *Life Cycle Assessment: Principles and Practice.* National Risk Management Research Laboratory, Office of Research and Development. Cincinnati, Ohio: U.S. Environmental Protection Agency, 2006.

LCA Inventory Analysis

#process module: #LCAinventoryanalysis

Learning outcomes

By the end of this module, you should demonstrate the ability to:

- Setup a spreadsheet to organize an inventory analysis for a life cycle assessment (LCA)
- List all of the processes in the LCA and categorize the inputs and outputs for each process
- Collect information on the mass and energy flows for each processes in a simple LCA
- Correctly enter and convert the quantities in the inventory to reflect the functional units chosen and consistent units of measure (SI or British units)

Recommended reading

Before this module:

- **Design for X > The Environment > 3. LCA Goal Definition and Scoping**

After this module:

- **Design for X > The Environment > 5. LCA Impact and Improvement**

1. Inventory Analysis

Inventory analysis and data gathering is the second step in a life cycle Assessment (LCA). The method discussed in this module is adapted from the process described in *Life Cycle Assessment: Principles and Practice* [1].

Once the life cycle has been scoped you will need to perform an inventory analysis. This means quantifying all of the energy and mass inputs and outputs for each process. It is sometimes easy to find good quality data to use by doing some searching. However, more often it is necessary to estimate the amounts of mass, and energy going in and out at each stage of the life cycle. The quality of the LCA depends on the quality of the data and estimates that are used. Therefore it is important to use best practices in the estimating procedure and gather credible information.

When developing the inventory, keep in mind the laws of conservation of mass and energy: What goes in must come out at each process step.

1.1. Step 1. Start Your Spreadsheet

In order to keep all of the data you will collect for your inventory organized you should use a spreadsheet. The spreadsheet will help you keep track of all of the data, making sure you do not have overlapping data or gaps in your lists. It will also help you convert units, and convert your raw data to the correct amount for the *functional unit* you have chosen. Ultimately it will make it easier to add up the impacts of different design alternatives and compare the results.

Start your spreadsheet by grouping and listing all of the processes in your life cycle. Include descriptions of the processes. Grouping means grouping together individual processes that all occur at the same time or in the same location. For example, there are numerous processes that go into creating an aluminum can, but you would group these together into an "aluminum can production" process in your spreadsheet. A description would explain the type of process used. This is especially important if there are a number of different processes that are commonly used in industry for this type of production or construction activity.

1.2. Step 2. List and Describe the Inputs and Outputs

For each process, enter into the spreadsheet all of the inputs and outputs. The mass inputs and outputs should include all gases, liquids, and solids. It is also useful to add a column to your spreadsheet that characterizes the output mass flows. The output mass can be described as a:

- Product that is the intended useful output of the process
- *Co-product* that is a useful, but unintended, output of the process
- *Waste* (production or post-consumer) that is an unintended and not useful output of the process

All waste and co-products (i.e., *residuals*) should be listed in the spreadsheet for each process. This includes waste going into the ground, into a body of water, or released into the atmosphere. Carbon dioxide and other greenhouse gas emissions should be included in the inventory. A column in your spreadsheet should be used to describe each residual output. The description should indicate whether the output is a product, co-product, or waste. If it is waste it should be characterized as non-hazardous, or hazardous with a brief description of the hazard it represents.

> ### Waste
>
> Waste that is generated by the processes that produce the technology is called *production waste*. This could be waste generated during manufacturing or construction of the technology or system. The disposal of the production waste is part of the impact of the technology on the environment and needs to be accounted for in the LCA.
>
> The waste that results from the operation of a system or technology or at the end of the useable life of the design is *post-consumer waste* and your life cycle analysis must include the disposal of this type of waste as part of the assessment.

LCA Inventory Analysis

383

Water represents a common input and output mass flow from many industrial processes. If the water is used, treated, and returned to the environment in essentially original condition (i.e., clean) then the water doesn't need to be included in the inventory. However, the treatment process must be included in the life cycle to accurately count the cost of remediating this important resource. If the water is "used up" in the process (through evaporation, incorporation into the product, or polluted without treatment) then the water must be listed in the inventory both in its input form and as an output with a description of its condition when it exits the process. You need to include the disposal of polluted water in your LCA if it constitutes a potentially significant environmental impact that results from your design.

The energy inputs should also be categorized. The energy used in a process can be heat, electricity, or transportation (i.e., energy required to transport mass from one location to another). The energy outputs should also be categorized (i.e., described as heat, electricity, potential energy, or other types of energy). Fuel going into a process counts as both mass and energy in the LCA inventory. The fuel going into a process as a mass flow should be described both in terms of its mass description (e.g., m^3 natural gas, kg coal, or L oil) and in terms of its energy content (which is called the **heating value**). The description of the energy will be important when you try to quantify the energy going into and out of each process.

1.3. Step. 3 Quantify the Inputs and Outputs

You now need to quantify each input and output of energy and mass for each process. As you gather the data and put it into your spreadsheet make sure you note in the spreadsheet the source of the data. The data should be put in the spreadsheet as it appears in the source to make it easy for you (and your reader) to see where the data came from. Then convert the values in another column, as necessary, into the units you are using or convert to reflect the functional unit you are analyzing. The quality of the data should also be noted. Is this piece of data an estimate? Does it come from research, or from an industrial measurement? Is it specific to a particular type of equipment or is it a general average? The quality and credibility of the LCA depends on the quality of the data and the documentation of the data.

Remember that for **foreground processes** you are trying to find specific data that reflects as exactly as possible the process that you are proposing for this part of the life cycle. The data for mass should be expressed in units of mass or weight. Energy units tend to be more mixed (kJ, kW, Btu). Be careful in converting these units, and do not add them together. Adding a kilojoule of heat to a kilojoule of electricity is like adding a kilogram of apples to kilogram of arsenic. A kJ of heat is not equivalent to a kJ of electricity because the environmental cost of producing one is quite different from the other (just as the environmental cost of disposing of arsenic is very different than disposing of apples).

For the background processes you can gather average data. For example, the production of electricity in a region generally involves a mix of different technologies. It probably involves burning fossil fuels, some electricity production using nuclear power plants or hydro-electric plants, and perhaps some other power generating technologies such as wind turbines. The environmental cost of this electricity production will depend on this local mix of technologies. For example, as of 2006 the U.S. average was approximately 11.3 MJ of energy input for every kWh of electricity produced [1]. This takes into account the energy used to extract the fossil fuels, transport them, and the energy inherent in the fuel that is "used up" when they are burned to create electricity. This type of data is often available through government sources, such as the U.S. Department of Energy.

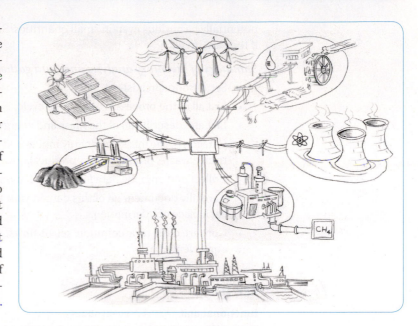

Data for **background processes** should reflect common practice in the region where the process will occur. You should not assume either the best practices will be used unless this is specified as part of your design, or the worst, such as illegal disposal.

When quantifying the residuals from each process it is useful to characterize the location of the outputs. Is the mass flow collected and stored? Or is it emitted into the environment? If it is emitted, is the impact local (for example, into the local soil), or national, or international (such as carbon dioxide gas)? This information will help you later in the LCA process when you try to characterize the impact of the life cycle.

1.4. What If You Can't Find or Even Estimate All of the Data?

It may not be possible to find all of the data you would want to have for your LCA. Some of the data may be **proprietary** (i.e., not in the **public domain**). Or it may take months of searching to find it, which is not feasible for an undergraduate project. It may not even be possible at this stage of your career, with limited experience, to create a meaningful estimate or educated guess for all of the values you need. Consulting with your instructor or a reference librarian may help you track down some of the information. However, where no information is available, you need to note this in your spreadsheet. Clearly indicate the data that is missing, and discuss this in your report.

1.5. When Can You Neglect Data?

You can ignore input or output data when the contribution is less than the error in the data for the substantial contributing mass and energy flows in the life cycle. Be careful that you document such omissions and that you have compared the data correctly before excluding them. Also, you should not ignore waste mass flows if the waste is potentially hazardous even in small quantities. Such emissions should be noted even if they are small.

Summary of inventory analysis and data gathering steps:

- Setup your spreadsheet.
- List all of the processes in your life cycle.
- List and describe all of the mass and energy flows for each process.
- Categorize the mass and energy inputs and outputs (e.g., heat, transport, or electricity; waste, product, or co-product).
- Collect and compile data for all of the mass and energy inputs and outputs.
- Carefully document all of this data in your spreadsheet including the source for each piece of information.
- Convert data to a common set of units, and to reflect your functional unit equivalents.

KEY TERMS

functional unit	co-product	waste
production waste	post-consumer waste	residuals
heating value	foreground processes	background processes
proprietary	public domain	inventory analysis

2. Questions and activities

1. Select one process in a life cycle of a product, such as the process of using a toothbrush, making coffee, eating cereal, using a washing machine, or some other process you are familiar with. Use a large number of repetitions of this process (e.g., brushing your teeth 1000 times) as the functional unit. Then create an inventory analysis for this process.

   a. Create a spreadsheet to inventory this process.

   b. Categorize the inputs and outputs for all mass and energy. What type of energy is needed? What types of energy are produced in the process? What types of mass (products, waste, co-products) are used and produced?

   c. Estimate the amounts of every energy and mass input to the process and output from the process. Are you able to find or estimate these quantities? Make sure to cite sources if you use published data (e.g., the back of the cereal box, or information from the Internet).

   d. Comment on the quality of the data. Is it from a published credible source? Or is it your own estimate from experience? Or have you done controlled experiments to determine this information?

   e. Convert the data into common units: use the functional unit selected to calculate the total input and output amount for a large number of repetitions of the process. And convert all units to a common system (SI or British units).

f. Create a brief report to accompany your inventory. Explain:

- How you approached researching the information
- What assumptions you made
- The range of quality in your data: which data are most accurate, and which are rough estimates
- What this inventory indicates in terms of waste production and environmental impact of this process

3. References

[1] Scientific Applications International Corporation. *Life Cycle Assessment: Principles and Practice.* National Risk Management Research Laboratory, Office of Research and Development. Cincinnati, Ohio: U.S. Environmental Protection Agency, 2006.

5

LCA Impact and Improvement

#process module: #LCAimpact&improvement

Learning outcomes

By the end of this module, you should demonstrate the ability to:

- Define the impact categories for a life cycle assessment (LCA) impact analysis
- Reorganize data from an LCA inventory analysis into an impact analysis worksheet
- Use midpoint assessment to draw conclusions about the relative environmental impact of alternative processes or technologies
- Use an LCA impact analysis to identify key areas of a lifecycle that could be improved to reduce environmental impact

Recommended reading

Before this module:

- **Design for X > The Environment > 4. LCA Inventory Analysis**

After this module:

- **Design for X > The Environment > 6. Sustainability**

1. LCA Impact Analysis

Impact analysis is the third step in a life cycle assessment (LCA). The method discussed in this module is adapted from the process described in *Life Cycle Assessment: Principles and Practice* [1].

By the time you have finished collecting and organizing the data for the inventory you probably already have some idea of what the most significant contributors are to the environmental impact of your proposed design. In the impact analysis stage you will aggregate your data based on scientific principles and begin the process of formulating conclusions based on the data. In the impact stage you will be describing the relative stressors on the environment that may cause impact based on the data you compiled in the inventory.

1.1. Step 1. Define Your Impact Categories

You will be organizing your inventory data into categories that describe the potential impact of the mass or energy flow (see Figure 1). Typical impact categories include human health, environmental health, and resource depletion. However, you can define other categories, or break these major categories down into subcategories, if there are specific impact issues that you want to consider in this analysis.

Start a new worksheet in your spreadsheet. In this spreadsheet list your impact categories.

1.2. Step 2. Organizing Your Data into the Categories

All of the data you have collected in your inventory will now be organized into one or more of the categories you have defined (e.g., human health, environmental health, and resource depletion). When you are finished your data will now be organized in two ways:

- First sheet: During the inventory analysis stage you organized the data by process according to the lifecycle diagram.
- Second sheet: During the impact analysis stage you will duplicate the inventory data but organize it by impact category.

You may find that the data fit into more than one of these categories, and the data should be copied into all of the categories that are relevant because the data in each

IMPACT ANALYSIS			
Mass and energy description	Amount	Originating process in the lifecycle	Impact: Midpoint assessment
Human health Aerosol particulates Carbon monoxide			
Environmental health Aerosol particulates Waste heat from process 1 Waste heat from process 2 Carbon monoxide Carbon dioxide			
Resource depletion Diesel fuel			

FIGURE 1 Example spreadsheet organized by impact category.

category will be used to derive a conclusion about a different type of potential impact. For example, nitrogen dioxide (NO_2) can both contribute to environmental degradation through the acidification of water (lakes or ponds) and also affect human health because it is a respiratory irritant, so it should be included in both categories. Make sure that you identify, the origin of the data in the new spreadsheet (i.e., with which lifecycle process the data is associated).

Advanced Material

Step 3. Determining Equivalencies

You might want to aggregate the data in the inventory so you can do a deeper comparison between two alternative designs. To do this you need a scientifically based method for comparing the potential impacts of different chemicals or other impact types. Engineers use the concept of equivalencies for this purpose. An *equivalency* is a way of converting data into a common set of units that can be added and compared. For example, methane (CH_4) is about 30 times more potent as a greenhouse gas (GHG) than carbon dioxide (CO_2). Therefore, every mol of methane can be considered the GHG equivalent of 30 mols of CO_2. You can use this fact as

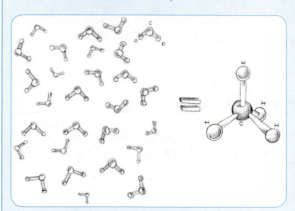

an equivalency to convert a mass flow of methane into the atmosphere into a CO_2 equivalent. You can use this information to identify the processes in the lifecycle that have the most significant impact in this category. It is important at this point to document your observations. Note the processes which are potentially the most impactful and consider whether there are alternatives that could reduce these impacts. For example, you might decide based on your analysis to concentrate your efforts on eliminating the emission of CH_4 from the design lifecycle rather than tackling the CO_2 emissions first.

Step 4. Aggregate the Data

Once the data has been categorized, and equivalencies have been used to convert the data into the same units, you can add up the impact in each category. This gives you a picture of the total impact of the lifecycle in each of the key categories. It also allows you to compare the totals of one lifecycle to another possible design to help you decide which one may represent a better choice in terms of environmental impact. Be careful not to add things together that are not equivalent. You may not be able to find equivalencies for everything, so some mass flows and energy flows will need to be left as is and not aggregated.

At this point you may want to perform a brief **sensitivity analysis**. Save your spreadsheet to a new file so you don't lose your original data set. Then try changing one of the significant inventory data values. How does the aggregate data change? Reset the data and try changing another input. You will probably find that the results are very insensitive to some of the data, which means that even if this data is not very accurate, the inaccuracy does not have a meaningful or significant impact on your results. However, you will probably also find that the results are very sensitive to a few key pieces of data. Go back and check on the accuracy of these key data. Doing a little more work to get better quality data for these key sensitive numbers will improve the quality of your LCA. Make sure you describe the sensitivity of the data to your audience so they understand it. You may also want to examine whether there are alternative processes that could be used to improve the impact of your design based on your knowledge of the key sensitive data points.

1.3. Step 3. Assess Midpoint Impacts

The data tabulated in the impact analysis spreadsheet can now be used to support statements, recommendations, and decisions. For example, you can use this information to explain why one design choice results in less impact on the environment compared to another.

In drawing conclusions based on this analysis it is useful to use the concept of **midpoint assessment**. In midpoint assessment you quantify and characterize the

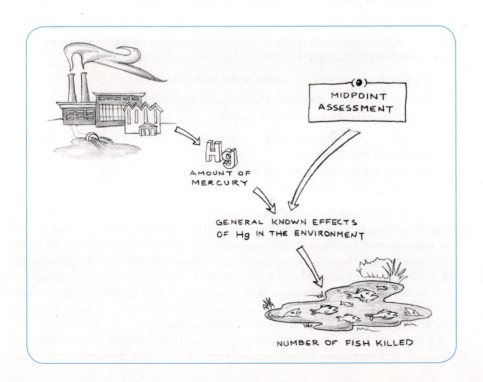

potential impact of the technology or system, but do not try to predict the precise consequences of this impact. For example, you could say that based on your lifecycle assessments, design A results in an overall release of toxic material into the local water that would be significantly less than the toxic material released by an alternative, design B. And you can characterize the type of chemicals that are released and the known effect of these chemicals on marine animals. However, in midpoint assessment you do not try to predict what the actual specific impact is on the water (e.g., number of fish that will be killed). A specific "endpoint" conclusion, such as the number of fish that might be killed, is really not possible based on the data and analysis. The conclusions you can reach are limited. A specific conclusion like the exact number of fish affected overreaches the validity of the information you have available, whereas a midpoint assessment that simply states and characterizes what is going into the environment is justified, credible, and valid.

Summary of impact analysis steps

- Define your impact categories.
- Using a new worksheet, or spreadsheet file, organized your data by impact category. Make sure you identify with which process the data is associated.
- Characterize the nature and impact of the materials and energy being extracted from environment, and going into the environment as a result of this lifecycle (i.e., midpoint assessment).

2. LCA Improvement Assessment

After completing the impact analysis you are ready to state the conclusions of the LCA. This is the fourth step in the LCA process and it is called the *improvement assessment*. Start by summarizing the analysis. This summary should:

- Remind the audience (i.e., reader) of the goal of the analysis
- Briefly remind your audience of the major processes in the lifecycle diagram
- Recall the major decisions and assumptions that went into scoping the lifecycle
- Describe the overall quality of the data used in the inventory, and note the consistency (or inconsistency) in the quality of the data
- Note any major estimations you had to make
- Summarize the results from the impact analysis

Now you can discuss the overall results and state your major conclusions. If you are comparing two alternative designs then it would be appropriate to have a section in your Improvement Assessment that compares and contrasts the results of the LCAs for the two or more design alternatives you are comparing. What are the advantages and disadvantages of each relative to the other? It may be clear that one design is much better, but usually the comparison demonstrates that all design choices have some advantages and some disadvantages (i.e., pros and cons). You need to weigh these results and decide what recommendations to make.

It is also important to note the limitations of your LCA. The quality of the data, the scope, and the assumptions and estimations you had to make in the analysis affect the value of the LCA. It is important to explain your level of confidence in the results and to acknowledge gaps in the analysis; for example, where data is missing from the inventory.

It is also worth noting that an LCA does not fully capture all of the potential environmental impact. For example, a typical LCA does not include disruption to sensitive habitats. It can quantify depletion of a resource, but it does not by itself indicate the importance of that resource for the people, plants, and animals that also depend on it. However, you can use the improvement assessment to point out some of these other issues and suggest changes to the location, or changes to the processes in the lifecycle that could potentially lessen the impact.

The improvement analysis should emphasize this last point: based on what you have learned through doing the LCA, what changes could or should be made to the design and the lifecycle of the design.

Summary of improvement assessment steps

- Summarize the analysis and the results
- Explain limitations of the analysis
- Compare and contrast the two alternatives on the basis of the LCA results
- Make recommendations

3. Services, Software, and Other Virtual Systems and Products

Virtual systems and services do not have a lifecycle that is comparable to a product. There is no need for the steady flow through of raw materials and the disposal of mass at the end of the lifecycle for a service or software package (if it is downloaded). In fact, some service companies were founded on this premise: that replacing a product with a service will reduce environmental impact. Here are some examples:

- Diaper services that replace disposal diapers
- Zipcar that allows consumers to drive a big car when they need one, or a small car when they don't

Some companies and organizations only produce virtual products or provide services with no physical product usage. For these organizations the most appropriate way to evaluate environmental impact is to consider the impact of their workforce and office buildings. This impact is called an ***environmental footprint***. To evaluate environmental footprint an LCA methodology can be applied to the organization's facilities.

Two examples of organizations that have taken this approach are Google and the Audubon Society. Google headquarters is in Mountain View, California. The office complex includes on-site electricity generation using photovoltaic panels, which supply about 30% of their electricity needs. They also grow on site some of the food that is served in the cafeterias. This particular company claims that they actively look for ways to reduce their load on the environment as part of their corporate culture. Similarly the Audubon Society has considered the environmental impact of their office building. The Audubon Society is a non-profit non-governmental organization (NGO) that works on nature conservancy issues. They

intentionally located their headquarters in a remodeled building in New York City. First, by remodeling instead of building a new facility they reduced the waste output of the construction project. The remodeled building is designed to make the most of natural light, reducing the need for artificial lighting. And because the building is located in an urban area, employees have easy access to a public transit system. These are just two examples; there are many more.

One additional example that you might consider is your college or university. Schools are service oriented, usually non-profit, businesses. Does your school have a policy on sustainability or environmental impact? How is this policy incorporated into the facilities design and services on the campus?

The fundamental conclusion here is that whether you are working on an engineering design project for a product, facility, service, or any other type of technology, LCA provides a useful approach to evaluating the environmental impact of the technology. And LCA is particularly useful if you are trying to distinguish between two possible alternative solutions based on "environmental friendliness" or a similar type of objective.

KEY TERMS

equivalency	**sensitivity analysis**	**midpoint assessment**
impact analysis	**improvement assessment**	**environmental footprint**

4. Questions and activities

1. Suppose one process in a life cycle inventory includes residuals (see table below) that are released to the environment as gas into the air.

   a. Using the impact categories human health and environmental health, organize these residuals into a worksheet to assess impact.

   b. Use research to assess the midpoint impact of the residuals on human health and environmental health.

   c. Summarize the environmental impact of this process.

TABLE FOR QUESTION 1 You can assume the total annual mass flow of gas is approximately 100 metric tonnes (1×10^5 kg) per year from this process. The table shows the mass fractions of gases that are in the residual flow into the environment from this process.

RESIDUALS	MASS FRACTIONS (%)
N_2	47.00
CO_2	9.80
H_2S	0.40
CH_4 (methane)	31.40
C_2H_6 (ethane)	5.60
CO (carbon monoxide)	2.60
NOx (nitrogen oxide compounds; NO and NO2)	1.60
SO_2 (sulfur dioxide)	0.80
CS_2 (carbon disulfide)	0.80

5. References

[1] Scientific Applications International Corporation. *Life Cycle Assessment: Principles and Practice*. National Risk Management Research Laboratory, Office of Research and Development. Cincinnati, Ohio: U.S. Environmental Protection Agency, 2006.

6 Sustainability

#process module: #designforsustainability

Learning outcomes

By the end of this module, you should demonstrate the ability to:
- Define and explain: sustainable design, industrial ecology, and preventative engineering
- Describe how these approaches can be applied to the design of a technology
- Analyze a case study to identify how the principles of sustainable design, industrial ecology and/or preventative engineering have been used to improve the design

Recommended reading

Before this module:
- **Design for X > The Environment > 5. LCA Impact and Improvement**

1. Introduction to Design for Sustainability

A *sustainable* design is one "that meets the needs of the present without compromising the ability of future generations to meet their own needs" [1].

Designing for sustainability means taking a longer view of the impact of a design. In part this is assessing the potential environmental impact of a project, and working to develop a design that minimizes this impact. However, designing for sustainability also means considering the impact on future generations and including this in the design process. The results can not only be advantageous for the environment, but can also result in innovations that are more economically, and socially effective.

2. Basic Principles of Industrial Ecology

Industrial ecology is one approach to sustainable design. It is a variation on the life cycle assessment method that looks at the life cycle of a product or technology as an *ecological system* analogous to natural ecological systems. Engineers use this perspective to try to find better, less environmentally impactful, approaches to the construction or fabrication of a technology, its use, and its disposal. This approach is based on the principles of conservation of mass and conservation of energy. You will recall that mass is cycled around within the earth's systems, almost always maintaining its original form (at the atomic level), whereas energy cascades through our planetary system. It arrives from the sun as electromagnetic radiation, and undergoes conversions from one form to another gradually losing its usability until it exits into space as low-grade heat.

Consider the waste products and post-consumer waste that are discarded in a typical life cycle for a technology. The waste materials are only considered "waste" because people have not found a use for them. If people recycle or reuse this waste then it becomes a co-product that has value. In nature, nothing goes to waste. All mass is recycled in one way or another. For example, leaves that fall from a tree are recycled into soil and the nutrients are returned to a state that the tree can use again. Or oxygen, a waste product from a tree, is used as a valued input product by animals. So you can look at your proposed life cycle and think about whether you can find a method for remediating the waste into a useable, or at least less toxic, material. The idea is to create technologies that are all part of a closed ecologic system where all mass is recycled into useful products. This is consistent with the three Rs: reduce, reuse, and recycle.

People can also improve the use of energy by matching the use of the energy to the source. For example, the generation of electricity from fossil fuels results in a large quantity of heat being discharged to the environment. However, this heat can be used before it is discharged. It can be used to warm greenhouses in winter for growing vegetables, or to heat other types of buildings. Using an industrial ecology perspective you can examine your life cycle for opportunities to make use of energy multiple times before it is discharged to the environment.

From an industrial ecology perspective engineers look at the design of a technology for use and reuse, rather than taking a "use it up" approach. In its purest form, all processes in the life cycle of a technology should only result in co-products or remediated waste that is environmentally benign. And all of the costs associated with energy production (e.g., remediation of coal mines, recapturing, and sequestering the resulting CO_2) should be included in the life cycle. By making sure that all of the "costs" of creating, using, and disposing of a technology (current and future costs) are explicitly included the true cost of the system can be accurately evaluated. This allows engineers to authentically compare all of the costs of one proposed design to another.

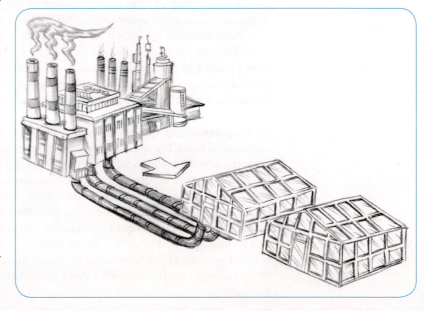

One criticism of the industrial ecology approach is that it does not actively promote reduction of waste. Industrial ecology might suggest that if you are able to turn all of the waste products in a life cycle into co-products then producing more residuals is better. Industrial ecology does not acknowledge explicitly that all production comes at a cost; a cost of energy and an *opportunity cost*, meaning the opportunity to leave the raw materials in their natural state, or for future use. A different approach is *preventative engineering* which makes this distinction and emphasizes reduce as the first key principle ahead of design for reuse or recyclability.

3. Designing for the Future

Considering environmental impact in the engineering design process is not just the right thing to do. The consequences of neglecting this important consideration can have serious consequences both in the near term and into the future. Neglecting potential impacts of this nature can result in lawsuits now and on-going liability in the future. In addition, it can impact negatively on the reputation of the company and your personal professional reputation as an engineer. Technical failures and environmental disasters can also result in substantial public disapproval, which is sometimes translated into intensive legal regulation of an industry. While the intention of regulations is generally positive, to protect the environment and public health and welfare, the implementation of regulations, or an overreaction in the wake of a disaster, can constrain technical enterprise significantly. Although these negative consequences may be important motivators for engineers to consider the environment in the design process, there are also some extremely positive reasons for including this in your thinking.

There are many positive reasons for taking broader issues into account in the design process. First, it adds value to what you do in the world as an engineer. Not only are you creating a technology that will have use, but you are also doing it in a way that you know will have the minimum possible negative impact on the earth. Some companies actually choose to reduce their profits in order to adhere to a corporate dedication to environmental ideals. This is not because they are economically inept. They have done the projections and realize that sacrificing some profits in this way has an overall positive impact on the company's reputation which they believe will have a positive benefit on their sales. This is another example of a reason for designing for sustainability. And there is a growing sector of *for-benefit companies* that seek to maximize benefits for society, the environment, and economics rather than being driven by profit motives alone.

The potential payoffs for thinking through the long term consequences of a design can be substantial. It was estimated, for example, that Toyota was losing about $10,000 on every new Prius they sold when they first marketed this hybrid vehicle. However, the result was that Toyota was the first company to grab a large share of a potentially very lucrative market in hybrid vehicles. Their strategy counted on increasing energy prices and growing regulations on CO_2 emissions that would lead to increased consumer interest in fuel efficient, low carbon emission transportation. Their push into this technology led to other companies bringing hybrids to market, and now consumers have a variety of hybrid vehicles to choose from. In this case the economics incentivized a design alternative with a lower environmental impact.

Sometimes a good design alternative from an economic perspective is also a good choice for the planet. However, whether the "greenest" alternative is the best economic choice or not, environmental impact is always an important aspect of the design process. As a profession, engineers have an obligation to examine the long-term environmental, social, and human consequences of design projects and factor these considerations into decision making because future generations will have to live with the outcomes of what is being designed now.

4. Innovation Opportunities

Taking into account environmental and societal issues in design can be difficult because there are competing stakeholder interests to consider. A business that is renovating their offices may want to use a recycled flooring material to reduce environmental impact, but if the material costs much more than standard flooring they may decide that it isn't worth it to go with the "green" product. As engineers you will face these types of decisions frequently and you will need to balance the competing interests of the stakeholders on a project. It can feel like environmental considerations, and issues brought forward from a local community are just adding more constraints to already difficult technical problems.

However, if you view these issues not as negative constraints but rather as opportunities for innovation, then you have the opportunity to conceive of creative solutions that you might not otherwise look for. An example of this is H-3 in Hawaii, a highway that cuts across the island of Oahu. The construction of this highway was delayed by decades because of the impact it would have had on the natural environment and because of community concerns. Only after the highway was reconceptualized was the project continued. Instead of trying to push through a plan that would have substantially impacted the land, an innovative design was developed that reconceptualized the highway as a bridge spanning the sensitive areas.

By listening to the communities that are impacted by a design, and by including sustainability in your thinking process, you have the opportunity to develop innovative solutions that work better from multiple perspectives. When this is possible you are doing a better job as an engineer for your clients, for the planet, and for future generations.

© Douglas Peebles Photography / Alamy

Sustainability

sustainable design   **industrial ecology**   **ecological system**

opportunity cost   **preventative engineering**   **for-benefit companies**

5. Questions and activities

1. In your own words describe and explain:

  a. Sustainable design

  b. Industrial ecology

  c. Preventative engineering

2. In the last century it was discovered that calcium oxide could be used to reduce sulfur emissions from coal-fired power plants and other industrial combustion processes. Some types of coal and other fossil fuels contain sulfur compounds. When the fuel is burned the sulfur in the flue gas is in a variety of forms such as H_2S, which is an environmental and human health hazard. Calcium oxide slurry is sprayed into the flue gas. This is called flue gas desulfurization. The resulting chemical reaction produces a synthetic form of solid gypsum, a mineral used in a wide variety of technologies. For example, standard wallboard used for houses and offices is often made of gypsum. As a result of this discovery, there was a reduction of gypsum mining and wallboard manufacturing facilities were retooled to use the synthetic gypsum produced in the desulfurization process to produce wallboard.

  a. Analyze this case from the perspective of sustainable design. Is the desulfurization process an example of sustainable design? If so, why? If not, why not?

  b. Analyze this case from the perspective of industrial ecology. Is this an example of industrial ecology at work? Explain. How could the mass recycling be improved, or the energy be better used?

  c. Analyze this case from the perspective of preventative engineering. Industrial ecology alone would suggest that the invention of the desulfurization process means that society should increase its use of high sulfur energy sources, such as high-sulfur coal, because we now have a means for turning the sulfur into a useable product. Is this an appropriate conclusion? How would a preventive engineering perspective differ from this conclusion?

3. Try applying the concepts of sustainable design, industrial ecology and preventative engineering to your own design project. Can you use these concepts to reduce the environmental impact of your design? Can you use these concepts to motivate innovation in your design process?

6. References

[1] World Commission on Environment and Development. *Our Common Future*. Oxford University Press, 1987.

Design for Flexibility:
Introduction

1

#process module: #designforflexibility

Learning outcomes for this section

By the end of this module, you should demonstrate the ability to:
- Evaluate the opportunities for flexibility in a design
- Adjust your design and process for design for flexibility

Recommended reading

After this module:
- **Design for X > Flexibility > 2. Managing Flexibility**

1. Introduction

Design for flexibility is a technique for incorporating into the design potential changes in the requirements that may occur during the design process. Change in the requirements is often expected, but what that change will be is usually not fully known. This approach to design allows flexibility; it allows the design process to adjust to some change in requirements without a huge increase in cost.

Of course flexibility is *not* a good option for all projects. Consideration of design flexibility will take design time and resources, so if there are no changes in the requirements then the additional cost is unwarranted. This approach to design is used for projects that are innovative and not routine, particularly when flexibility may result in a design that has a high profit margin thus offsetting the added cost. Or flexibility may be a priority when establishing a **product line** where future products in the line will build on the established **platform**.

1.1. The Place in the Project Cycle to Add Flexibility

Flexibility in the design can be added at two stages.

In the requirements phase. Flexibility can be viewed as a function or an objective if it is a goal of the design. When flexibility is identified as an objective it is often through client consultation or through a ***use case analysis***. When brainstorming use cases (aspects of use of the design) a variety of possible design features may be discovered. However, the design team may decide some features are not immediately necessary, and leave them for a later iteration in the design process or for the next generation of technology. The team will add an objective to the requirements such as "should accommodate the addition of new features" to acknowledge the need to keep the design flexible.

In the implementation phase. There may be opportunities to add flexibility at the detailed design or implementation phase of the project. If a change to the design makes it more flexible while still meeting other requirements such as cost and safety, then the flexibility can be added, and it may later prove to be useful in some unanticipated use of the design.

2. Tools to Create Flexibility

There are many tools that engineers use to create flexibility. The following examples, which may be of use to you at this point, include examining unconstrained values, designing in modules, using a general solution or a software solution, minimizing the effects of change, and incorporating flexibility in the design process itself.

2.1. Examining Unconstrained Values

It is worthwhile examining every unconstrained or virtually unconstrained aspect of the design and determining if it can be used to advantage for flexibility (or for any other DFX consideration, for that matter, such as increasing safety or reliability).

> **Example**
>
> If the design will be used in an open field, the size and weight of the design may not be constrained. Can the size or weight be varied to increase the flexibility of the design? For instance, if this is a power substation, can the concrete support pad be built larger than what is needed immediately, anticipating that larger power transformers may be required in the near future?
>
> Building in extra capacity is a very standard strategy for design for flexibility, and it often has a very high pay-off.

Here is a list of some of the choices you might make to increase the flexibility of a design:

- Change the size, strength, bandwidth, or other characteristics of the design to accommodate alternative uses or future needs. Doing so may make the design useful in other situations than the one it is specifically being built for.

- Extend interfaces to allow additions. Such extensions could be electrical, physical, or network and software. Putting a few extra USP ports on a device, for example, could be very helpful.
- Use generic interfaces rather than designing your own. There are many standards for connectors and cables, and meeting those standards can allow the extension or replacement of the equipment later at lower cost. For example, it would be frustrating if your new cell phone used a different charging cable than your old one if the phones were in the same product line.

2.2. Design in Modules

A second strategy for identifying opportunities for flexibility is to use ***decomposition***. Try to decompose the design structurally into a set of somewhat independent parts (see Figure 1). These modules may be useful in other designs, and may allow local changes to a module rather than the entire design when scope changes occur. This type of decomposition may also assist in project management because you may be able to assign the development work for each module to a different team allowing parallel work flow.

For example:

- Designing moveable units rather than permanent units. This allows reconfiguration of the layout and potentially alternate uses of the space involved. Many stadiums are now designed this way, so that many different sports can be run in the same facility.
- Building modules, such as software and electronics, in pieces with well-defined interfaces. Later, modules can then be replaced or extended without major redesign. For example, the use of subroutines in software programming allows for modularity of the development work.

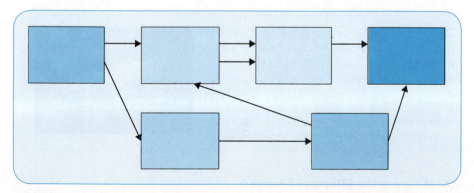

FIGURE 1 Modular design allows changes in modules without affecting the others.

2.3. Use General and Software Solutions

A third strategy is to use design decisions wisely to increase flexibility; choosing a more general solution or one that is easy to change over a solution that is specific or narrow.

In the early 1970s there were typewriters that did word processing. They were basically like traditional typewriters, but included some word processing features. Several companies that dominated the dedicated word-processing market in the early 1970s went out of business when general-purpose personal computers were introduced. The personal computers were able to do not only the word processing, but other functions as well.

Courtesy of the Authors

Software and soft techniques are easier to change than hardware. For example, screen-oriented user interfaces are easier to change than a panel of mechanical switches, and it is possible to have "soft" switches whose function changes depending on circumstance. Software-based process controls are more easily modified than dedicated units. More complex user interfaces can have controls and monitoring intermixed to maximize operator convenience and efficiency. Embedded control software allows many functional changes to be made without costly hardware alterations. In addition, software does not wear out the way mechanical controls do.

Sometimes this means looking at your design requirements and differentiating the parts that are necessarily physical (holding things, moving things) and those that involve control, where the need is only to impart a control decision.

Timing the electric pulses to the sparkplugs in a car engine was done mechanically for a long time. Now this is usually done using processors, eliminating a significant amount of size, weight, and complexity from the mechanical parts of the engine and improving the timing possible under varying operating conditions. The timing is a control function, not a mechanical function.

Courtesy of the Authors

2.4. Minimize the Effects of Change

A fourth strategy is to make design choices that reduce costs or problems when unforeseen changes occur. For instance, if the new design uses the same parts as other designs produced by a company, then there can be cost savings. This is called **design by inventory**. Next best is using **commercial off-the-shelf (COTS)** parts, where the cost is lower and the potential for resale is high should the parts no longer be needed.

You can also look for hooks in the design that could be used later to expand functionality. Examples of hooks that could be used for later expansion:

- Uncommitted inputs/outputs on devices, or other unused functionality.
- For software: Tables, linked lists, databases, and other expandable constructs (it is easier to add to a table than to totally redesign for an additional function). Similarly in the physical world; extra capacity, extra storage or workspace, unused electrical distribution.
- Connection methods that allow additional connections to be made of new devices or other equipment. Networks, for example, allow this as does the "cloud." Standard communication methods, such as universal serial bus (USB), can do the same.

2.5. Use Flexibility in the Design Process

Not only can you increase flexibility by decisions about the design, but you can also decrease the impact of changes during the design process by adopting flexible techniques. Note that, like the flexible choices for the design itself, these techniques can add unwarranted cost and time to projects if change does not occur. Two of the techniques that incorporate flexibility in the design process are promotion of early change requests and incremental design.

Promoting the Early Production of Change Requests

Use of prototyping, including models and user interfaces mock-ups, will help to encourage any changes to happen early in the design cycle when the impact is less. A more general method is "delayed commitment," which means choosing a design scheduling that delays commitment to a design decision as long as possible, thus allowing changes to occur before costly adjustments are required.

It is usually advantageous to heavily involve the client and users of the design during the early stages, because scope changes often come from these sources.

Incremental Design

In incremental design techniques, part of the design is implemented before the full design work is complete. The intent is that the design development will continue while the execution of one part is done. In some circumstances the part of the design produced early may be tested, accepted, and used by the client while the other parts are still being designed. The early delivered design works to some degree, but with reduced functionality. This will only work in designs where a partially-functioning system is of use.

> Note that producing part of the full design is different than producing a prototype. Here we have produced something intended to go into service, whereas for a prototype the intent is to discard the prototype after it delivers the decision information for which it was created.

For this to work there must be advantage in having a partial-design. Perhaps the project requirements are not fully specified or the design work is long and involved and the timeframe to deliver a partially functional product is short, or there is monetary advantage in delivering a partially functional design. In these cases it will be less costly to execute part of the design than to wait until a full design is ready.

Under some circumstances it may be possible to start the execution of part of the design before the exact details have been fully determined. For instance, construction of a building may start before the exact details of, say, the water distribution system, have been fully determined. The building site must be cleared, the foundation poured and so on, and these can be done with minimal knowledge of the water distribution system. This process is called *fast-tracking*.

In the software development field, some projects are done without a firm commitment to the exact details of what the final product will be. In many situations the project requirements will change often. For these types of projects product features are added in small steps. This approach has several variations; many are grouped under the name *agile design*. When used with proper precautions in suitable situations, this technique has been shown to be very effective.

There are other descriptions, names and techniques of incremental design that can be explored by designers interested in the method. A few of these are called helical or spiral development, extreme programming, and rapid application development. Although these were largely spawned to handle software development, they have application in other areas as well.

All incremental techniques must be done carefully. An incremental step might require rework of some results of previous steps where these previous results are incompatible with the new addition. Clearly if the costs of such reworking exceed any advantage of adaptation to scope change, the techniques should not be used. These techniques usually require careful planning, careful management, and consideration by knowledgeable designers.

KEY TERMS

product line	platform	use case analysis
decomposition	design by inventory	commercial off-the-shelf (COTS)
fast-tracking	agile design	

3. Questions and activities

1. Consider how flexibility can be added to a design through:
   a. Unconstrained or virtually unconstrained parameters
   b. Building in modules
   c. Using general solutions
   d. Leaving "hooks" that can be used for adding features
   e. The design process

Using these strategies, describe how flexibility might be applied to the design of the following items:

- Tablet
- Automobile
- Candy bar production equipment
- An air traffic control system
- A computer operating system
- Office desk
- Office lamp
- Stretch of new highway
- Disposable pen
- Security system
- Railway switch

2

Managing Flexibility

#process module: #managingflexibility

Learning outcomes

By the end of this module, you should demonstrate the ability to:

- Be able to determine if design for flexibility is appropriate for a design

Recommended reading

Before this module:

- **Design for X > Flexibility > 1. Design for Flexibility:** Introduction

1. Why Flexibility May Be a Good Idea

There are a number of reasons that flexibility may be a strong design consideration. Flexible design processes may address issues of **scope creep**, design mistakes, and rapid evolution or technical advances in the field. It allows the design team to take advantage of new ideas and features quickly, and adjust if the client's needs evolve during the project.

1.1. To Manage Addition of Function and Handle Scope Creep

Scope creep. The phenomenon of expanding project requirements (i.e., project scope) cannot always be controlled. Sometimes it may be a good idea to expand the scope, and we want to design with this in mind. Sometimes we will design specifically with the intent of allowing later additions of functions to the initial design.

People, and clients in particular, always develop better ideas as time progresses. This is due to a number of factors. They may have had more time to consider and to research their needs more deeply as the project progresses. There may be new features in a competing product that has just gone to market. The detailed design, prototyping, and implementation may suggest ways of further improving the design. Similarly, working with or observing a partly designed or partly built product may spawn new ideas. The marketing department may have gotten new ideas from users or market data.

The people controlling the design process will decide when the scope of the design must change in response to the new ideas being proposed. This will typically be the project manager, company managers, or the design team itself. The designers can expect to spend time determining the impact of the changes, changing the design plans accordingly, and implementing the changes. All of these will add cost of time and materials to the original design. (We won't consider here the inevitable expectation that change can be made at no cost or loss of time.)

> Many product releases, even from large companies, have had "last-minute" software changes and have suffered from unexpected bugs because of these "small" changes when insufficiently tested. A small addition to communications software, for example, could erroneously cause blocking or delays through the communications link, and failure of all software components using that link. The small change has major effect, but these may not be found if inadequate testing is done on the software because the change was "small."

Building a flexible design will often allow additional function to be integrated into the design at less cost and in less time than it might otherwise. A smartphone is designed in this way, allowing new operating system software to be downloaded to accommodate changes in behavior and addition of new features.

> *Example*
>
> The iPhone® is designed with a type of changeable semi-permanent memory (called "flash memory") that holds the program for the operating system. When upgrades to the system are available and the user agrees, the phone will enter a programming mode where the new operating system will be put in place of the old. The designers used flash memory in a special arrangement that added flexibility to the phone, and thus the ability to acquire a revised operating system with new features and functionality when it becomes available.

1.2. To Manage Mistakes in Design

Similar to scope creep, mistakes in design will result in unanticipated, potentially costly changes later in the design process. It is almost inevitable that some mistakes will be made. Checking and double checking work will help keep these issues under control.

> *Example: How flexibility can help manage mistakes*
>
> A building is designed but the funding falls through and the building it is never actually constructed. Later the design company gets a contract for a very similar building and proposes using the existing plans with a few small alterations to save money. Plans for
>
> *(continued)*

a moderate-sized building can cost millions of dollars to produce, so using an off-the-shelf solution is very economically appealing. Only after construction has started do the engineers realize that they need to change the design of the HVAC system that heats and cools the interior because the design was originally intended for a building located in Arizona, but now it has to handle the climate of upstate New York. In this case the engineers would:

1. Evaluate whether the existing system will support a larger heating unit. In other words, will it fit the space allocated for the unit and will the existing ducting allow any change in flow rates and other factors?

OR

2. If the existing system will not support the larger unit, then redesign or augment the original design to handle the increased heating-load requirements.

Where the original design provided for adding capacity (option 1), the cost and time loss will obviously be more manageable. This is an excellent example of how design for flexibility in the original plans could pay off when the design plans are reused. Huge costs and project delay may result if the existing system will not support a larger heating unit.

1.3. Considering the Future

Flexibility allows a design to serve longer into a future that is not fully mapped out.

- Products have new models introduced periodically. If a product is designed looking forward to the next model in the product line, it may be possible to reuse more of the parts in that next model. For example, in the case of a smartphone, the next version may use the same charging cable.

- Behavior changes. If a product or process can be designed that anticipates that change, then less time and effort will be required to make the change. For instance, it might be well worthwhile to purchase large road allowances for a highway, anticipating that traffic will increase over the years and additional lanes may be required.

Example: The Bloor/Danforth Bridge
In Toronto, Canada, growth of the city to the east was delayed by a large valley (ravine) separating two areas of development land. In 1911 construction began of a bridge (called the Prince Edward Viaduct) to link the two sides of the ravine. With significant foresight, the design engineers included a second level in the bridge design below the road surface with enough strength to carry railway cars or other traffic. There were no plans at the time for a railway, subway, or second level of traffic, but the bridge designers realized that the growing city would eventually need to accommodate a significant increase in movement across the valley.

Courtesy of the Authors

FIGURE 1 **The Prince Edward Viaduct under construction.** The lower level, eventually used for a subway, can be seen below the top level road. (Photograph reproduced with permission: City of Toronto Archives, Series 372, Sub-Series 10, Item 840.)

Almost 50 years later a subway project was started in Toronto, including a line running under the bridge using the second level. The city saved enormous amounts of money and time because it did not have to replace or reconstruct the bridge when the subway line was constructed.

2. When Flexibility May Be a Bad Idea

Like all DFX (Design for X) areas, design for flexibility will involve trade-offs. In other words, increasing flexibility will affect other parameters that may be objectives of the design.

- Size may increase.
- Power consumption may increase.
- Longevity or reliability/safety could be reduced.
- The design may have to be different than industry standards.
- Cost may increase.

So like all objectives, design for flexibility must be balanced with the other goals of the design.

3. Conclusion

Because design engineers are creating designs that do not exist yet, they must determine what functionality the design will need to have in the future. This is often not simple to determine definitively. As a result, design engineers often use techniques to enable their designs to adapt to what they cannot predict. These techniques allow designers to adapt to the evolving needs of clients and users, changing circumstances, and possible errors that may require late changes to the design.

KEY TERMS

scope creep

4. Questions and activities

1. You are designing software and have a piece of repeating software with a counter that increments by one every time this particular piece of software is executed. The piece stops repeating after the counter reaches 100. You could check for exactly the value 100 to cause the repetition to stop. Argue that checking for 100 **or greater** could aid recovery of program function and avoid system failure if the software piece was entered mistakenly.

2. Consider the following designs and:

a. Discuss whether design for flexibility would be/is suitable, and if so how it might be done or has been done.

b. Discuss why flexibility in the design might not be a good idea. Pick a specific reason or two from the section When Flexibility May Be a Bad Idea in this module to support your answer.

- Tablet
- Automobile
- Candy bar production equipment
- An air traffic control system
- A computer operating system
- Office desk
- Office lamp
- A stretch of new highway
- Disposable pen
- A security system
- Railway switch

Design for Human Factors: Introduction

1

#process module: #designforhumanfactors

Learning outcomes

By the end of this module, you should demonstrate the ability to:

- Define terms such as user-centered, experience design, participatory design, and universal design
- Explain the levels of human interaction with technology and provide examples of these levels
- Describe how the levels influence the design process
- Analyze existing design from a human-centered perspective
- Apply the principles of universal design in the engineering design process

Recommended reading

After this module:
- **Design for X > Human Factors > 2. Task Analysis**

1. Introduction

The field of **human factors** has rapidly evolved. Historically operators were viewed as mechanistic parts of the technology system: they served the system. Users were people who sought to minimize their interaction with technology. The primary goal of the designer was to create technologies that increased efficiency and convenience for the average user.

In the current view, technology serves people: individual, whole, thinking, feeling, and social human beings. Designers recognize the value of considering and including a wide range of people in their design process. There has been a realization that when technology is designed to work with the whole person, it works better because the person is put at the center, and the design functions to support their needs. The design world has moved from a "one-size-fits-all" model to recognition that customized or flexible design works better for a wide range of people. In addition, when people like using a technology they interact with it more and use it creatively and this can build economic and social capacity.

1.1. Levels of Human Interaction with Technology

Human beings operate at multiple levels with technology: physically, psychologically, socially, organizationally, and politically [1]. These levels are not independent. They operate together and affect each other. Successful technology must work effectively with people on all of these levels. Designing in this way is referred to as ***ergonomics***. Originally focused on designing efficient effective workplaces, ergonomics now includes all types of human-technology interaction design.

The physical level describes people according to the range of physical size, shape, and characteristics present in a population. For example, how much weight people can lift; the size and shape of a human foot; or how large a hand is when the user is wearing mittens. Information on human physical characteristics is called ***anthropometric data***. There are huge volumes of anthropometric information available to assist in the design of human-machine interfaces and systems that work effectively with human physical characteristics.

The psychological level describes people's cognitive and emotional abilities. For example, how many items a person can store in short-term memory; how stress affects reaction time; or how people organize and make sense of information when finding their way using a map. The understanding of psychology has had a major impact on the engineering design field. It is now understood that, to work, technology must coordinate effectively with people's psychological abilities and needs.

The social level describes the ways people interact with each other. This level has been called "team" [1], but people operating a technology collaboratively may or may not view themselves as a team. There are many examples of technologies that must function effectively to support collaboration. An air traffic control system, for example, has to be designed to work at the physical level, psychological level, and social level. It must support the protocols and communication practices that allow air traffic controllers to effectively collaborate with pilots. There are many examples of technologies designed to enhance or facilitate social interaction and collaboration. There are also designs that give people the choice of social interaction or isolation. In all of these cases the design engineers are trying to take into account the users' social needs and abilities to support effective collaboration and coordination.

The organizational level describes the way groups of people operate collectively or communally so that the efforts of one part of an organization are supported by other parts. A technology that works effectively in one company or organization, may fail in another if the technology is not designed to take into account the culture and norms of the organization. An improved design approach would take into account the differences that can exist in organizational structure and norms and build in this flexibility, or the ability to adapt the system to the organization. Trying to adapt the organization to an inflexible design is generally a far less effective strategy.

The political level could also be called the cultural or societal level. It describes the society and cultural norms in which the technology is situated. Ideally a technology should be designed to work with these norms. This is, in many ways, the most difficult to incorporate into a design because it is so difficult to be aware of the cultural assumptions and norms that you live with continuously. Yet ignoring these assumptions can create dysfunctional technologies and safety hazards. Consider a few of the

Example

Consider two organizations that have a need for a warehouse inventory system: one is a supermarket chain that sells food; the other is a non-profit agency that runs a food bank that gives food to people in need.

The supermarket chain needs an inventory system to monitor the grocery items they have in stock in a warehouse. The company probably relies on full-time employees who may receive substantial training on the inventory system. If this same system is implemented in the non-profit agency it may be completely ineffective. The food bank probably relies heavily on part-time volunteer staff, and extensive training is not feasible. The inventory system technology designed for a supermarket chain is not "bad" or "good"; it just was not designed to take into account this variability in the organizational context.

Courtesy of the Authors

simplest societal norms common in North America that could affect the way a user (or users) interacts with technology:

- Turning a switch *up* is associated with turning something *on*.
- *Red* is associated with stop, and *green* is associated with *go*.
- People tend to drive on the right side of the road. This carries over to public sidewalks and stairs: people tend to walk on the right.
- An ambulance has the right of way when its lights are flashing.

There are many more examples and often cultural expectations are very subtle. However, they play a huge role in expectations about how a technology should work, and it can be very frustrating when systems do not work in accordance with the norms of a society. This level also explains why technologies sometimes become archaic or obsolete; a technology may work well but no longer fit the evolving norms of a society.

2. User-Centered Design

The levels at which people operate with technology all together are very complex. However, the most successful technology works "naturally" with the human being, physically, psychologically, socially, organizationally, and politically. This ease of use often comes from a huge effort on the part of the designer to create a system that is in tune with people. This focus on the people who interact with the design is called **user-centered** or **human-centered design** because it puts the users at the center of the process and designs the technology to fit them, rather than the other way around.

When the user has to substantially change to operate technology, the results can be problematic. The user may successfully adapt, such as learning to drive on the left side of the street if you move to Hong Kong. Alternatively, the users may decide not to use the technology or work around it. The result can be failure: commercial failure of the technology or failure in terms of damage and injury. This can be in the form of workplace injuries, such as repetitive stress injuries from using ergonomically poor or inflexible designs such as keyboards, or in the form of substantial disasters. There are

Affordance

Courtesy of the Authors

Originally **affordance** meant anything you could do with a design. So, for example, you can use a park bench to tether your dog. This is not the intended function of the bench, but it is an affordance: the bench affords this possibility. In this sense of the term, affordance encompasses all intended and unintended functions or actions possible due to the design. A bench affords the action of tethering, which could include a dog, a bike, a horse, a stroller, or anything else.

Psychologically, people associate specific types of affordances with particular objects or forms, sometimes to such a degree that these objects or forms become representations of the action. For example, a button suggests a pushing action (e.g., a doorbell or an elevator button) to such a degree that psychologically any object, real or virtual, that looks at all like a button implies pushing. The button-push ideas become almost synonymous.

Photos courtesy of the authors

In his book *The Design of Everyday Things*, Donald Norman suggested a change in the definition of affordance to recognize that the way people use things is generally based not on what is possible to do, but what the design implies is the usage [2]. So while a button might be pulled rather than pushed (e.g., old doorbells were operated by pulling), the user would not immediately perceive this option and therefore it is not an affordance. Good human factors' design, Norman argues, works effectively with the user's psyche such that the way to use a thing is intuitive from the affordances suggested by the design; no owner's is manual necessary. A handle implies pulling, a button implies pushing. Norman's ideas have informed the development of user-centered design, interaction design, and experience design.

many well documented cases of failures that have resulted from engineering designs that were misaligned with people [1].

Incorporating a user-centered approach in design is now well established. It has become an integral part of the design process and there are companies which specialize in user-centered design. This is particularly the case in the consumer products field, and also has had substantial impact in the design of workplaces and commercial products and services. User-centered design, when it works well, is almost invisible to people. When you make a cell phone call you don't think about having to select what communications tower you wish to link to, then providing validation and access information in order to make the link. It all happens automatically for you.

More recently user-centered design has evolved to address more broadly the psychological needs and feelings of people. One step in that process was **interaction design**, which has now evolved into **experience design** (abbreviated "UX"). Experience design more fully takes into account not just what the user is capable of doing, but also the quality of the experience they have with the technology. Using an experience design approach the engineer considers not only the functional aspects of the system, or product, but how to design it so the experience will be fulfilling for the user.

Tools such as **use case** and **snapshot** are used to help the designer envision the experience the user will have (#usecase). In addition, **story boards** or **scenarios** can be worked up to assist the designer in imagining the technology from the user's perspective. The goal is not necessarily to achieve greater user happiness, but instead to design a technology that supports a user's personal goals which could include happiness, accomplishment, a feeling of security, creative innovation, and productivity. Designing a product or service that people like to use can provide an edge over a competing technology.

User Stories Example

A **user story** is a verbal version of a **story board**. It describes, from the user's perspective, the way the user could interact with a design. The story is used to identify functions and objectives in the definition of a design problem. For example; a university is designing an information system that will give administrators the ability to combine information from student, academic course, and HR (human resource) databases. The user stories written up by the design team include:

- "As an administrator, I want to see what combinations of courses students use to complete our academic program and 'rank' how many used each pathway."

- "As the instructor of SOC300, I want to see how my current cohort of students performed in SOC100 and SOC200. In other words, I want to select my class and view their performance as a group in the prerequisites or recommended prerequisites, to have a better sense of my current class and how I can tailor the course material to best suit their needs/background."

- "As an administrator, I want to be able to look at a class (es; ECO100) and for those who earned 90–100% in the course see what programs (i.e., majors) they enrolled in; I then want to see what programs the students that earned 80–89% enrolled in, and so on."

[User stories derived from Next Generation Student Information System Strategic Planning project, University of Toronto, 2014, reproduced with permission.]

---

3. Participatory and Universal Design

Focusing the design process on the user can be enormously beneficial because it is more likely to yield a design that fits the user's needs. However, this can also be difficult to do, particularly if the user is very different from the designer: physically, psychologically, socially, organizationally, and/or politically. It is hard to understand how other people think and what their needs are if they are very different than yourself. Two strategies have emerged in design to address this difficulty: participatory design, and universal design principles.

Participatory design brings people from the target user group into the design process as members of the design team. This requires building relationships with the community and requires sensitivity to historical and cultural issues that may play a role in the situation. Very often other types of professionals, such as sociologists, ethnographers, and experts in the culture of the user community, will work with the team to facilitate this relationship. Participatory design is becoming more common, particularly for the design of technologies for developing communities, which have historically not had a voice in the design process. It is a strategy for bringing their perspective, in the first person, into the process.

A **universal design** approach uses a set of principles to design for a broad range of users. The intent is to create design that is accessible and useable not just by the average user (if such a person actually really exists), but instead to take into account the broadest range of users possible. The principles of universal design were adapted from architec-

Samuel Croome

ture, and increasingly have found their way into legislation [3]. Consider the example of a ramp placed at the front of a building. The ramp not only provides access to the building for people using wheelchairs, but also provides better access for people delivering supplies, people pushing strollers, and people using other adaptive devices such as crutches. This is called the "curb cut" effect: a design feature implemented for one user group that actually improves the design for many. Text messaging, for example, was originally developed for people with reduced hearing but has now become a commonly used feature by many people. The recognized principles of universal design were developed by the North Carolina State University Center for Universal Design [4]:

1. Equitable use: Access and usability is provided to all users in an equivalent or equitable way (e.g., the ramp is the primary form of access at the front of the building, not around back).

2. Flexibility in use: There are multiple ways of accessing or using the technology (e.g., the elevator includes buttons with braille, visual identifiers, and sound cues that alert the users when they have reached the desired floor).

3. Simple and intuitive to use: Access and usability are obvious from the design form (e.g., how to navigate through a website should be obvious to the user and there should be navigation tools that make it simple to return to previous pages).

4. Perceptible information: Users are able to access and use the technology regardless of conditions (e.g., the buttons on a cell phone can be identified in low-light conditions, or if the user has low vision).

5. Tolerance for error: Access and usability are not compromised by minor user error (e.g., if the "uninstall" function on a piece of software is chosen, it will ask the user to confirm before carrying out the procedure).

6. Low physical effort: Access and usability should not require substantial physical effort (e.g., power assist doors allow users to open a door without exerting substantial force).

7. Size and space for approach and use: The physical size and space should accommodate a wide range of users (e.g., kitchen utensils, such as vegetable peelers, with a larger-grip handle that allows users with reduced gripping ability to more easily use the utensil) [5].

These principles provide guidance to the designer for creating technology that will be accessible and usable by a wide range of people and in the process usually create better designed technology for everyone.

4. Conclusion

The field of user-centered design has dramatically evolved over the past half century. It has gone from a nascent idea to a philosophy that is imbedded in industry practice and legislation, such as the Americans with Disabilities Act [3]. In engineering design today, user-centered design is widely used and strategies such as experience design and participatory design are becoming more widely accepted. All of these strategies are intended to take into account the complete user at every level from physical to political. Design solutions never perfectly fit everyone's needs, but taking into account a broad range of people holistically can improve the design of technology.

KEY TERMS

human factors	ergonomics	anthropometric data
user-centered or human-centered	affordance	interaction design
experience design	use case	snapshot
story boarding	scenarios	participatory design
universal design	user story	

5. Questions and activities

1. Explain in your own words the key terms in this module.
2. Chose an existing piece of technology that you have:
   a. What affordances does this technology have?
   b. In what ways does it take into account the user?
   c. In what ways does it take into account a wide range of users? Explain the usability of the technology using the seven principles of universal design.
3. Apply the strategies learned in this module in the design project you are currently working on.

6. References

[1] Vicente, K. *The Human Factor*. New York, Routledge, 2004.
[2] Norman, D.A., *The Design of Everyday Things*, New York, Basic Books, 1988.
[3] Americans with Disabilities Act.
[4] Center for Universal Design, North Carolina State University, 2008. Retrieved September 28, 2013, from www.ncsu.edu/ncsu/design/cud/about_ud/udprinciples.htm
[5] About OXO, OXO mission and design philosophy. Retrieved September 28, 2013, from www.oxo.com/aboutOXO.aspx.

2

Task Analysis

#process module: #taskanalysis

Learning outcomes

By the end of this module, you should demonstrate the ability to:
- Explain the use of task analysis in the design process
- Describe at least two different means for conducting a task analysis
- Apply a task analysis method to gather information about an existing technology or task
- Use the results of a task analysis to identify opportunities for improvement in a task, technology, process, system, or service
- Use the results of a task analysis to generate requirements and/or potential solution ideas

Recommended reading

Before this module:
- **Design for X > Human Factors > 1. Design for Human Factors:** Introduction

After this module:
- **Design for X > Human Factors > 3. Use Case Method**

1. Introduction

There are many methods used in design for human factors to assess existing systems and create new, novel designs that work effectively for people. One of the most fundamental of these methods is **task analysis**. Task analysis is a technique for understanding how people approach a task. Once the task has been fully understood and described, an engineer can analyze the process and figure out if there is a way to create, or improve, technology to do the task better. Better could mean faster, easier, more reliably, more safely, or in some other way improve the experience of doing the task for the user.

The most obvious way to approach task analysis is to ask a person who does the task to describe what they do (i.e., "Could you explain to me the steps you take in performing this operation?"). This is usually accompanied by questions that ask the person to elaborate on each step and explain the obstacles they may encounter.

However, it has been shown that the way a person describes a task from memory is often quite different than what is seen in direct observation. Also, people often leave out key details that may inform the design process. For this reason, it is preferable to augment the interview with more direct forms of observation.

2. Using Video or Observation

One means for improving the description of a task is using direct observation or video. This allows the designer to get a much more complete view of the task and the way people approach the task. Ideally, you would have the opportunity to observe more than one person so you can make generalizations about the way many people approach the same task. For example, if you were asked to design a new bicycle lock system, you might observe people locking up and unlocking their bicycles. You would take notes on the position of their body during this task, the steps they take in the process, and the amount of physical and mental effort required. In addition, you could watch where they put the lock when it is not in use; how easy is it to carry on the bicycle? In some design companies engineers do this type of observation. Other companies employ psychologists, anthropologists, and/or ethnographers who work with the design team to gather this type of information.

Courtesy of the Authors

To improve this type of direct observation it is useful to have the person (the subject) explain out loud what they are doing, thinking, and feeling while they perform the task. The task may look easy from an observer perspective, but the subject may express frustration or annoyance at the difficulty of the task. You can use just a voice recording, but you will get more comprehensive information if you pair the voice recording with a video or screen capture. Modern task analysis of software systems often also includes eye tracking because this shows the researcher where the subject's attention is focused during a task. This is useful information that will help the design team identify opportunities for a new design approach.

Typically in an undergraduate design project you will not be allowed to enroll test subjects for this type of activity and you should not videotape people without their permission. However, it is reasonable to observe people in public settings, such as watching the way a person locks up a bicycle at a public location, or watching the way people use a transit system. You can take notes about what you observe. You can also try the task yourself, or observe one of your teammates performing the task. This is a useful way to understand first-hand the challenges of a task.

There are a few potential sources of error in this process. First, it is important to recognize that you are not the same as everyone. That is, your individual perception of the task is only one viewpoint. Getting additional perspectives from people different than you is important to developing a full understanding of the task. Second, people will sometimes perform a task differently when they know they are being observed. They may be less likely to take shortcuts or do things they know are improper when being watched. This change in behavior can affect the information you collect.

Example: The Development of the Swiffer Cleaning System

When Continuum Design set out to redesign a floor cleaning system for Proctor & Gamble, they deployed ethnographers to watch people cleaning their floors. Some of their observations included [1]:

Courtesy of the Authors

- People used a broom to sweep the floor first and then they used a mop to wash the floor.

- People perceived cleaning the floor as a dirty job and put on old clothes to perform the task.

- People used a lot of different products in the task that were not necessarily specifically designed to work together (i.e., the cleaning fluid was not formulated for the particular brand of mop).

- People spent the same amount of time washing the mop out after cleaning the floor as they did mopping the floor in the first place.

This task analysis led the Continuum team to reimagine the task of cleaning a floor. They created an entirely new approach, the Swiffer cleaning system, which has been a phenomenal economic success. When it was introduced in 1999 it had $100 million in sales in 4 months [1]. Proctor & Gamble has successfully continued since then building on the Swiffer product line.

Note in this case some of the observations that might not have been uncovered in a user interview. For example, if a person was asked to describe the steps they take to clean a floor, they might not start by saying, "Well, first I get into old clothes because this is a dirty job." When using task analysis, observation—particularly observation of the *whole* task—is important for uncovering key pieces of information for the design process. Gianfranco Zaccai, a Continuum co-founder and president, has said, "Consider not just the act of using the product but the total experience around it" [2].

3.   Task Sampling

Another common approach to data collection for task analysis is sampling. This is used to get information about the frequency of tasks and the amount of time spent on different tasks. Sampling is often more accurate than information gathered from an interview, but it relies on the subject (person) accurately describing their activities.

There are several different methods for doing ***task sampling***. One common method is to prompt the subject at random time intervals to give a snapshot (i.e., a brief description) of what they are doing (#snapshot). For example, suppose you create a system that randomly texts a warehouse worker 10 times a day and asks them what they are doing at that instant. They respond by texting back. You then collect data by compiling the information you get back from each text. By collecting information like this day after day you can build up a picture of the person's activities, and how much time they are spending on checking inventory, or interacting with shipping companies, or doing other activities. This information can be used to create systems, such as warehouse inventory systems, that help the person do their job more efficiently. However, if your data shows that the person spends only a few minutes a day on this activity, then it may not be worth the time and resources to create a new system.

Collecting this type of data often requires coding the activities, which means grouping them into common themes. Sometimes researchers will use tables to help collect this information. For example, watching people work in teams and noting every 3 minutes what is going on by code (see Table 1), it is possible to build up an understanding of how effectively the team is working together.

TABLE 1 This table shows a tick mark for each observation; observations spaced 3 minutes apart during a 1-hour team meeting. This shows that the team is spending a lot of time in discussion but is rarely making decisions. This may indicate that they are having difficulty moving forward on their project.

DISCUSSION OR DEBATE	INFORMATION TRANSMISSION	DECISION MAKING	ANALYSIS OF INFORMATION	IDEA GENERATION
IIIII IIII	IIIII	I	II	III

Sampling can also be done using a diary system. The subject (person) is asked to create a diary of each task they perform and how long it takes them; or note down every time they have to do a particular task. For example, suppose you ask an assembly line worker to make a note every time they have to redo a solder connection because one fails. This would help you find out if there is a flaw in the soldering process or if possibly there is a quality control issue that needs fixing.

This type of data collection provides an overall understanding of a person's time, or the frequency of particular tasks. This data can be very useful in setting the objective goals for a design. For example, you might set as a goal that the new system you are designing should reduce the number of soldering flaws by 50%. By knowing how many flaws there are using the current system, you can then accurately assess whether a new system meets the objective.

4. Analysis

Once the task data has been collected it needs to be analyzed. Video information is often translated into a set of observations, such as those shown in the Swiffer example. The researcher will also quantify the information and create other types of qualitative descriptions based on questions such as:

- How long did the subjects spend on each step in the task, or how long did they spend doing each activity?
- What was the position of their body at each point, or how many times did their eyes focus on a particular element on a screen? Did they make use of special attire or equipment?

- How frequently was an error made in performing a task, and were there similarities in the errors?
- How did the person interact with each element of the technology?
- Did the subjects employ similar strategies in to accomplish the task, or were some people more "expert" at it? What did the experts do that was different than the other subjects?
- At what points in the task did the subjects express frustration? Or what emotions were associated with different activities or parts of the task?
- Did the subjects always use technology or do the task as intended? Or did they take shortcuts or do things differently than expected?

Analyzing the data to create quantitative and qualitative data allows the information to be used in the design process. Instead of designing based on one person's experience, or the perceptions of the designer alone, a complete task analysis study produces more reliable, multi-perspective information to use in the design process. This type of analysis can also be done for different types of users, increasing the value to the designer (see Figure 1). For example, if you are designing a new type of bicycle you might analyze the interaction of the users with the bicycle, and also the interaction of

Courtesy of the Authors

Courtesy of the Authors

Courtesy of the Authors

FIGURE 1 It is important in task analysis to capture the perspectives of different users, and different user modes.

bike mechanics with the bicycle to see how the design works for someone trying to fix the bicycle. This would give you insight into the design for the user and also for maintenance.

KEY TERMS

task analysis   **task sampling**

5. Questions and activities

1. Watch people performing a task and create both a quantitative and qualitative description of the task. Here are some suggested tasks to analyze:

   a. Activities of people in a coffee shop or cafeteria (e.g., ordering food or beverages, consuming their food)

   b. Checking out a book at the library

   c. Locking up a bicycle and unlocking a bicycle

   d. Attending a lecture (best done through task sampling to determine time spent in different activities during a lecture)

   e. Doing laundry or washing dishes

   f. Using software (pick a particular common piece of software to analyze)

   g. Using a parking meter or some other type of technology in a public space

   h. Some other task that is common and easy to observe

6. References

[1] Continuum, "Swiffer: A Game-Changing Home Product." Retrieved September 7, 2013, from http://continuuminnovation.com/work/swiffer/

[2] FastCompany, "Why Focus Groups Kill Innovation: From the Designer behind Swiffer." Gianfranco Zaccai, posted October 18, 2012. Accessed September 7, 2013, www.fastcodesign.com/1671033/why-focus-groups-kill-innovation-from-the-designer-behind-swiffer

3

Use Case Method

#process module: #usecase

Learning outcomes

By the end of this module, you should demonstrate the ability to:
- Develop a list of snapshots for a design
- Create a list of use cases for a design
- Distinguish between a snapshot and a use case
- Identify the actors in a use case
- Draw a use case diagram and extend it to include snapshot information
- Use the case information and snapshots to further develop requirements

Recommended reading

Before this module:
- **Design for X > Human Factors > 2. Task Analysis**

After this module:
- **Design for X > Human Factors > 4. Concept of Operations**

1. Snapshot and Use Case

Snapshot and *use case* are methods used together to develop user based requirements. Unlike the requirements themselves, these two tools develop descriptions that are highly specific to the context and usage of the design. Requirements are then derived from these descriptions.

These two tools describe interplay between *actors* and the design, where the actors are people, organizations, or some other parts of the system. Note that an actor can be an institution (such as a bank) or other selection of people (such as a lobby group) as well as an individual person. An actor can also be an object, such as an airplane or runway or database.

Both snapshots and use cases were initially developed in the software realm, but have obvious applicability to other engineering disciplines. The use case has become an important part of universal markup language (UML), a descriptive language for software development, but can be used independently in other fields of design [1]. The two tools differ from each other in their scope; we will look at the snapshot first.

2. Snapshot

The snapshot is a description of a technology at a specific point in its use; this could be a point during implementation, operation, or disposal of the technology (i.e., a specific point in the design's "lifecycle") (#snapshot). It is like a photograph—it contains no information about the past use or future operation. A snapshot describes a single aspect of interaction between the design and an actor. The table shows some examples.

SNAPSHOT	ACTOR	TECHNOLOGY BEING DESIGNED
Toaster put into a standard kitchen cabinet	Kitchen cabinet	Toaster
User reads the display in direct sunlight	User	Display
Speakers placed on the beach (and thus subject to sand and moisture)	Beach	Speakers
The input supply to the power unit has brownouts and blackouts	Electrical grid	Power unit
Batteries are changed	Service personnel or user	Battery-powered device
The user cancels the order midway through the process	User	Online ordering system

Note that, unlike requirements that must be absolutely testable, a snapshot can be missing the metric and criteria (i.e., goal). So "The camera is carried on a long hike" is a good snapshot, but would not be a good requirement. Developed into an objective and constraint it might be: "The camera is as light as possible and shall weigh no more than 200 grams." Here the designer has decided that anything like a camera that is 200 grams or less is going to be "light enough to be carried easily on a walk" and this can be tested to see if a potential design meets the requirement.

Snapshots can be steps in a procedure ("The user is identified") or conditions ("The product is used in the rain.") or events ("The user enters zero"). A ***user story*** can also be divided up into snapshots to capture the steps in the story. Snapshots naturally come to mind as a designer is thinking through the operation of the design. It is important to capture this thinking, and a snapshot description is one method. Some designers will use pictures or even photos for this purpose.

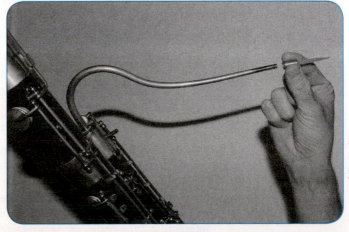

Courtesy of the Authors

User positions a reed to install in a bassoon

Use Case Method

427

Snapshots often can explain a requirement. Why must the camera weigh no more than 200 grams? Is it to avoid breaking the neck straps that it must work with? Is it because the assembly line can tolerate assemblies only up to this weight? Or is it a user requirement? The requirement alone does not include this information, but the snapshot makes the reasoning clear. If the designer had to revise this specific requirement because a camera could not be designed to meet the other requirements and still be less than 200 grams, having the reason for the requirement (i.e. the snapshot) would allow the designer to make better informed design decisions.

3. Use Case

The use case method is similar to a snapshot in that it describes the interaction between actors and the design, but it differs in complexity and scope. If a snapshot is like describing a picture then a use case is like describing a video. One of the actors will be considered the "primary" actor, and a use case is a description of how the design will provide the primary actor with a measurable, desired result consistent with the goals of the design [2]. The use case can be used further in the development of design solutions, and in validating the resulting design after development.

Consider a banking machine as the design and a person with an account who uses the bank machine for deposits, withdrawals, and other banking procedures as the primary actor. Making a deposit and making a withdrawal would be use cases for this primary actor. These are consistent with the goal of providing remote banking service without requiring a human service provider. If instead the design is the banking machine and the primary actor is the banking system, then "establishing the identity and account of the user and enabling user access" is a use case, since it also completes a goal of the banking machine. From this example we can see that there are many levels of use cases, depending on the selection of the primary actor. One can think of use cases as sections of a user's manual for the design.

A use case will describe the actions of the actor and the design that take place. Titles for use cases might include:

- Service people maintain the printer.
- A customer has a printer serviced. (This could be a use case for service center software.)
- The user installs a reed in a clarinet.
- An operator monitors the throughput of the transfer gate.
- The purchaser undertakes a formal acceptance procedure.
- A tsunami floods a nuclear reactor after an earthquake.
- A student registers for classes.
- A fire breaks out in the engine room.

The entire use case would have the step by step description of the steps in the interaction. This can be in the form of a flow chart, pseudo-code, text, or any other suitable representation. A use case must include not only the ideal interaction but also variants of the interaction that might happen. For the bank machine withdrawal example this would include, for example, the interaction if the machine was out of money, the interaction if the bank card was rejected, and the interaction if communication was lost with the central database. (Each of these could be snapshots.)

The use case can be used to generate requirements. High-level behavior information that is part of the description, including the variants of interaction, can also be used during requirements generation. Since the use cases indicate the behavioral goals of the system, these are also very important as the basis of system performance tests.

Often people confuse use cases, mode, and states.

Use Cases, Modes, and States

A *state* describes specific characteristics of the system at a particular point in time or at a particular location in the system. For example, on an aircraft the landing gear is normally in one of four states: fully up, fully down, moving down, or moving up. In a fluid flow system, such as a natural gas pipeline, the state at a particular location along the pipeline would be described by the flow rate, the temperature, and the pressure.

A *mode* describes a set of states implemented for a particular purpose. For example, to go into landing mode the pilot will generally lower the nose of the aircraft, reduce speed, lower the landing gear, and lower the flaps on the wings. Or a pipeline may operate in high-pressure mode or low-pressure mode by changing the set of operating characteristics of the compressor stations along the pipeline.

A *use case* is an interaction of "actor" and design for a "purpose." The use case for the plane landing scenario is pilot (an actor) lands (the purpose) plane (design). The full description of the actions needed for this use case would help the designer realize what control systems (flap control, landing gear control) have to be available to the pilot to implement this use case (i.e., what is needed to put the plane in landing mode).

3.1. The Use Case Diagram

Figure 1 shows an example of a use case diagram for a transit tracking system. In this system each bus communicates its current position and speed to the central system. In this diagram the bus is shown as an actor. Users are also actors. They can query

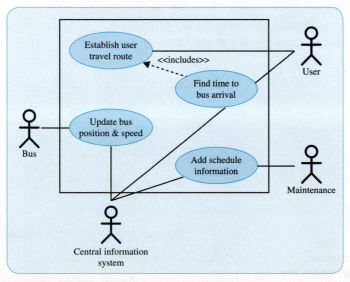

FIGURE 1 Use case diagram for a transit notification system. The technology being designed is designated by the box.

the central system to find out where a bus is, or how long it will be until the next bus arrives at a specific stop. A system maintenance person will have to interact with the system to update buses and routes.

In this example the bus is an actor, the user is an actor, and the maintenance person is an actor. The central information system is the technology being designed to provide the functionality. It is also shown as an entity in the diagram because it acts on and reacts to the actors. The use cases are shown as ovals with descriptions. Lines link each of the actors with the relevant use case or cases.

A use case diagram or use case description, such as Figure 1, has the following four important functions. In developing the use case:

1. You identify the interactions of people, organizations, and systems with the design.
2. You explore more than just the "ordinary" interactions, but also the upkeep, installation, and potential failure modes of the design.
3. You identify further requirements of the design, both directly and implicitly.
4. You can use it to develop validation tests, functional decomposition, anomaly analysis, and other perspectives on the technology.

4. Examples

Typically requirements, snapshots, and use cases are developed in parallel. Requirements will suggest use cases and snapshots; the development of use cases will suggest requirements and snapshots; and the development of snapshots will inspire requirements and use cases.

Consider an example. An altimeter (see Figure 3) is an instrument that tells a pilot the height of the aircraft above sea level. Some use cases include:

- Pilot reads the altimeter to determine the height of the aircraft.
- The airplane manufacturer installs the altimeter.
- The service crew maintains the altimeter.
- The manufacturer assembles the altimeter.
- The shipper ships the altimeter to the airline manufacturer.

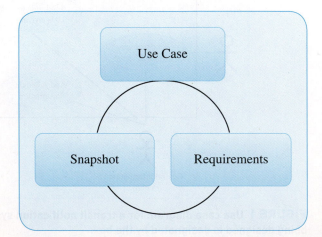

FIGURE 2 Use cases, snapshots, and requirements are developed iteratively and in parallel.

FIGURE 3 Altimeter [top center].

Courtesy of the Authors

Figure 4 illustrates the first use case on this list. This use case can be extended by showing some snapshots derived from this use case (see Figure 5). Also in this example the snapshot "blocked input tube" could suggest the use case "service crew cleans altimeter input" and suggest other snapshots such as "pilot must have a method of clearing the input or providing an alternate input to the altimeter during flight if tube is blocked."

Ideally you would work back and forth between snapshots, use cases, and requirements until you have a complete set of each. "Complete" means that you have described all of the usages of design by every actor that interacts with the design. This is, of course, ideal. In reality engineers try to capture the use cases that will have the most substantial impact on the design. In addition, use cases that impact safety and failure modes will be very important. As the design process progresses the design team will revisit the use cases and add detail as the technology takes shape. This will help the team identify possible failure modes, bugs, or weaknesses in the design as they approach the detailed design phase.

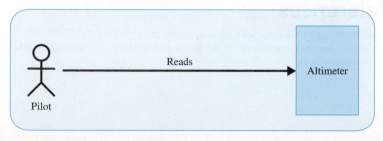

FIGURE 4 Diagram for the "pilot reads the altimeter" use case.

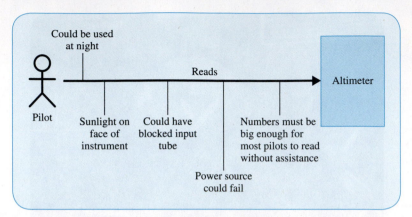

FIGURE 5 *Extension* of a use case showing snapshots.

KEY TERMS

use case	snapshot	actor
state	mode	extension
user story		

5. Questions and activities

1. For a toaster:

   a. Give two use cases.

   b. Name three actors.

   c. Give four snapshots.

   d. Give requirements that relate to your answers for a–c.

2. Do a diagram of the use case "person uses a credit card to pay for groceries."

3. There is a use case "person takes the subway to work."

   a. Name five physical items that would be involved in this use case.

   b. Explain how each item would be influenced by this use case.

   c. Give a snapshot for each item.

   d. Give tests that would indicate design compliance for at least one requirement for each item coming from your answers from b and c.

6. References

[1] Cockburn, A. *Writing Effective Use Cases*, Addison Wesley, 2001.

[2] Sharon, Yonat. "Extreme Programming." Object Orientation Tips. April 15, 1999. Retrieved July 10, 2006, from http://ootips.org/xp.html

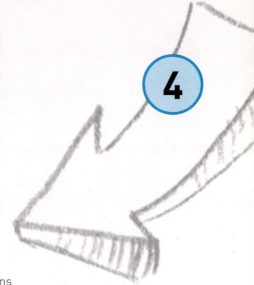

Concept of Operations

#process module: #conceptofoperations

Learning outcomes

By the end of this module, you should demonstrate the ability to:

- Explain the similarities and differences between a concept of operations document and other types of engineering design reports
- Develop a concept of operations report for the purpose of communicating with users and operators

Recommended reading

Before this module:

- **Design for X > Human Factors > 3. Use Case Method**

1. Introduction

A *concept of operations (ConOps)* document describes the design problem and conceptual design solution. In this way it is similar to a conceptual design specification document in content. The difference between a ConOps and a conceptual design specification, or other reports for the client, is that a ConOps is written for the future users and/or operators of the design. The relationship between these types of documents is shown in Figure 1.

Because the audience for this type of report is different, the way it is written is somewhat different. A ConOps document is written in accessible language the users can understand. It avoids the use of jargon and acronyms. A good ConOps document should allow the user group to understand clearly the problem that the designers have identified and be able to assess the quality of the proposed solution. In some projects the ConOps is used in place of a highly technical conceptual design specification type document to communicate with both the users and the client, particularly if the client is not immersed in the technical field.

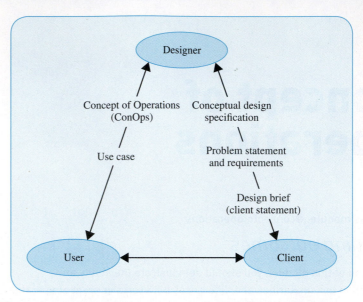

FIGURE 1 **The relationship between a concept of operations report and other types of reports developed during a design project.** The design brief is written by the client. The other reports shown are typically written by the design team.

This approach was developed to address the problem of users and/or clients being confused or frustrated by the highly technical language sometimes used by engineer's to describe problems and solutions. A user group that cannot understand the problem from the engineer's perspective is not likely to agree to the proposed solution, especially if the solution is written up in a manner that is also unintelligible. In addition, they cannot provide good feedback on the proposed design or problem definition. However, if the problem from the engineer's perspective is written up in a way that is clear and understandable to a professional non-technical audience, then the users can grasp the project and be full participants in the design process. Highly technical documentation may be necessary for proper implementation of a design, but this documentation does not always serve the purpose of communicating well with other audiences, and this is where a ConOps report is useful.

2. The Structure of a ConOps

There are formal standards for a concept of operations document, for example, IEEE Standard 1362-1998 (which has been superseded by IEEE Standard 29148-2011) [1,2]. These standards describe a very detailed, formal approach to developing a ConOps. When you work in industry you may be required to follow this type of standard in your ConOps documentation. Here we will describe an approach to developing a ConOps very loosely based on the standards, but at a level that is more appropriate for an introductory design project.

2.1. The Purpose

A ConOps should start with the purpose of the project. This is a short introduction that explains the main point of the design project at a high level. It gives the "big picture" view for the user and identifies the primary goals.

2.2. The Problem Statement from the User's Perspective

In a ConOps report the problem statement should start with a description of the current situation. This describes the existing system or the way the problem is currently being solved. If there is no existing system, the problem statement starts with a description of the current situation that is motivating the project request.

The problem statement would then go on to describe the motivation or justification for the new design. It would explain why the current system is inadequate, or what change in the situation has prompted the request for a new design. In the terminology used to describe a problem statement, this is the user "need" or the "gap" in the existing situation. It might explain how the world has changed in a way that motivates the development of a new system. The problem statement should also identify key functions and features that the new technology should have.

This description should be written from a user or operator perspective. In formal engineering requirements the use of phrases such as "the design should be reliable" are discouraged because this is not a measureable objective. In formal requirements this phrase would be translated into a set of specific measurable requirements. However, in a ConOps document this type of statement is permitted. You would probably add some detail to explain what "reliable" means in the context of the particular problem, but you would not include the very detailed requirements that will be used to test a prototype. The information provided should be sufficient for the users or operators to determine if you have correctly captured a full description of the problem. It should also help them understand the problem more comprehensively so they are in a better position to give you input on the design process and assess how well your proposed design meets the intended need.

2.3. Diagramming the Problem

It is very useful to include a diagram or schematic of the design problem. The diagram could be:

- A black box representation, showing the inputs to the design, the outputs from the design, and the service environment
- A use case diagram or diagrams, with snapshots
- A flowchart showing the inputs and outputs from the design
- A fishbone diagram (e.g., Ishikawa diagram) showing the root cause of a problem
- Some other graphic representation of the design problem

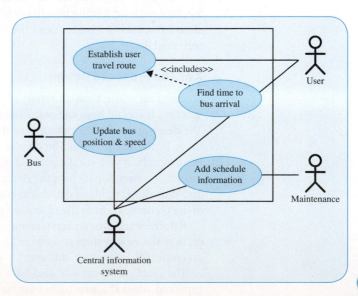

The diagram should not imply a particular solution (i.e., it should be solution independent). It should merely describe, visually, the design problem and the operating environment. For example, if the new design has to operate with existing equipment, protocols, utilities, or other service environment conditions, then these should be included in the diagram.

2.4. The Service Environment

The service environment section would be similar to the service environment you write up for a project definition; again the difference is that in a ConOps it is written from the user's or operator's perspective. The service environment section should clearly describe all aspects of the intended operating environment for the technology. It should also explain any existing systems that are expected to interface with the new system.

2.5. The Requirements

The detailed requirements in a ConOps report are often given in bullet point form. However, they are not typically organized into a function list, objective list, and a constraint list. Instead they are organized into categories that are more meaningful for the users or operators. These could by system requirements, operating performance, and constraints. Alternatively they could be organized by the attributes of the design and the capabilities. There may also be a category for the assumptions made by the design team. The design team is encouraged to choose an organization (set of categories) for the requirements that makes sense for the design problem, rather than holding to a specific formulation.

Ideally, this section of the ConOps would also indicate priorities; that is, which requirements have been given priority in the design process. This will assist the users in understanding the justification for the proposed design.

The requirements in this section can be more general than what is necessary in a formal set of engineering requirements. It is permissible, for example, to say that the design for an electronic file storage system must "accept content in all common formats" without getting into the details of the exact formats that are included. This allows the users to understand the requirements without getting overloaded with technical details.

2.6. The Proposed Design

This section of the ConOps report explains the proposed design from the user's perspective. It should have use case diagrams or other visualizations to support the text. The description should include how the design will operate and what features it will have. Typically this would be divided up into sections for each *mode* of operation. For example, if you are designing a materials processing system, this section would include a subsection on how the design operates in start-up mode, steady flow mode, and shut-down mode. It could also contain subsections on maintenance or servicing of the technology, from the operator's perspective.

If the new technology is replacing an existing one, then there would be information on how the new system is different than the current one. What are the new features and how is the operation different?

The ConOps report will also have a summary of the improvements that can be expected when the new technology is implemented. This should include reasonable quantitative or qualitative predictions of the outcomes that will be achieved using the new design. In particular, it should forecast the effects of the new design for users, operators, and other stakeholders. It may also have estimates of the reduced environmental impact, or improved economic outcomes that will be possible using the new technology. The reader should finish this report with an excellent understanding of the design problem, proposed solution, and potential impact of the solution.

3. Conclusion

A ConOps document is used to communicate the design process with the users and/or operators of the proposed technology. It shares many characteristics with a conceptual design specification, or other types of engineering design reports. The chief difference is that the ConOps is written for and from the perspective of user groups and operators. The purpose of this document is to inform users and operators about the design problem and proposed solution so they can actively participate in the design process, and understand the operation of the proposed design.

KEY TERMS

concept of operations (ConOps)   **mode**

4. Questions and activities

1. Explain how a concept of operations document is similar to, or different from, other engineering design reports.

2. Develop a concept of operations report based on the design project you are currently working on.

5. References

[1] IEEE Standard 1362-1998. *IEEE Guide for Information Technology: System Definition—Concept of Operations (ConOps) Document.* IEEE Standards Association, 1998.

[2] IEEE Standard 29148-2011. *IEEE Systems and Software Engineering:- Life cycle Processes—Requirements Engineering.* IEEE Standards Association, 2011.

1

Design for Intellectual Property: Introduction

#process module: #designforintellectualproperty

Learning outcomes

By the end of this module, you should demonstrate the ability to:
- List the different types of intellectual property
- Explain the benefits and drawbacks of intellectual property from a societal perspective
- Explain the benefits and drawbacks of intellectual property from the creator's perspective
- Identify when you can use an off-the-shelf item in a design without explicit permission and when you need to obtain permission or license the technology

Recommended reading

After this module:
- **Design for X > Intellectual Property > 2. Principles of Patentability**

1. Introduction

The term ***intellectual property*** (IP) refers to valuable things that are the products of the mind, rather than muscles. In order to encourage individuals and companies to invest time and money to create new products, services, and works of art and literature, the government allows the creators to benefit from the fruits of their labors by allowing them to own their IP. In fact, the creators of IP are allowed to put a "fence" around their creations, and to control access, creating the potential to profit from their work (see Figure 1). This is similar to the system used in to settle the American West: the early settlers who spent the time and effort to clear new land were allowed to stake a claim to the land and fence their property. Just as granting the right to own the physical property they developed motivated the early settlers, the right to own intellectual property creates a financial incentive to do the intellectual work that creates it.

Of course, the downside of allowing someone to own IP is that everyone else might be forced to pay to gain access to the work; whether it is a song on iTunes, a book, the latest toy, or something as important as a new heart medication. For example, people

FIGURE 1 Intellectual property is similar to physical property. It can be owned, leased, given away, or stolen.

might have to pay a high price for a drug because one company owns the IP and can prevent others from selling it, which reduces economic competition. In spite of this, virtually the entire developed world has decided that it *is* in the interest of the community as a whole to grant IP rights, but with some very specific limitations. Whereas a land claim often gave the early settlers a permanent right to use the land (and to stop others from using it), intellectual property rights are time limited.

We will focus on ***utility patents***, which are used to protect engineering designs. In most countries, a utility patent gives the inventor the right to "fence" their intellectual property for a period of 20 years from the date of the first patent application. After that period has elapsed the ownership right disappears, and anyone can use the technology. It is as if the fence collapses, and the private land becomes a public park. Patents cannot be renewed. Therefore, the patent laws promote innovation, but also set boundaries that are in society's best interest.

> This module contains general information about intellectual property as it pertains to engineering design work. Some details have been omitted for clarity and simplicity. Since laws vary from country to country and over time, and patent law is quite complex, this text is not a substitute for professional legal advice on these matters.

2. Types of Intellectual Property

Songs, movies, literature, art, and inventions are all created in the mind and actualized; developed into an actual thing, not just an unexpressed idea. Intellectual property laws protect these actualized creations. Just like real property, intellectual property can be bought, sold, rented, or stolen. Different types of intellectual property are protected in different ways:

- ***Utility patents*** protect the underlying function of inventions and of new engineering designs.

- **Design patents** protect the ornamental shape (but not the function) of physical things like cars, ballpoint pens, or toasters. There is potential for confusion here. The "design patent" is actually the protection for the artistic part of the design, not the technical part.
- **Provisional patent applications** are informal precursors to a full utility patent filing. In the United States a provisional patent application is simply a dated and filed document with a full description of the IP, and it is used to establish a **priority date** for the invention, useful for subsequent filings.
- **Plant patents** protect new strains of plants.

- **Copyright** protects writing, music, movies, software, art, and other expressions of ideas. Copyright protects the tangible expression of an idea, not the idea itself. The actual sequence of words in the book *The Time Machine* by H.G. Wells could only be reproduced with the permission of his estate. However, a copyright would not prevent the creation of an actual time machine, even if the book described all the details of a functioning device. To protect the underlying working principal of the device, a utility patent would have to be obtained.
- **Trademarks and servicemarks** protect names, symbols and specific graphical images.
- **Trade secrets** are simply things that are not disclosed publically and hence protected only because they are secret. The formula for Coca Cola® is an often cited example.

The University of Toronto seal is an example of a trademarked image.

In order to obtain protection from any of the patents (utility, design, or plant) a formal application to the government is required. In contrast, you automatically have copyright protection when you create a new piece of writing, art, or music. You are free to mark your work with the copyright symbol ©. However, a copyright symbol is not necessary to protect your work. In many countries, trademarks and servicemarks are also automatically protected once they have been in public use, although a formal registration is recommended to make it easier to defend against infringement. When a trademark or servicemark is formally registered, it is identified with the symbol® Trade secrets are protected by internal agreements with those individuals who know the confidential information. Within companies, such individuals are often required to sign a **non-disclosure agreement (NDA)**, which stipulates that they will not reveal the confidential information.

With the exception of a trade secret, all of these items represent a formal way of protecting the intellectual efforts of the creators for varying lengths of time. Engineering designs are focused on the fundamental workings of products and systems, and these ideas must be protected by a utility patent, so we will spend the rest of the chapter discussing this particular form of IP protection.

The Frisbee®

The protection of the Frisbee® flying disk toy provides excellent examples of both design and utility patents (#Frisbeepatents). The original plastic molded disk was developed by Fred Morrison and Warren Franscioni in the 1950s Because it was already common to play catch with inverted pie and cake plates, the basic concept of a flying saucer toy was not novel.

Courtesy of the Authors

The idea was already in the **public domain,** meaning that it was already known to people other than the inventor. However, the particular ornamental shape of an early plastic molded flying disc was eventually protected by a design patent, which was filed on July 22, 1957 (US Design Patent D183626). At that time, the term of a design patent was seven years, meaning that no one in the United States could copy the "Flying Toy" (as described by the two pictures in the patent document) for seven years.

After the Wham-O® company had negotiated the rights to manufacture and distribute the flying disc, their vice president, Ed Headrick, conceived of an important improvement. By adding some circumferential ridges on the top surface, the flow of air over the surface was modified and the flying disk, which had been renamed the Frisbee® by Wham-O®, flew much better. This represented a new functional improvement in the toy. Headrick applied for, and was granted, a utility patent for this innovation, which was assigned to Wham-O®. U.S. Patent 3359678 describes an aerodynamic toy including an "air flow spoiling means . . . comprising a plurality of concentric circular raised ribs." (Notice the very specific language used to describe the invention. This is typical of patents.) In the United States, for 17 years from the date of issue in 1967, Wham-O® could stop anyone else from selling a flying disc with raised circular ribs to modify air flow. In 1985, anyone could make and sell flying discs with this feature.

The design and first utility patent for the Frisbee® are publically available on-line through the U.S. Patent Office database. We have reproduced them in this text as an example.

Utility patents, as their name indicates, provide protection of the underlying functioning, or utility, of an invention. The invention could be one of several things: a physical product or machine that does something useful, composition of matter (e.g., a new drug molecule or material), a process for doing something, or an improvement to any of these. The actual range of "patentable" subject matter varies from one country to another. For example, methods of doing business are patentable in some places but not in others, and these rules evolve over time.

IP owners can prevent others from making, using, or selling the intellectual property for which ownership rights are granted in the form of a utility patent. That is, they can prevent others from *infringing* on the patent. It is not a criminal offense to use someone else's intellectual property without permission. Instead, the IP owner has a right to sue the infringer to stop them from trespassing, and to obtain compensation for any financial damage that resulted from the offense.

3. Why Do Design Engineers Care about IP?

There are basically two questions about intellectual property that will impact your design thinking:

1. Is it possible to protect the design that you are working on so that your client can profit from it without worrying about others copying the technology and competing for *market share*?

If a design can be protected with a patent, then your client will have more incentive to invest money to bring the technology to a state of commercialization where it can be of some benefit to society. The ability to protect intellectual work may, in some instances, be a necessary component of a successful design. In fact, it is quite possible that *patentability* trumps other considerations. The design direction chosen as optimal under the patentability objective might not otherwise be the best choice. However, you might choose the patentable design because it is more economically valuable. This could even be specified as a constraint in the requirements!

2. Does someone else already own all or part of the design you are creating? Are you trespassing, (i.e., infringing) when you are designing a new technology?

The second issue of concern in any new project is whether the design is trespassing on the intellectual property of others. In the IP world, trespassing is called *infringing*. When designing new products, it is important to know whether you will be infringing others' rights. Unfortunately, even if the product is not available online or in a local store, you cannot be sure that it is not protected. A patent owner does not need to commercialize the technology to retain protection on it. A thorough patent search is always required to be sure that a technology is unprotected.

In general, you cannot infringe on IP once it is more than 20 years old. Anything described in a printed publication or available for sale more than 20 years ago is no longer patent protected (although variations of it may still be protected). Also, you normally do not have to worry about the IP covering the commercially obtained components you buy to put into a system. For example, you might purchase a motor or a memory chip to put into your design. These are called *off-the-shelf* components. When you purchase an off-the-shelf item, you are also purchasing the right to use it. Although the law stipulates that you cannot use a protected invention without permission, it is usually assumed to be the responsibility of the manufacturer and *vendor* of these components to ensure that they are legally able to sell the item.

KEY TERMS

intellectual property (IP)	utility patent	design patent
provisional patent	priority date	plant patent
copyright	trademark	servicemark
trade secret	non-disclosure agreement (NDA)	public domain
infringe	market share	patentability
off-the-shelf	vendor	

4. Questions and activities

1. Give examples of each of the methods of protecting intellectual property described in the section Types of Intellectual Property.

2. Which method of protection is strongest for technical designs (where available)? Why is this the strongest protection?

3. How can you protect a design if a utility patent is not available—for example, if the design might have been in use for decades?

4. How long does a utility patent provide protection? What happens after the time limit?

5. What sort of protection does a utility patent provide?

6. Give an example of an off-the-shelf component you might purchase for use in a design, and give an example of a technology that you would need to explicitly license (get explicit permission) to use in a design.

2

Principles of Patentability

#process module: #principlesofpatentability

Learning outcomes

By the end of this module, you should demonstrate the ability to:
- Explain the principles of patentability, including novelty and non-obviousness
- Find a patent using a variety of search techniques
- Read a patent and provide a summary of its applicability

Recommended reading

Before this module:
- **Design for X > Intellectual Property > 1. Design for Intellectual Property:** Introduction

After this module:
- **Design for X > Intellectual Property > 3. Intellectual Property in the Design Process**

> This module contains general information about intellectual property as it pertains to engineering design work. Some details have been omitted for clarity and simplicity. Since laws vary from country to country and over time, and patent law is quite complex, this text is not a substitute for professional legal advice on these matters.

1. Four Criteria for Patentability

In order for a design to be eligible for protection by a *utility patent* (i.e., *patentable*), it must meet a few basic criteria: it must be a *non-obvious* advance, it must be new or novel, it must be useful, and it must be in a *patentable domain*.

Non-Obvious. Most engineering design is of a fairly routine nature, although it is nevertheless creative work of which the engineer can be proud. A chemical engineer might be responsible for laying out a new filtering unit as part of a fertilizer plant, and would make decisions about which commercial pumps to use, what method of filtering would be best, and so on. Any skilled chemical engineer could solve the problem, and although some would undoubtedly produce better designs than others, none would be surprising or vastly different. For a design to be **non-obvious**, it must be more than the result of the typical routine design process. In Europe, this extra something is called an **inventive step.** In practice, the inventive step does not have to be very big; even small advances resulting from fairly routine design activity are sometimes successfully patented.

Novelty. *Novelty* means that at no previous time, anywhere in the world, was the invention in the **public domain**. The specific definition of novelty depends on the laws of the particular jurisdiction. In the United States, under the America Invents Act, an invention cannot be patented if: "the claimed invention was patented, described in a printed publication, or in public use, on sale, or otherwise available to the public before the effective filing date of the claimed invention" [1].

The invention must not have been publically available prior to the filing of a patent. If you invent a new watch winding mechanism, but a description of the same mechanism was published in the *Obscure Watchmakers Weekly* in 1923, you are out of luck, even if you did not know about the article when you invented the winding mechanism. A patent can only protect things that are truly new. (There is a one-year grace period in United States and Canada after disclosure by the inventor under certain limited conditions.)

Only the inventor can apply for a patent. If your co-worker describes a great idea, you cannot steal it and file a valid patent application. However, if two people invent independently, either (or both) could file a patent application as long as the work has not become public. In this case, in the United States and most other jurisdictions, the first inventor to officially file for a patent would be given priority.

The sum total of information in the public domain is referred to as the **prior art** (#priorart). Of course, it might be possible to get a new patent for something that is in the prior art, if the inventor did not know about the previous work, and the **patent examiner** does not find it. However, a granted patent can be ruled invalid later if relevant prior art is discovered after the patent is issued.

For example, a U.K. patent assigned to Windsurfer International, Inc., was later challenged based partly on the fact that Peter Chilvers had invented (and ridden in public) a basic sailboard as a 12-year-old boy, at least 10 years before the first patent was granted for a stand-up sailing vessel [2]. Competitors who would like to operate in an area covered by a granted patent will go to great lengths to try to find prior art that will render invalid the patent they are trying to work around.

In order for prior art to invalidate the novelty of a patent, it must be an **enabling disclosure**. This means that it must contain enough detail to allow someone to reproduce the invention. H.G. Wells published a book called *The Time Machine* in 1895, but the book, of course, did not provide much information about how to actually build a working time machine! If you invent a working time machine today, you will still be able to file for a patent. However, a book that clearly describes how to build a device would be an enabling disclosure and you would not be allowed to patent the device it describes.

Patents cannot be granted for inventions that were on sale or in public use prior to the filing of the application. One purpose of the patent act is to make sure that the public has access to inventions as soon as possible. The government does not want a company to use an invention (and to make money selling it), but to hold off on applying for a patent until competition begins. The deal between inventors and society is simple: the inventor has 20 years to profit from the invention, and then everyone can use it.

Utility. In order for an invention to be patentable, it must have some practical use, or the potential for practical use at some time in the future (i.e., *utility)*. Exploding breakfast cereal might be new and non-obvious, but is unlikely to be granted protection by the patent office as it would have no practical, legal utility.

Patentable Domain. Depending on the country, certain domains (or areas) of interest may or may not be patentable. Business practice and existing chemical structures in living organisms are examples of domains where patents may not be possible in some countries.

2. Novelty: Searching for Prior Art

Since novelty is a requirement for patentability, a prior art search should be as broad and comprehensive as possible. Experts exist that will search the literature for a fee. Although this will cost a significant amount of money, it is still much less than the cost of a patent. Although it is upsetting for a designer to find that his or her invention has already been developed, it is far more upsetting to find out after substantial time and effort have been expended. Finally, if the relevant prior art is a patent applied for within the previous 20 years, it is even more critical to find it because missing it might lead to an infringement lawsuit by the owner of the prior art IP.

2.1. Internet

It is a good idea to start with a basic Internet search to determine if the design is on sale or widely known. One of the best resources to help you as engineering student with this process is the reference librarian at your campus library. Librarians are information experts who can be tremendously helpful, showing you how to conduct effective, efficient searches on the Internet. It is important to choose search terms carefully, and to think of all the various alternative terms that could be used in combination. An advanced search using Boolean operators (AND, OR, NOT) is a must if a very simple search does not turn up anything.

2.2. Commercial Distribution

A search of commercial companies in a particular field is best conducted with a specialized form of industrial database such as www.thomasnet.com or www.globalspec.com. These search engines report lists of companies that make products or provide services and don't clutter up the results with blogs, videos, and other Internet detritus. If you have an idea for a centrifugal pump, and you want to find companies in that business as part of your prior art search, industrial databases are very useful.

2.3. Existing Patents

When applying for a patent, it is obviously critical to thoroughly search the world's patent literature. At the time of writing, the U.S. patent database consisted of more than eight million granted patents. The U.S. patent database is freely accessible online at www.uspto.gov and can also be searched at www.patents.google.com and at various other online search providers. The U.S. Patent and Trademark Office (USPTO) provides access not only to granted patents, but also to patent applications that are in the process of examination. In the United States, a patent application is normally published (i.e., becomes available on the web) 18 months after the date of application. It can actually take several years between the date of application and the final issue date for patents to be granted, so a search of the Patent Application database at the USPTO is recommended for access to the very latest prior art.

The European Patent Server is accessible at www.espacenet.com and has a comprehensive database of patents filed in various places around the world. Since it includes the U.S. database, you can use Espacenet as a one-stop patent search engine.

A search through the patent literature can begin with a simple keyword search, similar to a general Internet search. However, some more sophisticated techniques must be used to be sure that all the important references are found. One method is to search backward and forward through linked reference lists (see the following section, Composition of an Individual Patent). On the bibliographic page of each patent, a list of prior art cited in the patent is provided. On the USPTO website, links to previous patents are clickable, so it is quite easy to jump back to an earlier patent from the reference list. With one mouse-click, it is also possible to find the list of patents that subsequently reference a published patent. This means that it is possible to skip back and forth through the literature using the links provided. This process can very quickly yield a huge number of patents, and it is helpful to go backward to find a key historical patent that the majority of subsequent patents in the field are likely to cite.

Classification numbers are also provided on the front page of modern patents. Patents are classified using an elaborate subject matter classification system, and each type of invention is given a number. In the United States for example, all pogo sticks like the one shown in Figure 1 get a U.S. classification number of 482/77. (The International Classification Number, used by Espacenet, is different.) Class 482 is for exercise devices. Subclass 77 is for bouncing devices. You can search the U.S. patent database to find pogo sticks by searching for the class 482/77, although you will also get other bouncing devices in the list. There you might find a patent that describes a "bouncing, spring actuated exercise device," and the words "pogo stick" might not be used at all. Hence the classification search would find the patent, whereas a simple keyword search would miss it. At Espacenet, a classification search will also turn up relevant foreign language patents that would be missed with an English keyword search. You can also combine keyword searches with class numbers in the advanced search forms of the various databases.

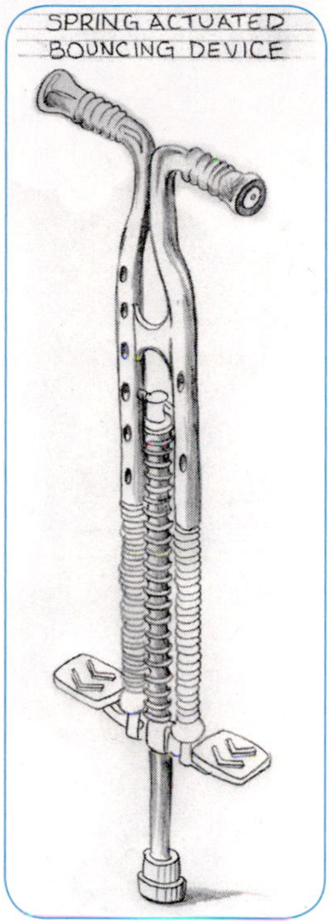

FIGURE 1 Patent titles and abstracts don't always use the words that you would expect to describe the invention.

3. Composition of an Individual Patent

We recommend that you go to the U.S. patent database and select a patent to look at while reading this information. This will allow you to identify how the bibliographic and other information looks in an actual patent. Or you may want to refer to the utility patent for the Frisbee® (#Frisbeepatents).

Bibliographic Information. The first section of a patent contains the bibliographic data. This includes details such as the assigned patent number, the title, the names and nationalities of the inventors, and so on. It also provides a list of references referred to in the patent, and the U.S. and international classification numbers. Both of these will be important during a patent search.

Abstract. The abstract summarizes the main features and use of the invention.

Drawings. Most patents include a set of drawings as an integral part of the description, although this is not mandatory. The various components of the invention in the drawings are given numbers, so that when they are referred to in the body of the patent, there is no confusion. Usually, the most representative drawing is selected for the front page of the patent.

Specification. The specification is the main part of the patent, and contains a summary of invention, some background about the prior art, and the need for the invention, then a description of the drawings, and a detailed description of the invention itself.

Claims. The *claims* are the most important part of the patent. The claims define the exact location of the "fence" around the *intellectual property* (IP) in very specific legal terms. The first claim is the most important, and usually describes the invention in the broadest acceptable terms possible given the prior art. That is, the first claim tries to fence off as large a piece of intellectual property as possible. Subsequent dependent claims will describe restricted subsets of the domain covered by the first claim. This is done in case prior art is subsequently found that invalidates the main claim. In this case, some of the dependent, more specific claims may still protect part of the intellectual property. There can be more than one independent claim in the patent; for example, a patent might claim a chocolate formulation suitable for injecting into ice cream, and might independently claim the injection process itself.

KEY TERMS

patentable	non-obvious	patentable domain
inventive step	novelty	public domain
prior art	patent examiner	enabling disclosure
utility	intellectual property	claims
utility patent		

4. Questions and activities

1. For a wheelbarrow:
   a. Go through the search process. Can you find an original patent? Patented aspects? IP that is now in public use?
   b. Given your results from the first part of the question, do you see any areas of the wheelbarrow where an inventor might be able to invent a novel wheelbarrow design?
2. Select another patent from the United States or European database. Identify the bibliographic information, abstract, drawings, specifications, and claims in the patent. Summarize the main claim in your own words. Explain how the other claims relate to the first claim in this particular patent.

5. References

[1] U.S. Patent Act. www.uspto.gov/web/offices/pac/mpep/mpep-9015-appx-l.html#al_d1fbe1_234ed_52.
[2] Van Dulken, Stephen. *Inventing the 20th Century: 100 Inventions That Shaped the World from the Airplane to the Zipper.* New York: New York University Press, 2002, p. 164.

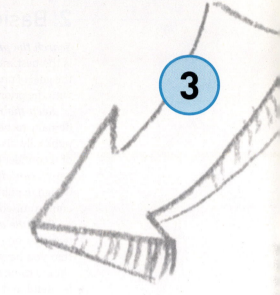

Intellectual Property in the Design Process

3

#process module: #IPinthedesignprocess

Learning outcomes

By the end of this module, you should demonstrate the ability to:
- Explain why designing for intellectual property might be a key objective
- Describe a design process that would lead to protected intellectual property
- Describe a few ways that an individual or company could benefit from intellectual property

Recommended reading

Before this module:
- **Design for X > Intellectual Property > 2. Principles of Patentability**

After this module:
- **Design for X > Intellectual Property > 4. Frisbee Patents**

> This module contains general information about intellectual property as it pertains to engineering design work. Some details have been omitted for clarity and simplicity. Since laws vary from country to country and over time, and patent law is quite complex, this text is not a substitute for professional legal advice on these matters.

1. Introduction

The ability to protect a design is an important competitive advantage, and may be a key objective in the design work. Inventions that are eligible for legal protection are *patentable* intellectual property. Without protected ***intellectual property (IP)***, competitors can copy your design and reduce your ability to profit from your work. Quite often, therefore, engineers work with the specific intention of creating commercially valuable intellectual property that can be protected. In order to do this, an engineer must do a few things that are not part of the conventional engineering design process.

2. Basic Methods

Search the prior art: It is important to stress that a very thorough ***prior art*** search is the best method to avoid wasting time and money. When you are designing with the goal of producing ***novel*** intellectual property, it is imperative that you are familiar with the prior art.

Keep the work secret: Since an invention has to be novel and not in the public domain to be patented, it stands to reason that inventors must take care not to publically disclose the work. The idea should only be discussed within the context of a confidential relationship. A formal confidential relationship can be established with a ***confidentiality agreement*** or ***non-disclosure agreement***. If you begin using a design in public or offer it for sale before applying for a patent, your ability to apply is compromised.

Find the alternatives: Suppose you have a great idea for solving a problem in a way that no one has before. You have kept your idea secret and you are convinced that you have the basis for a valid patent. Before you file a patent application, you should think about ***workarounds*** like the example shown in Figure 1. If your design is useful and starts making lots of money, competitors will probably try to find a way to achieve the same result without violating the specifics of your patent claims. This is referred to as a workaround. To ensure maximum commercial success, it is critical to try and anticipate workarounds yourself.

Fortunately, you already know how to do this because you have engineering design methods to use. For example, using a ***functional decomposition*** of the design, you can identify various alternative means of satisfying the sub-functions. Being an accomplished designer, you should have chosen the optimum means of accomplishing each sub-function. However, if you do not want a competitor selling one of the other (albeit suboptimal) versions, you should consider putting an IP fence around those alternatives too by broadening the patent or patenting the alternatives separately.

Courtesy of the Authors

FIGURE 1 **An example of a "workaround."** The variable length security bar was designed to wedge a door closed but to contract in length a fixed amount so that the door could be opened a bit, while still preventing an intruder from entering. The needed telescoping action is created using telescoping tubes (seen on the left), and this design was protected by US Patent 5988710. The design shown on the right contracts without telescoping tubes using a flap at the foot end, and hence "works around" the patent to achieve the same functionality without infringing.

3. Obtaining and Benefiting from IP

In order to formally protect the ideas, the inventor must file a patent with the patent office in the countries where protection is desired. While the inventor can, in principle, do the filing themselves as an individual, the process is quite complex and specialized help is highly recommended. In addition, the structure and language used in a patent specification, and particularly in the *claims* section, must be very precise in order to properly protect the IP. *Patent attorneys* are lawyers who specialize in IP. They often work for law firms that are entirely dedicated to this area of the law. *Patent agents* are not attorneys, so they cannot litigate IP disputes, but they are licensed professionals specializing in the patent application process. (Sometimes engineers become patent attorneys or patent agents because their technical expertise is excellent grounding for these careers.)

Since specialized help is recommended, we will provide only a very brief general outline of a typical sequence of events. A graphical representation of the process in the US has been provided by the USPTO www.uspto.gov/patents/process/index.jsp and is summarized below.

1. Invent something novel, non-obvious, and useful.
2. Seek assistance from a patent agent or patent attorney.
3. File a *Provisional Patent Application* in the United States (optional). A Provisional Patent Application is a simple filing that establishes a firm date of invention for the IP. It allows an inventor a one-year window in which to explore the commercial viability of an invention in public without sacrificing the ability to obtain a full utility patent [1].
4. File a full utility patent application.
5. File a *Patent Cooperation Treaty (PCT)* application. A PCT application is usually the first step taken to obtain patent protection in other countries.
6. Respond to the *patent examiner*. The patent office employs examiners to confirm that the patent claims are indeed novel and non-obvious. The patent examiner will do a detailed search of the prior art, and may object to one or more of the claims. The applicant then has an opportunity to respond in an interview or by correspondence.
7. Pay maintenance fees. There are fees payable at the time of filing, and also fees payable at regular intervals to keep the patent active. If the maintenance fees are not paid, the patent lapses, and the IP is available to anyone free of charge.
8. Defend against *infringement*. Competitors can choose to ignore granted patents and infringe on the IP holder's rights. The IP holder must take the infringer to court, and if successful the infringer will be ordered to stop making, selling, or using the invention, and may have to pay damages to compensate the IP owner for lost revenue. It is the patent holder's responsibility to defend their property from infringement.

4. Routes to Commercialization

There are three main methods by which IP is commercialized, depending on who did the work.

4.1. Employed to Invent

If the inventor works for a company, and his or her job entails the development of new products or services, then the ideas would normally belong to the company from the beginning. Although the individual would be named as the inventor in the patent, as required by law, the inventor would be obligated to assign the patent rights to the company. This means they give the rights to commercialize the invention to the company. In larger companies, development staff might be expected to sign a detailed employment contract stipulating this. However, these agreements will also sometimes give the inventor a fraction of the commercial benefits that result from the IP.

Note that if the inventor is employed as a janitor, for example, and develops a new product or process in his or her spare time, it is likely the inventor will be permitted to keep the patent and benefit directly even if it relates to the employer's business, since developing new technology was not part of the employee's job. However, as an engineer you will likely be required to sign an IP agreement with your employer giving up commercial rights to inventions. Professors at universities also often have this type of agreement with their employer. Even if you do not retain commercial rights, you can claim credit for your patented work on your resume and this accomplishment adds to your professional reputation.

4.2. Starting a Company

An individual who starts up a new company to market and sell a product or service based on IP is called an *entrepreneur*. Of course, it takes money to start a new company. This can be provided by the inventor and his or her family and friends, or can come from outside investors in exchange for *equity* (part ownership in the company). An *angel investor* is a wealthy individual who typically invests less than two million dollars in exchange for equity. For more money than this, a *venture capital* firm is normally involved. Venture capital firms are commercial ventures that pool money from a number of backers to invest in risky early-stage companies.

4.3. Licensing IP

It is also possible for the inventor to sell the rights to commercialize the IP to an existing company. This is referred to as *licensing* the IP, and is the dream of many home inventors. In this case, the inventor approaches a company with an idea, together with a patent or filed patent application, in hopes that the company will commercialize the idea in exchange for a *royalty*. A royalty is a percentage of the profits. For example, Fred Morrison reported that he licensed the Frisbee® to Wham-O® in exchange for 6.6% of their wholesale sales [2]. In this type of arrangement, the company typically assumes the cost of producing and marketing the product. A contract between an inventor and a company would be negotiated on an individual basis, and the company might assume the responsibility for defending the patent against infringement as part of the agreement.

It is possible to sell an idea to a company without formal IP protection, but many companies are not willing to discuss unprotected IP with an inventor. This is because as soon as a technology is on the market it can be copied if the IP is not protected.

KEY TERMS

patentable	intellectual property (IP)	prior art
novel or novelty	confidentiality	non-disclosure agreement
workaround	agreement	claims
patent attorney	functional decomposition	provisional patent
Patent Cooperation	patent agent	application
Treaty (PCT)	patent examiner	infringement
entrepreneur	equity	angel investor
venture capital	licensing	royalty

5. References

[1] Stim, R., and Pressman, D. *Patent Pending in 24 Hours,* 4th ed. Berkeley, CA: Nolo Press, 2007.

[2] Morrison, Fred. *Flat Flip Flies Straight. True Origins of the Frisbee.*® Wethersfield, CT: Wormhole Publishers, 2006.

4

Frisbee Patents

#resource module: #Frisbeepatents

Recommended reading

Before this module:

- **Design for X > Intellectual Property > 3. Intellectual Property in the Design Process**

1. Introduction

This section shows reproductions of the patents for the toy commonly known as Frisbee.®

The protection of the Frisbee® flying disk toy provides excellent examples of both design and utility patents. The original plastic molded disc was developed by Fred Morrison and Warren Franscioni in the 1950s. Because it was already common to play catch with inverted pie and cake plates, the basic concept of a flying saucer toy was not novel enough to warrant a utility patent. The idea was already in the public domain, meaning that it was already known to people other than the inventor. However, the particular ornamental shape of an early plastic molded flying disc was eventually protected by a design patent, which was filed on July 22, 1957. At that time, the term of a design patent was seven years, meaning that no one in the United States could copy the "Flying Toy" (as described by the two pictures in the patent document) for seven years.

After the Wham-O® Company had negotiated the rights to manufacture and distribute the flying disc, its vice president, Ed Headrick, conceived of an important improvement. By adding some circumferential ridges on the top surface, the flow of air over the surface was modified and the flying disk, which had been renamed the "Frisbee®" by Wham-O®, flew much better. This represented a new functional improvement in the toy. Headrick applied for, and was granted, a utility patent for this innovation, which was assigned to Wham-O®. U.S. Patent 3359678 describes an aerodynamic toy including an "air flow spoiling means . . . comprising a plurality of concentric circular raised ribs . . ." (Notice the very specific language used to describe the invention. This is typical of patents.) In the United States, for 17 years from the date of issue in 1967, Wham-O® could stop anyone else from selling a flying disc with raised circular ribs to modify air flow. In 1985, anyone could make and sell flying discs with this feature.

Dec. 26, 1967   E. E. HEADRICK   3,359,678

FLYING SAUCER

Filed Nov. 1, 1965   2 Sheets—Sheet 1

INVENTOR.
EDWARD E. HEADRICK
BY
Christie, Parker & Hale
ATTORNEYS.

Dec. 26, 1967   E. E. HEADRICK   3,359,678

FLYING SAUCER

Filed Nov. 1, 1965   2 Sheets—Sheet 2

INVENTOR.
EDWARD E. HEADRICK
BY
ATTORNEYS.

United States Patent Office

3,359,678
Patented Dec. 26, 1967

1

3,359,678
FLYING SAUCER
Edward E. Headrick, La Canada, Calif., assignor to
Wham-O Manufacturing Company, San Gabriel,
Calif., a corporation of California
Filed Nov. 1, 1965, Ser. No. 505,864
3 Claims. (Cl. 46—74)

ABSTRACT OF THE DISCLOSURE

A saucer shaped throwing implement. A series of concentric discontinuities are provided adjacent the rim on the convex side of the implement. The discontinuities provided on the convex side of the implement exert an interfering effect on the air flow over the implement and create a turbulent unseparated boundary layer over the top of the implement reducing aereodynamic drag.

This invention relates to aerodynamic toys to be thrown through the air and in particular to flying saucers for use in throwing games.

Over the past several years toys resembling saucers have become quite popular as throwing implements. In the usual embodiment the implement is made of a plastic material in a saucer shape with a rim located around the edge of the saucer, the rim having a somewhat greater thickness than the saucer portion of the implement. The rim curves downwardly from the saucer and has a configuration such that the implement when viewed in elevation approximates the shape of an airfoil.

The toy is used in throwing games and is normally gripped by placing the thumb on the convex side of the saucer and one or more of the fingers on the concave side. Throwing is usually accomplished with a wrist snapping motion wherein the thrower assumes a stance approximately at right hangles to the intended target and retracts his arm across his body. By uncoiling his arm and snapping his wrist, momentum and a spinning motion is imparted to the saucer to cause it to fly toward the target. The direction of flight from the thrower depends upon the thrower's skill and the type of flight path (e.g. curved or straight) depends upon the angle of the saucer relative to the ground when it is released by the thrower. Its appeal as a toy appears to reside in the fact that it exhibits definite aerodynamics characteristics, can be made to do maneuvers of various kinds depending upon the skill of the user, and is relatively easy to master.

The present invention provides an improved version of this well-known flying saucer. In this invention, means located on the convex side of the flying saucer are provided for interrupting the smooth flow of air over this surface. In aerodynamics this action is described as "spoiling" the air flow and the means by which this is accomplished are described as "spoilers." As applied to the present invention, this disruption of airflow is thought to create a turbulent unseparated boundary layer over the convex side of the saucer and to result in a reduction of drag especially in high-speed flight and an increase in stability while in flight. This means that a novice thrower can learn to throw the flying saucer more rapidly, that more expert throws will result with less experience, that better accuracy can be achieved and that a reduction in the skill required to use the saucer is made possible.

The invention contemplates an aerodynamic toy. The toy comprises a central portion and a rim circumscribing the central portion and curving downwardly from the central portion. The central portion and the rim together form a concave side and a convex side of the toy. In addition, means are located on the convex side of the toy

2

for interfering with the flow of air over this side of the toy when it is thrown. The toy is of a size to be readily gripped with one hand for throwing and for this purpose has a rim height which permits convenient gripping of the implement with fingers placed on the first side of the implement and the thumb placed on the opposite side.

The invention will be better understood by reference to the following figures in which:

FIG. 1 is an elevational view of the flying saucer;

FIG. 2 is a view of the top or convex side of the flying saucer;

FIG. 3 is a view of the bottom or concave side of the flying saucer; and

FIG. 4 is a cross-sectional view taken along the lines 4—4 of FIG. 2.

Referring now to FIG. 1 there is shown an elevational view of a flying saucer **10** of this invention. As can be observed from FIG. 1, the toy resembles an inverted saucer having a central portion **12** and formed integrally therewith is a rim **14**. To provide a smooth transition from the central portion **12** to the rim **14**, a curved surface **16** is provided. The central portion **14**, surface **16** and rim **14** together form two sides of the toy which will be referred to herein as the concave and convex sides of the toy.

As can be observed from FIG. 1, the bottom edge **18** of the saucer **10** together with the convex side of the toy resemble an air foil. A plateau or crown **20** is formed in the central portion **12** on the convex side of the toy. Although its contribution to the flight of the saucer has not been definitely determined, it is believed that the crown **20** also contributes to the stable flight of the saucer through the air.

Also shown in FIG. 1 are a plurality of ridges **22** superimposed or raised on the curved surface **16**. As will be discussed in more detail below, the placement of ridges on the convex side of the saucer has been found to produce a beneficial effect on the stability of the implement when thrown. This stability is thought to be due to the ridges causing an effect which is analogous to the effect of a "spoiler" as that term is used in aerodynamics.

In FIGS. 2 and 3 are shown plan views of the convex and concave sides of the implement **10**. As can be readily observed from FIGS. 2 and 3, the implement is circular in form. The rim **14** is placed so that it circumscribes the circular central portion **12** of the saucer. Further details of the relationship of the ridges **22** to the flat central portion **12** and the rim **14** can be seen from FIG. 2. The outermost ridge is located on the curved portion **16** in the area where the rim **12** and central portion **14** merge. The outermost ridge and any others which may be provided circumscribe at least a major portion of the periphery of the saucer. The other ribs or ridges **22** which are provided are located in concentric arrangement interiorly of this outermost ridge and in close juxtaposition with adjacent ridges as shown.

Another discontinuity **26** in the convex surface of the toy, also in the form or a rib or a ridge, is located interiorly of ridges **22** and forms a closed circle on the central portion **12** concentric with the center **28** of the saucer. Interiorly of ridge **26** is a ridge **30** marking the beginning of crown **20**. Crown **20** resembles a circular plateau and is located on the central portion **12** of the saucer with its center coincident with the center **28** of the implement.

FIG. 3 depicts a view from the concave side of the saucer and as shown therein has a relatively smooth surface **32** extending from the rim **14** interiorly to the crown **20**. Circle **34** corresponds to the ridge **30** located on the convex side of the saucer. In this view the crown appears as an indentation in the concave side of the

3,359,678

3

In FIG. 4, a section view taken along lines 4—4 of FIG. 2, further details of the relation of the rim 14 to the central portion 12 are shown. As shown the lower side of toy is concave and merges into the rim at 31. On the upper side the circular portion 12 is connected to the rim 14 by means of the curved transitional area 16 on which a plurality of ridges 22 have been placed. In addition, as can be observed, the thickness of the rim portion 14 is substantially greater than the thickness of the central portion. In the preferred embodiment the rim 14, ridges 22 and central portion 12 are formed integrally from plastic or other lightweight material by means of a molding operation.

The height of the rim 14 is selected such that the implement may be conveniently gripped by placing the thumb on the convex side of the saucer with the finger or fingers of the hand extending around the rim and being placed on the concave side of the saucer. It has been found that when the implement is thrown in a manner such that the saucer is approximately horizontal with respect to the ground as it leaves the hand that it displays definite aerodynamic properties and tends to "fly" in the direction in which it is thrown. It is believed that the saucer flies because the saucer approximates an airfoil and hence its flight through the air is enhanced by aerodynamic lift. Depending on the skill of the throwers the angle of the saucer with respect to ground can be varied to obtain greater eccentricities in flight such as causing the saucer to curve in one direction or another. Similarly, the angle of attack with respect to the air can be varied such that if the saucer is thrown at a high angle of attack relative to the wind or airflow, the saucer can be made to demonstrate an action similar to a boomerang.

As indicated in a preceding discussion, the ridges 22 provided on the convex side of the saucer have been found to result in an improved flight of the saucer regardless of the skill of the thrower. This improved flight is not completely understood but is thought to be due to an effect analogous to a spoiler on an airfoil which interferes or interrupts the smooth aerodynamic flow over the top of the saucer. Put another way, the ridges or spoilers cause a disruption of the normal airflow pattern over the top of the saucer. This interference with the smooth airflow pattern results in a reduction in drag and an increase in stability, especially under high-speed flight conditions. The increase in stability under high-speed flight conditions is highly desirable since slight mistakes in execution of a hard throw tend to be exaggerated under these conditions. Due to the fact that the spoilers are provided and stability is thereby increased, the chances of a poor flight due to a hard or high-speed throw are reduced.

In addition to the various methods of throwing heretofore discussed, the saucer can also be used to perform other maneuvers. For example, a high, easy throw of the saucer causes it to stall and float downward gently as lift forces resist the pull of gravity. This floating action enables someone participating with the thrower to easily catch the toy as it settles vertically. Similarly, the throw can be executed by a throw in which the thrower begins with the toy behind his back and carries it in a vertical plane past the thigh and then rotates the wrist such that the saucer is brought to a nearly horizontal plane and released with a flip of the wrist in a relatively horizontal orientation causing it to fly away from him.

4

Although the spoilers which for the basis of the improvement in the saucer with which this invention is concerned have been shown to be circular ribs or ridges circumscribing the periphery of the saucer, various modifications and embodiments of such spoilers are believed to be possible. Provision of other means for interrupting the normal airflow pattern over the convex side of the saucer are possible without departing from the scope of the invention as determined by the following claims.

What is claimed is:

1. An aerodynamic toy comprising:
   a circular central portion having a center surrounded by a substantially flat circular surface area whose boundary is defined by a predetermined radius;
   a rim circumscribing the circular central portion;
   a surface of curvature extending from said boundary and curving downwardly to a point of juncture with said rim to form an upper convex surface and a lower concave surface of the toy; and
   air flow spoiling means located on the convex surface of curvature and extending substantially from said boundary to the point of juncture at the rim, said spoiling means comprising a plurality of concentric circular raised ribs being concentric about said center.

2. A toy according to claim 1 wherein said plurality of raised ribs are evenly spaced.

3. A toy according to claim 1 wherein the rim has a greater thickness than the body portion.

References Cited
UNITED STATES PATENTS

D. 183,626	9/1958	Morrison	34—15
356,929	2/1887	Cruttenden et al.	273—105
2,659,178	11/1953	Van Hartesveldt	46—74
2,835,073	5/1958	Dame	46—74

RICHARD C. PINKHAM, *Primary Examiner.*

ANTON O. OECHSLE, F. BARRY SHAY, *Examiners.*

T. ZACK, *Assistant Examiner.*

United States Patent Office

Des. 183,626
Patented Sept. 30, 1958

183,626

FLYING TOY

Walter Frederick Morrison, La Puente, Calif.

Application July 22, 1957, Serial No. 47,035

Term of patent 7 years

(Cl. D34—15)

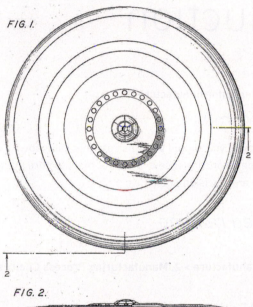

FIG. I.

FIG. 2.

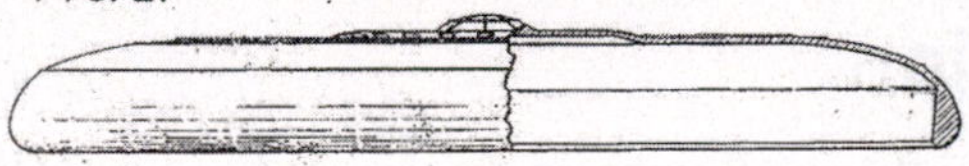

Fig. 1 is a top plan view of a flying toy, showing my new design;

Fig. 2 is a side elevational view thereof, partly in cross section taken on line 2—2 in Fig. 1.

I claim:

The ornamental design for a flying toy, as shown.

References Cited in the file of this patent

UNITED STATES PATENTS

D. 137,521	Davidson	-----	Mar. 28, 1944
1,404,132	Manes	-----	Jan. 17, 1922
2,690,339	Hall	-----	Sept. 28, 1954

FOREIGN PATENTS

890,001	France	-----	Oct. 25, 1943

Design for Manufacture:
Introduction

#process module: #designformanufacture

Learning outcomes

By the end of this module, you should demonstrate the ability to:
- Evaluate design decisions for ease and cost of manufacture in the wider context of the full design and decision outcomes

Recommended reading

After this module:
- **Design for X > Manufacture > 2. Manufacturing Process Choices**

1. Introduction

Courtesy of the Authors

FIGURE 1 Soft drinks are an example of a product manufactured in very high volume.

With any product, a very large influence on the design decisions is how that product is to be produced, developed, or constructed (i.e., brought from a design idea into a real technology that can be used). This is called the execution of the design. Design for manufacture is concerned with reducing the time and/or the cost to execute a design, in particular a product. For a mass-produced commercial or consumer product in particular, the effects of the decisions will be apparent long after the design process is complete, because any avoidable manufacturing costs will be incurred with every copy of the design produced. For example, soft drink consumption in the United States is over 150 liters per person per year [1]. At this level a very small decrease in the cost of manufacture would result in enormous savings.

Therefore, most of the emphasis of this module will be manufactured goods produced in ***high volume***,[1] such as washing detergent or cell phones. In these products there are significant savings to be made through manufacturing decisions. Note, though,

---

1 "Volume" in this context does not refer to the size of the product. Volume refers to the number of units (i.e., copies of the product) produced.

that even low-volume or single-copy products can benefit from the same design approach. They often have repeated elements in them (such as multiple floors in a building) and decisions about manufacture that could significantly increase the cost and/or the time to completion. Sometimes, even though only a single *unit* (i.e., a single copy of the product) is being produced, a design decision can be made that will significantly reduce the time to fabricate the technology. For this reason, many design teams include an engineer who specializes in manufacturing.

Checking a design for manufacturing feasibility is essential. Wonderful ideas that cannot be built (or that cannot be built for a reasonable price) are plentiful. For instance, the facsimile machine (the "fax" machine) was invented in various forms many years before modern electronic technology allowed the device to be manufactured at a reasonable cost. A design engineer must eliminate or modify those potential product solutions that cannot reasonably be manufactured.

2. Choices of Material and Parts

Designs can often be produced from many different choices of parts and materials. The choices of materials and parts are determined through the design process and this process may include input from a marketing and sales department. These choices, in turn, can determine the manufacturing time and cost. The study of materials is important in itself, and understanding materials is one of the basic skills of the complete engineer. Here we will briefly look at some of the characteristics of materials and parts that are important to manufactured designs.

- **Cost**. Manufacturing cost is a general issue; the costs of the raw materials and parts will be part of this concern.

- **Availability.** Parts or materials that are *single-sourced* (*sole-sourced*) or that come from sources that could be interrupted would interfere with manufacturing should the source(s) be unable to deliver. These interruptions can include strikes, natural disasters, or political disruptions. The major earthquake and tsunami in Japan in 2011 had a significant impact on the deliveries of car parts and integrated circuits. This is particularly important since a large number of companies now limit the inventory they have on hand; an economic strategy called running lean.

- **Handling the materials and parts.** Some materials and parts require special handling because of their delicacy, environmental hazard, weight, size, and composition. For example, special equipment is needed for handling materials that could be hazardous if inhaled by humans. Where alternate choices exist, these alternatives could prove to have lower total cost. In other cases, choice of materials may open up processing options. Many ferrous metals, for example, can be manipulated using magnetic fields and magnetized tools.

  In addition to cost factors, choice of materials and parts will also affect manufacture time. Depending on the parts and material choices, steps may be required in the manufacturing process that may not be required if different materials or parts were chosen. Some of these factors include:

- **Finishing.** Some materials have an acceptable finish when they are used in manufacturing (e.g., molded plastic). Other materials may require polishing, painting, or other operations to improve the aesthetic appeal or to protect the surface. Some materials may allow labels, logos, serial numbers, and such to be built into the part; others may allow these to be embossed or printed onto the part, and others will require a separate labeling operation.

- **Inherent wait times during processing.** Paint must dry; concrete must harden; fermenting takes time; baking and cooling can sometimes be sped up through process decisions, but will always take some time. In each of these cases there must be an allowance in the full manufacturing procedure to hold product during these processes. It may be advantageous to change the parts and material selections or to alter or augment the process to shorten or eliminate these waits if possible.

2.1. Designing or Purchasing Parts to Speed the Manufacturing Process

As a general rule, products that can be assembled using fewer parts will be faster and less expensive to manufacture than alternative designs that use more parts or components. In addition, sometimes parts can be purchased or designed to speed the assembly process. Some examples include:

- **Pre-assembled parts** or parts that replace assemblies only need to be mounted in the final assembly. This saves steps in the final assembly effort. Such parts may cost more than what they replace, but can lower the all-over unit cost by reducing manufacturing time.

- **Parts with built-in guides and housings** for other parts will make assembly faster and can decrease the number of mistakes made in the manufacturing process. With some types of parts, molded parts in particular, such guides and housings are possible.

- **Components that reduce programming time.** Some devices with embedded control require programming. This can be a time-consuming task, done serially for devices. Using devices that do not require programming, or that are pre-programmed, may speed production of the units.

Most consumer hard goods will have evidence of these decisions. Figure 2 shows the lower part of a computer speaker. Figure 2a is of the completed assembly. Figure 2b is of the rear case part. This part shows several examples of built-in guides and housings to improve manufacturability. The grey arrows point to slots for the circuit board assembly and the white arrow points to one of the posts for the screws that hold the case together. Figure 2c shows the circuit board in place in the slots. Figure 2d shows the other half of the case, which has posts that accept the screws and further guides for the circuit board.

2.2. Using Testing during Manufacturing to Reduce Time

If the design requires that the product be set up or tuned during manufacture, or if the product must be tested before shipping, this will increase the time of manufacture for all units and could warrant other design choices that reduce or eliminate these steps.

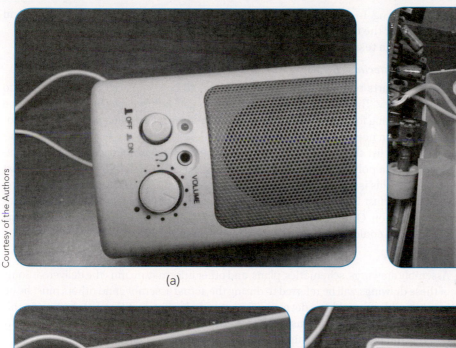

Courtesy of the Authors

(a)

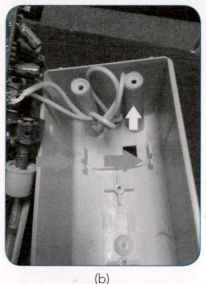

Courtesy of the Authors

(b)

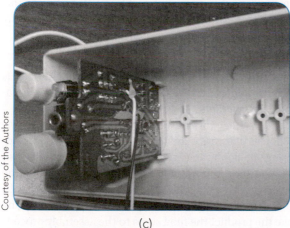

Courtesy of the Authors

(c)

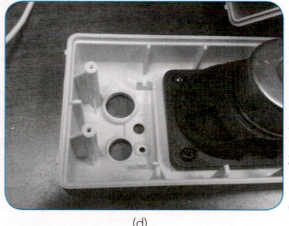

Courtesy of the Authors

(d)

FIGURE 2 **Parts of a disassembled computer speaker housing.** The arrows point to guides for the printed circuit board and the screw posts that accept the screws that hold the two halves of the housing together.

For products with a large number of assembly steps, such as an automobile, testing is done frequently during the process, and an automobile will not be released from the end of the assembly line until all tests are passed. The efficiency of the test process design will, in part, determine the manufacturing cost and time.

3. What the Design Engineer Delivers to Manufacturing

Once a detailed design is complete the design engineers have to transfer the information to manufacturing. The manufacturing may occur within the same company as the design unit, or may be sourced from another company that specializes in manufacturing. In either case, the information about the design needs to be clear and complete for the manufacturing to be performed properly.

The design team will generally deliver:

- ### Bill of Materials (BOMs)

The list of parts to be used in manufacture should be completely specified, down to the last screw. The fact that screws are "easy to get" still means that delivery will be delayed waiting for the screws to be located, purchased, and delivered if they were not at hand when needed.

The BOM list should include at least one *source* (*vendor*) for each part (some companies have a policy that two or three sources must be listed). The description must be specific; this is called a *specification* or *spec* for short. For instance, a specification for a screw must include the material, length, thread size, head type, drive type, hardness, and so on. Often acceptable alternatives for the part are also specified, since the original part could be discontinued or become hard to obtain.

- ### Drawings

Assembly drawings, schematics, plans, and other information must be delivered. Some of these drawings will be referred to during the actual assembly, and others must be on hand should any problems develop—for example, if a product that fails testing. In this case the drawings would be used to track down the possibly defective parts or identify differences between the manufacturing process and the specified design.

- ### Manufacturing Process or Procedures

Where there is an order of assembly that must be followed, or special handling is required, then this information must come from the design group. This documentation is sometimes called *assembly instructions*.

Properly done, the design team will be working with manufacturing during the design process to make design choices that ease manufacturing processes. The "hand-off" of the design to manufacturing will not be an abrupt event, but a gradual process. For very low-volume items, the design team will usually be heavily involved until the end of the project including working with technicians on the assembly of the units. For a high-volume product, there may be engineering teams devoted to working with the design engineers to take the product the final steps to the assembly process. Engineering teams may be involved in the design and implementation of specialized manufacturing equipment designed specifically to support the manufacture of the product, such as a specialized conveyor belt to carry the partially assembled units from one area to another. Design and manufacturing engineers will also frequently review the manufacturing process to circumvent problems and improve efficiencies.

4. The Interaction of Design for Manufacturing Decisions and Other Issues

Ideally you would design every product for optimal manufacturing. However, the ideal is rarely possible and good design is about choosing an optimal balance (or as good a balance as can be managed) between many factors. Figure 3 shows some of these for manufacturing decisions.

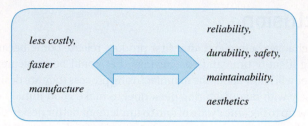

FIGURE 3 Possible trade-offs in manufacturing.

There are some common factors that are balanced with manufacturing or support decisions that favor reductions in manufacturing cost and time. It is useful to be aware of these as you develop a design:

- **Cost to the User**

There is a direct relationship between the cost that must be passed on to the consumer or client and the cost to build. Removing costs in manufacture gives the opportunity to reduce the cost to the purchaser, or to increase profits for the seller.

- **Aesthetics**

Manufacturing decisions can change the aesthetics of a product. Building a product in a plastic case versus a metal case changes the **look and feel**. Packaging has a large impact on production and product costs and often represents a significant portion of the overall unit cost. Packaging needs to be practical to allow the product to be easily shipped and sold, but also serves a marketing and aesthetic purpose. Engineering design teams will often work with **industrial designers** to achieve a balance between **aesthetics** and manufacturability.

- **Reliability**

Reliability of the product could be increased or decreased by decisions involving manufacture. Parts made of plastic could be less expensive to manufacture, but may be more prone to failure under temperature extremes or ultraviolet radiation when compared to parts made of metal.

Soft drinks often come in plastic bottles. The technique used to make the bottles is called "blow molding," where softened plastic is forced by air into a mold, which then opens to expel the bottle after the plastic cools. Although the technique produces very inexpensive containers, the designers must balance the use of minimal material to reduce costs, and the strength and durability requirements of the product. This is particularly important for carbonated drinks, where the bottle is under internal pressure and must survive being subjected to hard handling and extremes of temperature by the supply chain and by the user. For example, soft drink bottles must work reliably in airplanes at altitude under low ambient pressure. If you have ever opened a carbonated beverage on a plane you may have noticed the effects of the pressure difference. Imagine if one of the bottles explodes onboard; it would be a mess, and might cause the airline to switch to a different drink brand.

5. Conclusion

Many design changes before and after first product release come because people in manufacturing identify time and cost savings that could be implemented. A good design engineer will know about the manufacturing process they are designing for, and will work closely with manufacturing personnel to make good initial design choices and then follow up to become aware of and to implement better design choices during production. A design team may even include an engineer who specializes in manufacturing for just this reason.

Like many design decision spaces, design for manufacture is heavily dependent on the context. For manufacture, these decisions can depend on the item(s) being manufactured, the resources available including the labor market, the technologies available, and the quantities being manufactured. Further, they will depend on the cost-effectiveness of the methods that are already in use by the company that could be used or extended to manufacture the new product. Like all objectives, any design for manufacture objectives and decision must be weighed against other objectives and goals.

> **Examples of Savings Estimates Resulting from Manufacturing Changes**
> Very small savings from manufacturing decisions at the unit level can multiply into very large savings. iPhone sales were estimated at 100M by the end of 2011 [2], meaning a savings of a dollar per unit, much less that 1% of the consumer purchase cost, would result in remarkable savings, and even a penny a unit would be worth many engineering salaries for a year!
> In the consumer goods industry, savings in the packaging required has a significant impact; box size can increase shipping costs if it is not designed to fit a standard shipping container.

This module has given a very quick overview of a topic that is quite industry-specific. Only some of the specific techniques from one manufacturing design can be transferred to another, particularly if the industries are very different from one another. For a design engineer, it is important to be able to evaluate the options available, not just in terms of the design effort, but even more importantly in terms of their effects on the production of that design.

KEY TERMS

high volume	manufacturing	unit
single-source or sole source	source	vendor
bill of materials (BOMs)	specification (spec)	assembly instructions
look and feel (aesthetics)	industrial designer	

6. Questions and activities

1. Examine the speaker in Figure 2. List and specify the order of a set of manufacturing steps to produce the speaker. Or pick a simple product you own and develop a simple set of assembly instructions for the unit.

2. Give examples where giving design for manufacture priority over other considerations might negatively affect:

   a. Reliability

   b. Durability

   c. Safety

   d. Maintainability

   e. aesthetics

3. Consider the following pairs of materials for cost, availability, manufacturing method, safety concerns during manufacture, manufacturing time, and equipment/labor requirements. State any assumptions. You may have to research each of these materials to gain enough knowledge.

   a. Fiberglass versus steel

   b. Plastic versus aluminum

   c. Titanium versus aluminum

   d. Wood versus steel

   e. Glass-wool insulation versus spray foam insulation

7. References

[1] Martinne, G. *UPDATE 2-U.S. soda consumption fell faster in 2011,* March 20, 2012. Retrieved from www.reuters.com/article/2012/03/20/drinks-idUSL1E8EK1P620120320

[2] Ian, P. *iPhone Sales Forecast to Hit 100 Million by 2011,* June 18, 2010. Retrieved from www.pcworld.com/article/199237/iphone_sales_forecast_to_hit_100_million_by_2011.html

(2)

Manufacturing Process Choices

#process module: #manufacturingprocesschoices

Learning outcomes

By the end of this module, you should demonstrate the ability to:
- Make advanced decisions about the manufacturability of your design

Recommended reading

Before this module:
- **Design for X > Manufacture > 1. Design for Manufacture:** Introduction

1. Manufacturing: Designing the Process

In this section we will look at specific decisions that can be made in the area of design for manufacture and the processes available that will assist you in design. The type of material selected or the parts selected can affect the processing methods that can be reasonably used, and the processing methods available can affect the material and part choices. Like design overall, this is seeking the best balance, and to understand that balance we must look at process decisions in greater detail. In industry, manufacturing analyses are often done by Industrial Engineers trained in the area.

2. Process Management

Manufacturing time is an important factor because it relates directly to cost and to response time. As a result, management of the manufacturing process is important. Using the methods of ***project management*** an engineer can determine which operations can be done in parallel (i.e., at the same time concurrently) and which operations must be done in series (one after the other). For example, parts of an assembly may be able to be produced in parallel, but packaging of the assembly must be done only after the parts have been built. This specialized area of project management is called ***process management***.

Consider the production of a toaster. Modern toasters have small circuit boards that control the timing of the process. This is shown at the center-right in Figure 1. Production of the circuit boards, of the heating elements and of the outside case could be done in parallel, with final assembly of the parts, testing and packaging done after these subassemblies were built. Each subassembly and the final assembly are manufactured on an **assembly line**, often just referred to as a **line**.

Manufacturing process design and management often includes transport and storage. Transport, such as a conveyor belt, is used to move the parts from the subassembly manufacturing line to the main assembly line. And storage is required when parallel production results in subassemblies that need to be stored for use later. For example, it would likely be complex and/or costly to simultaneously produce jelly beans of the various colors in a typical mixed bag of jelly beans at exactly identical rates. Instead, each color is produced in a separate production line and stored until needed in a mixing and packaging line. They are then fed into the line at the right rates to produce a perfect mixture.

FIGURE 1 A disassembled toaster.

3. Other Factors in Process Decisions

Besides trying to match the timing of parallel operations, other decisions go into choosing the manufacturing methods for each step. One essential question is: will the manufacturing require specialized resources? These resources could be human, a specially-trained operator or one with years of experience, or it could be a specialized piece of equipment. This might save time or money or deliver a better product, but also could be a risk should the resource become unavailable, even temporarily. Related to special resources is the choice of whether to mechanize a manufacturing process, or use people (hand-assembly). There are many interesting videos online about robotic assembly, for example, which is the ultimate automated assembly method [1]. Table 1 shows some of the other key factors that affect manufacturing decisions.

At the end of the module is a substantial example of process-level decisions applied to the manufacturing of a box.

Courtesy of the Authors

Courtesy of the Authors

Manufacturing Process Choices

TABLE 1 Factors influencing choice of manufacturing process.

	NOTES
Use of existing facilities, tools, processes, and expertise	It will generally be less risky and less expensive in start-up to use existing knowledge and facilities, but this could be less than optimal in the long run
Scalability	If the expectation is that the output of the facility will vary, then the method should be appropriate for all anticipated levels of output
Energy requirements	Lower energy requirements could prove very important in the cost of the process
Waste reduction: recycling, reuse, and waste	Manufacturing choices that reduce waste, and particularly those that recycle or reuse failed product will reduce the overall cost and increase community good will
Human requirements	Avoiding operations that increase the hazards to people, both workers and the public, will often decrease the cost of manufacture overall
Special environments	Clean rooms and other environments that must be strictly controlled will increase the cost of manufacture

4. Design Specifications that will Ease Assembly

In this section, we will look at some specific design decisions that will speed assembly and lower costs.

4.1. Decisions about Connections and Mating Pieces

While the design engineer tends to concentrate on the functions and user-related objectives of a product, any assembly of parts will necessarily have connections that are made during assembly. The designer should consider these carefully with manufacture in mind. The designer may also be asked to create assembly instructions to guide the manufacturing process. Some considerations are:

- Where possible, specify fasteners that are quick to use: welding or glue instead of fasteners; snaps or rivets rather than bolts or screws; or screws that allow power tools, which are preferred over those that require hand tightening (see Figure 2).
- Specify color-coded wires and keyed and different connectors for each cable so that incorrect assembly is less likely or not possible. This simplifies the job of the assembler and reduces the chance of error.
- Specify in the design the guides, mating pieces, markings, labels, and so on, that will help the assembler to build the product correctly. Many of these can be designed into the parts themselves, such as molded guides and labels in a plastic part, or printing on a circuit board. If the design makes it is easy and obvious how the parts go together the correct way and hard to put together the wrong way, there will be fewer assembly errors.

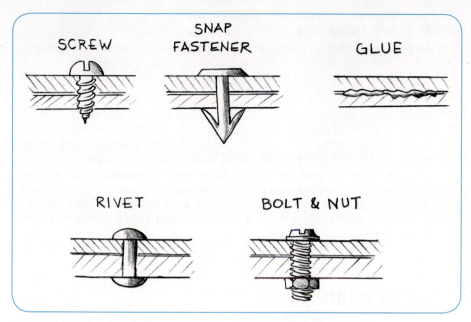

FIGURE 2 Examples of fastener types for joining two parts together.

One of the authors was involved in the design and production of automated test equipment with many different pieces of test equipment linked together using multiple cables for data and power. The connectors were designed to be absolutely identical, which had many advantages in terms of ease of assembly. However it also made it possible to hook up the equipment in a way that would damage the components.

The problem was solved by the use of keys. A key is an insert that changes the mating profile of the connector. So only the correct plug would mate with its matched connector. Incorrect plugs would encounter mechanical interference from the keys and could not be connected to the wrong outlet.

4.2. Decisions for Automation and Speed of Assembly

We have already talked about molded guides and housings for parts. Here are some other considerations having to do with automation and speed of assembly:

- Design with parts that allow automatic insertion, automated handling, and so on. The range of parts that could by handled by automated equipment in the past was quite limited, but with advances in robotic vision and manipulation this is no longer the case.
- Build jigs and guides that the assemblers can use to more quickly handle parts.
- Kit parts: that is, bring the parts already grouped to the assembler so they can be accessed quickly during the assembly process.
- Design parts, subassemblies, and assemblies with built-in test features, or the ability to link to automated testing equipment. Since the testing has to be done, automating it will speed the assembly process. For example, design in easy access for multimeter probes.

> ### Offshoring and Outsourcing
>
> **Outsourcing** means having the work done outside the home company. For example, a company that does some of its own manufacturing may choose to outsource a new product line if the cost is less or if they do not have the internal capacity to handle the work.
>
> **Offshoring** means having the work done in another country, typically by another company. It is a form of outsourcing. Many companies do high volume manufacturing offshore, primarily due to the low labor rates available elsewhere. For a design engineering team this produces additional challenges. When manufacturing is in-house the design engineer can visit the assembly line frequently in person and easily meet with the manufacturing team. Offshoring makes this more difficult and may introduce communication, cultural, and time-zone differences. The engineering design team will have to document their work more closely to guarantee the product is manufactured to spec.

5. Conclusion

Consideration of a design from a manufacturing viewpoint is not a trivial concern. Huge amounts of money can be gained by appropriate decisions that involve in-depth knowledge of the materials and fabrication processes involved. This is particularly the case for high-volume mass-produced technologies. However, even for low-volume designs, knowledge of manufacturing processes will improve the detailed design and prevent the design of un-makeable parts or assemblies.

KEY TERMS

project management	**process management**	**assembly line** or **production line** or **line**
outsourcing	**offshoring**	**job shop**

6. Questions and activities

1. Bring in a consumer item that can be destroyed, like a toaster or hairdryer. Often you can find these at yard sales or second-hand stores (e.g., thrift shops, Goodwill stores). Take it apart and make notes on the design-for-manufacture decisions made.

2. Consider the connection methods shown in Figure 2. For each one, propose a situation where it would be the best choice. Besides design for manufacture, consider safety, testing and maintenance, and product lifetime.

3. If overstress happens, it is often important that components fail in a manner that minimizes damage. This is called failsafe design. Explore the strength ratings of bolts to see how bolt strength might be specified. How are bolt characteristics noted on the bolt? You could go to a local hardware store to look at different bolts for this activity.

7. References

[1] Freightliner Robotic Manufacturing. *Cab and Sleeper Panel Assembly*, July 7, 2009. Retrieved from www.youtube.com/watch?v=xMqlgtI3Tbg

8. Substantial Example: Process-Level Manufacturing Decisions

We will look at the effects of material and of manufacturing method by looking at an example. In the following we will consider making a simple box in a number of ways:

- Build it from individual pieces, like one might build a packing crate out of pieces of plywood.
- Build it from sheet metal, where the metal is folded into a box.
- Mill (cutting out) the box from a block of metal using software that goes from a design on a computer through to a computer controller that guides the cutting blades.
- Sand-cast the block in a sand mold. In this method molten metal is poured into a void in a packed sand form made by pressing the sand around a model of the box we want to build, then removing the model. The molten metal will then harden into a replica of the model.
- Mold the box in plastic. A metal mold is made and molten plastic forced into it. We will assume an injection-molding process is being used, where the form is similar to the form for sand-casting above, but has been machined in two halves from metal.
- Use 3D printing, a deposition technique where a part is built up through rows of deposit, and layers of these rows.

Table 2 lists some of the issues that differentiate the various methods of manufacture. Note that these are not the only methods available.

> Summary of the chart: For simple boxes that are designed for a unique situation, using individual parts or sending a design to a metal-folding *job shop* would make the most sense. If a development/production group already had the software and automated machinery in place, using an automated milling machine or 3D printing would work for small quantities. Sand casting would likely take too much time and cost too much to make a simple box; it would make sense for small runs of boxes with very complex mounting tabs, guides and so on, and for boxes with very thick walls. For large volumes of boxes, the choice would likely be injection molding. Each copy made from the mold is extremely inexpensive, but the mold itself can be costly. For high volume units, the cost of designing and producing a mold is worthwhile because the cost can be spread over the number of units produced.

TABLE 2 Comparison of manufacturing methods for a box.

	PIECES	SHEET METAL	MILLING (AUTOMATED)	SAND-CASTING	INJECTION MOLDING	3D PRINTING
Major strength of method	Inexpensive parts and simple tooling	Short design cycle	Design to finished product can be all digital	Less waste; complex shapes possible	Large volumes at low unit cost	Small volumes, complex shapes
Major weakness of method	Small runs of simple items only	Large volumes difficult; many designs impossible due to tooling restrictions	Wasteful of material; large volumes difficult	Labor intensive; large volumes difficult	Must produce expensive mold	Significant time in production. Requires specialized software
Tooling required	Minimal; could be done with hand tools	Cutting, folding, punching machines	Automated machines and compatible software (expensive)	Must melt metal and be able to handle it	Sophisticated injection molding machine	3D printing machine and software plus materials
Preparation required	Minimal; rough plans only	Plans where automated machines do cutting and punching	Detailed design for downloading into machine software	Must prepare model of what is to be produced to make image in sand	Detailed mold must be machined before method can be used	Detailed computerized drawings
Comparative cost for run of a single unit	Very low	Very low	Medium	Low	High	Low
Comparative cost for multiple units	Very high	High	Medium	High	Very low	Low (except time problem)
Guides/mounts possible?	No	A few types possible; might be difficult	Some types possible	Yes	Yes	Yes. Specializes in shapes that may be impossible to mill
Finish (color, smoothness, etc.)	Must be painted; many corner joins	Must be painted	Must be painted (unless milled of colored plastics)	Generally requires final machining or other finishing and coloring	Could be made of colored plastic; no finishing required	Rough but acceptable for some parts

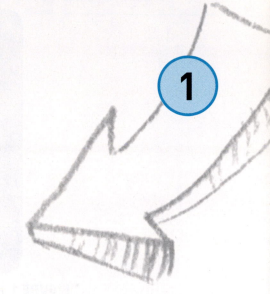

Design for Safety:
Introduction

#process module: #designforsafety

Learning outcomes

By the end of this module, you should demonstrate the ability to:
- Describe the role of safety in the overall design process
- Describe the use of safety engineering at various levels
- Explain the difference between risks and hazards
- Explain the relationship between safety engineering and codes and standards

Recommended reading

After this module:
- **Design for X > Safety > 2. Identifying Hazards**

1. Introduction

The engineering profession considers the health, welfare, and safety of the public to be of paramount importance. Doctors generally help one person at a time, while engineers design things that can benefit many. The converse of this is also true: whereas a medical mistake generally harms one person, a faulty engineering decision that leads to a bridge collapse or an explosion may cause widespread losses. For this reason, a consideration of safety is normally considered to be the most important of the *design for X (DFX)* principles.

Good engineers design things that work as intended, and do not unintentionally damage property, the environment, or human health. They anticipate circumstances that might lead to such losses, and weigh the risks carefully. They design protective and warning systems, and create training and operational *protocols* to make losses unlikely. A diligent engineer is not expected to be perfect, but is expected to be competent and careful. Nevertheless, the history of our modern technical society is littered with examples of failures: airplane crashes, toxic waste releases, massive software crashes, and so on (see Figure 1). In principle, of course,

FIGURE 1 In 2010, the failure of the blowout preventer on the Deepwater Horizon drilling rig in the Gulf of Mexico led to the deaths of 11 workers and an environmental catastrophe. (Photo courtesy of the US Geological Survey.)

all of these failures are preventable, provided the circumstances leading to them are anticipated.

You may be getting worried about being "responsible," perhaps even ***liable***, for your design work. In fact, anyone paid to do professional work is responsible for outcomes, not just engineers. Doctors, lawyers, plumbers, and tattoo artists are all responsible for what they do. Engineers need to be familiar with the basic principles of designing for safety, and to be diligent in applying these principles where appropriate. They need to follow all ***regulations*** and ***industry standards***. It is not necessary to be perfect. All that is expected is that you will act with at least the care and competence of an engineer of ordinary skill. It is only when engineers disregard common practice, ignore regulations, and take shortcuts that they irresponsibly create trouble for themselves and others. Here we will focus on the basic principles underlying the field, and tell you when it is appropriate to seek more information.

The basic principles of safety engineering can be applied to design in a variety of ways. Engineering failures can be caused by something as simple as the failure of a bolt or o-ring, or as complex as a poorly designed organizational structure in a large corporation. In the book The Human Factor, engineering design is discussed at five levels: physical, psychological, team, organization, and political [1]. There are examples of design successes and failures spanning this entire range. For example, the specific cause of the Space Shuttle Challenger disaster was a faulty o-ring, but the organizational structure and culture at NASA were contributing factors. We will focus primarily on the physical and psychological aspects of design for safety, but it is important to remember that these are only two of the levels of engineering design that can affect safety.

2. Concepts of Risk and Hazards

When engineers talk about safety, they are talking about avoiding undesirable losses. Here we will quickly review the concepts of risk with an orientation to safety (#failure&risk).

A *hazard* is a situation where there is the possibility of damage to health, property, or the environment. The result of exposure to a hazard is the *consequence*. Unintentional exposure to the hazard is an *accident*. Gasoline vapor, in the right concentration, can explode. This potential for explosion is a hazard. The destruction that results from an accidental explosion is the consequence, and is characterized by its *severity*.

Our concern about a particular hazard depends not only on the severity of the consequences, but also on the probability there will be unintentional exposure. The combination of these two factors is referred to as the *risk*, and the simplest form of an equation for risk would be expressed as follows:

$$\text{Risk} = (\text{severity and cost of consequence}) \times (\text{probability of occurrence})$$

$$R = S \times P$$

For a gasoline vapor explosion, the risk would be acceptable only if the probability of an explosion was very low, because the explosion would have serious consequences. For example, filling your gas tank at a gas station is considered an acceptable risk because although the consequence is potentially very serious the probability of an explosion is very low. However, you are warned not to use a cell phone while pumping gas to eliminate a source of ignition, and therefore to keep the risk as low as possible. A paper cut is quite likely for office workers, but because the consequence is not serious, the risk is judged to be acceptable: high probability, minor consequences. Office workers do not have to wear special protective gloves when handling paper. Here the risk is low because the damage is relatively inconsequential. A design might be considered safe, therefore, if the level of risk is an *acceptable risk*.

It is not necessary for risk to be zero for a design to be considered "safe." Children are taught to use cutlery, and even though they could put out an eye with their fork, people deem the risk to be acceptable once the child has had suitable training. An individual's perception of acceptable risk, and society's decision about what constitutes an acceptable risk, is complicated and evolving. In North America seatbelts in cars were optional 60 years ago, then they became a standard feature in cars, and now most

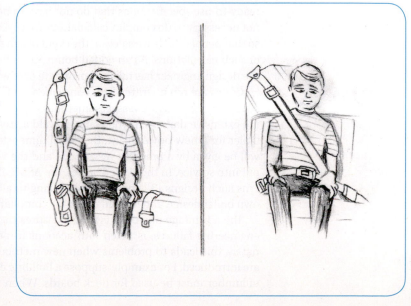

jurisdictions require the driver to wear a seatbelt while driving. What is perceived as acceptable risk also varies substantially between cultures.

From an engineering design perspective the goal of design for safety is to reduce risk to something less than the threshold of acceptable risk. There are a number of ways to do this:

- You can eliminate the specific hazard or risk.
- You can reduce or mitigate the risk.
- You can guard against the risk.
- You can warn against the risk.

In all cases you, as an engineer, must first and foremost follow all legal regulations, codes, and standards that pertain to the design you are developing.

3. The Role of Codes in Safety Engineering

When the failure of a design would lead to serious consequences, more often than not it will be governed by *regulations*, *codes*, and standards, which we will refer to jointly as "codes." A code is a written rule, enforceable in some way, that dictates exactly what a design must or must not be. It may be imposed by government, or it may be simply agreed upon by a representative industry group. For example, part of the Ontario, Canada FIRE PROTECTION AND PREVENTION ACT, 1997, REGULATION 213/07, Section 2.7.1.2 stipulates that the main aisle in an office space must be at least 1100 mm wide, in order to ensure an adequate route to safety in the event of a fire [2]. The person designing the office space is not allowed to choose a width they think is safe; the width must be 1100 mm or wider.

Many codes, like the Fire Code, are put in place to ensure safety, while others are simply intended to ensure that designs are adequate for their intended purpose in other ways. In both cases however, the code embodies best practice, and provides ready-to-use specifications that do not have to be designed from first principles. It is not necessary to do complex calculations to work out how wide a corridor needs to be so that people can leave safely in the event of a fire. The Fire Code embodies the results of such calculations. As an added bonus, compliance with the code is evidence that the design engineer has taken appropriate care with the design.

Codes are often written as a consequence of previous failures. For example, early steam engine pressure vessels sometimes exploded violently, killing people, and causing extensive damage. If you want to build a steam pressure vessel today, perhaps a boiler for a new power plant, the vessel dimensions, materials, and wall thicknesses will be given by a pressure vessel code and the vessel will be inspected before being put into service. In the United States, the ASME Boiler and Pressure Vessel Code governs such designs [3]. Society is not willing to allow individual engineers to do their own boiler design and instead demands compliance with this code.

The advantage of a code is that it ensures that past experience, particularly with engineering failures, is taken into account in new designs, reducing risk. Unfortunately, this leads to problems when new methods of construction or new materials are introduced. For example, suppose a building code specifies that a particular grade of lumber must be used for deck boards. When a new plastic composite product is introduced, builders cannot use it until the code is modified or special exemptions are arranged. This is the downside of code-based engineering design: codes ensure safety, but can slow innovation.

In upper-year engineering courses, you may be introduced to specific codes governing the design projects you are working on. In this introductory text, we are interested in teaching design for safety from first principles. What you must understand now is simply that codes exist in all fields of engineering. A *critical* part of any engineering design exercise is to determine what codes pertain to the design you are developing.

> You probably realize that civil engineers deal most often with building codes and electrical engineers deal with electrical codes. Perhaps less obvious is the fact that database engineers and website designers must be familiar with the codes and regulations governing the collection, storage, and use of personal data.

4. Conclusion

A successful engineering design should first and foremost perform the function for which it was intended. It must do so while satisfying all the constraints set out, and optimizing the performance with respect to the objectives. Whether it is explicitly stated in the objectives or not, safety is paramount in engineering practice and in engineering design. An objective concerning safety—"The design should be as safe as possible"—always comes with an implied constraint. "The design must be safe" means that the risks associated with the use of the design are judged to be acceptable.

KEY TERMS

design for X (DFX)	**protocols**	**liable**
regulations	**standards**	**codes**
hazard	**consequence**	**accident**
severity	**risk**	**acceptable risk**

5. Questions and activities

1. Examine an electric kettle. Think about the various hazards: electrical fire, scalding water, and contamination of drinking water.

   a. Rank the probability and severity of each of these hazards, and hence rank the risk associated with each

   b. Try to determine what the designers have done to combat each of these hazards

   c. Flip the kettle upside down to see if there are any markings on the bottom to indicate compliance with codes

   d. Try to design a feature, to be added to an existing kettle, that would reduce the risk of scalding

2. Describe an example of an instance in which a risk is tolerable when taken voluntarily, but not when taken involuntarily.

3. A risk that is acceptable for one or two individuals may not be acceptable for a large collection of individuals. For example, no laws prevent you from driving your car for 24 hours straight, but bus drivers are not allowed to do this. Describe two other examples.

4. Serious accidents usually involve the failure of physical parts or systems, but also a failure of the systems and management that created the environment where the accident happened. Review the failure of the Deepwater Horizon oil rig to find out if there were systems or management failures, or if the failure could be strictly attributed to a mechanical failure.

5. Group discussion: If a growing small airport replaces its manual radio-based aircraft control system with a software-based system and there is a crash linked to aircraft scheduling, who might be liable? What might determine this? Consider:

    a. The controller (the person running the software)

    b. The airport owners (the purchaser of the software)

    c. The software designers/software company

    d. The software installers

    e. Others?

6. Pick a code or standard in your field of engineering or that pertains to the design project you are working on.

    a. Find the code online or in the library

    b. Pick a specific project or the design project you are working on for your course. What parts of the code will pertain to the project?

    c. How does the code help guide an engineer toward safe practice?

6. References

[1] Vincente, K. *The Human Factor*. New York, Routledge, 2004.

[2] Retrieved December 16, 2011, from www.e-laws.gov.on.ca/html/source/regs/english/2007/elaws_src_regs_r07213_e.html

[3] Retrieved December 16, 2011, from www.asme.org/kb/standards/publications/bpvc-resources/boiler-and-pressure-vessel-code—2010-edition

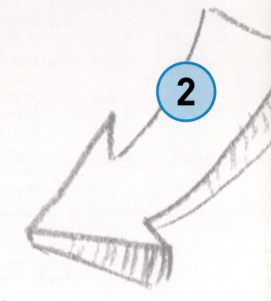

Identifying Hazards

#process module: #identifyinghazards

Learning outcomes

By the end of this module, you should demonstrate the ability to:

- Take an inventory of possible hazards in a design
- Identify the storage of energy and ways in which it might be released
- Describe the special hazards that pertain to small children, and a wide range of adults

Recommended reading

Before this module:

- **Design for X > Safety > 1. Design for Safety:** Introduction

After this module:

- **Design for X > Safety > 3. Safety in the Design Process**

1. Introduction

In order to produce safe designs, you must be able to search for and identify potential *hazards*: situations where there is possibility of damage to health, property, or the environment. Hazards can arise from many possible sources. We will discuss a few of the more common causes here, but you should also actively investigate your potential design ideas to uncover hazards that arise from other causes.

2. Identifying Energy Release Hazards

A common factor in many accidents is sudden energy release. You climb up a ladder, storing potential energy, and this is converted to kinetic energy when the ladder tips over. Eventually, the energy is transferred to your body, which has a limited capacity to absorb it. The failure of a pressure vessel full of compressed steam is another example. Catching your finger in a running lawnmower blade is a painful third example. Here is a partial list of ways in which energy can be stored, and consequently could be transferred in ways that could cause harm:

- Kinetic energy: a moving car, a spinning lawnmower blade
- Gravitational potential energy: a bucket of paint on a high shelf, your body weight, a piece of ice on a roof
- Chemical bond energy: a flammable, explosive, or corrosive material
- Strain energy: a stretched bungee cord, a winch cable under tension
- Compressed gas energy: a pressure cooker, a gas cylinder, a vacuum flask
- Electrical energy: a capacitor, battery
- Thermal energy: boiling water, a hot light bulb

A faulty main cable supporting a bridge is a hazard, but it is really the strain energy stored in the cable and the gravitational potential energy stored in the bridge that are going to do the damage. You should examine your design to identify where energy is being stored and how it might be released. In subsequent sections, we will discuss how to control and reduce the risk associated with these hazards.

3. Identifying Chemical and Biological Hazards

In addition to being a storehouse of potential energy, chemicals can also present a hazard for humans and the environment in the form of toxicity. As a result, whenever solids, liquids, and gases are used, their potential for human or environmental damage must be considered. Some important considerations are summarized below:

- Obviously, care must be taken when using toxic elements like beryllium, or *carcinogenic* compounds such as formaldehyde. A *toxic material* causes immediate harm, while a carcinogenic material increases the incidence of cancer later in life.
- Special care must be taken for materials that come into contact with food, or compounds used in food itself, and formal codes cover these situations. In the United States, the Food and Drug Administration (FDA) sets the regulations.
- Baby soothers, toys, materials for cribs, and playpens are more closely regulated than things intended for adults, because babies and toddlers put things in their mouths, and their lower body weight and immature internal organs make them more susceptible to toxins.
- *Biologically active materials* such as drugs are heavily regulated.
- Materials that can harbor biologically active materials such as viruses and bacteria need careful consideration in hospitals, food processing plants, and restaurants. Non-porous, sterilizable surfaces are often required in these situations.

In a workplace setting, *toxicity* information is supplied by the manufacturer of the material in the form of a *Materials Safety Data Sheet*, or *MSDS*, and for liquids, solids, and gases supplied in containers, a warning label is used (see Figure 1). Consumer products such as household cleaners, solvents, paints, glues, and so on, are also labeled with warning information if they pose a health hazard.

---

2. HAZARDS IDENTIFICATION

Emergency Overview

Product is a clear, colorless liquid with a sour odor. It is irritating if it comes in contact with eyes or skin. If vapor is inhaled it affects the central nervous system causing drowsiness and dizziness. Repeated exposure may harm kidneys, and liver. Repeated external exposure may cause skin dryness or cracking. If swallowed it causes damage to stomach, lungs.

Health hazards

Eyes	Irritating to eyes.
Skin	Irritating to skin. Repeated exposure may cause skin dryness or cracking
Inhalation	May be harmful if inhaled. Inhalation affects the central nervous system. Repeated exposure may harm kidneys and liver. May cause irritation of

---

FIGURE 1 Part of a Material Safety Data Sheet (MSDS) showing information related to safety considerations.

Common materials like steel, aluminum, and most common polymers (i.e., plastics) do not present significant toxicity hazards when in solid non-powdered form, and unless they are being used in food contact or medical situations they are probably fine. (They may also be fine in food contact situations, though more careful review is needed.) However, it is always a good idea to consult an MSDS to see if special precautions are needed in handling or use of the material.

The MSDS summarizes all the potentially hazardous aspects of a material including the toxicity, flammability, and explosion hazards. The MSDS also lists protective measures needed when handling the materials, and provides disposal information. A sample MSDS is presented in the appendix, and online MSDS databases are easy to locate through a simple Internet search. Workplaces where chemicals are in use are generally required to keep an MSDS file readily available for their workers.

Many countries, including U.S. and Canada are moving to a new Safety Data Sheets (SDS) system under the guidelines of the Globally Harmonized System of Classification and Labelling of Chemicals (GHS).

4. Identifying Hazards to Young Children

Young children present a special set of challenges for the design engineer. Children climb things, touch things, and put things in their mouths, creating additional and often unanticipated hazards. If you are designing a display unit for a retail store, and there are low shelves with convenient handholds, then you can be sure that some child will eventually try to climb to the top. If this causes the display to tip over, the design might be considered faulty. Also, small children cannot read warning labels and lack the experience to recognize a dangerous situation.

Children can choke on small objects that fit in the mouth, strangle themselves on scarves or dangling cords, such as those from drapes or blinds, and suffocate by putting their heads in plastic bags. Window blinds often have two cords that need to be raised or lowered simultaneously, and these cords used to be tied together. When the blinds were raised, the resulting loop of cord could be at the height of a toddler, and many tragic strangulation deaths resulted. Modern blinds usually have a fastener between the cords that supports very little force before separating and hence reduces the risk of strangulation. An engineer designing a technology that is meant to operate in a household environment must consider the possible presence of small children and the possible hazards that arise from their interaction with the technology.

Even if a design is intended for adults, the possibility that children might come into contact with it must be considered. You must ask the question: Will children interact with my design? What additional hazards should I consider? If your design is a plastic bag for packaging something, you might punch holes in it, and add a printed warning: "Not a toy. Suffocation hazard. Keep away from small children." These are examples of reducing risk and warning against risk, which are two strategies for handling risk.

5. Identifying Hazards to a Wider Range of People

A community is made up of people with a wide range of mental and physical abilities. Engineers have had a tendency to design for a fictitious "average" person who really does not exist. All of us have unique differences and at times will find some physical or mental tasks challenging. It is important to consider the broadest possible set of users when designing a technology. **Design for accessibility** is an important engineering discipline, which encompasses both the elements of design for basic usability and design for safety for those who have accessibility challenges. Related to design for accessibility is **universal design**: a consideration of the needs of a broad range of users in the design process. A broad range of individuals need to be considered in design for safety:

Courtesy of the Authors

1. How will a person in a wheelchair exit a building in case of a fire?
2. How will a visually impaired person navigate a subway platform?
3. How will a hearing impaired person know that a fire alarm has been activated?
4. How will an older adult with a poor memory manage their medications?

All of these situations represent potential hazards that need to be considered in the design process. There are volumes of material on dealing with these design challenges, but if you do not think to ask the appropriate question, you are obviously not going to incorporate appropriate accommodations. Many jurisdictions now have extensive code requirements related to accessibility and safety for a wide range of people. In the United States this falls under the Americans with Disabilities Act and other subsequent related legislation.

6. Identifying Hazards in the Virtual World

Even software engineers are not immune to the need to design for safety. When software controls things in the physical world, like airbags in an automobile, aircraft landing gear, or pressurized valves in a refinery, an ordinary analysis of safety includes an analysis of the software and interfaces. But hazards can also exist in the purely virtual world.

You are probably familiar with computer viruses, and you may have occasionally read newspaper stories about the accidental release of personal information such as banking or credit card records. While these might not pose an immediate threat to human health, they can cause economic damage, and hence may be treated like other hazards. As always, when the consequences of a hazard are severe, regulations are not far behind. For example, in Ontario, the safeguarding of personal data is regulated by FIPPA, the Freedom of Information and Protection of Privacy Act. This act stipulates what can and cannot be stored, how it may be used, and how it is to be protected.

An engineer producing software should consider what would happen if data was lost, corrupted, or copied in ways causing damage to people trusting the system. These could happen because of hardware malfunction, software errors, or malicious intervention (viruses and the like).

7. Conclusion

Identifying hazards is the first step in design for safety. There are many different types of hazards; we have discussed a few important ones in this section. However, a design engineer is obligated investigate all the potential hazards that might be associated with the technology they are developing. Once the hazards have been identified there are a number of strategies that can be deployed to reduce the risk associated with these hazards to an acceptable level.

KEY TERMS

hazard	carcinogenic	toxic
biologically active material	toxicity	Materials Safety Data
design for accessibility	universal design	Sheet (MSDS)

8. Questions and activities

1. Take an inventory of all hazards in your kitchen or room. How many can you identify? Start by looking for stored energy.

2. After completing Question 1, design the steps needed to prevent a curious toddler from being exposed to these hazards (i.e., "baby proof" your kitchen or room).

3. A cruise ship has stairs and elevators to move between floors. The elevator is not operable during a fire emergency. Consider people in wheelchairs having to move to an evacuation deck. What would you recommend?

4. Your grades are likely kept on computers: computers of individual instructors and a central computer with a "learning system" on it perhaps also linked with other computers. Each of these computers will be communicating to the world over the Internet. What precautions would you expect to be taken to preserve your grades and transcript against failures and malicious intrusions?

3

Safety in the Design Process

#process module: #safetyinthedesignprocess

Learning outcomes

By the end of this module, you should demonstrate the ability to:
- Describe the various ways in which engineers can design for safety
- Describe the four-part hierarchy of safety engineering
- Describe the various methods of designing to prevent a hazard from materializing
- Describe the various methods of guarding
- Be able to write a simple warning for a common household appliance such as a power tool

Recommended reading

Before this module:
- **Design for X > Safety > 2. Identifying Hazards**

After this module:
- **Design for X > Safety > 4. Workplace Safety**

1. The Hierarchy of Safety Engineering

Suppose you have identified **hazards** inherent in a design project that you are working on, and evaluated the **risk** these hazards pose by anticipating combinations of circumstances that could lead to an accident (#handlingrisk).

What should you do to address these risks? The general methods of dealing with risk are dealt with in specific ways when safety is an issue. Risk is a combination of the probability of an accident and the severity of the resulting consequences. The first strategies engineers use to design for safety is to eliminate or reduce the potential for damage caused by an accident, that is, reduce the severity of the consequences. Or to reduce the probability of an accident occurring. The primary strategies used, in order, are:

1. Reduce or eliminate hazards, thus reducing the potential for consequences. Related to this is the use of safer alternatives.
2. Reduce the probability of failure. This is often done through overdesign to prevent accidents.

3. Guard humans and property to protect them from hazards.

4. Warn and train the users operating the technology.

In order to make use of this hierarchy, it is necessary to consider safety from the very beginning of the design process. When safety is an afterthought, the resulting design will probably be more expensive and less effective than it should be.

2. Reducing or Eliminating Hazards through Design

The best solution for reducing safety risks is to use less hazardous systems or to eliminate the hazard altogether [1]. So design engineers search for safer alternatives or ones that eliminate a specific risk. This is best illustrated by some examples:

- A saw for removing plaster casts in a hospital can be designed to vibrate rather than spin; cutting brittle plaster but not skin. The skin does not need to be shielded, and the doctor needs little training to ensure patient safety.

- Pressure vessels are proof tested by applying a pressure greater than the rating on the vessel. Compressed air can contain a great deal of energy, so if a weld fails during a proof test with compressed air, there is a rapid energy release, leading to damage. Proof tests are therefore done by pressurizing with a relatively incompressible liquid, such as water or oil, so that little energy is stored during the test.

- Benign chemicals can be used instead of toxic ones. Toxic sulfur dioxide gas was used in refrigeration systems until safer chlorofluorocarbons (CFCs) were introduced. CFCs were eventually shown to damage the ozone layer, an environmental hazard, so they have largely been replaced with safer HCFCs and HFCs.

- Biometric companies want to provide access based on fingerprints, but could do this by storing a single number characterizing a few key features of the fingerprint. By not storing the actual fingerprint, the possibility that this highly private information is lost or stolen is eliminated, and elaborate safeguards are less necessary.

Courtesy of the Authors

In each of these examples, an early consideration of less hazardous alternatives eliminates the expense of guarding against the hazard later or dealing with the damage caused by a failure. There are often safe or at least less hazardous alternatives in a design, and good practice dictates that they be used if possible.

3. Reducing Probability of Failure

Sometimes energy must be stored or toxic chemicals must be used in a design. If the hazard cannot be eliminated, the next layer of protection is to ensure that accidents are unlikely to occur by designing against the possibility. If you know that the failure of a part is an unacceptable risk, you should ensure that a design or part of a design is highly unlikely to fail by **overengineering** and/or putting backup systems in place in the event that it does fail. There are some commonly used techniques for accommodating hazards.

Safety factor (SF). A safety factor (or *factor of safety*) is a multiplier that is used to ensure that materials are sized to more than adequately carry a load. If you are designing a crane cable for a construction site, and you have 3000 kg buckets of cement on a construction site, you will make sure that the cable can carry at least 30,000 kg to reduce the chance of failure. In this case, you have used a safety factor of 10 (i.e., 30,000/3000 = 10). The size of the safety factor depends upon circumstances, and may be specified by *code*. The more serious the consequence of failure, the higher the safety factor must be [1]. (Note: Always check the code!)

It should be noted that over-engineering in this way is expensive and can also affect performance, which puts a limit on the acceptable safety factor. Safety factors are particularly challenging when used on things that move, since they often add mass. In F1 racing cars, airplanes, and rockets, safety factors are highly constrained. Instead, very careful engineering and very tight tolerances on materials and parts are employed.

Overspecifying. Overspecifying is a more generic term for a safety factor applied to things other than material strength. If you require a 5000 W electric motor to lift the expected loads on a winch, you might choose a 10,000 W motor for the job to be sure that it never fails and does not wear out prematurely. Once again, the specific choice may be governed by code.

Redundancy. It is also possible to ensure that failure does not lead to disaster by putting in redundant systems or load paths. Commercial airliners are fitted with redundant control systems so that they can still fly if the primary system fails. Again, the more serious the consequences, the more important it is to build in redundancy. One flashlight is enough to go into a dark closet in your house to change a light bulb, but not enough to go cave exploring.

Redundancy is also routinely built into software. A control decision made at one point by a human or computer is checked by an independent method. This helps guarantee that, even under circumstances unforeseen by the first decision maker, a hazardous condition will not be produced. For instance, software code may be used to set the radiation level for a radiation therapy device. The software code that set the level is probably independently checked by another program to ensure that the level is appropriate before the treatment is started.

Specific service life. Sometimes detailed calculations can be done to compute the expected life of a part, called the *service life*. For example, aircraft engines are completely overhauled after a specified number of flight hours, regardless of whether or not there are any signs of trouble. This is done long before any parts are expected to wear out or fail. The concept of service life is related to the idea of a safety factor, although in this case the safety factor is being applied to a calculated lifetime.

Self-warning systems. Some systems are designed to provide warning before failure, making them easy to maintain. This is called a *self-warning system*. The brake pads in an automobile can be designed to make a loud squealing noise before they completely wear out. An example of a more sophisticated system is fiber optic cables and strain sensors being built into modern bridges. These sensors monitor the condition of the materials in the bridge and report on the condition of the bridge remotely.

Design for inspection and maintenance. Engineers can make systems safer by designing the technology so that it is easy to inspect and maintain. This is a *DFX* (*design for X*) consideration. If parts are easy to inspect and maintain, then it is more likely that this will be done, and parts that are about to fail can be identified and replaced prior to failure.

Failsafe design. If all other design approaches still cannot prevent failure, a component can be designed to fail in a safe way rather than an unsafe way. This is called a *failsafe* strategy. For example, it is possible to design pressure vessels so that they leak before they catastrophically fracture, releasing their pressure safely, and warning the users about the presence of a

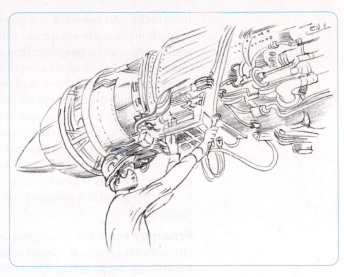

crack [2]. An automobile thermostat operates a valve that controls how much radiator coolant circulates through the engine. A failsafe thermostat is designed to fail in the open position, since an engine that is too cool is much safer than an engine that overheats.

4. Guarding

Sometimes exposure to a hazard is unavoidable. A saw-mill, which processes whole logs into lumber, contains saws that could cut a person in two just as easily as a log. There is no way to eliminate this hazard. The answer in this case is to keep people away from the blades using *guards*. The guards can be physical (put the saw inside a cage), or they could be electronic (shut the saw down when a sensor detects that a human is approaching). Even common household items are often guarded in these ways.

Courtesy of the Authors

Physical guards. Physical guards prevent access by separating humans from the hazard with a physical barrier. They must be made of a material that is strong enough to contain the chemical or stored energy should failure occur, and the choice of material will normally be specified by code. If wire mesh or fence is used, the gap between the hazard (e.g., a saw blade) and the fence is related to the size of the holes in the fence. If the holes are large enough to admit a finger but not an arm, the gap must be longer than the longest finger on earth! Again, it is not necessary to work this out on your own: the gap sizes are generally specified by code.

Ventilation. Exposure to hazardous dusts and vapors is controlled through ventilation. For example, toxic chemicals are handled inside fume hoods with forced ventilation carrying gases away through a ventilation stack. Paint spray booths perform a similar function.

Interlocks. An *interlock* is a power switch that is only activated when a guard is properly in place. The equipment cannot be turned on unless the interlock is engaged. Some things can be permanently guarded. For example the slot on a paper shredder is only big enough to admit a few sheets of paper but not a finger. Other things require intermittent access. A household food processor must be loaded with food prior to use, and emptied and disassembled after use. The guard on a food processor is not much good unless the user puts it in place, but experience shows that users make mistakes and frequently take shortcuts if they can. Hence physical guards are either locked in place permanently so that they cannot be removed by the operator without a special tool, or are interlocked, so that the machine will not operate without the guard in place. Food processors and coffee bean grinders have an interlock feature. Putting the guard in its proper place activates a switch that allows the machine to be turned on.

Proximity sensors. Sometimes it is impractical to provide a physical guard. An alternative is a proximity sensor or light curtain that will detect if humans are near the hazard, and trigger action if so. Automatic garage door openers and modern elevator doors are equipped with an "electric eye" that consists of an infrared light on one side of the door and a detector on the other side. If the beam is interrupted while the door is closing, it will automatically reverse. (Note that a garage door that closes on a child or a pet can be fatal. Modern automatic garage door openers therefore have redundancy built in: in addition to the "electric eye" they can sense the closing force and will reverse if the force is too high while the door is partially closed.)

Control placement. One method of ensuring an operator has hands clear of a hazard is to force them to press two buttons simultaneously that are spaced more than a hands length apart. An alternative is to place the control panel some distance from the hazard. However, for obvious reasons, emergency "Stop" buttons are always large and prominent and placed close to the hazard.

Redundant safety measures. Since guards are critical to engineering design, they themselves must be designed with safety principles in mind. You must ask the question: What if the guard fails? Machine guards are therefore carefully engineered with safety factors, redundancy, and self-checking features. For example, a modern computer-controlled milling machine on a production floor is used to cut blocks of metal into finished parts. There is a hazard from the moving stage, as well as from the cutting tool. This machine is designed to have an interlocking door, so that the machine will not start until the door is closed, and the interlock is detected to be in the "on" position. Furthermore, the interlock is designed to not open until the machine is completely de-energized, and contains no stored energy that could injure the operator. This includes the kinetic energy in the rotating head driving the cutter. Sensors ensure that the cutting head is no longer spinning before the door lock is released.

Many home washing machines have a similar type of interlock on the cover, not allowing operation unless closed, and being physically locked until the machine has stopped moving. Microwave ovens similarly will not operate with the door open for safety reasons.

Lockout. Perhaps you have been in a room when someone accidentally turned out the lights because they didn't know you were in there. Now imagine that you are repairing a large metal-stamping press when the operator comes back from lunch and starts up the machine, unaware that you are working on it. Industrial accidents of this type can be avoided with a *lockout* procedure, which is a very common procedure in industry. Maintenance staff have a personal lock, for which only they have the key. The lock is used to disable controls and other components of a system before the staff member works on it. If many people are working at the same time, each one uses a personal lock in parallel with the others so that all the locks must be retrieved before the machine can be operated (see Figure 1).

Personal protection. The last line of defence is personal protective gear, used when direct exposure to hazards is unavoidable. When working on the floor of a steel mill, for instance, every person must have steel-toed boots, a hard hat, and hearing and eye protection. When underway, the mill could generate flying debris and particles and excessive sound that cannot reasonably be contained with machine guards. Chemical workers must use respirators if it is not possible to provide effective control through external ventilation. Personal protection is also used as a redundant safety measure so that if the primary guard or safety system fails the operator will not be injured.

Courtesy of the Authors

FIGURE 1 A lockout system permitting multiple users to use individual locks in parallel. All locks must be removed before the master tag can be removed and the power to the machine turned on.

5. Warning and Training

Some hazards cannot be eliminated or guarded against. A car can run over a pedestrian, and while backup cameras and proximity sensors are used on some vehicles, it is unlikely that engineers will ever be able to design a perfectly safe car. Some guarding is built into the system by separating pedestrians and cars using sidewalks, crosswalks bridges, and so on, but the main safety mechanism is warning and training. Children are taught not to run in the road, not to cross unless the light is green, and not to walk behind a car with reverse lights on. Sometimes, this is the best you can do. If you are a carpenter or a machinist working with hazardous power tools, well designed tools will reduce the probability of an accident, but practice and training are essential for keeping you safe.

Human factors. Even after training users to do the right thing, accidents still happen. Some circumstances go beyond the training, and sometimes people get complacent or distracted. So our cars are equipped with seatbelts and airbags, even though all drivers have to pass a driving test. Design engineers can make a difference by designing so that it is easier for users to operate the technology correctly and harder for them to operate the technology in a way that is hazardous. This is an important general design principle, but is even more important when safety is involved. Emergency stop buttons on machines are red because in our society the color red is associated with "stop." It would be very poor design to use a green stop button on an industrial machine, because this would invariably lead to an accident.

As always, the answers seem obvious once you know them, but the critical skill is to know when to ask questions. What is a person likely to do in this set of circumstances?

→CAUTION←
KEEP FINGERS CLEAR!
windows are heavy
and may open rapidly.

What happens if they are tired, or intoxicated, or not wearing their safety glasses? What if they take a shortcut to make a highly repetitive task easier on a production line?

Warning labels. When hazards exist, it is important to put a warning label close to the hazard, at the point where the knowledge will have immediate impact. Warnings must be written in clear simple language, and must adequately warn of specific dangers inherent in the design. For industrial chemicals, the *Materials Safety Data Sheet (MSDS)* provides a complete description of hazards and control procedures, but is supplemented by a warning label on the container that summarizes the key hazards.

Quite often, hazards are indicated with warning diagrams, such as the International Standards Organization (ISO) standardized warning symbols. These symbols are intended to have meaning regardless of the language spoken by the user.

6. Other Techniques for Ensuring Safety

Benchmarking. Benchmarking means understanding the best existing and competitive solutions to the design problem you are working on (#benchmarking). These existing designs may have safety features built into them, and it is a good idea to understand and include such things in the new design if possible. In particular, if the industry is using an effective standard method to deal with a safety issue, new designs should incorporate this feature or the engineers risk being accused of incompetence or negligence. For example, if most ballpoint pen caps have ventilation holes to allow air to pass if a child swallows them, and you are designing a new pen top, you should include holes too, even if this is not mandated by code. (In fact, there has been a British standard in place since 1990!) The failure to include a safety feature well known in the market, and available at virtually no cost, would not only result in a poor design, but might even be considered negligent. Benchmarking is an important general design tool, but is particularly important when considering safety.

Prototyping/testing. Prototyping is a key method of testing design features and assumptions (#models&prototypes). Prototyping is particularly important for establishing the safety of a design, and for discovering hidden hazards. For instance, it may not be until the design team has a working prototype that they determine that accessing one part of the machine places the operator's ear unacceptably close to a source of loud noise, or that remotely accessing one part of a software program unnecessarily exposes the code to malicious attacks. Although modern software simulations reduce the burden of prototyping, prototyping can reveal issues that simulation will not and so is still considered an integral part of design education and the design process.

Test-based design. Test-based design is a technique common in the software industry, but is potentially useful in all design work. The design process starts with a set of tests that the design must pass before the design choices are made. In this way safety criteria can be specified very early and the design will then have the mechanisms to meet these criteria. Software often goes through several phases of testing (usually referred to as alpha testing and beta testing) before it is fully implemented.

7. Conclusion

In design for safety, we have identified several stages of engineering for safety: recognizing a hazard, recognizing the circumstances that will lead to an accident, controlling the risk through design, and warning and training when there is some risk remaining. Since the protection of the welfare of the public is of paramount importance, design engineers must:

- Think about safety as an integral part of any design that involves hazards
- Think about safety early in the design process
- Bring the risk to an acceptable level through standard engineering practices
- Follow codes, laws, guidelines, and benchmarked standard practice
- At a minimum, exercise the standard of care of a reasonable engineer of ordinary competence
- Carefully document the design and testing procedure

If you take the design for safety exercise seriously, you will produce superior designs, prevent wasteful product recalls or software updates, and avoid unnecessary harm to people, property, and the environment. This is a critical part of any design process.

KEY TERMS

design for safety	overengineering	safety factor (SF) or Factor of Safety
code	overspecifying	redundancy
service life	self-warning system	design for inspection and maintenance
guard and guarding	failsafe	test-based design
interlock	proximity sensor	lockout
personal protection	warning label	Materials Safety Data Sheet (MSDS)
benchmarking		

8. Questions and activities

1. The wood chipper is used to grind tree limbs into wood chips. It has a set of toothed in-feed rollers to drag the branches in, and sharp blades inside for chipping. In British Columbia in 2005, one member of a four-man crew was dragged through a large running wood chipper with tragic consequences [3]. The mechanical hazard and stored energy are clear in this case. How could you design the chipper to be safer:

   a. Can the hazard be eliminated or reduced?

   b. Can the device be guarded making it impossible for a worker to be dragged into the machine?

   c. What guards or safety mechanisms could be used in this case?

d. What warning labels would be appropriate for this machine?

e. Write a brief numbered list of warnings as part of the training manual.

There is a video about this accident available on YouTube. After you have completed the exercises, review the video here: www.youtube.com/watch?v=ji_-m_ylq10

2. Some cold medicines have warnings on their labels about operation of heavy machinery and driving. Examine the warning on a cold medicine. Does it seem appropriate? How might you make an industrial situation safer where an operator potentially was using one of these medicines to treat a cold? (Suggestion: Think of procedural solutions as well as physical ones.)

9. References

[1] Infrastructure Health & Safety Association. *Hoisting and Rigging Safety Manual,* August 2010. Retrieved March 30, 2014, from *www.csao.org/images/pfiles/4_M035.pdf*

[2] Ashby, M.F, *Materials Selection for Mechanical Design,* 2nd Ed., Butterworth-Heinemann, Oxford, 2003.

[3] WorkSafe BC Incident Report NI 2005134580241. Retrieved March 30, 2014, from www2.worksafebc. com/PDFs/investigations/IIR2005134580241.pdf

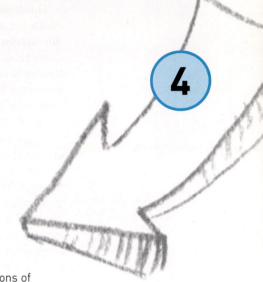

Workplace Safety

#process module: #workplacesafety

Learning outcomes

By the end of this module, you should demonstrate the ability to:

- Describe the features of a workplace that lead to special considerations of safety
- Describe some workplace related codes and standards

Recommended reading

Before this module:

- **Design for X > Safety > 3. Safety in the Design Process**

1. Introduction

The workplace is a special environment and is given special treatment in most jurisdictions for a number of reasons:

1. Workplace hazards are more complex and often have more severe consequences. No one operates an oil refinery or a nuclear reactor at home.

2. The risk is increased by the amount of time workers spend at their jobs, typically 1800 hours per year or more. Even events that are highly improbable may occur if many workers are engaged in an activity for many years.

3. The risk is taken involuntarily: workers need their jobs to support themselves and their families, and hence may do what they are told, even if the activity is not safe.

4. There is potential conflict of interest between safe work and quick work or cost savings. In the short term, work can sometimes be done more quickly or cheaply if safeguards are bypassed and cumbersome procedures are neglected. If you need to climb a ladder on a jobsite and your work partner is not there to stabilize the base, it is obviously quicker (in the short term) not to wait.

For these reasons, governments set up strict regulatory protocols to protect workers in the workplace. In the United States, the *Occupational Safety and Health Administration (OSHA)* is responsible for maintaining and enforcing workplace safety rules. If you are designing something to be used in the workplace, you must be familiar with the codes that govern industrial machinery, chemicals, materials, and processes. These regulations may be stricter than those governing designs outside the workplace. For example, you can purchase a band saw for home use that would not be permitted for repetitive use by unskilled workers on a production line.

In addition to specifying details about such things as machine guarding and chemical storage, workplace safety regulation places great emphasis on the systems and organizational structures that are in place to recognize and deal with potential hazards. The goal of such regulation is to ensure that management and workers treat safety procedures as a cooperative venture. This text deals primarily with the physical design of products, structures, and systems, but the same types of thinking can be used to design effective organizational systems to ensure safety within companies. Think about circumstances in the work environment that might lead to an accident, and the rules or organizational structure you would put in place to prevent it from happening:

- A production manager is behind schedule, and asks his workers to continue working on a machine, even though the guard is visibly cracked.
- A worker is too hot in a full-face respirator and decides to shovel dangerous silica dust without it.
- A press is jammed and the line worker decides to free it up herself, without waiting for a maintenance worker with a proper lockout system.
- A database operator backs up some highly private personal medical information on a USB key so that he can work on it at home.
- Device control software is released before full testing because the programmers are under pressure to meet a scheduled product release.

In each of these cases, the bypassing of safety procedures leads to a sharp increase in risk. The system of reporting and the culture of the organization must be designed to prevent these problems from leading to an accident. For example, it should be clear that workers cannot be fired for refusing unsafe work, and conversely that they will be penalized for doing unsafe work by bypassing standard safety procedures. The implementation of such rules is an example of design for safety at the level of the organization or government.

KEY TERMS

Occupational Safety and Health Administration (OSHA)

2. Questions and activities

1. Imagine you own a manufacturing plant with 1000 workers. Describe the organizational structure you would put in place, in the absence of any government regulations (which of course do exist), to prevent an accident caused by the pressure to meet production schedules.

2. Under the circumstances of question 1, how would you handle a worker who ignored a safety rule once? Repeatedly? What if the worker had specialized knowledge, difficult to find elsewhere, or was involved in work that needed to be done?

3. A press has two "Go" buttons that must be pushed simultaneously to make the machine activate. This means both hands must be used and will be out of the way of the press.

a. How could this system be defeated by an operator?

b. If you are a manager, how might you deal with these methods of defeating the system?

c. If you were designing the press, what methods might you employ to make the activation of the press safer?

5

Resources

Introduction to Estimation

#resource module: #estimation

Learning outcomes

By the end of this module, you should demonstrate the ability to:

- Recognize the situations where an estimate is preferable to a detailed calculation
- Form the underlying question in a given situation
- Be able to form an answer with appropriate significance to the underlying question

Recommended reading

After this module:

- **Resources > Estimation > 2. Estimation Techniques**

1. Introduction

Engineers often have to make quick, approximate calculations when they don't have the time or the information needed to make precise calculations. Except for simple counting operations and arithmetic, physical measurements and the calculations based on these measurements almost always involve approximations. The trick is to make the measurements carefully enough, and the approximations good enough, such that the resulting answer has the **accuracy** needed. If you are looking for new subatomic particles using a particle accelerator (atom-smasher), incredibly accurate measurements are needed. However, in the preliminary stages of a design process engineers can often make decisions using less accurate values.

In this set of modules, we will discuss the process of making **estimates**, with a special focus on quick but useful crude estimates called **back of the envelope** calculations. An ability to make such estimates is essential for practicing design engineers.

2. Why Do Engineers Need to Estimate? Why Not Just Compute the Answer?

In a typical engineering problem set or exam question in school, you will be provided with clear data, and you will use some equations (preferably the correct ones in the correct order) to compute an answer. In engineering practice however, you often won't have sufficient information and/or time to work out an exact answer. In such cases, it may be necessary to come up with a reasonable but approximate answer: an estimation. There are a variety of terms used to describe estimated answers:

- ***Ballpark*** answer (or to ballpark an estimate) is slang for an answer that is very approximately correct. The "ballpark" refers to a baseball playing field.

- ***Order of magnitude*** estimate is a value that is within a factor of ten of the true answer.

- ***Back of the envelope (BOTE)*** estimates provide very quick approximate answers, based on information that is already known or can be very easily found. BOTE estimation got its name from the quick calculations scribbled on any available piece of scrap paper such as the back of an envelope.

Even if you have all the required data, you may want to make a quick estimate instead of taking the time needed for a full calculation. For instance, in a meeting you might be asked how much processing time it will take to generate animated graphics

---

Example

Imagine that you are asked to determine if a chain rated for 3000 kg is sufficient to lift a 2000 L vat of cooking oil, but you are not given the density of the oil that would allow you to calculate its exact weight. If this were an exam question, you would probably raise your hand to ask the professor for the missing data, but what if you are on a job site, with a 3000 kg capacity chain, a 2000 L vat full of oil, and a winch? 10 people are standing around waiting for you to tell them whether to go ahead or not. In this case, you might remember that cooking oil floats on water, and that 1 L of water weighs about 1 kg (approximately, depending on the temperature). So because oil is less dense than water you can reliably estimate that the contents of the vat must weigh something less than 2000 kg. However, you also have to factor in the weight of the vat itself. You might guess that the vat weighs at least a hundred kilograms, and this means that the total is more than half the rated value and, therefore, too close to the chain rating to risk the lift. There is not a sufficient safety factor. You would decide to wait for a stronger chain. You have made a useful estimation even though you were missing some data.

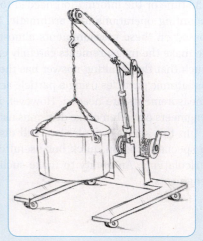

for a user interface. Given enough time, you could find the exact value, but the expectation is not that you compute the actual answer, but that you give a rough value on the spot. If you have knowledge of the processing speed of the computer in use, a rough idea of the complexity of the task, and knowledge of the processing times of other tasks, you can probably give an approximate answer relatively quickly.

A third motivation for estimating is that it provides an answer that is "in the ballpark" and can be used to test the credibility of a more detailed solution. In complex engineering computations, such as the determination of stress in an airplane wing, engineers often resort to numerical simulations where a sophisticated computer program determines the answer. An independent estimation allows them to catch the slip of a decimal place or the omission of a negative sign in the precise calculation and to pause and double check when the detailed answer doesn't seem right.

The ability to make quick, effective estimations is important to the engineer's overall efficiency and usefulness.

3. The Basic Estimation Process

The basic estimation process, much of which might be done informally and quickly:

1. Determine the underlying question, and work out how much accuracy the answer needs. This will guide your choice of input data and calculation approaches.

2. List the data that is either available or easily found.

3. Brainstorm one or more approaches to the calculation and choose the most promising one. This may not be the easiest calculation, but rather the one that is most feasible, given the data and knowledge that is available.

4. Find sufficiently accurate values for the input data required.

5. For the approach chosen, determine the key physical relationships (equations) that must be used in the calculation.

6. Perform the calculation.

7. Consider the potential error bounds for the calculation given the approximations made. Decide if the answer is "good enough" given the underlying question. Estimates should include a disclaimer and/or an estimate of the possible error (e.g., within ±10%).

4. The Underlying Question

When estimations are required, it is important to identify what question is truly being asked. In the last section we asked "How much processing time will it take to generate animated graphics for a user interface?" The real question to be answered is something like "Would adding animated graphics to the user interface take such a large amount of processing time that the application would appear sluggish?" This, in turn, might be more simply part of the answer to the question "Is adding animated graphics a good idea?" Often the real question is "Should the design team go further with this idea, or not?" The temptation for inexperienced engineers is to compute the most accurate answer possible, but using an estimate to make a decision can save time while delivering the same decision.

Introduction to Estimation

A knowledgeable and skilled engineer can produce realistic estimates for many types of calculations, particularly when the underlying or fundamental question is recognized and used as guidance for the degree of accuracy required. Some examples of underlying questions:

- What is the projected development cost for a new product? (Will the cost be grossly prohibitive, or should the team look more closely at the possibility of developing the product? The team needs a quick **go/no go decision**.)

- How many trips to the client should the team expect before the order is signed? (Is it 5 or 50? They need a rough idea to roughly determine time and costs.)

- About how much will it cost to modify the process to include the new chemical scrubbers? (The team needs to decide between alternatives; exact numbers may be impossible to calculate, but may not be required as long as there is a clear "best" strategy.)

- How long can the user expect this technology to last before repair is required? How long does it need to last? (The design team is attempting to determine longevity and thus future costs for repair and replacement.)

- What size of ladder is required? (There are probably only a few ladders available, and they only come in certain sizes. The designer is asking "what ladder is long enough and not too long?")

The 2010 BP Gulf Oil Spill

In the spring of 2010, a British Petroleum (BP) oil rig was destroyed by an explosion, killing 11 people, and oil from the well began to spill into the Gulf of Mexico. Initial estimates were 5000 barrels of oil a day coming from the well. This estimate was revised many times, and estimates of up to 70,000 barrels per day were reported in weeks following the disaster [1].

The estimate was very important to BP and to the people near the spill, since the response to the spill had to be proportional to the size of the disaster. First, the flow rate determined the total amount of oil that went into the gulf before the well was capped. The oil had to be removed through dispersants, skimming, and other forms of collection, with materials and teams determined by the size of the spill. Second, the flow had to be stopped, and the design of a method to do this would have been influenced by the flow rate.

The estimates themselves were made from the known flows before the accident, plus video and audio information from submersible robots at the scene. No direct measures could be made. Initial estimates were refined as more information was made available, and as estimation techniques were refined.

U.S. Geological Survey Department of the Interior/USGSU.S. Geological Survey/ photo by U.S. Coast Guard

(continued)

The estimates, even the ones that proved significantly wrong, were important. Important, because they formed the basis of the cleanup response being mounted. Also, important, because they formed the basis for the design of the devices and techniques created to try to capture or stop the flow.

5. Conclusion

A good engineer will be able to make appropriate estimations during various phases of design and other engineering work. This involves understanding the accuracy required to make the necessary decisions in each phase of the design process. It is important to start with a clear understanding of the underlying question and the data available (along with its accuracy). The engineer must then identify the other data that needs to be collected, or estimated, to answer the question and work out a solution plan. The results of an estimate should always be stated with the caveat that this is only an estimate, and ideally should be stated with an approximate error level so the reader can gauge the accuracy of the result.

KEY TERMS

estimate	accuracy	back of the envelope (BOTE)
ballpark	order of magnitude	go/no go decision

6. Questions and activities

1. Pick a landmark in your area, such as an airport or train station. Estimate how far in advance you need to leave your current location to catch a train or plane.

   a. What is the underlying question that is being asked in this instance? What are you actually being asked to estimate?

   b. What value did you get? What information did you use in this calculation? Explain how you arrived at your estimate.

   c. Is there information posted (or available by GPS) that provides another estimate that you can compare to? How does your estimate compare?

2. Estimate how much exercise you get in a typical day.

   a. What is the underlying question that is being asked in this instance? What are you actually being asked to estimate?

   b. What value did you get? What information did you use in this calculation? Explain how you arrived at your estimate.

   c. What measurements could you make that would help you arrive at a more accurate estimate?

7. References

[1] Suzanne, G. *Marine scientists study ocean-floor film of Deepwater oil leak*, The Guardian, May 13, 2010. Retrieved from www.theguardian.com/business/2010/may/13/bp-oil-spill-ocean-footage

2

Estimation Techniques

#resource module: #estimationtechniques

Learning outcomes

By the end of this module, you should demonstrate the ability to:

- Use methods to get data quickly for use in the estimations
- Use inexact calculation methods to speed the estimations

Recommended reading

Before this module:

- **Resources > Estimation > 1. Introduction to Estimation**

After this module:

- **Resources > Estimation > 3. Estimating Cost and Labor**

1. Basic Method

The basic method for estimating involves

1. Identifying the underlying question, and a method for answering the question that relies on approximate information
2. Finding or estimating the information needed
3. Calculating a result
4. Reflecting on the answer

The basic method of estimating will be illustrated in this classic **BOTE** (**Back-of-the-envelope**) example calculation, which is sometimes attributed to the physicist Enrico Fermi.

> **Example: Back-of-the-Envelope Calculation**
>
> An engineer invented a special low-cost tool for tuning pianos that is so good, she is sure every piano tuner in the United States would buy one. She would like to know if it makes good business sense to spend $5000 to $10,000 to obtain a U.S.
>
> *(continued)*

patent to protect the invention. The first thing she needs to do is estimate the size of the market. How many people in the United States are in the business of tuning pianos? Very detailed market research would likely provide an accurate answer, but a BOTE calculation could help decide if she should bother to spend the time needed to do the detailed calculation.

The calculation consists of a sequential set of educated guesses leading to the answer.

Courtesy of the Authors

- Approximately how many people are in USA? ~300 million

- Is there likely to be more than one piano per household? No

- How many households are there in the United States? ~100 million

- Does every household have a piano? No. Perhaps 1 in 25

- How many pianos are there in the United States? 100,000,000/25 = 4,000,000

- How many piano tuners are needed for 4 million pianos? Don't know

- How many piano tunings can 1 piano tuner do? 2 per day maybe

- How many piano tunings per piano? 1 every 2 years maybe

- How many piano tuners are needed in the United States?

- X = 4,000,000 pianos/(225 working days per year * 2 years/tuning)/2 pianos per working day per tuner = 4,444 tuners.

(The Piano Tuners Guild lists ~4,000 members [1], so the estimate has yielded a fairly good answer.)

So she might sell a few thousand low-cost piano-tuning tools and hence might decide that it is not a good idea to pursue a patent. In fact, she might be confident enough in the very rough answer to decide against spending time on a more detailed calculation.

Let's discuss the elements of this calculation:

- We used approximate numbers based on knowledge that we already possess, can easily acquire, or can guess. We know the population of the United States is about 300 million, and we guess that the mix of single people and large and small families leads to about 100 million households. Or we could use data from the U.S. census. We think about our friends to decide how many people own pianos, while recognizing that our friends might not properly represent the general population of the United States. And so on . . .

- Here the computation itself is a very simple equation. Notice that we used all of the units for each of the quantities to ensure that the units cancelled properly when the calculation was done. This is a simple form of *dimensional analysis*.

- Since the equation is simple, any errors are likely to come from errors in the assumed data. For example, a tuner could perhaps do three or four tunings a day in practice, but only if he or she had a full calendar of appointments. Obviously, we would have to do a survey of working tuners to do any better. Some people have their piano tuned annually, some never. And so on . . .

- Even if we know that the actual population is actually slightly higher than our assumed value of 300 million, but given the rest of the approximations, it is not really necessary to find or use a better number. As a ***rule of thumb***, you don't need data to be much more accurate than the least accurate piece of data you are going to use.

This is a very straightforward approximation; however, with practice it is surprisingly easy to make reasonable approximations in many circumstances. Some common techniques necessary for making these approximations are given in this module.

2. Finding the Information for an Estimate

Many pieces of information will go into a detailed calculation. However for an estimate, ignore the information that is of lesser significance and take an approximation of the information you do not have, perhaps using some of the techniques of this module.

2.1. Measuring Length, Area, and Volume

Physical dimensions are a common part of engineering estimations. Most people can make a reasonable estimate of length and area. You can make reasonable estimates of volume by breaking a complex volume into an assembly of regular shapes such as rectangles, spheres, and cones. (Of course, you need to know how to calculate the area of these!)

Courtesy of the Authors

You can make a more accurate estimate of area by imposing a grid on top of an area you wish to measure. Sometimes the area will already be covered in a grid—floor tiles, for instance. You might also superimpose a grid on a photograph as illustrated in the figure given below. You can count the squares to determine a complex area.

Courtesy of the Authors

2.2. Working from Known Information

Often engineers have some information about what they are trying to estimate, but the specific values are not available. For instance, Table 1 (from [2]), shows the coefficient of thermal expansion for steel in units of strain per degree Celsius. The table is only for a specific type of steel, type 1025. So at 200°C you would expect a meter long piece of steel to expand 6.3×10^{-6} m for every degree the temperature is raised.

But this rate changes with temperature, and say you were interested in the rate at 250°C. In this case you might estimate: $(12.8+13.3)/2 = 13.1$. This method of determining an intermediate value between known values is called *interpolation*. In the case where you extend the range of a set of values, say to estimate an expansion value for 500°C, the method is called *extrapolation*. Extrapolation is used to estimate a value that falls beyond the range of values given. Interpolation and extrapolation are exact for linear relationships, but are not exact for non-linear relationships.

TABLE 1 Rate of 1025 Steel Expansion with Temperature [2].

TEMPERATURE (°C)	EXPANSION (D/D/C° $\times$ 10^{-6})
200	12.8
300	13.3
400	13.9

Non-linear Scaling

It is often necessary to extrapolate from known data when estimating. If you need five bricklayers to do the bricklaying to build 1 home in 1 week, then 500 will be needed to complete 100 homes in the same time. You must be careful though, as many systems do not scale in this simple linear way. This is just another way of saying that there are many non-linear relationships (equations) and non-obvious relationships in engineering. A beam that is twice as wide as another is twice as stiff, but when a beam is twice as deep, it is eight times as stiff. Simple estimating needs to include these sorts of scaling relationships. This type of information is learned in your technical courses in engineering.

If you know (or can infer) the non-linear relationship between things then you can use this to interpolate or extrapolate a value from known data. This is simply a slightly more complicated calculation than simple linear interpolation or extrapolation.

Complex systems, involving interactions between the components, also scale in a non-linear way. Sometimes engineers can estimate using a ***complexity count***: a count of the difficult and interrelated parts of the system they are considering. Consider air traffic control. If there is only one aircraft near an airport, there is no problem with guiding it in for a landing. If there are two aircraft, then one also has to consider the interactions with the second aircraft when determining a landing route for the first aircraft. With three aircraft there are three pairs of interaction to consider; with four aircraft there are six pairs of interactions. The complexity of the problem grows geometrically rather than linearly (in this example related to n!, the factorial), so the cost of the people and equipment needed to control the system also scales in a non-linear fashion.

2.3. Reasoning Using Analogy

It is sometimes possible to use a comparison or analogy to deduce the value of missing data. People are much better at comparing things than judging in absolute terms. So it is easier for us to compare the weight of two items than to guess the absolute weight of an item in isolation.

If you need to pump a fluid through a pipe, and you would like to estimate the drop in pressure from the inlet to the outlet, you need to know the fluid viscosity (a property of the liquid). Your field engineer reports that the liquid is thick and gooey, but did not make the appropriate measurement of viscosity. However, the field engineer can probably remember whether it was as thick as (a) water, (b) honey, or (c) peanut butter. So you could ask her to give you a better description using one of these materials as a comparator. You might know or can easily look up viscosity data for these common foods or other materials.

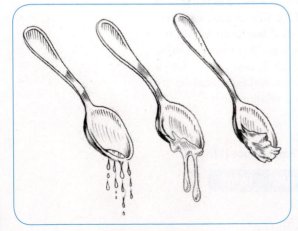

The following table provides some examples of simple things that you could use to make engineering estimates.

ESTIMATES FOR COMPARISON (APPROXIMATE VALUES)	
Air pressure	Soccer ball: 75 kPa
	Car tires: 250 kPa
Material modulus	Plastics: 1–3 GPa
	Aluminum: 70 GPa
	Steel: 200 GPa
Power	Incandescent light bulb: 100 W
	Athletic person: 400 W
	Car engine: 200 horsepower or 150 kW
Density	Wood: 0.6 Mg/m^3
	Water: 1 Mg/m^3
	Aluminum: 2.7 Mg/m^3
	Steel: 7.9 Mg/m^3

Estimate the weight of your cell phone by guessing its volume, and guessing its average density by comparing it to an equivalent volume of wood, water, aluminum, or steel. Do all the calculations in your head and then check the *specifications*. How well did you do?

Note that the term *specifications* is used in several different ways in engineering. It can be used as a synonym for requirements. However, it is more frequently used to refer to the detailed technical description of a system or product. You will typically find the specifications listed on the box, and on the instructions, for a product.

For example, if you were purchasing a laptop, the specifications you might look at would be display size, display technology, main memory size, size of hard drive, processor type, processor clock speed, battery life, outside dimensions, and so on.

The term "specifications" is used so widely in engineering that engineers have a short form they use frequently: *specs*. For example, a mil-spec is a military specification, and it tells a vendor exactly the technical properties a product or system must have to be purchased for military use (the term mil-standard is also used in this way).

2.4. Prototyping and Simple Experimentation

When data needs to be determined, it might be possible to do a simple experiment or make a rough prototype to obtain needed data (#models&prototypes). If you are calculating friction losses in a rotating shaft with a plain bearing, and you do not know the coefficient of friction for steel sliding on plastic, you can set up a very simple experiment placing a block of steel on a sheet of plastic and tipping the assembly until sliding motion begins. If you need to know the airflow from a fan, you can use a stopwatch and a garbage bag to collect a known volume of air over a known period of time. In fact, it is often possible to do simple experiments like these to fill in missing data.

Prototypes serve to validate design assumptions, including any estimations or calculations that were part of the design process. The basic premise is that they should be built with just enough care that the results of tests can be relied upon to represent to some degree the performance of the final design. So a prototype of a smartphone could be rendered to different accuracies, depending on the testing to be done.

TABLE 2 Examples of prototypes that could be used to gather data for estimates related to the design of a cell phone.

QUESTION TO BE ANSWERED BY SMARTPHONE PROTOTYPE TESTING	PROTOTYPE CONSTRUCTION
Will it fit in a pocket?	Block of wood
Is it comfortable to hold?	Properly weighted block of plastic with corners and edges matching the final design
Is the keyboard on the touch screen useable?	Properly sized prototype with working touch screen hooked up by cable to external computer

2.5. Using Non-Standard Information Gathering: Fermi Methods

As you gain more experience with engineering computations, you acquire more and more of the information needed to make estimations. You will remember data such as material strengths and densities, coefficients of friction, computer chip speeds and power consumption, and so on. You will also remember physical relationships and the equations that represent them. The famous physicist Enrico Fermi was so well known for his ability to make estimations that simple questions amenable to BOTE estimation are sometimes called *Fermi Questions*.

One of Fermi's strengths was to use non-standard data measures for his estimates. Fermi once estimated the explosive power of an early atomic bomb test by observing the path of some paper that he dropped while observing the test from a safe distance. This information was top secret at the time and he was not supposed to know it, but his estimate was amazingly accurate.

2.6. Approximating Data

Where you don't have the information, it is now relatively easy in many cases to find approximate information needed using an Internet search. For example, high-resolution satellite imagery can be used to estimate building sizes and land areas. Mapping programs produce accurate distances between places. Online encyclopedias, and other knowledge bases and engineering sites provide quick access to mountains of data, the equations governing physical relationships, and specifications for existing systems, services, and products. As more powerful computational knowledge search engines are developed and the amount of material increases, it will become even easier to find the information you need through on-line sources. As an engineer you will be expected to know how to use and make use of these tools.

Remember that for an approximation you do not need exact information. The general principles discussed in this section, such as interpolation or extrapolation, can be applied to the process of estimating to help you get around the problem of missing, inaccurate, or inadequate information.

3. Performing the Calculation for an Estimate

There are two elements to estimation: the input information and the calculation. In the previous section a variety of means of obtaining or estimating the approximate input information were discussed. Now we consider the calculation itself. It stands to reason that if the information is only approximate, it is reasonable to do rough calculations, particularly if they are much easier to do. Although modern scientific calculators make complex arithmetic, trigonometry, and many other calculations trivial, it is still useful to develop an ability to do rough calculations on the fly, without the need for computational help!

Approximate arithmetic computations can often be done with an acceptable degree of accuracy. If you must multiply 13×17 in a hurry, we can estimate the result as $15^2 = 225$, and this is a reasonable approximation to the true answer of 221. Quite often, you can simply round the numbers so that the simple calculation can be managed in your head. 9850/125 is about 10,000/100 less about 20%, so we might guess ~80. The true answer is 78.8, so our approximate is certainly good enough.

Much more complex computations can also be done without a computer, with a bit of practice. Integrals may be approximated as areas under the curve, and curves can be approximated using straight lines. The integral $\int_0^3 x^3 dx$ must be less than the area under the straight line linking (0,0) to (3,27). If you visualize or plot this function as a series of points (0,0), (1,1), (2,8), (3,27) you recognize that the area under the line between 0 and 1.5 is rather small, and take the area under a straight line from (1.5,0) to (3, 27) = ½ * 1.5 * 27 ~ ½ * 1.5 * 30 ~ 45/2 = 22. The actual answer: $\int_0^3 x^3 dx = \left.\frac{x^4}{4}\right|_0^3 = 20.25$. Again, good enough!

In a similar vein, derivatives are approximated by finite differences, so that the acceleration, $a = dv/dt$, can be approximated by $a = \Delta v/\Delta t$. Many other simple computational approximations are possible, and in fact there people that can do seemingly impossible calculations in their heads. Arthur Benjamin gave a TED talk in which he did a variety of calculations, and ended the talk by squaring a five-digit number out loud by breaking the problem down into manageable pieces (available at http://www.ted.com/talks/arthur_benjamin_does_mathemagic.html).

Punching numbers into a calculator is easy, but it is also easy to make a mistake. Just as the entire BOTE can serve as quick double-check on the feasibility of a design decision, a quick approximate calculation can serve as a check of a more detailed computation. A willingness to practice approximate calculations can pay dividends on a job site or in a business meeting when a calculator is not readily available.

4. Conclusion

Estimations involve finding values quickly, so they often deal with missing or inexact data and do calculations that are inexact compared to the detailed calculations that might be done. A good engineer will be able to recognize the opportunities to exploit these methods to produce quick estimates. A further extended example of this type of problem solving is given below.

KEY TERMS

back-of-the-envelope (BOTE)

interpolation

specification (spec)

dimensional analysis

extrapolation

Fermi question or Fermi problem

rule of thumb

complexity count

5. Questions and activities

1. Estimate the number of bicycles in the town or city you live in. If you decided to open a bicycle shop this would be important information.

2. Estimate the volume of the building you live in.

3. Estimate the square area of some part of the campus, such as a quad, without using a measuring tape. Try doing this on the ground, standing in the area. And then try doing it using Google Earth or another mapping program.

4. Practice interpolating and extrapolating information using the following table:

Viscosity of water as a function of temperature.

T (°C)	VISCOSITY (μPA·S)
0	1793
20	1002
40	653.2
60	466.5
80	354.4

Data source: J.V. Sengers and J.T.R. Watson, "Improved international formulations for the viscosity and thermal conductivity of water substances," *J. Phys. Chem. Ref. Data* 15, 1291, 1986.

a. Estimate the viscosity of water at 25°C and at 62°C.

b. Estimate the viscosity at 87°C.

c. Estimate the viscosity at 100°C, and compare your estimate to the published value: 281.8 µPa·s.

6. References

[1] *Piano Technicians Guild.* Retrieved from www.ptg.org

[2] efunda (Engineering Fundamentals). *Carbon Steel Carbon Steel AISI 1025* Retrieved from www.efunda.com/Materials/alloys/carbon_steels/show_carbon.cfm?ID=AISI_1025&show_prop=cte&Page_Title=Carbon%20Steel%20AISI%201025

7. Examples

Example 1. Say we are to estimate the power use of a home theater circuit. We ignore the parts of the circuit that consume almost no power, such as the initial amplification circuits that work will small signals, concentrating on the output power amplifiers that drive the speakers. By looking at the power consumption requirements of the speakers, we can roughly determine the peak output power of these amplifiers and from this approximate the average power at full output of the circuit. While this may not be very exact, it may be close enough to determine (for a standard North American home with 120 V AC supply voltage) whether we need a 20 amp or higher supply circuit for this amplifier, or whether a standard 15 amp circuit would be enough.

Example 2. Microcontrollers are mini-computers that are much cheaper than the processors in typical personal computers. Can you use a microcontroller to do real-time refresh of a computer video screen, where every pixel of the screen must be updated from memory? Typical refresh rates are 50 to 60 refreshes a second. (Pixels are the "dots" on a screen that compose the picture.)

Input: (from screen properties) Screen resolution 1280 × 800 pixels, each pixel is defined by a word of memory. Say 20M machine instructions per second, from your knowledge of the processor.

Quick assumptions:

1280 × 800 -> say about 1,200,000 or 1.2 M pixel locations and corresponding memory places to manage.

You might assume one access to read or to write all the data for one pixel.

As a reasonable guess, figure on the order of 10 instructions per pixel to do the processing required. (This could include transfers of pixel data and addresses, the updates to allow us to do the next pixel values, and a check to see if you are done specifying every pixel in the screen for a refresh cycle.)

Figuring:

Time to do one screen refresh using the microcontroller

$$\frac{(1.2 \text{ M pixels}) \times (10 \text{ instructions per pixel})}{20 \text{ M instructions per second}}$$

Or 0.6 seconds per screen refresh, meaning under 2 screens per second.

Estimation Techniques

This is far under the required refresh rate, therefore you need a special processor or special hardware to update the video. The method you chose will not allow you to use a microcontroller for the refresh.

Consider the estimate of 2 screens per second of video using a microprocessor in this example. Despite the fact that you know it is not exact, you can make the decision about standard microcontrollers. You may be able to find microcontrollers that work much faster than the one used in the estimate, but they would have to work more than 30 times faster to reach the levels you want. You may think that 10 instructions per pixel is a few too few, or a few too many, but this number would have to be significantly different to affect your decision. A more likely way to make a microcontroller useful would be to change your expectations. You could use a microcontroller if lower refresh rates or a lower screen resolution were acceptable. Or you might want to increase the processing being done by the video unit, perhaps having the video unit making high speed memory accesses to refresh the video, while the microcontroller only did updates of the information in the memory to make changes. Regardless, you can easily, and without exacting calculations of values, selection of a specific microcontroller, or any programming, reject the original idea as unworkable without some modification.

> In modern personal computers there are specialized graphics computers, called GPUs, outside the main processor to put up the video screens. These graphics computers have many, many multiple cores, working in parallel to reach the required refresh rates. These graphics computers are those found on "video cards."

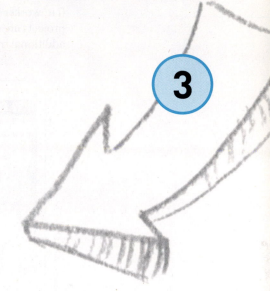

Estimating Cost and Labor

3

#resource module: #estimatingcost&labor

Learning outcomes

By the end of this module, you should demonstrate the ability to:
- Identify two methods for estimating the cost of a project
- Apply these methods to estimate the cost and other resources needed to complete a simple project

Recommended reading

Before this module:
- **Resources > Estimation > 2. Estimation Techniques**

After this module:
- **Resources > Estimation > 4. Estimation Confidence**

1. Estimating Cost and Labor

We have used the word "estimation" in a generic sense, to represent an approximation to the exact answer. Estimation has a more specific meaning with respect to planning a large project. In this case, an estimate refers to a prediction of the time, cost, and resources needed to complete the project. In the installation of a chemical plant, or the construction of a new bridge, engineering firms are invited to **bid** on the job in response to a **request for proposals (RFP)**. The bid includes an estimate of what the job will cost. The bid that promises acceptable quality at the least cost will typically be awarded the contract. The firm responsible will sign a contract to complete the work in exchange for payment. Depending on how the contract is structured, the bidding company will generally have to deliver what they promised at the price they promised and in the time they promised. If the project is very big, they better be careful about what they promise! At the same time, the competitive nature of the bidding process means that the bidding companies have an incentive to keep their bids as low as possible. An accurate estimate of how much time, money, and **manpower** (i.e., worker-hours) a project will take is critical. For this reason, estimations for large projects are usually done by experienced engineers, who may even specialize and have additional training in doing these estimates.

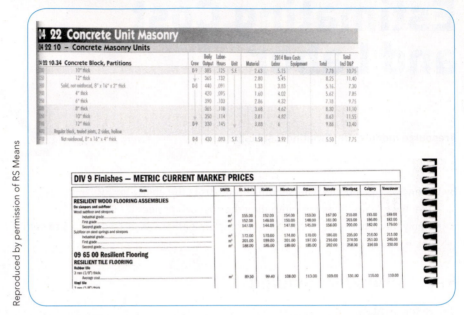

Reproduced by permission of RS Means

FIGURE 1 Samples of tabulated costs that can be used for estimation purposes [1,2].

Estimations done for projects are often based on historical information. Some of this will come from previous projects done by the company or the people in it. For instance, if you know of a software project that is of similar complexity to the one you are designing, you can use the cost and time information from that project to help you estimate cost and time for the new one. If you know typical turnaround times for part fabrication from other projects, you can use that information to estimate the turnaround time for the parts in a new project.

However, there are also published guides for various types of work. Auto mechanics, for example, can refer to a guide to help them provide a binding estimate before a major repair job such as the replacement of a transmission.

Examples of information used by engineers for these estimates are shown in Figure 1. Labor costs can be estimated using tables such as the ones found at: www.bls.gov/ooh/business-and-financial/cost-estimators.htm

Of course when using tabulated estimation values, care is needed to avoid certain pitfalls.

- Many estimating techniques will assume that doubling the number of workers, or doubling the available equipment, will halve the time required. This is almost never exactly true and sometimes the effect of these actions on time is negligible. Learning some *project management* strategies, and about the effect of critical path on a large project, will help you develop better estimates (#projectmanagement).

- Specific buildings, apartments, computer programs, or kitchens are not identical to one another, and so some adjustments must be made depending on the specifics of the project.
- Tabulated data could be out of date. Inflation increases the cost of labor and materials, while changes to legislation can dramatically influence costs. Conversely, new materials and techniques may reduce costs. Try to base your estimate on a current publication or up-to-date online information.

2. Conclusion

Estimating the resources necessary to complete a project in terms of time, cost, and work-hours is an essential skill. You have probably started to be able to do this effectively for your own school projects so you complete them by the due date. However, once you are involved in more complex projects that involve multiple people (who might need to be paid), supplies, deliverables, and delivery dates, estimating becomes more difficult. Using a similar project as a baseline, or published information, is helpful. However, people new to estimation almost always under-estimate the amount of time and resources needed to complete a project. So, when you are new at estimation you should probably substantially overestimate the time and resources needed to get close to an accurate prediction.

KEY TERMS

bid   request for proposals (RFPs)   manpower
project management

3. Questions and activities

1. Select a simple project such as:
- Building a bicycle from parts with professional assistance
- Creating a complex flower arrangement
- Baking a cake decorated using professional methods
- Planning a trip including hiring a guide to provide assistance touring the region
- Have a simple part of your residence (such as the bathroom) remodeled

a. What is the cost if you do this yourself? Estimate the hours and cost of supplies and tools (if necessary).

b. What would be the cost if you hired a person or people to do this for you? Estimate the hours, supplies, and cost using the methods explained in this section.

4. References

[1] *RSMeans Building Construction Cost Data, 72*nd annual ed., Reed Construction Data, LLC, Norwell MA, 2014.

[2] *Hanscomb Yardsticks for Costing*, RSMeans, Toronto ON, 2014.

Estimating Cost and Labor

Estimation Confidence

#resource module: #estimationconfidence

Learning outcomes

By the end of this module, you should demonstrate the ability to:
- Apply basic techniques of bounding estimates and checking using multiple methods to improve the level of confidence in an estimate
- Explain how a factor of safety is used in the estimation process

Recommended reading

Before this module:
- **Resources > Estimation > 3. Estimating Cost and Labor**

1. Introduction

When an engineer makes an estimate there must be some level of confidence in the final information produced. Even though it is an estimate, it is not the same as a guess, and far from being a random number. Strategic decisions will be made based on the estimate, so a statement about the level of accuracy should be part of the estimation.

The most important steps in generating this confidence are embedded in the basic estimation process. By determining the underlying question, the engineer can determine the level of accuracy required of the answer. This module is concerned with determining whether an answer is likely to have the accuracy required.

2. Is the Estimate Good Enough?

With every estimate engineers make, whether it is expected to be very close to the "real" answer, or just in the ballpark, they must determine if they have answered the underlying question well enough. There is great danger in working with the estimate if engineers misjudge its accuracy. A ***go/no go*** decision will often not require an accurate estimate, provided the estimate clearly indicates what the decision should be.

A planning estimation, on the other hand, could lead to huge budget shortfalls if the accuracy is poor.

2.1. How Far Off Might You Be?

Depending on the information that went into the estimate, the result will vary in accuracy. As you are developing the estimate, also consider the accuracy of the solution. You might, for example, estimate an upper and lower value for the answer to see how wide the range is. Even if you don't know the true answer, it is sometimes possible to estimate the potential uncertainty in your answer by keeping track of the uncertainty in the information used to compute this answer.

The more reliable the sources of information you use, the more reliable the estimate will be. If you base your information on data from an extreme political website, or from an obscure Wikipedia page, or from another estimate of unknown accuracy, there is a significant chance that you will end up with a poor result. On the other hand, data from sources such as industry-standard reference compendiums are likely to be very accurate and thus increase the confidence you might have in an estimate derived from these.

2.2. Does the Estimate Make Sense?

You know from experience that your computations can sometimes be wildly wrong. Your model could be incorrect, your input data could be faulty, your method may be in error, or you may have made a simple calculation error. It is important to apply common sense to your answer and if in doubt, to recheck your calculations, preferably with a different approach to the problem that should give the same answer if things are done correctly. Unfortunately for young engineers, common sense accumulates with experience: an experienced mechanical engineer might know by comparing to existing commercial solutions that a certain gauge of tubing is not adequate for a high pressure pipe, whereas a new engineer less likely to have the background needed for this judgment. You need to be more careful when you begin working in a new area. To double check your calculations you may need to do more research to find more accurate data, rather than relying on experience.

2.3. What are the Consequences of a Bad Estimate?

When engineers make an estimate, they must consider the consequences of an error. Are there serious financial, environmental, or safety issues if the estimate is wrong? It is common sense that you must be surer of the result when the risks are greater.

If you are estimating how much water you should carry on a trip, your effort to develop an accurate estimate will depend on whether the trip is a four hour trip in a car, a three day hike in the desert, or a four-month mission in space.

2.4. What Is the Cost of Making a Better Estimate?

It costs more time, energy, and resources to make a more accurate calculation than it does to make an estimate. These costs must be weighed against the benefits of a more accurate result.

Estimation Confidence

3. Techniques to Raise the Confidence of an Estimate

3.1. Checking by Multiple Methods

One method of increasing the certainty of an estimate is to find a second, independent way of estimating the answer. If two very different ways of estimating produce a similar answer, then you can have more confidence that your calculation is correct. For example, you might estimate the weight of a load of grain in the hold of a ship by estimating the volume of the hold and the density of the grain. To check this estimate, you could also simply record the change in the waterline when the ship is loaded, and estimate the weight of the load knowing that it is equal to the weight of the additional displaced water (i.e., using the principle of buoyancy). If the two completely different calculations yield similar results, your confidence in the answer is greater.

3.2. Bounding Estimations

If the information needed for an accurate estimate is unknown, it might still be possible to compute upper and lower bounds for the true answer using known information. For example, assume that you have to estimate the number of gumballs in a 5 L jug. The gumballs are 1 cm in diameter, so their volume is known, but you do not know the fraction of space filled by the solid balls (known as the packing fraction) when the balls are randomly dropped into a container. However, if you have taken an introductory course on metallurgy, you will remember that the packing fraction for a hexagonal close-packed array is 0.74. Balls cannot be packed more closely than this, so you can estimate that:

$$\text{Upper Bound for the Number of Balls} = 0.74 * V_{container}/V_{ball}$$

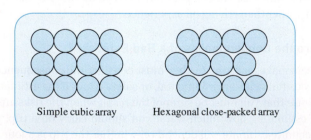

Simple cubic array   Hexagonal close-packed array

On the other hand, a simple cubic array has a packing fraction of 0.52. Common sense tells us that randomly packed balls must be more densely packed than this, so you can estimate that:

$$\text{Lower Bound for the Number of Balls} = 0.52 * V_{container}/V_{ball}$$

Experiments show that randomly packed balls have a packing fraction of about 63%, which is actually halfway between our upper and lower bounds.

4. Factor of Safety

In any estimate it is important to use a ***factor of safety*** when the safety of people or the environment is at stake. This means that any estimate that will be used to make a decision about hazardous situations or safety should be adjusted so that the resulting decision will be appropriate even if the estimate is substantially inaccurate. Critical systems such as elevators generally have very high factors of safety (between 3 and 10) so they can hold many times their rated load without plummeting to the basement. This doesn't mean they will work properly when overloaded, only that they will not catastrophically fail. Factors of safety of 2 or more are not uncommon in engineering, and should be factored into your estimate.

4.1. Example: Calculating the Slope of a Ramp

Suppose that you determined the maximum angle allowed in a car ramp using the coefficient of static friction of rubber on dry concrete. You would realize that any oil, water, ice, or other substance on the surface, or any other forces acting on the contact area would cause the car to slip off the ramp, and so would realize that the determination was likely too steep for a real ramp. As a check, you might find the slopes of the steepest roads in the world and compare your angle to these.

The accuracy of your estimation will depend on the model you use (here, that the friction force required depends only on the materials in contact) and the numbers you use with the model (here, that the substances were new and dry). You must recognize the potential flaws in these assumptions and the effects on your estimate. So in this case it would be appropriate to apply a factor of safety to your estimate.

5. Conclusion

The level of confidence is of huge importance when engineers make estimates. With all calculations, and particularly with estimates, an engineer should reflect on the answers to make sure that sufficient care has been taken to come to them so that sufficient accuracy is attained to answer the underlying question. The confidence in the accuracy is often lower for estimates because the data is sparse or less reliable. Therefore, using techniques such as bounding the estimate, using multiple methods to check, and applying a factor of safety are important to improving the level of confidence in an estimate and the decisions based on the estimate.

KEY TERMS

go/no go   factor of safety

6. Questions and activities

1. Using multiple estimation methods or bounding techniques to improve the level of confidence for the following:

   a. Estimate how many calories you burn in a day.

   b. Estimate the amount of time it will take to get from where you are to an airport or train station.

   c. Estimate the cost of your lunch.

   d. Estimate the number of ping pong balls that fit in the trunk of a car (pick a car to use for this).

   e. Estimate how many steps you need to walk to get from where you live to your engineering design class.

   f. Estimate how deep a pool should be for you to safely dive into it.

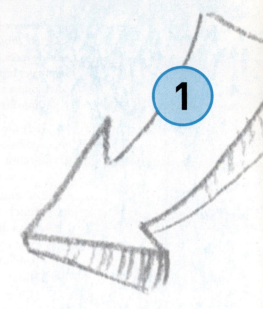

Introduction to Economics

#resource module: #economics

Learning outcomes

By the end of this module, you should demonstrate the ability to:
- Estimate the costs of the design activity of your project
- Account for the time value of money in a purchase decision

Recommended reading

After this module:
- **Resources > Economics > 2. Time and Money Calculations**

1. Introduction

Engineering economics is the study of costs and revenues as they apply to engineering projects. Engineering creates benefit. Engineering creates products and services and the infrastructure upon which industries depend. Understanding economics allows you to assess the benefits of engineering activity to a company, to its people, to its clients and customers and to society.

Economics is not perfectly understood, nor well controlled. Economic trends and outcomes cannot be exactly predicted. However, engineers must be able to estimate the cost of projects and the value of the project to the company, clients, users, and society. Projects must align with goals of the organization; company, government, or non-profit organization at a reasonable cost. And the products and services produced must add value for a project to be implemented. For commercial projects this means that the technology produced must be something that clients and customers are willing to pay for, such that the company has a surplus of money (profit) to be used to expand the company and to reward investors and employees.

The profit is calculated using a very basic equation, which will be referred to as the foundational equation:

$$Profit = Revenue - Cost \qquad \text{Equation 1}$$

Revenue is the monetary benefit coming into the company before any costs are subtracted. ***Profit***, also sometimes called ***net revenue*** or ***income***, is the monetary benefit after the deduction of the ***costs***. Costs are the costs borne by the company to operate or to make a project happen. The equation can be used for a company, a subsidiary, a department, a division, or any other entity, including a project. In this module we will focus on projects.

Three factors make the simple foundation equation a large, complex area of study:

1. Time: When costs or revenues are delayed or accelerated in relation to each other, complexity grows. The simplest way this happens is through borrowing, but it can also happen through bulk purchasing, inventories, investments, and many other ways. All these make the foundational equation harder to evaluate.

2. Evaluations: When costs and revenues come from many different sources, evaluation of the foundational equation becomes complicated. For example, does the cost of a telephone system at the engineering company get ignored or partially or fully applied to a particular project when there are several different projects underway at the same location?

3. Influences: When selling a product or service the market situation can change. Changes in demand for the product and service may influence the revenue. Changes in the cost of providing the product or service will influence the profit. These influences are often difficult to predict.

The ultimate basis of economics is the study of money coming in versus money going out. Profit is the difference of these two and a business or project usually is intended to maximize this difference in a positive way. Even non-profit projects, such as new roads, are measured according to this equation; the goal is typically to minimize cost relative to the positive economic and social impact of the service the road provides.

2. Cost-Benefit and Accumulated Benefit Curves: Project Focus

When and what revenues come in, and what and when costs are involved, depends on the project and on the organization's viewpoint. A consulting engineering firm that is working on a wind-generator installation will get consulting fees long before the wind generator begins to provide power, perhaps even before the actual installation begins. To the owner of the wind power company, this payment to the consulting engineering company will be an expense. The users will have no costs for this energy until it begins to be supplied; their costs will be the power company's revenue.

Consider a simple, representative example of a company that does work to develop a product, and then uses production and sale of that product to create revenue. This example represents many development situations, including those situations where the "product" is a service. Figure 1 shows the costs and revenues over the startup period for this example project. Each bar is a uniform periods of time, a monthly period for this example. The blue bars below zero are the total cost and the black bars above zero are the total revenue for each month.

During the early months of the project there are only costs. This is typical during development: You cannot sell smartphones you have not yet designed or sell condominiums you have not yet built. During this design and development period, costs will

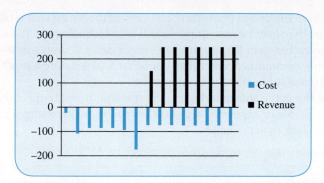

FIGURE 1 Cost and revenue over time for an example company that is designing, producing, and selling a product.

have to be paid from savings, from revenue from other sources such as selling other products, from loans, or by using money from investors. When the product is ready for market, it is used to generate revenue, shown by the positive lines. For our example we have constant cost and revenue amounts after the development period. However, this would be unusual. Usually sales and costs will vary month to month throughout the life cycle of the technology.

We can look at this in a different way, as shown in Figure 2. In this figure we show the profit on a month-by-month basis in blue and the accumulation of profit (or loss), shown by the black bars. The accumulation of profit shows how much *total* profit has been made to date.

In the early months of the project when there are only costs and no revenue, this profit is negative (i.e., it is *loss*). Eventually revenue pushes the profit up (for a commercially successful product). Within a few months of having positive profits development costs are covered and the accumulated profit becomes positive. In many instances even very successful products will take years to reach this *break-even point*, which is often referred to as a *payback period* or *payback time*.

We can apply this economic analysis to any project and technology. The accumulated profit graph is effectively a curve, shown in Figure 3 for the lifetime of a

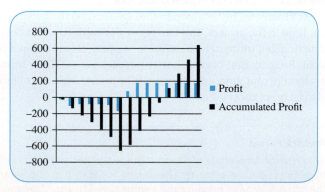

FIGURE 2 Monthly profit and accumulated profit for the example.

technology: from the design phase through to end of the life cycle. It is often useful to consider more than just the monetary rewards for a technology, so the **accumulated benefit** curve, rather than the accumulated profit curve is shown in this figure. For example, a new community center may not be built to generate a profit, but it does provide measureable benefit to the community.

Note that the curve shown in Figure 3 is smooth. In reality the curve may have bumps and dips. Major expenses, large bulk purchases, sudden surges in sales could all cause irregularities in the graph.

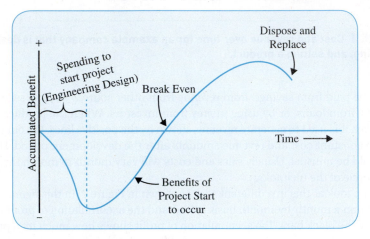

FIGURE 3 The accumulated benefit curve.

2.1. True Cost: Putting It All Together

Ultimately, our economic study will bring us back to a generalization of the original foundation equation:

$$Excess\ Benefit\ [Profit] = Incoming\ Benefits\ [Revenue] - (All)\ Costs \qquad Equation\ 2$$

Equation 2. A more general expression that measures the net benefit of a technology by the difference between benefits and all costs.

For any project there will be times when the costs exceed the benefits coming in at that particular time in the project. This can occur during design and development, or during construction, but ultimately the success of a project is measured by the excess benefit or profit. Projects that have a shorter payback period (time from start of the project to break-even) and more excess benefits or profits over their lifetime are generally judged to be more successful.

> **Example: Payback Period**
>
> In the early 1990s the University of Toronto made the decision to purchase a gas turbine engine to produce electricity. The engine is located in the steam plant on campus. The initial purchase price for the engine was $10.4 million and the engine produces 6 MW of electricity using natural gas as the fuel. The university calculated

at the time that there would be a 7-year payback period. That is the difference between purchasing 6 MW of electricity from the local power utility for 7 years minus the cost of purchasing natural gas to produce 6 MW of electricity for 7 years was $10.4 million. They also factored into this calculation the cost of maintenance and other expenses. An additional important factor in this calculation was that the exhaust heat from the turbine is used to produce steam to heat the campus.

So they projected at the time that after 7 years the cost of the gas turbine would be recovered (i.e., the accumulated cost savings equals the cost of the turbine). After that the university would be effectively getting 6 MW of electricity for less cost, thus producing a significant cost savings for the university after the payback period.

In actuality the price of electricity in the 1990s went up faster than anticipated after the gas turbine had been installed. As a result the payback period was substantially reduced. The university continues to use the gas turbine co-generation system, which covers about one-third of the campus electricity demand.

KEY TERMS

revenue	profit	net revenue
income	cost	loss
payback period (or payback time)	accumulated benefit	break-even point

3. Questions and activities

1. Create a simple budget of your personal expenses. What are some of the major costs? Where does the money come from to pay for these expenses?

2. Consider the purchase of a new computer, printer, smartphone, or tablet. Itemize the costs and benefits of this purchase including:

- The expense of the personal time you put in to researching which model to buy
- The cost of the device
- The cost of any supplies, such as ink cartridges or service contracts, that will be needed to operate the device
- The benefit derived from having the device

a. Itemize each of these costs and benefits month by month for the first year you own the device.

b. Explain how you quantified the benefits. For example, how did you assign a dollar value to the benefit derived from owning and using the device?

c. Draw the accumulated benefit curve for this purchase.

d. What is the payback period?

3. List some of the non-monetary costs and benefits of a company and comment on the net benefit. Consider the benefits and costs to society of the industry. You could choose a local company or industry for this question or you could choose a restaurant, a saw mill, a bicycle shop, a food manufacturing plant (such as a plant that makes breakfast cereal), a software company, or the construction of a new condominium building.

Introduction to Economics

527

2 Time and Money Calculations

#resource module: #time&money

Learning outcomes

By the end of this module, you should demonstrate the ability to:
- Determine the costs of interest on money borrowed or held
- Take account of the time value of money in a purchase decision

Recommended reading

Before this module:
- **Resources > Economics > 1. Introduction to Economics**

After this module:
- **Resources > Economics > 3. Project Decisions**

1. Introduction

Money coming into a business and money going out are almost never occurring simultaneously. If you are running even a small business you will make commitments for future expenses, borrow money and give credit to customers. This borrowing and lending does not come for free, and so usually the first step in understanding economics is the understanding of the relationship of money and time.

> **Examples**
> Commitment for future expenses:
> - You rent office space for your company; the rental agreement commits you to paying rent every month.
> - You hire employees; you commit to paying them every month.
> - You start a business; you commit to paying taxes annually on the business activities and property.
>
> Debt and borrowing money:
> - You get a loan or line of credit from a bank to start your business; this loan will cover your expenses until your business becomes profitable.

- You take out a mortgage; this is essentially the same as a loan guaranteed with a piece of property.
- You buy supplies and services on credit; you get the supplies now and the bill (or *invoice*) will come to you later, usually at the end of the month.

Credit:

- You give your customers products or services and bill them for the work at the end of the billing period which is often at the end of a month.
- You allow your customers to buy your products or services using credit (such as a credit card) with the expectation that later the funds will flow into your account.

2. Money and Time

The value of a dollar, euro, yen, or peso changes with time; due to factors such as inflation, the value of a unit of currency declines with time. If you have ever watched an old movie, or read an old magazine or newspaper, you probably have been amused by the references to money. Cars could be purchased for what is now the cost of a month's rent of an apartment. A good dinner was less than you now pay for a piece of candy. However, at that time, wages were also low, and so many common items cost you fewer hours at work now than they would have at that time.

Courtesy of the Authors

If you hide money in your house for a year, instead of investing it, you have lost money in two ways. First, the worth of the money has declined because of *inflation*, the reduction over time of money's purchasing power. Second, you have lost the *interest* that the money could have generated. If instead of hiding the money in your house you lend the money to a bank, say putting the money in a bank account, then the bank will give you a certain percentage *interest* for the use of the money.

In order to reasonably compare future costs and revenues and past costs and revenues, the value of money needs to be adjusted to a common baseline. This adjustment is based on an evaluation of inflation and interest rates. This adjustment of the value of money is used by businesses to make decisions about when to borrow money, when to make purchases, and when to give credit to customers.

2.1. Evaluating Future Costs in Current Dollars

Changes in purchasing power, interest rates, and inflation have to be taken into account when determining the economics of a project. To do this the costs and revenues needs to be adjusted for time so a purchase made 5 years from now can be appropriately compared to a purchase made today. Engineers do this by using a calculation called finding the *time value of money*.

For example, if a project will require $1000 a year from now, and the interest rate is currently 3.5% per year, then how much present-day money needs to be in the bank to cover this future expense? The answer is $1000/1.035 = $966.18, since $966.18 invested at 3.5% will produce $1000. So the ***present worth*** of $1000 payable a year from now is $966.18.

Important: Note that $966.18 is NOT $1000 minus $35! Even though $35 = $1000 × 3.5%. This is a very common mistake that people make in calculating the present worth of money.

2.2. Present Worth Factor

Economists use a general factor to calculate present worth over a number of years, called the ***present worth factor (PWF)***:

$$PWF = P\!\big/\!F = \frac{1}{(1+i)^N}$$
Equation 1

Where: P is the present value (the value of the money today)

F is the future value (value of the money N years from now)

i is the annual interest rate

N is the number of years in the future

So $P = F\dfrac{1}{(1+i)^N}$ and if you substitute $N = 1$, $i = 0.035$ and $F = \$1000$ you will get $966.18.

This equation also works using months, days or other time units when i is the monthly or daily interest rate and N is the number of months or days accordingly.

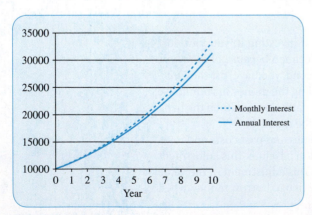

FIGURE 1 Cost of borrowing $10,000.

2.3. The Costs of Borrowing

The present worth factor can also be used to determine the effects of various borrowing rates. For instance, what is the difference between 1% interest per month and 12% interest per year? Initially, it may look like these are the same, but in the case of the monthly interest the factor is

$$\frac{1}{(1+i)^N} = \frac{1}{(1+0.01)^{12}} = \frac{1}{1.127}$$

And in the case of yearly interest:

$$\frac{1}{(1+i)^N} = \frac{1}{(1+0.12)^1} = \frac{1}{1.12}$$

These are left as ratios so that the effect can be seen: The interest in the monthly case amounts to an effective increase in interest of $\frac{127-120}{120}$ or over 5% additional interest over the year. Similar sorts of equations are used to determine payments on loans and mortgages, to help determine bank rates, and to find other values where one has to determine future value in present and future measures of worth.

The net result of the interest analysis provides mathematical support for most people's realization that it is better to receive income sooner and pay costs later. This is provided, of course, there are no extra charges for doing so, charges which are often there to compensate creditors for the losses with "buy now, pay later" incentives.* This also assumes you invest your income, and that the interest rate is greater than inflation.

3. Conclusion

The ultimate basis of economics is the study of money coming in versus money going out. Profit is the difference of these two and a business or project usually is intended to maximize this difference in a positive way. What makes this more difficult is that the value of money changes over time. So money that has been loaned or borrowed (directly, or through the provision of goods) makes the profit calculation more complicated. It is important that engineers adjust for the time value of money when projecting costs and revenues in a project so they can make well informed decisions and recommendations.

KEY TERMS

inflation	interest	time value of money
present worth (present value)	present worth factor (PWF)	

4. Questions and activities

1. Check out the fine print of an online ad from, for example, a furniture store that offers no interest or payment for a period of time. Work through an example to find out the final costs, using the current interest rate to evaluate future costs of money. Include any charges added to administer the purchase.

2. You company uses 6 industrial ink cartridges per month for label production. Each cartridge costs $603.00 if you buy between 1 and 9 cartridges. However the supplier offers a volume discount. Each cartridge costs $473.50 if you buy 10 or more up to 24 and $430.56 if you buy 25 or more. If the inflation rate is 0.15% per month, how many should your company purchase each time in order to save the most money on ink cartridges? (Assume the price rates don't rise, and the volume discount remains constant.)

3. In question 2, what other factors besides the cost of the goods might you consider when deciding how many to purchase in order to keep the costs minimal to your company?

4. In groups, create a graph that shows how $966.18 invested at 3.5% interest per annum is not the same as $966.18 + $35 per annum. How much will $966.18 be worth in 10 years if invested at this interest rate?

5. Find the present worth factor at 2% interest over 10 years.

   a. What is the present value of $5,000 in 10 years at this interest rate?

   b. How would you change the present worth factor calculation if the interest over the 10 years was not constant, but was 1.5% for 5 years and 2.5% for 5 years?

   c. Can you generalize the PWF calculation to manage different interest rates every year?

---

*Often sales promotions will claim "No Interest or Payments for 3 Years!", for example, but one can get a discount for paying immediately that is equal to or greater than what you save in interest.

3

Project Decisions

#resource module: #economicprojectdecisions

Learning outcomes

By the end of this module, you should demonstrate the ability to:

- Explain the decisions made at the different stages of the accumulated benefit curve
- Describe events and circumstances to which a particular project may be cost sensitive
- Identify circumstances that may lead to decisions that do not maximize total accumulated benefit, or even instantaneous accumulated benefit
- Define and determine a design project's cost commitment

Recommended reading

Before this module:

- **Resources > Economics > 2. Time and Money Calculations**

After this module:

- **Resources > Economics > 4. Types of Costs and Revenues**

1. Economics of a Project over Time

This module looks at the costs of a project over the lifetime of the project in greater detail. The graph in Figure 1 shows the ***accumulated benefits*** during the lifetime of an engineering project. Time proceeds to the right from the initiation of the project. Expenses in the design and construction stages push the curve down. Engineering design work costs money. And the construction, manufacturing, development, or implementation of the design will also contribute to the cost of the project. Once designed and executed, the benefits of the project will begin to be realized and the curve will rise. For a successful project the accumulated benefits will exceed the costs at the break-even point. After this point benefits will continue to accumulate. There will typically be operating expenses associated with the technology, but these will be much less than the benefits accumulating. Once the operating expenses or maintenance costs begin

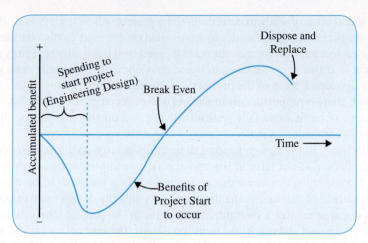

FIGURE 1 The accumulated benefit curve.

to rise, or the benefits begin to wane, a decision will be made to dispose of or replace the technology. Engineering companies will use projections of costs and revenues to draw curves like this in order to evaluate projects and decide which projects to develop.

The accumulated benefit curve applies to most engineering projects. We will use the terms "company" and "clients" in this discussion, although the curve and the concepts apply to government agencies, non-profit organizations, and other groups, and to projects that are internal to a company and may not have an easily identified client. Exactly how the curve looks will depend on the project circumstances and the perspective of the company.

There are often non-monetary benefits and costs in engineering projects. However we will use only money as the cost and benefit values to simplify the discussion. Any point in the graph shows the net (or total) of all the revenues less all the costs accumulated to that point in time. If the point is below the x-axis, the project has not made money (that is has not achieved a net benefit) at that time. If the point is above the axis then the company has made money or benefited from the project.

2. The Accumulated Benefit Curve

All projects will start with only costs, so the curve will start downward. This is because early negotiations between company and client, and work such as bid preparations, will cost the company, and there is no immediate compensation. This negative flow is particularly evident in companies doing custom projects such as consulting engineering companies. A consulting or construction company may research and bid on several projects which require significant preliminary preparations. However, not all of the projects will be awarded to the company. These companies will have to spend money long before any payments come in. And when they win a bid they will need to cover the expenses accumulated from unsuccessful bids.

All projects, even internal projects, have a similar evaluation and commitment stage. In situations where a product is being produced for manufacture the projects have an initial stage of study to determine if the project will be undertaken. The salaries and support for the persons doing this evaluation are costs to the company.

As the project starts in earnest, the costs increase and there are not yet any positive benefits. It is at this stage that research and development is generally done and the design is developed. Most of the primary design engineering is done in this stage of the project. There is proportionally much less design accomplished in the later stages. The amount of profit eventually realized will depend on the engineering decisions made at this earlier design stage.

The decisions in the design process determine, in large part, the **costs committed** that will be realized later in the construction or implementation phase of the project. For example, if today a design team decides a part should be made from titanium instead of aluminum, the decision costs only the design team's time today. But this decision creates a committed cost later in the project when the material has to be purchased and machined to fit the design. The cost of buying and machining titanium is much, much higher than for aluminum. So the decisions like this during the design process will significantly affect the costs and therefore the profits of the project later.

It is especially important to consider committed costs for projects with very large expenses over time. Figure 2 shows the costs incurred in designing and building a bridge, as an example. Early in the project the costs are relatively low. The cost of the design work may be several million dollars, but this is very low compared to the cost of actually building the bridge, which will involve more workers, materials, and other resources. At the time the design is given the go-ahead to proceed to execution, orders go out for materials and contracts are signed for resources. There is a commitment to a huge expenditure that cannot be stopped without considerable expense in penalties if at all. For this reason the design work, and decisions in the design phase, are very carefully considered (or should be) because these decisions create huge committed costs.

Figure 2 also underscores the importance of careful engineering design, and of taking the time to verify the correctness of design calculations. Small mistakes in the

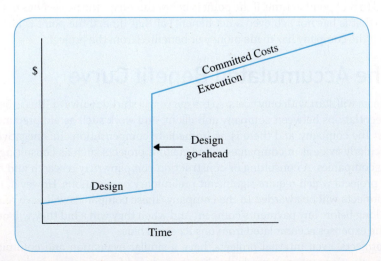

FIGURE 2 **Example of costs for designing and building a large infrastructure project such as a bridge.**

design stage can result in considerable costs during execution when the mistakes need to be corrected.

Once the technology is for sale or in operation, benefits start to accumulate. The accumulated benefits curve (Figure 1) reaches its lowest point and begins to rise as the on-going costs decline and benefits are realized. These benefits from the project are in the form of use or sale of the design, or in the form of payment for services.

At some point, if the project is successful, the accumulated profit will move from negative to positive; that is from a net loss to a net gain. This is the "break-even" point and the time it takes from the start of the project to this point is called the payback period. For many projects, such as purchasing a new piece of machinery for production where the purchase will support other benefit generation, this is considered to be good enough. For other projects, such as the introduction of a new product, the curve is expected to positive for a considerable time, as this accumulated positive benefit provides the funds that pay the shareholders and that will allow the company to finance other projects through the design and implementation stages.

Every project will have a unique accumulated benefits curve. Some projects will have a short payback period and, others a long payback period. Some projects require a very large upfront investment of resources, and others will require very little to get started. Often the accumulated benefits curve is very bumpy and uneven, particularly after the payback period as profits go up and down during the lifetime of the technology. A company uses economics to predict the benefit curves of possible strategies and usually selects projects which have the best **return on investment (ROI)**.

2.1. Curve Sensitivity

Many engineering project cost and timing decisions are made based on economic expectations. Typical contributors to these decisions are:

- Existing competition; alternatives available to the potential customers or clients
- Anticipated competition; future alternatives likely to become available to potential customers or clients
- Current or anticipated interest rates on loans
- Changes in currency valuation
- Global and national economies, and the discretionary purchasing power of potential customers or clients
- Labor unrest (strikes and other actions); loss of critical staff for any reason
- Changes in government regulation, or patent expiration
- Environmental events (earthquakes, hurricanes, temperature extremes)
- Accidents (causing human deaths, repair or replacement costs, time losses)
- Interruption of necessary supplies or equipment due to problems with suppliers or shipping
- Technical impediments that extend development time or that make delivery impossible
- Technical innovations that reduce the value of a technology, or open up new areas of design
- Warranty charges and product recalls

If the cost estimates are done properly, these and other contributors to the accumulated benefit curve will be analyzed and the most likely contributions will be integrated into the calculation. However, the behavior of the actual accumulated benefit in response to any change in circumstances should be evaluated and risks avoided or mitigated appropriately.

2.2. Choosing a Curve with a Lower Net Accumulated Profit

Consider the two curves shown in Figure 3 for two alternative projects: project blue and project black. For the blue project there is a higher accumulated profit over the lifetime of the technology. The black project produces less accumulated profit. If the sole consideration were highest accumulated profit, the choice would be simple. The company should choose to engage in the blue project. However, accumulated profit is not the only consideration.

The blue project requires higher initial investment and has a longer payback time. This is called *exposure*, because it exposes the company to financial risk. The black project has a lower initial investment and has a shorter payback time. These issues factor into the decision of which project, or approach to a project, a company might choose. A new start-up company may need to choose the black project because they do not have the financial means to undertake the blue project, and they need the quicker return on investment to keep the company going. That is, a new company may not want the exposure that comes with the blue project.

In another scenario, the black project might make use of existing equipment, an existing design approach, or lower cost technology. Perhaps it is a manufacturing process that produces few units per hour than other ways of manufacturing the product. The blue project may require new equipment, a completely new design, additional training for workers or higher cost implementation technology. For example, a new assembly line in a manufacturing plant that produces more units per hour of the product. This might lead a company to choose the black project over the blue project to simplify the implementation.

There might also be additional considerations associated with the choice between the blue project and the black project that relate to social, environmental, or

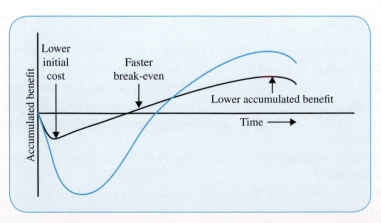

FIGURE 3 Comparison of two possible projects with different accumulated profit curves.

ethical concerns. The black project might be more environmentally benign, give jobs to people in a community with high unemployment, or be better aligned with ethical considerations and company values. For these reasons a company may choose to engage in the black project instead of the blue project.

2.3. Profit Issues

Fixed costs are the costs routinely incurred by a company that are essential to the business. This could include rent on the office space, the cost of the information technology (IT) and utilities, and basic administrative costs. The *variable costs* are costs that arise from individual projects or unique circumstances (e.g., a new piece of equipment to replace one that has stopped working). Variable costs are dynamic and will rise and fall over the course of a year. Because of fixed costs, the costs of operating a business will never drop to zero. Projects undertaken by the company must cover both fixed costs and variable costs of the business in order for the business to remain viable.

Figure 4 shows the accumulated benefits curves for three scenarios. The blue curve shows a successful project. The accumulated revenues exceed the initial cost and the project results in net accumulated profit. In the second scenario, shown in black, the project does not make enough revenues to cover the initial costs. The accumulated profit never becomes positive and the project does not break even. The initial costs, like design development, have exceeded the revenues from the production or construction of the technology. Likely the project will continue as long as it is generating net revenue over a period of sales because the company has an interest in recovering as much of the development cost as possible.

In the third scenario, shown in gray, costs continue to exceed benefits long after the initial investment in the development. This is essentially a failed project. Unless there is reason to believe the trend will reverse, there will be strong pressure to discontinue production long before expected end date for the project. If the company engages in too many failed projects it is likely that the company will exhaust all sources of funding, such as lenders and investors, and will go bankrupt.

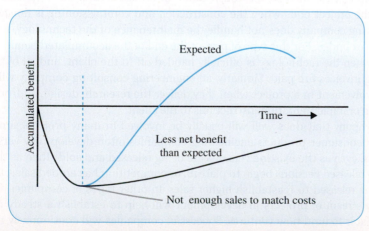

FIGURE 4 Accumulated benefit curves for three different scenarios.

2.4. The Curves of Clients and Suppliers

The scenarios discussed in this module consider the economics of a single engineering company. In actuality, engineering companies are heavily influenced by the economics of their suppliers, customers, and clients.

A client contracts an engineering company to do work for them. The client expects the engineering company's work to contribute to their own profit. A client will want to know:

- How does the engineering company's work contribute to the client company's accumulated benefit curve? What will it cost the client and what will be the effect on the revenues generated? What is the sensitivity of the client's profits to the cost of the engineering product or service?
- What additional revenues or future cost reductions might come from more features, higher quality, faster delivery, and so on?
- Can the engineering company adjust their work to be better suit the client's needs and motivations?

Similarly, the suppliers to the engineer's company have benefit curves that will be affected by the engineering company's work. It is important that valued suppliers receive appropriate net benefit from their dealings so they can stay in business. Engineering companies exist in an economic network of co-dependent businesses. It is important to consider not only the economic well-being of the company you work for, but also the economic health of the businesses your company relies on. How well an engineering company works with their clients, customers, and suppliers will determine, in large part, the success of the business.

3. The End of the Technology Lifecycle

Eventually the project benefits slow, and the accumulated benefits start to decline. This could happen when the repairs and upkeep costs outweigh the positive benefits, or if new development results in an updated product model that undercuts sales. Perhaps there is declining production from a mining project.

In some cases, such as the construction of infrastructure, the engineering portion of the project ends when the construction and commissioning is finished. The engineering company does not handle the maintenance of the technology, such as a highway or bridge. For this type of project the end of the accumulated benefits curve comes when the technology is officially handed off to the client, and all of the outstanding invoices are paid. Similarly, an engineering consulting company will finish their involvement in a project when they deliver the research, design, and/or manufacturing or implementation instructions to the client.

A company that does well will usually be involved in many projects simultaneously. In consumer products engineering companies often develop new versions of products, even as the existing versions are being released and sold. Then as the sales of those released versions begin to plateau, or competition has eroded sales, the new version is released to reestablish higher sales. In other types of companies, keeping a steady stream of projects in development will help to establish a steady revenue stream later to fund new ventures. Successful companies will continuously monitor the costs and revenues, both realized and expected, from every project so they can make decisions on strategic goals for the business.

KEY TERMS

costs committed	**return on investment (ROI)**	**exposure**
fixed costs	**variable costs**	**accumulated benefits**

4. Questions and activities

1. In groups, consider what the selling price should be for widgets your company produces. Assume the cost of production is $100 per widget.

    a. What would go into the decision beyond the production cost?

    b. What would be the effect of raising the selling price?

    c. What would be the effect of lowering the selling price?

    d. Assuming the widget project is successful, draw an accumulated benefits curve that might result, and explain, based on this curve how lowing or raising the price might affect the payback time and accumulated benefits.

2. Suppose you are working for a petro-chemical company. What economic sensitivities can you see to profits from the end product? Make a short list of key circumstances that could affect profits.

3. What are the possible bad outcomes of forcing a supplier to sell you product at a very low price?

4. Discuss both sides of the Ford Motor Company's decisions regarding the Pinto gas tank design [1].

5. References

[1] Wojdyla, B. *The Top Automotive Engineering Failures: The Ford Pinto Fuel Tanks, Popular Mechanics*, May 10, 2011. Retrieved from www.popularmechanics.com/cars/news/industry/top-automotive-engineering-failures-ford-pinto-fuel-tanks.

Types of Costs and Revenues

#resource module: #typesofcosts&revenues

Learning outcomes

By the end of this module, you should demonstrate the ability to:
- Divide costs by stage of a project
- Divide costs between fixed and variable
- Allocate proportions of fixed costs
- Distinguish between capital costs, operating costs, and disposal or decommissioning costs

Recommended reading

Before this module:
- **Resources > Economics > 3. Project Decisions**

After this module:
- **Resources > Economics > 5. Payback Calculations**

1. Introduction

Economics is important to a business for many reasons. Chief among these reasons is determining the current viability (health) of a company and whether it will continue to make a profit in the future. Economic forecasting allows an engineering company to make choices about which projects to pursue. A company analyst may ask:

- Is the business going to survive a year? A few years? Or more than a decade? Answering these questions requires us to look into the future at the anticipated profits the company will acquire from their projects and the benefits they provide to society. It also requires anticipating the costs to develop and produce those technology services and products.

- Does the company have enough money readily available and cash input to keep up with current expenses, anticipated expenses, and expenses related to new projects? This is a shorter-term **cash flow** outlook. A company that cannot keep up with the purchases, wages, rent, parts for production, loan payments, and other expenses on a month-to-month basis will go bankrupt.

- Is a current project going to make a profit? It is not enough just to cover the cost of materials in a project or wages for a service. Determining whether the company as a whole makes money on a certain project involves a more complex calculation. A project that does not provide net benefit to the company and is not likely to do so should be discontinued.

To calculate the answers to these questions costs and revenues are divided into categories and summed up in different ways. There are standard balance and income sheets (which we will not go into here). Often each project will have its own budget and balance sheet. Within a project, further divisions are made so that the numbers can more easily be understood and manipulated.

In this module the terms used to identify different types of costs and revenues will be defined.

2. Costs from the Perspective of Project Lifecycle

Costs of an engineering project over its lifetime can be split into three phases: Initial costs, on-going or usage costs and final or disposal costs. In some projects an engineering company will be responsible for all of these phases. For example, a mining company may be responsible for researching an ore deposit, designing a mining operation, managing the mining process, and decommissioning the mine at the end of production. In other types of projects the engineering company will be responsible for only one or two phases of the project and other organizations will handle the other phases.

Initial Costs—Initial or start-up costs are the costs incurred during the design and development phases of a project. They are the costs for development or design, procurement, installation, implementation, and/or construction. These occur before the technology can be used or before it begins to generate benefit. For a product or service, these costs come before sales begin. For infrastructure or plants these costs come before the project is functional. These are one-time costs and do not include the costs of operating the technology.

Initial costs are sometimes referred to as capital costs. A **capital cost** is a one-time-only cost to buy a piece of equipment or system, build a building or create a design or development plan. The thing that is purchased is called capital or a capital asset which adds value to the company or organization. Capital assets can be held, and depending on the type of capital they can be sold, licensed, rented, or traded to other organizations.

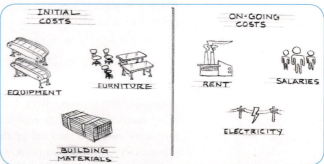

Initial Costs and Ongoing Costs

The design of a new product might require the development of a new assembly line to produce the product. The cost of buying the equipment, customizing it, setting up the assembly line, and testing the manufacturing processes would be considered an initial, or capital, cost for the project. Paying the workers who will make the product would be considered an on-going, or operating, cost for the assembly line.

Ongoing costs. Ongoing costs include the maintenance, and *operating costs* of the design. These are recurring costs that are required during the operational life of the technology. For a manufactured product, this would include the costs of running the assembly line including the materials, electricity and other utilities, and worker salaries and benefits. For a plant it would be the costs incurred during regular plant operations (e.g., a water treatment plant, or a chemical manufacturing plant). For a software company it would be the cost of providing technical support for their customers and continuing to monitor and maintain reliability and security.

The on-going costs, particularly maintenance costs, may include some capital expenditures. For example, if there is a need to replace a piece of equipment that has failed during the operation of an assembly line or plant. The cost of the new piece of equipment is a capital expenditure.

Final costs. Final costs are also called *disposal costs* or *decommissioning costs*. These are the costs required to decommission the design and to dispose or disperse the physical parts. Depending on the business, product, or project, these can include the costs of dismantling, disposal or treatment (in the case of hazardous material) and may include the costs of change to production facilities. These might also include employee costs such as retraining, moving, and/or severance. For a product assembly line decommissioning would include reconfiguring, repurposing, selling, or disposing of the manufacturing equipment. For a plant it would include the costs associated with shutting down operation, and disassembling the plant systems. For a software package or IT system decommissioning would include helping customers and clients transition to a new system or uninstalling the existing system.

Any parts or capital assets still of value that could be sold or repurposed would not contribute to final costs, and in many cases the reuse of the parts or sale of assets can offset the final costs.

3. Fixed and Variable Costs

A different way of categorizing costs is by identifying them as *fixed costs* and *variable costs*. The total cost will be the sum of these two.

$$Total\ Costs = Fixed\ Costs + Variable\ Costs \qquad \text{Equation 1}$$

Fixed costs are those costs that occur regardless of whether the business engages in engineering projects. Rent, basic telephone charges, accounting costs and equipment maintenance are typical fixed costs. Though these may change over time (e.g., rent may increase) they are predictable and virtually independent of the particular projects in which the company is engaged. These costs are also sometimes referred to as *indirect costs* or *overhead costs*.

These costs are often assigned to a project on a proportional basis. For example, if a project is taking 15% of the area of a building, it would be assigned 15% of the heating costs and the rental costs. If a project has 25% of the people employed, then it might be assigned 25% of the telephone system cost, and internet charges, and 25% of the front office clerical costs and human resource office costs. It is not unusual for indirect costs to be 30% to 70% of the cost of running an engineering business and can be even more.

Variable costs are those costs that depend on the particular projects in which the engineering company is engaged. It is easiest to identify these costs by imagining what costs would disappear if a project was stopped: if a design project was discontinued, if a contract was not pursued or undertaken, if the plant or facility stopped producing a product, if a product line or IT system was discontinued. The variable costs are dependent on quantity of work, unlike the fixed costs. One way to understand this concept is to consider the difference between the fixed cost of basic telephone service and the variable cost of long distance phone calls. If you are making sales all over the world, the long distance cost will go up. However, whether or not you are making sales, there will be a cost for basic telephone service. For manufacturing operations variable costs are often divided into cost of the manufacturing process and the cost of materials and supplies (i.e., *consumables*).

4. Internal and External Costs and Revenue

Costs can also be divided into internal and external costs. Similarly profit (or more generally benefits) can be internal or external.

Internal costs and revenue. *Internal costs* are the costs of running the business and *internal revenue* is the money the company earns from its work and investments. Supplies, equipment, labor, rent, and costs of borrowing are examples of internal costs. Internal revenue comes from clients or customers paying the company for goods and services. Typical corporate financial statements deal almost exclusively with these internal profits where profit equals revenue minus costs. The benefits a company acquires from clients and customers include non-monetary things such as reputation, customer loyalty, and brand recognition. However, corporate balance sheets generally deal only with the costs, revenues, and profits involving money.

External costs and revenues. The production, use, or disposal of a technology can lead to costs not paid by a company directly. For example, if a product indirectly leads to a lifestyle that decreases society's health and well-being, the costs of dealing with these resulting problems are handled by society—by governments or health care systems—and not by the company. There may also be costs to the environment such as pollution that is within government regulations (but still has some effect on the environment) or carbon dioxide emissions that are unregulated. Or there could be social costs, for example, improved automation reducing low paying jobs and increasing the number of jobs for highly skilled workers requiring more training. These kinds of costs are called *external costs*; there effect may be local, national, or international depending on the range of the impact.

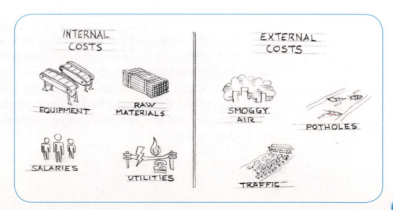

Historically, external costs were largely ignored. Take the internal combustion (IC) engine, for example. As the numbers of IC vehicles in urban centers increased in the 1900s, the amount of smog increased and the number of health-related problems because of this pollution rose dramatically. Further, there is considerable evidence that the carbon dioxide released into the air from vehicles contributes significantly to the buildup of greenhouse gases and that the buildup of these gases will result, and has resulted, in climate change. The cost of these environmental issues is external as the cost has not, historically, been borne directly by vehicle manufacturers or operators.

How do we measure this external cost? For the smog: is it strictly the medical costs of those with respiratory problems, and how can we be sure that the smog was the cause? Or is there penalty for "suffering," and is damage being caused to others also exposed to smog but not needing medical care (yet)? For the carbon dioxide: is it the cost of maintaining forests to absorb the carbon dioxide, or the cost to build scrubbers that will remove carbon dioxide from other contributors, and do we only concern ourselves with the human problems only in our own country, and assess the cost as what air conditioning would be needed for the elevated temperatures, or do we include the damage to other countries and to plant and animal life? Do we somehow absorb the costs, or some of the costs, of those countries that are not running so many vehicles? There are many differences in opinion on these subjects and no clear answers.

Calculations of some of these external costs have led to stricter government regulations that try to reduce the external costs, or taxes that try to recover the costs or penalize people for creating the costs. Taxing the sale of cigarettes is an example of this strategy. Regulations or taxes can have a negative impact on the economy, but they can also incentivize technological innovation; for example, work on electric vehicles and other propulsion methods that emit fewer pollutants, provide lower long-term external costs and a more sustainable future for the private vehicle.

Of course there are often positive **external benefits** of engineering work. A corporate decision to build a new manufacturing plant in an area will provide work for more than just the people that the plant employs, as these plant workers will require housing, food, education, and other products and services. The corporate taxes from the business will support local and national government work and projects, including education. The company products and services may enable a net social benefit. Many companies support social (e.g., scholarship programs, or local arts programs) and environmental programs. A company will benefit indirectly from these long term positive community and global relationships. Many engineering companies also produce goods and services that have a positive net impact on society. People's lives and well-being depend heavily on technology; modern health care, social programs, and educational systems all rely on technology to operate effectively. The external benefit, for example, of a life-saving technology can be far greater than the internal revenue (money) paid to the engineering company that designed it.

KEY TERMS

cash flow

variable costs

disposal (decommissioning) costs

external costs

external benefits

ongoing costs

operating costs

overhead (indirect costs)

internal costs

capital costs

consumables

fixed costs

internal revenue

5. Questions and activities

1. Consider yourself in your home or residence, and the economics relating to your being there. Identify expenses that are fixed and expenses that are variable. For the fixed expenses. If you share your residence with other people, determine what percentage of the fixed costs might be assigned to you. Explain how you calculated this.

2. Consider "going to school this year" as your project. Characterize the following expenses as fixed/variable, and initial/ongoing/final.

   a. Residence costs

   b. Meals every day

   c. Internet costs

   d. Renting a robe for graduation

   e. A laptop computer

   f. Travel to campus

   g. Buying Books

   h. Tuition

   i. Renting a van to move at the end of the school year

5

Payback

#resource module: #paybackcalculations

Learning outcomes

By the end of this module, you should demonstrate the ability to:

- Calculate the economic break-even points of alternatives
- Calculate the lifetime costs of alternative

Recommended reading

Before this module:

- **Resources > Economics > 4. Types of Costs and Revenues**

1. Comparing Alternatives Using Payback Period

The time from the start of a project to the break-even point is called the ***payback period***. This is because at the break-even point the total costs of the project to date will be matched by the total benefits to date and further use of the project design will result in net positive profit.

The payback period is one of the simplest economic metrics used to compare alternative designs that meet the functions and constraints of the requirements. Often one alternative will have a higher initial cost, but a shorter payback period.

1.1. Example

Consider the following example: you need to purchase a machine that will be used to wash parts in an industrial production process. There are two competing machines on the market that have the same capacity. The machine will be used to wash one batch of parts every day. You can use economic payback time to compare the two machine choices:

Choice A:

Initial cost: $10000

Electricity use: 6 kWh per load

Start Delay Feature: No

Choice B:

Initial cost: $11000

Electricity use: 6 kWh per load

Start Delay Feature: Yes

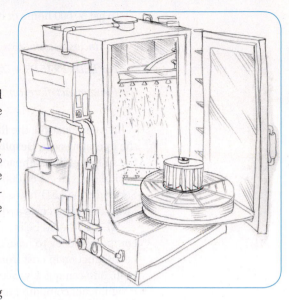

The benefit of the washing machine is considered the same in both cases, so this does not enter into the calculation.

The cost of electricity in the location of the plant is 15.7 cents per kWh (kilowatt-hour) during the day, with a 55% discount for the period 7:00 pm to 7:00 am. By using the delay function of Choice B the company could take advantage of the discounted electrical rate for by washing the daily batch of parts at night.

The savings per load would be

$$15.7 \times 0.55 \times 6 = 51.8 \, cents \, per \, loa$$

So the payback on the price difference (without considering the time value of money) would take

$$\frac{\$11000.00 - \$10000}{\$0.518} = 1931 \, loads$$

Or about 5 years at one load per day. This may be a sufficiently short payback time to justify selecting option B, especially if you expect electricity prices to go up or an increase in the discount.

2. Comparisons of Purchase Alternatives with Lifetime Considerations

When doing payback comparisons of alternatives, the product lifetimes also come into the calculations. A product that costs more to purchase, but that lasts longer might be a better alternative.

Imagine we have three alternatives for lighting an area: Incandescent bulbs, LED bulbs, and compact florescent bulbs. 60 watt Incandescent bulbs cost $1.63 each and last for 2000 hours. Equivalent output Light Emitting Diode (LED) bulbs are 9.5 W, cost $15.97, and last for 25,000 hours. Equivalent output compact florescent bulbs are 13 W, cost $7.97 and last for 10,000 hours. The price of electricity we will assume is $0.055 per kWh.

Again the positive benefit of lighting is satisfied by all of the alternatives so we can ignore the benefit received and look only at the costs. These costs can be divided into capital cost (or purchase cost) and operating cost (or the cost of operating the unit).

Courtesy of the Authors

To get 50,000 hours of operation we can then calculate the costs:

For incandescent: $25 \times 1.63 + 50000 \times 0.055 \times 0.060 = 40.75 + 165.00 = 205.75$

For LED: $2 \times 15.97 + 50000 \times 0.055 \times 0.0095 = 31.94 + 26.13 = 58.07$

For compact florescent: $5 \times 7.97 + 50000 \times .055 \times 0.013 = 39.85 + 35.75 = 75.60$

In each equation, the first part of the sum is the capital cost and the second part is the operating cost. Despite the large difference in the purchase price of a single unit, the LED and the compact florescent are better than the incandescent lamp both in capital costs to get 50,000 hours of operation and in cost of operation for that period. Similarly, the LED bulbs come in ahead of the compact florescent because of their lower operating costs and longer life.

We also should be including the costs of replacement of the bulbs—it will cost the company money to have someone replace bulbs, so if the bulbs last longer there will be a reduction in cost from this point of view. If it takes, say, ten minutes to replace a bulb, one can quickly calculate the employee-cost to replace the bulb, and may find it significant compared to the other costs involved. Again, the incandescent bulb and the compact florescent bulb will have more of these replacement costs for 50,000 hours of operation.

This example also does not include the costs to society. For the 50,000 hours of operation 25 incandescent bulbs will need to be disposed of, but only five compact florescent bulbs, and only two LED bulbs. Looking deeper, however, compact florescent bulbs contain mercury, and thus require special handling during disposal, which increases the cost of handling. Recent studies show that LED bulbs may have handling problems as well [1]. Whether this tips the balance in favor of incandescent bulbs is something that would require further effort to find the true costs of disposal. Increasingly, governments are attempting to impose disposal fees that make the disposal a realistic and real cost that must be included in the analysis.

Finally, the effect of inflation on the value of money has not been included in this example. Usually when costs are low, inflation is low and where costs can be expected to rise with inflation, this is ignored, at least for an initial rough calculation.

KEY TERMS

payback period

3. Questions and activities

1. If an employee costs you $30/hour, what is the revised cost of the incandescent bulbs in our example?
2. Redo the calculations for the incandescent, LED, and compact florescent bulbs if the cost of electricity is
   a. 10 cents per kilowatt-hour
   b. 2 cents per kilowatt-hour

3. Select two alternative methods for accomplishing the same task and calculate the cost differences over time. For example:

a. Driving to work versus taking public transport

b. Eating at the cafeteria versus packing your own lunch (remember to factor in the cost of your time.

These are simple examples but they relate to industrial decisions such as what form of shipping to use, or when to perform a task in-house versus when to use a contractor.

4. References

[1] Lim et al. "Potential Environmental Impacts of Light-Emitting Diodes (LEDs): Metallic Resources, Toxicity, and Hazardous Waste Classification." *Environ. Sci. Technol.* 45 (1):320–327. DOI: 10.1021/es101052q. Online publication date (Web): December 7, 2010.

Introduction to Failure and Risk

#resource module: #failure&risk

Learning outcomes

By the end of this module, you should demonstrate the ability to:

- Define the basic terminology in failure and risk, including hazard, consequence, and fault
- Explain how risk is related to failure and consequences
- Explain "good will" and decisions relating to it

Recommended reading

After this module:

- **Resources > Failure & Risk > 2. Handling Risk**

1. Introduction

This module will provide the basics for understanding and managing failure and risk. It is an introduction to the subject that will allow you to apply basic concepts in an introductory engineering design project. A complete understanding of this subject is highly technical and requires specific industry knowledge, which is beyond the scope of this text.

This area of study is known formally as reliability. **Reliability** is the expected lifetime of usefulness, and is a complicated field based on statistics. The study of reliability underpins forensic work on downed airplanes and collapsed bridges, on determinations of road and pipeline lifetimes, and on reducing the chance that a space mission or a smart phone will fail due to a component failure. Your future work may well require specialization in this area, but here we will take a more general approach to the field.

Ultimately we expect all technology to fail. If you buy a new laptop computer, you do not expect it to fail within the first month, but it may. The probability of that is quite low, but not zero. Over the first year, you might expect the probability of failure to remain low. A couple of years after the purchase, you know that there is higher likelihood of something going wrong. Perhaps it is only a battery failure, or a hinge breaking. Perhaps your friend accidently stepped on it when you left it on the floor. Some failures are just annoying, others costly. However, in the first 2 years, it is likely the laptop is still working well if it has not been stepped on.

Most people will replace their laptop within a few years, whether it works or not, to get a newer, lighter, faster model. For people designing and building computer components, it does not make sense to create expensive components that last centuries, because few consumers would be interested in such a product.

So reliability is governed by probability and statistics. Failures can be due to accidents, purposeful acts of sabotage, and/or **wear and tear**. Wear and tear means expected deterioration due to routine use. There are a wide variety of different kinds of failure modes, and the cost will also vary widely. If your computer fails it may not be very costly, compared to the failure of a computer controlling a banking system, for example. And there are often trade-offs between reliability and other objectives such as cost, weight, and speed.

> ### Vocabulary of Failure
>
> A situation where there is a possibility of damage to health, property, or the environment is known as a **hazard**. The result of exposure to a hazard is the **consequence**. Unintentional exposure to the hazard is an **accident**. These terms are used extensively in design for safety.
>
> Formally, a problem is known as a **fault** (such as a weak spot in a component), which may or may not result in a **failure**. A failure means the design no longer fulfills the requirements of the design.

2. Risk and Risk Management

The combination of the probability of a failure occurring and the cost of the failure if it occurs is known as **risk**. Risk of technology failing virtually always exists. Generally engineers try to design technology to reduce the risk of failure to an acceptable level. There are many factors that influence individual people and society's decisions about what constitutes an acceptable risk. In particular, perception of risk and tolerance for risk will change based on circumstances [1].

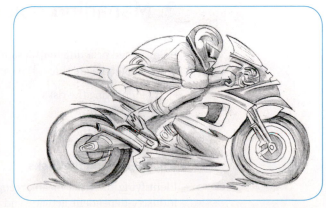

- A risk taken voluntarily might be acceptable, but the same risk is not acceptable if it is involuntary. For example, in some places you can choose to ride a motorcycle without a helmet. This is a voluntary risk. However, it would be unacceptable for a company or the government to require you to ride without a helmet.
- A risk might be perceived to be acceptable if only one person at a time is affected, but

Samuel Croome

unacceptable if many people are affected simultaneously. In the US 33,808 traffic fatalities were recorded in 2009, the equivalent of a large plane crashing every two days [2]. However, air travel would be perceived as unacceptably risky if a large plane crashed every other day.

- A risk is more acceptable if the person is involved in a common activity. For example, people rarely worry about the risks involved in walking down a sidewalk because this is a common part of everyday life in society. They worry more about the risks associated with skydiving because this is an uncommon activity.

- A risk is more acceptable if the person taking the risk is well informed. Medication packages generally include information about the risks associated with taking the medication. Society considers it acceptable for a person to take a medication that has risks associated with it, as long as the person is well informed and chooses to take the medication voluntarily (i.e., ***informed consent***).

There is often a trade-off between risk and reward. Engineers cannot design a car that does the job of getting people from A to B without also having designed a machine that can injure or kill people. Society tolerates this risk because of the valuable function cars serve. Society would not tolerate this risk if the reward were less important. Automobile fatalities are an unfortunate part of life; amusement park fatalities are not and are not accepted.

It is also the case that tolerance for risk is decreasing over time. In the 1950s and 1960s, seatbelts in cars were optional, and children were allowed to freely move around in cars. Today of course, both seatbelts and airbags are mandatory in passenger cars in most places, and special child car seats are required. This reduction in risk tolerance is happening everywhere, including the workplace, where modern standards are continuously tightened to protect workers. Improved safety is enabled by better technology, and the interest in improved safety has motivated innovations in technology.

3. Managing risk

Many design decisions are made depending on the risk calculations. The design might be modified or a different design chosen to reduce risk to an acceptable level. At a basic level there are three steps to managing risk:

1. Identifying the risks
2. Assessing the risks
3. Managing the risks

3.1. Identifying the Risks

Identifying the risks means recognizing what problems could occur. This is a very industry-dependent activity, and having industry knowledge is crucial to doing this well. Mainly this involves examining the tasks, parts, or interactions that occur when a technology is operating or when a project is in progress. For each aspect of the system ask, "What can happen here that would cause this aspect of the system to be delayed, to cost more than expected, or to cause a system failure?" For project management, this is most important for those items on the critical path of the project plan, or critical to the design or involving high costs, where the problems will have high significance. For technology it is most important when the technology is operating under extreme conditions or when failure has serious consequences.

3.2. Assessing the Risks

Assessing the risks requires estimates of the probability of occurrence of a problem and the *consequences*. Consequences can include time lost, cost, and health and safety. All resulting costs, time losses and damages should be included in this assessment. For example, in a building the failure of one support beam could cause overloading and failure of other beams, and consequently the failure of an entire structure. The risk is not simply the loss of one beam in this case, but the damages caused by the collapse of the whole structure.

3.3. Managing the Risks

Managing the risks involves taking action to address those risks that are found to be significant. The cost of an event multiplied by the probability of the event, the formal definition of risk, is used to determine how the response will be managed. So an event with a high cost but a very low probability might be ignored, as might an event with a reasonable possibility of happening but very low associated cost. Note that damage to *goodwill* must also be considered as being part of the cost.

> *Goodwill* describes the attitude of others toward the company. It is your reputation. Delivering projects on budget, on time, and in a manner that builds goodwill with the client, with the suppliers, and with the community adds to your reputation. People value technology that works well and enables them to fulfill their goals and aspirations. Creating products or systems over budget, that are faulty, or disruptive to people's lives will hurt your reputation.

4. Conclusion

Failure and risk are two important considerations of a design. An engineering designer needs to be particularly aware of failures affecting safety, but the thought and analysis is also necessary for other longevity considerations.

Risk is based on probability and statistics, since there is no method of determining exactly how and when a failure will occur. Instead, the engineer must rely on data from past events (statistics), derived from similar designs and from testing of materials and products. The other component of risk is the cost of the event, should it occur. The failures can be product, process, or plan failures. Once the risks are identified they can be managed through design decisions, process control, and other methods. The goal in any project is to reduce the risks to an acceptable level that safeguards human health and welfare and keeps the project within time and cost objectives.

KEY TERMS

reliability	wear and tear	hazard
consequence	fault	failure
accident	risk	informed consent
goodwill		

5. Questions and activities

1. Consider a product that is in operation. The product could fail in multiple ways.

    a. Is it possible for the risk to be near zero? Under what circumstances?

    b. Where do you see similar situations possible in every-day products?

2. Consider the risk posed by routine usage of a paper clip versus a lug nut that holds a truck wheel on. Both have a similar functional basis: to hold together objects. Discuss the differences in risk using the terminology discussed in this module: hazard, failure, consequences, accident, fault, and so on.

3. Make a list of possible failures associated with a computer keyboard. Which of these are precipitated by accidents and which occur through normal wear and tear?

4. What are the costs involved if you skip one of your classes? How can this risk be reduced?

5. The "low fuel" warning light on an automobile will indicate when fuel reserves run low.

    a. How costly is the failure if the light burns out?

    b. How is this linked to a more costly failure that might otherwise have been averted?

6. References

[1] Dieter, G.E. *Engineering Design,* 3rd ed., McGraw-Hill, 2000.
[2] U.S. DOT HS 811 363, Highlights of 2009 Motor Vehicle Crashes, August 2010.

Handling Risk

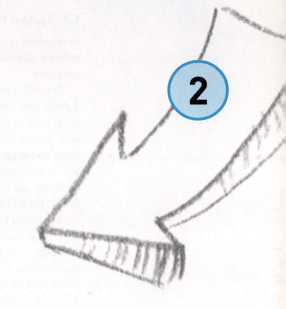

#resource module: #handlingrisk

Learning outcomes

By the end of this module, you should demonstrate the ability to:

- Use risk in design decisions
- Manage risk in design executions

Recommended reading

Before this module:

- **Resources > Failure & Risk > 1. Introduction to Failure and Risk**

After this module:

- **Resources > Failure & Risk > 3. Why Things Fail**

1. Assessing Risk

Identifying the possible failure mechanisms is the first step in incorporating risk-based methods into design. These methods will ultimately guide the choices of the design used, and "design for" such as design for durability, design for testing and maintenance, design for safety, and so on.

Risk is the product of the severity of a **failure** (often quantified as a cost) and the probability of the failure. The total risk is the sum of all the individual risks of all the potential problems which are called **faults**:

$$Total\ Risk = \sum_{All\ Problems} (Cost) \times (Probability) \qquad \text{Equation 1}$$

In many industries models and simulations exist to help evaluate the expectations of faults. Engineers also use prototypes and run "accelerated lifetime" tests to find failure modes and probabilities.

1.1. System Failures and the Fault Tree

In complex systems there are often interactions of circumstances and faults to cause failures. Methods have been developed to organize and understand this complicated structure.

One of the ways to look at systems in terms of potential failure analysis is a fault tree [1]. A ***fault tree analysis (FTA)*** links various events or problems to consequences, and is often used in forensic analysis of accidents. For example, "antenna hit by space debris" and "power supply connection failure" which are both faults could both lead to the same consequence: "unable to communicate with satellite". A fault tree will link these, and more complex combinations of faults and circumstance, to various failures. The fault tree can be used in lifetime analysis for durability, maintenance, safety, and other design for X (DFX) considerations.

The trivial example in Figure 1 shows how the interaction of a pet cat and a computer keyboard might result in erroneous capitalization of input at a later time. The rectangles are consequences and the rounded rectangles show "and" (simultaneous occurrences of items that must happen to give a certain consequence). This is a type of Boolean operator also known as Boolean logic. Other logical combinations, such as "or" or "not" can also be used to link elements.

The diagram shows, for instance, that the cat walking on the keyboard is only important when the computer is on, and then will result in the Caps Lock being on. This is not an issue, unless keys are pressed, in which case we have the failure of INPUT THAT IS CAPITALIZED. (shoo, cat!) Clearly this problem has low risk. Although the probability of occurrence may be high, the cost is very low since the problem is easily reversible at little cost. The example shows how a visual fault tree can represent complex interactions.

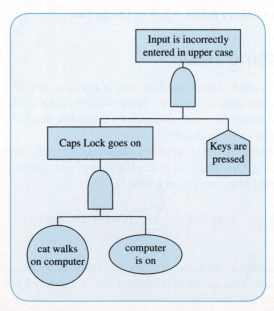

FIGURE 1 **An example of a fault tree diagram.** In this example the symbol ⊃ means "and," which is a Boolean operator.

Other types of analyses in this field include:

- **HAZOP analysis:** A *hazard and operability analysis* is a qualitative method used in the chemical process industries to predict the outcome of deviations from intended operation parameters.
- **FMEA:** A *failure modes and effects analysis* is a formal method of tracking the higher level consequences of low level failures. The effect of the failure of each individual component is tracked upwards to determine the global result.

2. Handling Risk

We have looked at the first two aspects of managing design failure and risk: determining where failures could occur and determining the risk involved (the probability and the cost of the failure involved). The last step is the set of actions taken to **mitigate** (i.e., reduce) or eliminate the risk.

There are a number of ways this can be done:

- A specific risk can sometimes be *eliminated*. For example, in a product design, if there is a part that may fail in a design that makes the failure risk unacceptable, you may wish to choose an alternate design that does not have the part. Note that this eliminates one specific risk, but will not eliminate the entire risk for the technology.
- The risk can sometimes be *reduced*. In a project execution, there may be risk that parts will be delivered late and delay the completion of the project. By ordering early or possibly by finding alternate sources this risk can be reduced.
- The risk can often be *passed along*. Often companies will reduce the costs of a subcontractor being late through penalty clauses in the agreement with the subcontractor. This will share or off-load the cost of failure.
- The risk can be *shared*. A common method for sharing risk is **insurance**. For example, you might purchase car insurance that will help reduce the cost of repairing the car if you are involved in an accident. Insurance works because all of the policy holders pay for the shared risk they have collectively. This reduces the **liability** for any one of the individual policy holders.
- If the potential failure is that the project will run over cost or time, then increasing the estimates for cost and time mitigates the risk.

2.1. Factoring Risk into Cost for a Project

Suppose your company is considering bidding on a project. There would be a "most likely" set of circumstances, and an associated cost (say $MM). There would also be an optimistic estimate if everything went the best way possible which will cost less; say $OO. Similarly, there may be events that have low probability of occurring, but a substantial risk because they would be costly. If these events occur then the worst case scenario will cost more, say $PP.

The company could bid using the most pessimistic cost, $PP. But in this case they would likely lose the contract because others would bid less. The company could bid using the expected cost, $MM, but this is risky because the project could cost more if some of the low probably events actually occur. Clearly bidding using $OO might work, but likely would not, as the whole project would have to go perfectly for the company to profit.

Although not foolproof, one simple formula used in industry is:

$$(\$PP + \$OO + 4 \times \$MM)/6 \qquad \text{Equation 2}$$

This weights the expected cost heavily, but also factors in the optimistic case and the pessimistic case. If the pessimistic case is very, very much more expensive then the formula will suggest a higher bid. However, this formula is very crude and depends substantially on the quality of the cost estimates for the three cases. In dealing with risk there is no perfect formula and this should always be kept in mind.

2.1. Reducing Risk

Reduction or mitigation of risk is done by altering the design or project plan in engineering projects. There are a number of common strategies that are employed for accomplishing this.

1. The engineer can reduce the probability of the failure mechanisms by using better parts or construction techniques. This will reduce the probability of failure and thus reduce risk.

2. The engineer can introduce detection and shutdown mechanisms into the design that prevent the failure mechanism from propagating. For example, there is a detector that monitors the landing gear on an airplane. It tells the pilot when the landing gear is fully down or fully up. This information lets the pilot know that the gear is ready for landing, thus possibly preventing a catastrophe. If the landing gear is malfunctioning, the pilot has the opportunity to try to fix the problem before a larger failure (e.g., a crash landing) occurs.

3. The engineer can design for testing and maintenance to reduce risk by detecting failure potential and dealing with it before failure occurs. This would include the testing and maintenance of failure detection and shutdown mechanisms introduced into the design.

4. The engineer can change the architecture to remove the failure mechanism or cause of the event. For instance, for the landing gear example, the aircraft can be designed such that the landing gear is always down. This has other ramifications, of course.

5. The engineer can introduce redundancy into the design. This increases the number of events that must occur in sequence for the failure to happen. For example, most commercial aircraft have three instruments for each of the key information points. If one instrument fails the pilot will still have two others that provide correct information.

6. The engineer can use appropriate *factors of safety* in the design and other industry standards to guide the design process. *Regulations, codes,* and *standards* are often developed in response to previous failures. Adhering to up-to-date standards of practice and current regulations in the design process will mitigate risk.

3. Example of Working with Risk

Say you need to design a circuit involved in sequencing explosive charges that will bring down an old building. Part of the design involves a firing circuit. The electrical energy from the firing circuit is used to start the explosive chemical reaction. How do you design that circuit? If there is a 1% chance that the circuit will fail and detonate the explosive prematurely, is that acceptable? Is 0.1% acceptable? 0.01%? Is 0% the only acceptable figure, and can that be achieved?

Figure 2 shows the diagram of a simple firing circuit. In actual industry practice the way this circuit is designed and component used would be specified by a *code* and the firing of an explosive charge will be conducted following a strict *protocol*. As an engineer you must follow industry codes and protocols. They embody past experience and guide you to safe practice. However, to illustrate the points in this module we will assume here that you are free to design the circuit without the benefit of guidance from a code.

We can analyze the failure rates of the components, predict the outcomes, combine all these probabilities of failure causing unwanted detonation and come to some conclusion as to how probable it is that the circuit will fail prematurely. In Figure 2 there is a switch (in a fancy plunger box) which closes to complete the electrical circuit and blow up the dynamite. This analysis can be simplified by assuming that failure of the switch is the only way the dynamite could explode prematurely. In reality there are a number of other possible faults, including build-up of static electricity, which could cause failure. However, for purposes of illustration we will focus on the possible failure of the switch; specifically the possibility that the switch is stuck in the closed position such that it detonates the explosive as soon as the explosive is connected to the circuit.

The risk is the cost of the failure times the probability of the failure. What is the cost if the dynamite blows up prematurely? There could be deaths, probably lawsuits, property damage, loss of your company's reputation, and possibly the end of the company.

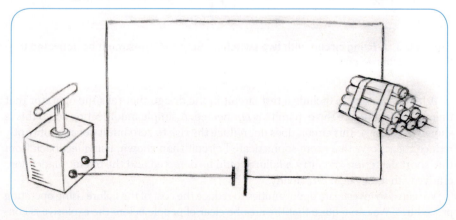

FIGURE 2 Example diagram of a simple firing circuit for an explosive charge.

The cost could be measured in the millions, some of which would be covered by insurance, some not. And, of course, there is no amount that will remedy a loss of life. If the final cost of the failure is $500,000,000 (considering only the cost of damages) and the probability of failure is 0.1%, then the risk is 0.1% of $500,000,000 or $500,000. This would have to be added to the cost of the circuit to the customer, meaning the customer would be paying over $500,000 for a switch and a few wires.

This is likely unacceptable. Assuming the cost of failure will not change, we need to drive down the probability of failure to make the circuit less risky. One strategy is to duplicate the circuit. Using our figures from before, the circuit of Figure 3 has a probability of failure of 0.1%* 0.1% or 0.0001%, so our risk would come down to $50,000. This is because both switches would have to fail to get premature detonation.

We could also use chemical dynamite fuses instead of electrical detonation and completely eliminate the risk caused by a faulty switch. Using fuses would, however, create other risks and may not be appropriate for the problem of bringing down an old building.

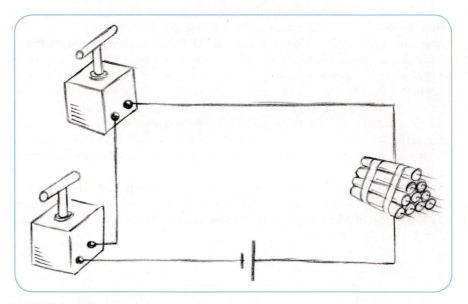

FIGURE 3 **A firing circuit with two switches.** Both switches must be activated to fire the explosive charge.

A better route is to include a test circuit in the design that tells the operator that there is a short before the dynamite is connected. A simple modification to do this is shown in Figure 4. This circuit does not reduce the risk to zero but it does significantly reduce the risk. With a more sophisticated circuit than shown, using feedback from the "short detection" circuitry, a failure could be detected and the circuit made inoperative with a very high probability.

We can also investigate opportunities to reduce the cost of the failure using operating procedures. We can require that the area be clear of people before the explosive is connected. The risk is less because the cost of failure is less. We can also purchase insurance to cover any damages that occur. However, insurance will not cover criminal negligence (e.g., if the engineer did not adhere to the design code), and it will not remedy a loss of life.

FIGURE 4 Firing circuit with a detector that warns of an impending explosion.

There is no method of reducing risk to zero, but using safety procedures helps to reduce the risk. This simple example shows how we can deal with risk through design, both of the circuit and the operating procedure. It should be clear that no method is going to be risk-free. It should also be a reminder that, as an engineer, you must prioritize the health and welfare of the public. A protocol that requires the area to be clear before the explosive is attached to the circuit is an obvious step toward safeguarding life.

4. Conclusion

Failure and risk are two important considerations in design. An engineering designer needs to be particularly aware of failures affecting safety (design for safety), but the thought and analysis is also necessary for other longevity considerations (design for durability; design for testing and maintenance).

Risk analysis is based on probability and statistics, since there is no method of determining exactly how and when a failure will occur. Instead, the engineer must rely on past data, derived from similar designs and from testing of materials and products. Simulations, modeling and prototyping are also important tools in the evaluation of risk and probability of failure. The failures can be product, process or plan failures. Once the risks are identified, methods can be used to reduce and sometimes eliminate the higher risk failures in a project.

KEY TERMS

risk	failure	faults
fault tree analysis (FTA)	hazard and operability analysis (HAZOP analysis)	failure modes and effects analysis (FMEA)
mitigate	insurance	liability
factor of safety	regulations	codes
standards	protocol	

5. Questions and activities

1. Look at the circuits in Figures 3 and 4.

   a. How could multiple failures cause the risk to be non-zero, even with the assumptions made?

   b. Where do you see similar situations possible in every-day products?

2. What failures are possible with a computer keyboard? What methods could you use to reduce the probability of these failures?

3. Choose another simple household device. Identify the possible failure modes associated with this device. What methods could be used to reduce the probability failure and the risk associated with these failure modes?

6. References

[1] U.S Nuclear Regulatory Industry, *Fault Tree Handbook*, Washington: U.S. Government Printing Office, 1981. Available online at www.nrc.gov/reading-rm/doccollections/nuregs/staff/sr0492/sr0492.pdf

Why Things Fail

3

#resource module: #whythingsfail

Learning outcomes

By the end of this module, you should demonstrate the ability to:

- List and explain common modes of failure
- Analyze a design situation for failure modes
- Analyze existing designs for failure modes
- Explain how failure provides the motivation for design for safety, design for testing and maintenance, design for durability, and design for human factors

Recommended reading

Before this module:

- **Resources > Failure & Risk > 2. Handling Risk**

1. Introduction

When they envision technology failing, people often think of bridges collapsing, aircraft crashing, or a laptop no longer booting. Technology failing is often perceived as a material failure (i.e., an object breaking). Design engineers must look deeper than this, to the reasons behind these material failures and to consider other types of *failure*.

A failure is a situation where a design no longer entirely meets the needs of the user and other stakeholders. This could be because of a material failure, but could also be because of needs beyond the original scope of the design, or a design flaw, or a misuse of the design.

In this module we will look at the ways engineers measure the probability-based background to failure. Then we will look at common reasons for technology failure.

2. Technology Lifespan: The Bathtub Curve

The chance of failure during the lifespan of a technology often follows a ***bathtub curve*** (see Figure 1). It is called this because it roughly resembles the cross-section of a bathtub. Just after a technology enters operation (time equals zero) there is a relatively high probability of failure. This is known as ***infant mortality*** in industry. It is caused by assembly and manufacturing process problems, design flaws, bugs, and other failures that result from the design or fabrication processes. Manufacturers of consumer products will often ***burn in*** their products to get them through this period and sort out the units that fail. Early failures can also be caused by the shipping (transport) and mistakes in installation or implementation of the technology.

As time progresses the probability of failure decreases and remains fairly low for a period of time (#designfordurability). The expected time before a problem occurs is called the ***mean time to failure (MTTF)***. As the product ages, it will again become more probable that failure will occur as parts wear and show the effects of interaction with the environment, normally called "aging." A bathtub curve analysis like this one does not include accidental damage or failure due to abuse or purposeful sabotage. It is meant to map the lifespan of a technology under normal use conditions.

Designing for reliability means making design decisions that extend that central area of the curve where the risk of failure is low. This can be done through risk analysis and risk reduction.

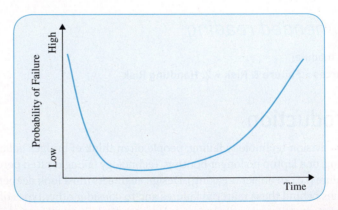

FIGURE 1 **An example of a bathtub curve describing typical technology lifespan.**

3. Classes of Failure: Anticipated Failure and Accident

A sewer line is designed to last decades. A backpack is designed to last years. This does not mean that these technologies are designed to fail, but that the cost versus lifetime trade-off has been made in different ways by the people who are designing these technologies. If a sewer line, designed to last at least 50 years, fails at 70 years, it is not unexpected. These failures are anticipated by the designers and the managers of the sewer line (typically, a city) can take steps to guard against significant damage to users of the line and to people and vehicles that use a road built over the sewer line.

When a there is unintentional exposure to a hazard, engineers call this an **accident**. Accidents are not caused by designs operating in the expected or customary way. They normally occur when an unusual and unanticipated combination of circumstances occurs. In 2010, Toyota recalled many vehicles after a fatal collision in which it was first suspected that the floor mat had interfered with the accelerator pedal [1]. This is a very simple example of an unanticipated set of circumstances leading to a serious problem.

The goal of good design is to anticipate, to the extent possible, these types of problems, which are called **faults**, and design them out or mitigate the risk they pose. This starts with an inventory of potential faults in a design. The next step is to consider what combination of circumstances would lead to an accident. Anticipating accidents involves, fundamentally, a process of repeatedly asking "What if..."

4. Failure Cause: Human Error

What if the operator does not do what he or she is supposed to do?

Human error is a common cause of accidents. The operator of a power boiler is supposed to turn a switch to the left in the event of a pressure spike, but instead turns it to the right. A worker is not supposed to go up a ladder without a co-worker to stabilize the base, but takes a shortcut and does so anyway. A consumer is not supposed to stick his hand in the food processor to dislodge a stuck potato, but is in a hurry to make dinner and reaches in anyway.

Courtesy of the Authors

FIGURE 2 Hopefully no one would try to retrieve this coffee cup from the inlet of the wood chipper! It is difficult to guard a wood chipper to prevent humans from being killed, since large tree branches must be fed into the blades. This unit is equipped with two "last chance" safety straps in the chute. If these are pulled, the chipper feed reverses immediately.

Why Things Fail

Sometime people do things they are not supposed to, even when they know it is wrong. There is time pressure, or the effort required to do the wrong thing is much less than doing it right. Engineers try to prevent this type of human error through design for safety (#designforsafety). However, sometimes the design itself affords the human error. For example, the engineer has put the switch to extend the landing gear of an aircraft right next to the switch to adjust the wing flaps and they are close together and the same shape [2]. This type of design enables error, and even encourages it. Engineers try to prevent this type of human error through design for human factors (#designforhumanfactors).

5. Failure Cause: Material Failure

What if a critical cable, beam, or bearing suddenly fails?

Materials failure, by corrosion or fatigue, is often the cause of an accident. When engineers design a load-bearing structure, such as a building or a walker, the primary calculations are based on short-term material properties. The design is stiff enough, strong enough, and tough enough when the material used has sufficient strength, modulus and fracture toughness to handle the loads imposed. In this case, the very first time the design is loaded, it will work as expected. Safe operation of the technology depends on correct calculations and design.

Four long-term material failure modes are at the root of many accidents in older components: *fatigue*, *corrosion*, *creep*, and *wear*. Of these, fatigue and corrosion are the two most common causes of unexpected failures.

- *Fatigue* refers to the initiation and growth of cracks under the influence of a load that does not cause immediate failure. In 1988, a Boeing 737-200 that had been flown successfully by Aloha Airlines 89,680 times, failed on its 89,681st flight. The failure was attributed to a fatigue crack that grew gradually as the plane was pressurized and depressurized with each flight, until it reached a critical length [3]. Only one flight attendant was killed when a large section of the fuselage skin separated at altitude, leaving five rows of passengers flying in the equivalent of a convertible car with the top down.

- *Corrosion* is the electrochemical process that leads to material loss and transformation in metals. A bolt rusts and gets stuck or breaks suddenly, leading to an accident. On June 23, 2012, the roof collapsed on the Algo Centre Mall in Elliot Lake Ontario. An engineering forensic report concluded that the cause was decades of corrosion of the support beams due to road salt and water infiltration [4]. The report identified one particular weld that had been so badly corroded it could hold only 13% of its original capacity as the precipitating failure that brought down the roof.

- *Creep* refers to the slow deformation of a material over time. It is the failure mode for the filaments in incandescent light bulbs. A related process, stress relaxation, refers to the gradual reduction in stress in a component subjected to a fixed strain. The sag in a bookcase shelf over time is an example of creep that might eventually lead to failure. The permanent compression in a polymer gasket between two pipe flanges is an example of stress relaxation.

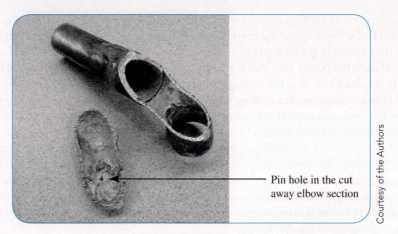

Pin hole in the cut
away elbow section

Courtesy of the Authors

FIGURE 3 Cutaway of a copper pipe elbow that has been corroded and eroded, causing a pinhole to develop in the pipe.

- *Wear* is the loss of material where two solid surfaces slide over each other. Automobile brake pads "wear out" and thus eventually fail, although this failure is gradual and anticipated. The related phenomenon of *erosion* refers to the loss of solid material because of a flowing fluid. Erosion created the Grand Canyon, but can also result in pipe failure over long periods of time.

There are other failure modes associated with material failure, such as embrittlement and UV degradation of polymers. Engineers need to be familiar with both the short-term and long-term properties of the materials they use in their designs to prevent or plan for material failures. In critical components, where failure has severe consequences (e.g., airplanes, nuclear reactors), great effort is taken to design against long-term failure. In these cases, the materials are chosen not only for strength and stiffness, but for corrosion resistance, fatigue resistance, creep resistance, and wear resistance, as the situation demands. Also, the design would include a rigid maintenance and inspection schedule.

In less critical components, a routine replacement schedule may be suggested. Or in some cases, such as brake pads, the component is designed for easy, inexpensive replacement and is designed to provide a warning to the user before it fails.

6. Failure Cause: Circumstances Outside Design Expectations

Natural Disasters and Other Extremely Unlikely Events

We can also have failure due not to age, but due to infrequent environmental or use problems that stress the design beyond the design limits. At the time that this section was first being written, engineers were struggling to regain control of the Fukushima Dai-ichi nuclear power plant in Japan. Because the consequences of a nuclear plant failure are so serious, there is extensive design for safety, including a careful consideration of earthquakes. So when a 9.1 magnitude earthquake struck on March 11, 2011, the three operating reactors at Fukushima I responded properly and shut down by inserting the control rods fully into the reactor core. However, the plant was cut off from the power grid, and the huge tsunami that struck about 15 minutes later led to the failure of all of the backup generators. Once the battery power ran out, the ability to circulate needed cooling water was compromised, and there were a variety of leaks, fires, and explosions. An evacuation zone around the plant was set up to reduce exposure to radiation.

This was one of the strongest earthquakes ever recorded, so it is not surprising that it caused extensive damage. However, some of the backup generators and other critical components were in a vulnerable position when the tsunami struck, and failed in the resulting flood. Extreme events are generally ignored in design unless the consequence of failure is also extreme. In routine design where the consequences are not extreme engineers usually design:

- To meet all regulations and standards, which often include possible natural disasters.
- To the 95% or 99% events: the conditions that occur 95% or 99% of the time rather than trying to create a system that would work under 100% of all possible conditions.

The probability of the natural disaster that an engineer designs for is proportional to the consequences.

7. Failure Cause: Design Failures

Designs that do not perform as intended

Even designs that meet regulations, pass verifications that show they meet the requirements, and have been carefully checked and carefully tested may still have design flaws, particularly when subjected to loadings and event sequences that were not anticipated by the designers. Perhaps one of the most famous failures of this type was the Tacoma Narrows Bridge, which opened for traffic in July 1940 and collapsed four months later. The bridge experienced an unexpected and extreme resonant vibration due to wind loading in the area where the bridge was constructed.

There is no formula or process that guarantees a perfect design. There are good engineering practices that help a design team design for safety, durability, and human factors. But perhaps the most important practice is thinking carefully about "what ifs." The design team needs to constantly be stepping back from the design and considering how it might fail.

KEY TERMS

failure	bathtub curve	infant mortality
burn in	mean time to failure (MTTF)	accident
faults	human error	fatigue
corrosion	creep	wear
erosion		

8. Questions and activities

1. Recall the worst accident that you have been involved in, and describe the chain of events that led to the accident.

    a. Analyze the accident and chain of events. Could it have been anticipated? What measures could have been put in place to avoid the accident?

    b. Repeat this question for a major engineering disaster such as the hydrogen cyanide release in Bhopal, India.

2. Try bending a paper clip back and forth until it fails. What type of failure is this?

3. List and explain common examples of material failures. Try to find an example for each of the four common failure modes. If possible find examples of these modes at work around your home or school and include a photo. (e.g., a rusted bolt, a sagging bookcase, etc.).

4. Find an example of a technology that might fail under each of the following extreme conditions.

    • For each case, describe a technology and the circumstances under which it would experience this type condition (e.g., motor oil in the arctic will experience over-chill in the winter).

    • Describe how the technology might fail: Would it break? Would it bend? Would it stop working (freeze)? For example, when standard motor oil becomes very cold it becomes extremely viscous. It doesn't freeze, but it also won't flow.

    • What would be the consequences associated with the failure you describe?

    a. overheat, overchill, overstress

    b. different input than expected or designed for

    c. use for longer than intended

    d. lack of or bad maintenance

    e. alpha hits* and magnetic or electrical noise disruptions in electrical circuits

    f. accidental and deliberate damage

    g. unanticipated inputs, unanticipated circumstances (interaction between tasks, timing of events), untested error/problem handlers

    h. incorrect installation

9. References

[1] Retrieved December 16, 2012, from www.toyota.com/recall/videos/floor-mat-entrapment.html.

[2] Vicente, K. *The Human Factor*, New York, Routledge, Taylor & Francis Group, 2006.

[3] National Transportation Safety Board Report: AAR-89/03, June 14, 1989.

[4] NORR Report, Forensic investigation of the collapse of the Algo Centre Mall in Elliot Lake. NORR Ltd. Retrieved August 17, 2013, from www.elliotlakeinquiry.ca/exhibits/exhibit_lists/ELI_Exhibit-List_March-12-2013_NORR.pdf

—————————

*An alpha hit is the collision of a cosmic sub-atomic particle with the design.

6

Case Studies

1

Aerial Photography

#resource module: #aerialphotography

Summary

This case study was written by a third-year engineering student in 2011 (reproduced with permission of M. Chang). It is a first draft; the requirements are still incomplete and the research done on the problem is limited. It should give you an idea of the type of work required to develop the definition of a design problem.

The client, Tori Noir, is the owner of a small photography studio that specializes in providing wedding and event photography in the Chicago, Illinois area. He has approached the design team requesting a way to take aerial photographs of the guests at the events he shoots.

The client statement is short, but does a good job giving a context to the problem and explaining how the problem has arisen. There are various problems in the client statement, however, including ambiguous and irrelevant information, solution-driven statements, and erroneous information. The work developed here illustrates how to dissect a design brief and create a solution-independent set of requirements.

Design Brief

The following is the client statement from Tori Noir, a Chicago photographer:

I am the owner and head photographer of Noir Photography, a small photography studio serving Chicago and the Chicago area. We specialize in wedding and event photography, and also work on portraits and fashion photography.

A client has recently asked me to take a photograph of all 400 of the guests at her wedding and I have realized that I have no way to do this. I cannot take a group photograph of all of those people without taking it from high above the ground. I would like to provide my client with the option of aerial photography and so I am requesting that you design for me something that will allow me to take photos of parties from above. In the future I plan on making this an option for all of my clients. I think this will be most popular at weddings and other large parties, though we also take photos of smaller events.

Like I mentioned, we cover parties and weddings throughout the year, both indoor and outdoor. Usually there are 100 or more guests at these events. When I shoot an event I will normally have my photography assistant with me to help with the shots. The camera I use is a Canon EOS-5D with an EF 24-70mm f/2.8 L USM lens and so the device will have to allow me to lift it up high enough to take pictures of the guests; this would probably be about 10–15 ft. However, for my convenience it needs to be collapsible so it can fit in my car.

In addition to fitting all of the guests in, I must also maintain a certain quality of photo because this is, after all, my profession. I need to be able to adjust my camera to get the best shot possible. An extremely tall tripod could do it, but it is essential that I have a way to control the camera.

In terms of cost, we're a fairly small business and so I don't want to spend a fortune on this device unless I know for sure it is worth it and will last long enough so that the cost is justified.

I hope you will be able to help me with this project. You may contact me if you need anything else.

Best,
Tori Noir

Problem Statement

The client, Tori Noir, is the owner of a small photography studio that specializes in providing wedding and event photography Chicago and the surrounding area. He has approached the design team requesting a way to take aerial photographs of the guests at the events he shoots. In addition to providing this main function the solution should also be easy to use, low-cost, and able to accommodate different types of cameras other than the DSLR he uses primarily. In addition to these objectives there are some constraints that must be met. The solution is required to be operated with the aid of at most 2 people, to take photos of areas of $50\,m^2$ to $500\,m^2$, and to accommodate the Canon EOS-5D and a wide variety of lenses. This solution will be used at both indoor and outdoor events; as such the physical and virtual characteristics of the service environment will be taken into consideration, along with the characteristics of any people or animate beings to come in contact with it. The users and operators of the solution are the client and any other photographers employed at Noir Photography. Their concerns, along with other stakeholders of the project, will play a large role in design decisions. In addition, human, safety, and environmental factors will be considered throughout the design process.

Requirements

Functions

Primary Function: The fundamental functionality of this design is to support mass; as such, its primary function is to operate a camera remotely at a specified elevation above the subject matter.

Secondary Functions: The secondary functions are what the design must do in order to accomplish the primary function. The secondary functions are to:

- Hold the camera
- Elevate the camera above the subject
- Frame the image
- Adjust zoom
- Adjust aperture

- Adjust shutter speed
- Focus the image
- Take the photograph
- Return the camera to the photographer

Objectives

The objectives, desired attributes, for this design are outlined in Table 1, with specific goals for each objective. The objectives are ranked from top to bottom, from most important to least important, as determined by a pairwise comparison.

TABLE 1 Objectives.

RANK	OBJECTIVE	DESCRIPTION	GOAL
1	Durable	Maximizing the solution's lifetime	Minimum 5 years or 300 uses
2	Easy to use	Maximizing intuitiveness in operation	All parts correspond to similar parts familiar to photographers
3	Low cost	Imposing minimal expense to the client	Less than $500
4	Easy to set up	Requiring two or fewer people to assemble in a minimal amount of time	Less than 10 minutes
5	Easy to maintain	Using standard spare parts in as many cases as possible	No custom parts are required

Constraints

The **Constraints:** This design must be:

- Able to take photos of areas from $50\,m^2$ to $500\,m^2$ at a minimum
- Portable—fits within a space of $1m^3$ maximum for travel
- Able to accommodate daytime and nighttime shooting
- Compliant with all applicable local and federal laws, bylaws, and regulations
- Accommodating to the Canon EOS-5D
- Accommodating to standard, wide-angle, zoom, and telephoto lenses

Service Environment

Physical Environment: The physical environment is a description of the physical characteristics of the location in which the design is expected to be used. Based on the client's needs, the design will be required to function both indoors and outdoors.

Indoor events:

- Will most often be banquet halls, churches, or restaurants.
- Have average room temperatures of around 72°F, although this could climb in summer.
- May have height restrictions depending on the height of any ceilings.

Outdoor events

- Most usual weather conditions are possible in the Chicago area. This includes sunshine, clouds, rain, snow, wind, and hail.
- Normal temperatures range from 20°F in winter to 80°F in summer; however, lows can reach sub-zero to 90+°F in summer [1].
- Average annual rainfall is 36.9 inches [1].
- Average annual snowfall is 36.7 inches [1].
- The average wind speed is 10 mph [1] (although gusts and sustained higher winds can be expected).
- The landscape of the outdoor environment may include grassy, muddy, or paved areas and uneven ground.

Living Things: There will be many crowds present at the weddings and events that the photographer will be attending. The sizes of these crowds range from 50 to 750 guests. These guests will include a diverse range of individuals. The ages of the guests can range from newborn babies and young children to elderly seniors (ages 0 to 100+). Outdoor events must also take into account small animals that may be present, such as squirrels, raccoons, or various species of birds.

Virtual Environment: Because the environment in which the device is being used changes from venue to venue, cell phone service, wireless networks, and satellite networks are unpredictable and thus unreliable. In addition, the availability of electricity is unreliable as well, due to some venues being located outdoors, while others are indoors.

Users and Operators

The users and operators of this design are the client, Tori Noir, and any other photographers and photography assistants associated with Noir Photography.

Physical: The physical description of the users and operators include the physical characteristics that should be taken into consideration during the design process. The users and operators are assumed to be all photographers and photography assistants in the age of adulthood. They may be either male or female. The height of an average North American male is 69.4 inches, and the height of an average North American female is 63.8 inches [2]. There is a possibility for the photographer or assistant to be using a wheelchair or other mobility assistance device.

Psychological: The psychological description of the users and operators include languages spoken, level of education, along with any significant psychological norms. It is assumed that all users will have the ability to read and speak English. It will not be assumed that all users have all received a post-secondary degree or even a secondary school degree, as there is no mandatory level of schooling or specific special training required to be a photographer. As for special skills and knowledge, it is assumed that each user and operator has an expert knowledge of photography. This leads to some psychological norms involving the use of photography equipment like cameras, lenses, lighting, and tripods.

Social: The social description includes any social norms or cultural commonalities exhibited by the users and operators. As the users and operators are all photographers, or photography assistants, a common social norm between them would be the belief in the value of time. Taking photos is all about capturing the correct moment at the right time. As such, they highly value not wasting time on things that distract from their photo capturing time.

Organizational: The organizational description includes any organizational ethics or values shared by all users and operators. Each of the users and operators are employees of Noir Photography. As it is a professional photography studio, the employees want to portray this image to their clients and make a good impression. In addition, it is their highest priority to deliver high-quality photographs to their clients in order to continue the growth of the business. This device could allow Noir Photography to advertise a unique service that sets them apart from other photographers. Much like most companies, they wish to maximize their income and minimize their expenses.

Stakeholders

Stakeholders are parties outside of the client, users, and design team that have a stake in the design decisions. The interests of each of the stakeholders have been categorized into either economic, social impact, ethics and morality, legality, or environmental impact concerns in Table 2.

TABLE 2 Stakeholders.

STAKEHOLDER	INTERESTS	
Client of Noir Photography	Social impact	• Wants the event to go as well as possible • Concerned with keeping the guests happy
Client of Noir Photography	Economic	• Concerned with the overall cost of the event
Guests of client's event	Social impact	• Want to have a good time • May or may not like having their picture taken • Want to look good in any memento of the event
Venue coordinators	Economics	• Wants to maintain physical integrity of the venue
Patent regulators	Legality	• Want to maintain protection of existing intellectual property
Environmental NGOs	Environmental impact*	• Want to maximize positive environmental impacts and minimize negative environmental impacts

*This may become important if the design impacts the environment or requires alterations to the environment to operate.

Design for Safety

Designing for safety involves considering the safety of all the design's stakeholders during the design process and addressing all safety concerns. As the solution is to be used at large events with many people, it is important to consider safety in order to ensure no one is harmed. This includes taking into account aspects of the service environment such as wind. Also the weight of the device and height (depending on the solution) could pose a potential energy hazard.

References

[1] Illinois State Water Survey: State Climatologist Office for Illinois. "Climate of Chicago: Description and Normals." Retrieved March 30, 2014, from www.sws.uiuc.edu/atmos/statecli/General/chicago-climate-narrative.htm#tables.

[2] Centers for Disease Control and Prevention. "Body Measurements." Retrieved July 7, 2011, from www.cdc.gov/nchs/fastats/bodymeas.htm, April 2, 2009.

Glossary

3D-printing 3D-printing (also called additive manufacturing) is a form of rapid prototyping. Computer-aided design (CAD) drawings are printed layer by layer into an actual 3-dimensional physical object made out of plastic, starch, metal, or some other specialized materials.

5 whys see why, why why

Acceptable Risk A risk is generally considered "acceptable" if either its probability is very low or its consequence is not serious. It is not necessary for risk to be zero for a design to be considered "safe." An individual's perception of acceptable risk, and society's decision about what constitutes an acceptable risk, is complicated and evolving. Acceptable risk also varies substantially between cultures. Acceptable risk is often formalized as regulation and codes.

Accident Unintentional exposure to a hazard. An accident may result in negative consequences.

Accumulated benefit The total amount of benefit gained from a project less the total amount spent on the project at any point in time, summed up from the start of the project. When this is zero, the project is at the break-even point; when it is above zero, there is net benefit from the project. The accumulated benefit of an enterprise is often in the form of money, but more broadly it can be a measure of benefit to the enterprise (company or non-profit organization) or other entity such as a community, society, or the environment.

Accuracy Precision or exactness, or a value of maximum expected error. Also, in agreement with objective facts or findings; i.e., an accurate statement of fact.

Action item A task assigned to a team member at a meeting or through another type of communication. For example, in the minutes of a meeting, action items will identify who is responsible for a task that was assigned at the meeting, what the task is, and when it should be completed.

Active listening Concentrating on what the speaker is saying and listening attentively without interrupting. Active listening strategies include taking notes as someone is speaking and asking questions that will clarify what you have heard and confirm your understanding.

Active voice A sentence in which the subject (or agent) precedes the verb (or action).

Activity on the arrow A way of depicting a PERT chart. It is an ordered graph of tasks in which the nodes are numbered and the tasks are shown on the arrows.

Activity on the node A way of depicting a PERT chart. It is an ordered graph of tasks in which the task is shown in the node bubbles between the arrows.

Actor A term used in the Use Case method. An actor is a component within the design of a system. The actor can either be a human or a system interacting with the design.

Actual cost In project management, the cost of a project or part of a budget spent, in contrast to budgeted cost, which was the predicted cost.

Actual cost of work performed (ACWP) In project management, the actual cost of the work done to a certain time point or other time interval of a project; in contrast to Budgeted Cost of Work Performed.

Adjourning The fifth and last stage in the Tuckman team model. Adjourning involves the dissolution of the team, generally at the end of a completed project. However, adjourning may also occur with a permanent team if a member leaves or there is significant change in the team structure or purpose.

Aesthetic The look and feel of a design. The aesthetic is the artistic aspect of the design, which is often created in collaboration with an industrial designer, architect, or artist.

Affordance Originally any function of a design, both intended and unintended. The more current definition of affordance recognizes that the way people use things is generally based not on what is possible to do, but the usage implied by the design. Thus affordance is the function of a design that is intuitively obvious: A button suggests pushing, a handle suggest pulling, and an arrow suggests directionality or pointing.

Agenda The list of items that will be addressed during a meeting. Often a certain amount of time is budgeted for each item; this is also noted in the agenda.

Agile development (Agile design process) An iterative, incremental, adaptive development process where requirements are developed as the project progresses. The emphasis is on short design cycles involving user and/or client feedback and reevaluation at each cycle.

Alpha testing The testing of a complete or almost complete version of a product by developers before it is given to users to test. See also Beta testing.

Analogy An idea-generation tool that uses solutions in other realms to inspire design ideas. For example, to find a solution to a filtration problem, you might look at the ways plants and animals filter fluids. This is called biological analogy. Other types of analogy are technology analogy, which uses technologies in other fields for inspiration; fantasy analogy, which uses fiction (e.g.. futuristic science fiction) as inspiration; or synectics, which uses symbolism as inspiration. See also Technology analogy and Biological analogy.

Angel investor A wealthy individual who typically invests a few million dollars or less in exchange for equity in a project.

Anthropometric data Information on human physical characteristics. This includes data such as the range of physical size, shape, and characteristics present in a population; for example, how much weight people can lift, the size and shape of a human foot, or how large a hand is when the user is wearing mittens.

Appendix (pl: Appendices) An extra section of a document, added after the end of the document (i.e., after all required sections, including reference list, if appropriate). The appendix or appendices contain additional material that provides more detail to supplement the document.

Assembly instructions An order of assembly that must be followed or special handling that is required for manufacturing a product. The documentation is called assembly instructions. Assembly instructions describe precisely and clearly the steps needed to correctly assemble a product from the component pieces and materials.

Assembly line (also called production line or just a line) A set of manufacturing processes that typically occur in the same location. In an assembly line component parts or pieces are processed (e.g., cleaned, etched, cut, soldered, joined, painted, and so on) and assembled to create a product. This may be a finished product ready for sale, or it may be a sub-assembly that will be used in the assembly of a larger system.

Assessment of future technology Making an educated guess about the future technology that will be available in a particular field.

Audience assessment (often called audience analysis) A structured method to determine characteristics of the person receiving a communication. The audience analysis enables the writer or speaker to design the communication for greatest effectiveness. An audience analysis is useful in preparing any type of communication, including written, ora,l and graphic (drawings, illustrations).

Axis (pl: axes) The number line on a graph. A two-dimensional graph has an x-axis, which is a horizontal line, and a y-axis, which is a vertical line. The axes should be properly labeled, and the data depicted on the graph should be placed accurately relative to the x and y values.

Back of the Envelope (BOTE) Calculations that provide very quick approximate answers based on information that is already known or can be very easily found. Back-of-the-envelope calculations are often used to do a preliminary quick "reality check" on an idea.

Background processes A term used in Life Cycle Analysis (LCA). Background processes are common, generic processes used for multiple processes aside from the one you are analyzing and for which you may gather average data. For example, you may use regional average data on the mix of fuels used to produce electricity when calculating the impact of the electricity consumed for a life cycle process.

Ballpark answer (or to ballpark an estimate) Slang for an answer that is a very rough approximation (related to order of magnitude approximations and back of the envelope calculations).

Bathtub curve The probability of failure of a piece of technology over its normal lifespan is high at the beginning (due to early failures), lower in the middle section, and high at the end due to age. The curve is similar to the cross-section of a bathtub, hence the name.

Benchmarking The analysis of an existing technology. This process enables the designer to identify requirements, learn about existing systems, and compare design ideas to currently available technology.

Beta testing The testing done by a small group of users to find problems in a product before it goes into commercial production. Usually the beta test is followed by a short period of final development work before the product is released for full production. See also Alpha testing.

Bias The tendency to judge things in a certain way due to the values and pre-existing ideas a person brings to the situation rather than to the details of the situation itself.

Bid An estimate of the cost of a job submitted by contractors in response to a Request for Proposal or other form of call for proposals. A bid may also include information about the amount of time, materials, or other resources needed. It may also include a preliminary design plan to show the client that the bidder has the expertise to undertake the job. A bid must match exactly the specifications in the Request for Proposals (in terms of formatting, length, content, etc.) and must arrive before the deadline to be accepted.

Bill of materials (BOM) A list of materials used in the manufacture of a design. This includes the various pieces and connectors, usually showing an item and a quantity. The BOM may also include other information such as source (where it was purchased), part number, cost, and/or location in the company inventory.

Biodegradable The quality that the mass that makes up a product can be returned to an essentially natural state in a reasonable amount of time (on the order of days, months, or years), through the action of living organisms (microbes, animals, fungi). A product that takes hundreds or thousands of years to degrade is considered nonbiodegradable.

Biological analogy (related to Biomimetic design) Using knowledge about the natural world to inspire design solutions. This is a creativity method that relies on looking at natural systems related in some way to the design problem. Research on how the natural system works is used as inspiration for creating a new technology.

Biologically active materials Materials capable of having an effect on an organism. Drugs are an example of biologically active materials.

Biomimetics The process of adapting solutions from the biological world to engineering design problems. See also Analogy and Biological analogy.

Black box method A method used to discover (uncover) functions of a design by connecting the components going into a design to the components that result from the design.

Blue sky thinking A creativity method. In this method people imagine what is possible with no constraints. In blue sky thinking, any preconceived notions about what is possible or impossible are ignored and ideal solutions are imagined as if there were no limits.

Body language Physical demeanor while communicating; both physical stance as well as tone of voice.

Body text The primary font and paragraph style in a document, which is the style of the "normal" paragraphs in the document. This is in contrast to other types of text in a document such as headings, figure captions, references, footnotes, or boxes that may use a different formatting.

Boilerplate clause A clause (paragraph or statement) that is reused in contracts. Many contracts are constructed using a combination of boilerplate clauses and new, original language.

Brainstorming A term usually used to describe the process in which a group of people have an intensive session of idea generation. The concept of brainstorming was developed by Alex Osborn and is used to generate ideas to address a specific question or defined problem. The main focus of this technique is to generate a large quantity of ideas without worrying about their quality. See also Structured brainstorming and Free brainstorming.

Break-even point The point in the history of a project when the costs are covered by revenues; that is, when the accumulated benefit is zero. Successful projects will attain net positive profits (positive accumulated benefit) after the break-even point. See also Payback period.

Budgeted cost A project management term for the estimated amount of money with allowances for potential deviances needed for a project or part of a project in a certain time period. See also Actual cost.

Budgeted cost of work performed (BCWP) A project management term for planned cost for work done in a specified time interval, in contrast to the Actual Cost of Work Performed.

Budgeted cost of work scheduled (BCWS) A project management term for planned cost for work scheduled to be done in a specified time interval. Note that the work may not have been completed, nor may the costs be as estimated. See also Budgeted Cost of Work Performed and Actual Cost of Work Performed.

Burn in A time at the end of manufacture of a product when the product is powered; used to check for faults and premature failure. See also Bathtub curve and Infant mortality.

Capital Costs The costs associated with the purchase of items needed to initiate a project, in contrast to items continuously purchased to maintain the project (such as power and rent). Capital costs often include equipment and furniture; e.g., delivery trucks or other vehicles, desks, computers for staff, manufacturing equipment, communication equipment (phones), and printers. Capital costs can also include the cost of building a plant or other facility.

Capital equipment Tools, machinery, or structures purchased for the implementation of a design. The capital equipment is paid for at one point in time and then is used over time to support the implementation or operation of a design; for example, a company may purchase manufacturing equipment to build an assembly line.

Caption A title for a figure, such as a diagram, photo, graph, or chart. However, unlike a title, the caption is placed below the figure. It typically identifies the figure number and gives a short description of the figure. A figure caption may be one sentence or more long.

Carcinogenic Increasing the incidence of cancer.

Cash flow The money moving into and out of a project, department, or other entity, in contrast to the assets tied up in inventory, equipment, and similar capital investments.

Cell In the context of an information table, a cell is one box in the table. It typically contains information relating the column header item to the row header item.

Chart A pictorial representation of statistical data. Common chart types include pie charts and histograms.

Check points Places in a problem-solving process that allow stopping to reflect on the process and how it is going. Milestones are check points with associated intermediate results: calculated or derived pieces of information that will be used later in the process or form part of the solution. See also Milestones.

Citation information (also called bibliographic information) The details of ownership and publication or distribution of ideas, documents, articles, images, and books. Such information includes (but is not limited to) the name(s) of the author(s), the name of the book, website or article in question, the name of the journal, collection or larger body that the article belongs to, the date and place of original publication or distribution, any company (such as a publisher) responsible for the distribution. Specific details of citation or bibliographic information necessary for ethical documentation of information sources are given in style guides or handbooks associated with the organization to whom you are submitting your work.

Claim 1. An assertion that requires supporting explanation and evidence to be persuasive. A credible statement includes a claim, an explanation of the claim, and evidence to support the truth and validity of the claim.
2. The most important part of a utility patent. The claims define the intellectual property (IP) being claimed for protection in the patent in very specific legal terms.

Client An individual, a company, or some other organization that commissions the design. The client gives the task of designing the system to the design team, and the client also typically funds the project.

Client statement An initial written statement from the client that describes a design problem. The statement may be short, only a paragraph or two, and vague, or it may be very long. Some requests for proposal (RFPs) are thousands of pages long and describe the design problem in detail. See also Design brief.

Closed question A question designed to elicit a specific, short answer. For example: Did the company make a profit last year? (yes/no) How much profit did the company make last year? (dollar figure) What is your best-selling product? (specific short answer). This is in contrast to open questions which are meant to elicit longer, more complex responses.

Closed-ended problems Problems having one correct answer. The goal of solving this type of problem is to find that one answer, or an approximate solution that is close enough to the right answer, in contrast to open-ended problems.

Code A set of requirements that follow from government regulations. Typical codes include building codes, fire codes, and health codes. A code is a set of rules intended to ensure minimum levels of safety and performance. Codes may be defined by governments or by professional associations and may be enforced by government agencies or by the professions involved. Therefore, codes will vary between jurisdictions and a design must adhere to the local code requirements.

Column In a two-dimensional table (matrix) made up of rows of cells, a column is a set of cells along a vertical path in the table.

Commercial-off-the-shelf (COTS) (also called custom-off-the-shelf) Proposed designs that use parts, components, and other systems that are readily commercially available (or open source). COTS design is often less expensive. In addition, off-the-shelf items have already been tested and certified if this is necessary. Many innovative engineering designs are the result of putting together off-the-shelf technologies in new and unique ways.

Commission To sponsor, authorize, or support (fund) a project.

Commissioning Getting a plant or system up and running after it has been constructed so it is ready for the customer to use without further setup work.

Community consultation (also called public participation) The process of involving the community in the design process. This is particularly important for public works and large infrastructure projects. Community consultation includes both informing people about the project as it goes through the design process and also listening to their views and perspectives on the project and using this information as input to the design process.

Complex 1. A system in which a variety of parameters interact to create events that are not perfectly predictable or may not be fully anticipated. Chaotic systems, such as weather, are one example of a complex system, as well as sociopolitical or economic systems that behave in ways that are not fully predictable.
2. A sentence in which the parts (clauses) have a specified relationship to one another; for example, one part may be the result of the other, in which case the word "because" connects them, or one part may contradict another, in which case the word "but" connects the clauses.

Complexity count A count of the interrelated parts of a system and the number of interactions these represent. Complex systems, involving interactions between the components, often scale in a nonlinear way. Sometimes engineers use a complexity count in their estimations to take this into account. A system with five interacting parts, for example, represents 10 possible interactions. This type of complexity count helps the engineer correctly estimate values such as the degrees of freedom in a system.

Complicated system A system that has many interacting parts, which can be modeled. The behavior of complicated systems can be predicted accurately with sufficiently powerful modeling, in contrast to complex systems in which modeling, particularly modeling behavior far in advance, is not possible.

Components 1. In functional basis a component refers to mass, energy, or information. 2. In detailed design, particularly product or plant design, a component refers to a piece or part of hardware or software.

Compound sentence A sentence made up of two parts (clauses), each of which could stand on its own as a simple sentence; the two parts are joined with the word "and."

Compound-complex sentence A sentence that has clauses connected with the word "and" (as in a compound sentence) as well as clauses in relation to one another, connected with such words as "but," "because," or "if...then" (as in complex sentences).

Compromise A dispute resolution method characterized by all parties giving up some aspect of their individual desires in order to achieve a group decision.

Computer aided design (CAD) A process in which a software package is used to create a detailed two- or three-dimensional representation of a physical structure. In a modern CAD program, each physical part in a full assembly is represented by a detailed description of its mass, volume, and composition in a virtual 3D space. A 3D CAD model serves the role of a virtual prototype and can be easily modified.

Concept of operations (ConOps) A document that describes the design problem and conceptual design solution from the user's perspective. A good ConOps document should allow the user group to understand clearly the problem that the designers have identified and be able to assess the quality of the proposed solution.

Conceptual design alternatives A primary set of ideas for possible solutions

Conceptual design specification A document that describes the alternative design solutions considered and recommends a design for further detailing. It may include documentation of decision making, information that has been gathered that informs the project, and results of prototype testing, modeling, or simulation activities. The document should give the reader (supervisor, client, or instructor) confidence that the team has explored the solution domain and that the team's recommendation for a design solution is sound.

Confidence level A term used in probability and statistics. It is the degree to which the samples used in a statistical analysis represent the true distribution.

Confidentiality agreement (also called a non-disclosure agreement) An agreement that legally establishes confidentiality. In the field of intellectual property, it is intended to prevent the use of a design in public or its offer for sale prior to a patent being applied for.

Consensus A team decision-making strategy that requires group assent to a decision formed through discussion. Some members of the team may disagree with a choice, but agree to go along with, and support, the choice.

Consequence A result; for example, the result of exposure to a hazard.

Conservation of energy A physical law stating that energy is neither created nor destroyed. This means that all energy going into a system must either be stored in the system or flow out of the system. This law is essential for the design of many technologies.

Conservation of mass A physical law that states that mass is neither created nor destroyed. This means that all mass going into a system must either be stored in the system or flow out of the system. This law is essential for the design of many technologies.

Constrained problems Problems that have limits to possible solutions. Most design problems are constrained by physical laws, government regulations, budget, availability of current technology, and so on, so they are basically constrained problems. Some problems are very constrained, thus only having a few existing solutions rather than providing wildly diverse options. In the limit as constraints increase, the problem becomes more like a closed-ended problem that has only one possible solution.

Constraints Absolute limits of a design. If a potential solution violates a constraint, then it must be dismissed.

Consulting company A type of engineering company that works exclusively through collaborative projects or contracts. A consulting company usually takes on numerous projects for a wide variety of clients. They do not necessarily get involved in the full design process or in the full implementation of the design. They are contracted to provide a specific outcome or set of deliverables.

Consumables Resources that are used up in the activity (or task). Examples are fuel, printer ink, and cleaning supplies.

Continuous distribution A probability distribution where the set of outcomes or possible events forms a continuous curve. For example, if you measured the precise height of every person in the world and put this data on a graph, you would have a distribution that could be well described using a continuous mathematical expression (i.e., a curve).

Co-product A term used in Life Cycle Analysis (LCA). It refers to a useful, but unintended, output of a process. For example, the generation of electricity using fossil fuels produces heat. This heat can be considered a co-product if it is put to use heating buildings or supplying process heat for a plant.

Copyright Legal protection for some types of intellectual property such as writing, music, movies, software, art, and other expressions of ideas. A copyright protects the tangible expression of an idea, not the idea itself.

Correlation The process whereby two variables appear to change together, i.e., a change in one variable can be used to predict a change in the other. A correlation, however, does not necessarily imply cause and effect.

Corrosion A failure mode in materials due to chemical processes that lead to material loss and/or degradation. One example of corrosion is rust.

Cost What must be paid or given up in order to accrue a benefit. Costs may be determined financially, as in the payment of money for a product or service. Costs may also be determined in relation to time or other resources.

Cost variance (CV) A project management term that is equal to Actual Cost of Work Performed minus Budgeted Cost of Work Performed. When positive, this is the amount the work came in under budget; when negative, it is the amount overspent to get the work actually done.

Cost-plus contract A contract to do work or services where the cost of the actual work or services provided is paid for, with some additional amount (the plus) representing the profit to the contractor.

Costs committed The costs associated with work or services agreed to or contracted before those costs are actually incurred. For example, if you sign a lease for an apartment for six months and the rent is $600 per month, then the cost committed is $3600 even before you pay the first month's rent.

Cover page Part of the front material of a report; the top-most page that contains all relevant information identifying the report, its authors, client, and other significant information.

Creativity methods Systematic techniques to generate imaginative, unique, and unconventional solutions. There are a wide and growing number of creativity methods used by engineers and other professionals for addressing design problems.

Credible statement See Statement.

Creep A failure mode due to the slow deformation of a material over time.

Criteria (sing: Criterion) Qualities or characteristics defined in order to evaluate something; i.e., assessment measures. In the Roe, Soulis, and Handa framework for requirements, the term criteria is used to describe the preferred direction of a goal (e.g., _more_ is better, or _smaller_ is better).

Critical path In project management, the sequential set of tasks in a project where any increase in the expected execution time of any of the tasks will result in a change in the overall time for the project. Hence, in the critical path method, project management efforts concentrate on those particular tasks to keep the project running on time.

Critical thinking The ability to analyze a situation, idea, or problem to come up with unique ideas. Critical thinking emphasizes objectivity, the importance of recognizing bias, and other factors that affect the validity of information. Critical thinking involves judgment based on evaluation of multiple perspectives and evidence weighed against criteria.

Cross-functional team A team whose members come from various disciplines and bring varied skill sets. For example, a cross-functional team may have members from civil, mechanical, electrical, and other engineering disciplines, financial personnel, and architects or industrial designers.

Custom off-the-shelf (COTS) Designing using components or items that are already in production and available to buy or license. See also Commercial-off-the-shelf.

Datum Singular of data. A datum also refers to the standard solution used for comparison in the Pugh method of decision making.

Deciding how to decide The procedures by which a team determines how to come to agreement. Two of the more common decision-making methods are consensus and voting. See also Team decision making.

Decision-making methods Methods used to try to remove bias to ground decisions firmly in engineering reasoning. They also assist in developing the justification for decisions when the design process and reasoning are explained.

Decision tree (decision protocol) A sequence of decisions where each decision influences the next decision or set of decisions that need to be made. Take the example of going out to dinner: Where to go to dinner? What will you wear? What will you choose to eat? The decision you make in response to the first question will then dictate, to some degree, the answers to the other questions.

Decommissioning The processes needed at the end of the useful life of a product, structure, or plant. This typically requires shutting down the system and disposing or recycling the components.

Decomposition Breaking the design or design problem into smaller parts. Decomposition is a method used for generating requirements, such as secondary functions. It is also used for generating solutions to a design problem. See also Morphological charts, Functional decomposition, and Structural decomposition.

Deliverable A piece of documentation or other tangible result from the design process (e.g., a set of drawings, a working model of the design, a report, a cost estimate, a set of operating instructions, or construction of one part of the project). In a course, the deliverables are submitted to the course staff (e.g., the instructor or teaching assistant). For an actual industrial project, the deliverables would generally go to the client. Deliverables are usually specified in the agreement (contract) between the company performing the design work and the client or customer.

Dependence A term used in probability analysis meaning that for two trials, the result or outcome of one trial influences the result or outcome of the other.

Dependent clause A clause that cannot be understood on its own, but requires another clause to complete the idea. Clauses, or groups of words, in a single sentence become dependent on one another when they are connected by a word that signals relationship, such as "because," which signals a cause-effect relationship.

For example, in the sentence "The results were inaccurate because the equipment was faulty" the clause "because the equipment was faulty" is the dependent clause.

Design brief The initial written statement from the client that describes the design problem. As an engineering student, this design brief (or client statement) will often come from your instructor in the form of project instructions. The design brief may be short, only a paragraph or two, and vague. Or it may be very long. See also Client statement and Request for proposals.

Design by inventory Making design decisions in favor of parts that are already in company inventory (presumably already being used in other company products).

Design for X (DFX) Design for a specific consideration (X) that is of particular importance to the project you are working on. Some common DFX considerations are:
– Design for assembly
– Design for accessibility
– Design for human factors
– Design for durability
– Design for flexibility
– Design for the environment
– Design for testing and maintenance
– Design for disassembly
– Design for safety
– Design for intellectual property
– Design for aesthetics

Design patent Legal protection for the ornamental form (but not the function) of physical products such as cars, ballpoint pens, or toasters. A design patent is actually the protection for the artistic part of the design, not the technical part. It is one of several types of legal protection for intellectual property. See also Utility patent, which provides protection for the functional aspect of technology.

Design space An abstract concept of a space that encompasses all of the design solution ideas that are being considered at any point in time during a project. Developing requirements defines the boundaries of the design space. Expanding the scope of the project expands the space. Discarding ideas, i.e., removing solution ideas from consideration or reducing the scope of the project, shrinks the design space. Building and testing prototypes or models explores (investigates) the design space.

Design, build, operate, maintain (DBOM) In a DBOM contract, the contractor (i.e., company holding the contract) is responsible for designing and building the technology, then is also responsible for the operating and maintaining the system after it is built.

Design, build, operate, transfer In this type of project, the company is responsible for designing, building, and then operating a system for a short period before the design is transferred to the client. This means that any early problems are taken care of before the client takes over operation.

Design/build project A turnkey solution in the construction industry. The design engineering company is responsible for the design work and the construction of the design. In construction this would involve creating the design then building it. The technology is then transferred to the client for operation.

Detailed design In general terms, a plan for implementation of a design that is so fully detailed it does not require further involvement of the engineer except in an oversight role. See also Final design.

Determinate (determinacy or fully determined) problems Problems that have one unique correct solution. Often fully determined problems have a routine solution, e.g., the area of a triangle that has a known base and height. See also Closed-ended problems.

Diagram A visual depiction of a concept or model that captures the essential features or characteristics of the system from an engineering perspective while generally not trying to depict realistically the actual visual look of the system. For example, a vehicle may be shown simply as a moving mass: a rectangle labeled "m" in a diagram. Diagrams are used in problem solving to help engineers think through a problem. They are also used in documents and presentations to communicate ideas to others.

Diagrammatic communication Using diagrams, schematics, flowcharts, graphs, charts, and other pictorial elements to communicate ideas. These are common elements in engineering documents and presentations.

Dichotomy A two-sided idea, often simplified to an idea with two, diametrically opposed sides.

Dimensional analysis A method for analyzing a problem, a model, or a system. In this method the dimensions (e.g., mass, time, length, voltage, current, viscosity) that are relevant in the problem, model, or system you are analyzing are examined. Terms should "make sense"; for example, terms in an equation that are being added together should all have the same dimensions, or if you are constructing a model for the resistance in a circuit, then the result of the model should be in ohms (Volt/amp).

Dimensions Related to units and indicate the fundamental measure of a quantity. Dimensions include mass, time, and length and other measures such as viscosity or voltage. For example, velocity has dimensions of length/time, and it can be expressed in many different units such as m/s, ft/minute, or km/hr.

Disciplines Engineering fields, each of which has particular training and methods, i.e., chemical engineering, civil engineering, computer engineering, electrical engineering, industrial engineering, mechanical engineering, materials engineering, or mineral engineering.

Discrete distribution A probability distribution that results from an experiment that has discrete (i.e., countable) number of possible outcomes, for example, the distribution of marks on an exam in a class of 50 students. Typically marks are integers between 0 and 100, so the number of possible outcomes for each student is countable: 101. The result, when graphed, is a discrete distribution.

Disposal (decommissioning) costs (also called Final costs) Costs associated with disposing of a product or facility at the end of its usable life. Decommissioning a plant, for example, may require dismantling the systems, selling off any reusable parts, and recycling or disposing of everything else.

Divergent process Any process that increases the number of possible solutions to a problem or increases the breadth of a problem. Divergent thinking means imagining ideas that are very different from existing solutions or solution pathways. See also Lateral thinking.

Document (v.) To explain what has been done in writing and pictures (i.e., graphic communication and drawings). The act of documenting the work also helps engineers develop ideas and clarify their own thinking.

Documentation The documentation of the design process can take many forms: written, graphic (drawings, pictures), and recorded oral (presentations). It is an essential part of the design process that generally forms the bulk of the work that is produced.

Drawing Engineering drawings are formal documents that show the exact specifications of a design in detail. Drawings typically follow a standardized format that conveys information in a clear, precise manner. Drawings are used to communicate the design details to the people executing the design, i.e., people involved in the construction, development, or manufacturing of the technology.

Due diligence Conscientious work practices; taking reasonable care to ensure that all relevant standards, codes, and statutes are adhered to, including pertinent engineering ethical codes. Also, making sure calculations have been checked and that you are following the current, modern practices in your discipline.

Dummy task In project management, tasks used in PERT charts to correctly represent a project plan. A dummy task is a dotted line arrow added to an Activity on the arrow graph to indicate a dependency that would not otherwise show up using this graphical method.

Duration In project management, the amount of work time it takes to complete a particular task.

Ecological system (ecosystem) The system of interrelationships between organisms that share a habitat and between these organisms and their environment.

Effort driven In project management, a task where increasing resources will result in the task taking less time, theoretically in proportion to the resource increase. For example, in painting a building, doubling the number of painters will reduce the time by as much as half. See Fixed Duration.

Elapsed time In project management, time assigned to tasks that do not require the efforts of a person, such as waiting for paint to dry. Elapsed time can happen at night or on weekends or other times when workers would not be scheduled.

Embodiment design A stage in the design process that comes after a preliminary design solution is selected. In this stage of the work the design is "roughed out." Essentially, the engineer is trying to get all of the major implementation decisions made, so that in the last phase of work he or she can focus only on the details. The same basic design may take on different embodiments depending on the application: A portable travel hairdryer will look different but have many of the same features as a hairdryer for home use. This is an example of different embodiments of the same design. See also Intermediate design.

Emotive language Language that is used to excite emotions. It is often used in advertising.

Enabler One type of worker behavior that can create problems with teamwork. Enablers want to help everyone out, at the expense of their own time, health, well-being, and other work. At the extreme, enablers will promise to do too much and will be unable to deliver. As a result, a team that relies on the enabler to do the work for them will fail.

Enabling disclosure A term used in the field of intellectual property. In order for prior art to invalidate the novelty of a patent, it must be an enabling disclosure. This means that it must contain enough detail to allow someone to reproduce the invention.

Engineering design process A systematic method to help engineers develop design solutions that have the best chance of being effective from multiple perspectives.

Engineering notebook A common method for keeping track of design work. Each person on a team will keep his or her own notebook to be used as a journal to record work as it progresses. Starting an engineering notebook is the first step to beginning a new design project. The information documented in a notebook is important to claiming the intellectual property developed during a project.

Entrepreneur A person who starts up a new company to market and sell a product or service.

Environmental footprint The environmental impact of an individual, a device, product, or system over its life cycle.

Equity Part ownership of a company.

Equivalency A term used in life cycle analysis (LCA) meaning a way of converting data into a common set of units that can be added and compared. For example, methane (CH_4) is about 30 times more effective as a greenhouse gas (GHG) than carbon dioxide (CO_2). Therefore, every mol of methane can be considered the GHG equivalent of 30 mols of CO_2. You can use this fact as an equivalency to convert a mass flow of methane into the atmosphere into a CO_2 equivalent for purposes of comparing the environmental impact of a flow of methane to a flow of CO_2.

Ergonomics The applied science of design in relation to the human user. It is intended to reduce operator fatigue and discomfort. Ergonomics is also referred to as biotechnology, human engineering, and human factors engineering. Originally focused on designing efficient effective workplaces, ergonomics now includes all types of human-technology interaction design.

Erosion The loss of solid material because of a flowing fluid.

Estimate (estimation) To estimate is to determine an approximate answer. The approximate answer (called an estimate) should be close enough to the true answer for the purpose for which it is being used.

Evidence Part of making a credible statement. Evidence is information that supports a claim or statement. It may be a scientific method or principle or data generated through rigorous experimentation or gathered from research of previously published work.

Evolutionary design See Incremental design.

Executive decision A decision made by one person, usually the team leader, project manager, or other type of leader. The person making the executive decision may use discussion and consultation to get other people's opinions and ideas, but ultimately the decision is made by a single person.

Executive summary A brief summation of all the most important points in a document. Its purpose is to give a busy reader a complete understanding of the most significant aspects of the document so that he or she can make reasonably informed decisions, perhaps about funding, continuation, mitigation, or cancellation of a project. This is a stand-alone summary; the reader should understand all essential information, conclusions, recommendations, time lines, and budget issues from this summary.

Expected cost The amount of money that the projections indicate should be required to complete each stage of the project. See also Budgeted cost.

Experience design A design approach that more fully takes into account the quality of the experience the user has with a technology. Using an experience design approach, the engineer considers not only the functional aspects of the system, or product, but how to design it so the experience will be fulfilling for the user.

Experiment A test devised to assess or characterize a system, technology, or natural phenomenon. Each performance of an experiment is called a trial.

Explanation Part of making a credible statement: information that defines terms and fills out the details of the ideas in statements to enable others to understand them.

Explicit meaning The literal or dictionary definition of a word. For example, the explicit meaning of the word "smart" is "intelligent." However, depending on the tone a person uses, the word can take on an implied meaning that is just the opposite. When someone says, "Oh, yeah, that's a smart idea" with a sarcastic tone, the person is implying that the idea is actually stupid.

Exposure Financial risk as a result of a decision.

Extension Term used in a use case to identify system requirements necessary to account for users who deviate from the behavior expected. Designers determine what common deviations (or errors) might take place and design the system to accommodate them—this accommodation is known as an extension.

External benefits Benefits not received directly by a company, but which positively impact the community or environment. For example, a company's activity may result in employment and growth of service companies, attraction of other industries to the area, and increased tax revenues to local governments.

External costs Costs not paid directly by a company, but which impact the community or environment. These may include health, environmental, or social costs that are borne by society, governments, or healthcare systems. External costs may be local, national, or global.

Extrapolation A method to estimate a value that falls beyond the range of known or given values.

Fact Statement accepted as "truth," including well-proven scientific laws or axioms or statements supported by tangible evidence.

Factor of Safety See Safety factor.

Factory-follow work (also called Sustaining engineering) Engineering work that arises from problems during manufacturing. The engineer may be called in to solve a problem or approve a change during the manufacturing process.

Failsafe A design that fails in a safe rather than an unsafe manner. For example, in automobiles, a failsafe thermostat is designed so that if the valve that controls how much coolant goes to the radiator fails, it fails in the open position. This is because an engine that is too cool is much safer than an engine that overheats.

Failure Degradation of part of a system such that the part no longer fulfills its full purpose in the system.

Failure modes and effects analysis (FMEA) A formal method of tracking the higher-level consequences of lower-level failures. The effect of the failure of each individual component is tracked upwards to determine its final result. See also Fault tree.

Fallacy An invalid argument. While the conditions may be true, their relationship or consequences have not been logically proven.

Fast tracking A project management method where parts of a design are detailed in parallel with the execution of other parts of the design, thus reducing the overall time to execute the design when compared to methods where all the design is detailed before execution begins.

Fatigue A failure mode that develops over time due to the initiation and growth of cracks under the influence of a load (often a transient load). If you bend a paperclip back and forth until it breaks, the failure is due to fatigue.

Fault An aspect of a design or product that was unintended and has the possibility of detracting from the worth or of interfering with the intended use of the design. A fault may not be detected or cause issues, or it may result in a failure (see Failure). In manufacturing, there are often quality assurance processes built in to try to detect a fault before the product leaves the assembly line.

Fault tree analysis (FTA) A means of calculating the probability of some undesirable consequence, based on a computation of the probability of all the events that could lead to the consequence. The information is put in a tree structure that shows the necessary circumstances (human, environmental, etc.) that bring a fault or set of faults through to the consequence or failure.

Feasibility analysis (Feasibility check) An analysis of a preliminary design or design idea to make sure that it meets all the functions and constraints, including economic constraints, and significantly addresses the objectives. The design is also analyzed to make sure that it is safe and can be implemented. The design idea may also be analyzed to assess if it is physically possible (this is related to a reality check).

Fermi questions This term is derived from physicist Enrico Fermi who was well known for his ability to make estimations. Fermi questions are very difficult to solve exactly, but can be solved approximately using creative estimation methods. See also Back-of-the-envelope estimation.

Figurative language Language that uses figures of speech such as similes and metaphors. These are inexact and poetic comparisons intended to allow readers or the audience to bring their own subjective experience into understanding an idea. These should be avoided in engineering communication and should serve as a warning to use caution when they are found in an information source.

Figure A type of diagrammatic communication. A figure is a picture (illustration), a diagram, a photo or a graph, or any other type of visual element in a document or presentation. Figures are typically identified in a document by a number and a caption below the figure.

Final Costs The costs to decommission the design and/or to dispose or disperse the physical parts. See also Disposal costs.

Final Design In general terms, a plan for implementation of a design that is so fully detailed it does not require further involvement of the engineer except in an oversight role. See also Detailed design.

Final design specification This report deals with the details of the design solution, usually including detailed drawings, or schematics and the embodiment of the design (sketches or renderings of the technology). It may include economic, environmental, or other types of analysis of the design solution. This type of report is usually developed near or at the end of a design project. Along with the detailed drawings, it may be the last *deliverable* for the client.

Finish-to-start In project management, an association of two tasks where the first task must finish before the second task can start.

Finite element analysis (FEA) A solid object (component in a design) is represented by a large number of discrete elements in a computer program that can compute the full three-dimensional stress field resulting from imposed loads or displacements. FEA can be very accurate and is used widely in industry to analyze systems before they are built or to investigate failures.

Fixed costs Those costs that occur regardless of the quantity of activity in a company or other enterprise. Rent, basic telephone charges, and basic equipment purchases are typical fixed costs. Whether the company is making sales or not, these costs must still be paid. This is in contrast to variable costs, which will depend on the quantity of activity.

Fixed duration In project management, tasks that are of fixed duration cannot be made faster by the addition of resources, such as extra workers. See Effort Driven.

Fixed-price contract A contract where work, goods, or services are provided for a set amount of money regardless of the time taken or the problems encountered. See also Cost-plus.

Float In project management, the difference between the duration of a project path and the duration of the critical path. The critical path has zero float. See also Slack.

Flowchart A type of diagrammatic communication that depicts the flow of information in a system. Flowcharts are a common way to visually represent the logic pathways in computer code, but can also be used to represent protocols, procedures, and other information processes.

Flow Refers to mass, energy, and/or information. The mass, energy, or information does not need to be actually moving to be called a flow. Flow simply refers to the presence of these components, which are operated on by a technology.

For-benefit company A company that may include in its corporate culture and policies means to maximize benefits for society, the environment, and the local economy rather than being driven by profit motives alone.

Force-fitting Forcing your mind to make lateral jumps to novel concepts. One method for doing this is ruling out a typical, known solution to a problem. For example, what if you had to transport a cup of coffee (i.e., a cup of liquid) without a cup? What if you had to transport a cup of coffee using only materials you have in your school bag or backpack?

Foreground processes In life cycle analysis (LCA), the processes that are unique to the lifecycle being analyzed. The data collected for foreground processes will be specific to the technology being analyzed.

Form (also called embodiment) The physical manifestation of a design idea. For example, all pens serve the same essential purpose of enabling a person to translate thoughts into writing. However, pens take many different forms: long, short, ballpoint, fountain pens, permanent markers, etc. The form, or embodiment, of a design should align with the function (i.e., form follows function).

Forming The first stage in the Tuckman team model. In the forming stage, the team is coming together but still operating as a set of individuals. This stage is dominated by individual team members thinking more about themselves than about the team.

Frame of reference A person's perspective; it will affect that person's observations and perception of reality.

Free brainstorming An idea-generation method used to develop the design requirements for a project or generate solution ideas. Design team members call out ideas, which are recorded for future discussion and sorting. In free brainstorming all team members contribute ideas simultaneously. See also Brainstorming.

Free riding The lack of participation by a team member, particularly in a brainstorming activity. See also Hitchhiking.

Functional basis A powerful technique for dissecting an engineering design problem and understanding it from a functional perspective. Functional basis identifies a few fundamental types of operations (i.e., functions) that a technology can perform on a set of components (flows). Virtually all technologies that exist can be characterized by their functional basis.

Functional decomposition Splitting up the functions of a design into a set of sub-functions in order to analyze each separately. This can include identifying several primary functions or identifying a set of secondary functions that together enable a primary function. Functional decomposition can be used to identify requirements, such as secondary functions, or it can be used to generate solution ideas for a design. Related topic: Morphological chart method. See also Decomposition and Structural decomposition.

Functional unit In LCA, a unit of technology that has a defined quantity of functionality. Two technologies can be compared based on their functional units. For example, if comparing the environmental impact of bar soap to liquid hand soap, pick quantities of each that provide the same number of washings—i.e., that are functionally equivalent—for comparison.

Functions What the technology must do. The functions are part of the project requirements. A function describes what the design must do in order to work, i.e., in order to be functional. Basically, any idea that meets the functional requirements could conceivably be a possible design solution for the problem.

Gantt chart A project management tool that shows the project schedule visually as a list of tasks illustrated as a bar graph on a calendar.

Gated project A project that includes evaluation points (gateways) where the project is assessed to see if it is proceeding well enough or if it is costing too much or taking too much time or resources to continue. Decisions made at gateway points are often tied to continued funding of the project.

Gateway (also called a gate) A decision point in a gated project. Gateways may require deliverables or a project review, and they may result in a decision to further fund the project, discontinue the project, or take the project in another direction. Funding installments for a project are often contingent on gateway deliverables.

Go/no go decision A decision to proceed or not to proceed (or continue) on a project. It is a quick and simple yes or no; not a decision with conditions.

Goal definition and scoping The first stage in a Life Cycle Analysis (LCA). In this stage the goal and scope of the analysis are defined.

Goals The particular characteristics of the design solution. In the Roe, Soulis, and Handa framework for requirements, goals include functions (what the design will do) the design will perform and objectives (what the design should be).

Goodwill A positive attitude of others toward a company, an organization, or an individual.

Graph The visual depiction of data, typically on two or more axes (e.g., Cartesian coordinates, polar coordinates). There are many different types of graphs: for example, scatter plots, line graphs, bubble plots.

Graphical decision chart (also called graphical decision matrix method) A decision-making method used to compare ideas to two criteria; for example, to compare design solution ideas to the top two design objectives. This method works particularly well if there are two objectives, or criteria, that are clearly more important than the other evaluation criteria for the project.

Group processing Activities, such as discussion, that promote synergy in a team and characterize real teams and high performance teams. In group processing the team members build synergistically on ideas from others, creating outcomes that would not be possible from a group of individuals working independently.

Guard (guarding) A barrier between people, such as operators, and a hazard. Guards can be physical (such as a fence), or they can be electronic, such as a sensor that automatically shuts off the equipment when detecting a person nearby. Even common household items are often guarded; e.g., a food processor that won't start unless the lid is securely in place. In this case the lid acts as a guard.

Hazard A situation where there is the possibility of damage to people, property, or the environment.

Hazard and operability analysis (HAZOP analysis) Qualitative method used in the chemical process industries to predict the outcome of deviations from intended operation parameters; in other words, to predict high-risk hazards.

Header (also called a column header) The first row in a table. The cells in this row announce the title for each column.

Headings Short titles at the start of sections, usually separated by a space and differentiated from other text by font or format, such as bold and/or italics. They break up a document and signal readers about the information they are about to read. This makes reading faster and documents easier to navigate.

Heating value The energy content of a fuel that can be converted to heat when the fuel is combusted.

High performance team A group of people who come together to work in an interrelated manner toward a common goal; they operate in a highly communicating, cooperative, and synergistic manner that fully actualizes the potential of every team member. Their performance together goes well beyond the capability of the individual members.

High volume In this context "volume" refers to the number of units (i.e., the number of copies of a product) that are produced. A high-volume product is one that is produced in large quantities. This would be typical for a successful consumer product. There is no exact number for what constitutes "high" (versus "low") volume; it is a subjective term.

Hijacker A type of behavior that can create problems with teamwork. A hijacker takes over control of the team. This may not be the team leader, but is often a person who wants to be the team leader.

Hijackers have very little trust in other people's work and will often redo it themselves. This behavior is corrosive to collaborative teamwork and often results in a dysfunctional situation.

Histogram A type of bar chart used for depicting statistical data.

Hitchhiker A type of behavior that can create problems with teamwork. A hitchhiker is a person who contributes significantly less to a project than everyone else for no legitimate reason.

How-why tree A graph that organizes information from high-level global objectives down to measurable local objectives. The how arrow points down the tree: How will the general objective be realized? (i.e., by one or more measurable local objectives) The why arrow points up the tree: Why is the sub-objective required? (i.e., to fulfill the high-level global objective) How-why trees are used to organize and operationalize objectives in many fields including design, business, education, and government. They are used when high-level objectives (e.g., maximize profit for the business) need to be translated into specific executable objectives (e.g., increase individual customer sales).

Human error (also called operator error) When an accident is caused by a person doing something incorrectly, rather than a failure of the technology per se.

Human factors The physical, psychological, social and political circumstances that affect the successful operation of a design.

Human-centered design See User-centered design.

Idea generation The stage of the design process when the design team generates possible solutions to the design problem. This stage is characterized by use of creativity methods to develop as broad a set of solutions as possible.

Impact analysis The third stage in LCA. In this stage the impact of the life cycle on the environment is assessed.

Implied meanings Messages that are carried in words, independently of, or in contrast to, the words' explicit meaning. For example, the word "crippled" explicitly means having an inability. However, over time the word "cripple" has taken on an insulting sense, as if a physical challenge reduces a person's overall worth. When writing an engineering document, it is important to be aware of both the explicit and implied meanings of the words chosen to transmit information accurately and respectfully.

Improvement analysis The fourth stage in an LCA. In this stage the results of the LCA are used to suggest improvements that will reduce the environmental impact of the design.

Income (net) The monetary benefit after the deduction of the costs. It is basically equivalent to Net revenue, or Profit. Note that often in business "Income" is used where "Net Income" is meant, and it sometimes refers to the revenue before deduction of the costs.

Incremental development Stepwise development of a system: implementation and testing of parts of a design before the whole system is completed. Effectively this is like the development of partial prototypes, each building on the last, until the final result is the full design.

Independent A term used in probability analysis describing experiments or trials where the outcome from one is not influenced by another; e.g., the result of the first trial of an experiment does not change the probability of the outcomes of a second trial.

Independent clause A clause, or group of words, that contains a whole idea and is grammatically complete; on its own, an independent clause is known as a simple sentence.

Indeterminate problems Problems that do not have enough information to be solved uniquely, e.g., the area of a triangle where the base length and height are unknown.

Industrial designer An artist who specializes in the aesthetics and human factor aspects of products. Virtually every major consumer product, from shampoo bottles to cars, is designed with the assistance of an industrial designer, who will be involved with the development of the form (embodiment) of the design. An industrial designer may be part of the design team from the beginning of a project or may join the project later during the intermediate design stage.

Industrial ecology A variation on the lifecycle assessment method; it looks at the lifecycle of a technology as analogous to natural ecological systems where everything is reused and recycled. The objective is to identify opportunities for turning waste products into co-products and overall reduce the environmental impact of a technology life cycle.

Industry standards See Standards.

Infant mortality A failure of a product very shortly after manufacture due to manufacturing variances or faults. See Burn-in and Bathtub curve.

Inflation Reduction of purchasing power of a unit of money due to rising prices of goods and services.

Informed consent It is acceptable to subject a person to a risk under some circumstances. However, ethically the person must be informed of the risk and consent to it. Informed consent procedures are often overseen by ethics boards and/or legal experts.

Infringe (infringement) Using or marketing protected intellectual property without permission. For example, if you market a device that utilizes a patented technology in the design without licensing the technology from the owner (i.e., without getting permission), this would be considered infringement. You would be infringing on the owner's intellectual property rights.

In-house design team A design team that works for and designs technology for the company that employs it.

Insurance A contract with a company to manage the risk associated with a costly event. The company will pay part or all expenses associated with the event should it occur. The contractee periodically (monthly or annually) pays a sum based on the risk regardless of whether the event occurs.

Intellectual property (IP) Anything humans create through processes of thought and/or creativity become the property (a tangible thing owned) of the creator or the organization for which the creator works (because organizations often have agreements for rights). Works that are considered intellectual property include anything written, either artistic, academic or journalistic; any visual representation, again, whether artistic or scientific; and/or any patentable designs. To use intellectual property without following the appropriate guidelines, rules, or laws is considered the same as stealing.

Interaction design A design approach that considers the interaction of the user with the technology as a key focus of the design process. Interaction design stems from user-centered design and has evolved into experience design.

Interest 1. Stakeholder interest: An aspect of an organization or person's ongoing welfare that might be affected by a design. An interest implies that there may be benefit or loss due to the implementation of the design. The kinds of interests that the stakeholders have in a design project generally fall into categories related to economics, ethics or morality, legality, human factors, social impact, and environmental impact. 2. Interest as used in economics: The cost of borrowing money, generally assessed as a percentage of the amount borrowed (which is called the capital). Also, if a customer "lends" his or her money to a bank by putting it in a savings account, then the bank will pay the customer interest. Similarly, investments will pay interest to the person who has invested the money. Interest is usually paid daily, monthly, or annually depending on the type of loan or investment.

Interlock A type of safety system. A design with an interlock has a power switch that can only be activated when a guard is properly in place.

Intermediate design A stage in the design process where the objective is to "rough out" the final design. This means different things in different disciplines, but essentially, the engineer is trying to get all of the major implementation decisions made, so that in the last phase of work, he or she can focus only on the finer details. See also Embodiment design.

Internal costs Costs borne by the company or enterprise alone, not including costs to society or the environment.

Internal revenue Describes income to the company alone; does not include benefits to society or the environment.

Interpolation A method of estimating an intermediate value that lies between two known values.

In-text citation A note, put into the body of a report or other communication, signifying that the material being presented has ideas from a source other than the presenter. The in-text citation may be a number in square brackets, a superscript number or a name or names and numbers in round brackets, depending on the referencing standard being used. The number or name will relate to a reference list, normally located at the end of the document. The reference list will have the full bibliographic details of the source.

Invention The process of solving a problem in a new way that would not be obvious to engineers practicing in the field. See also Inventive step and Inventor.

Inventive step For a design to be non-obvious (for purposes of patentability), it must be more than the result of the typical routine design process. In Europe, this extra something is called an inventive step. In practice, the inventive step does not have to be very big; even small advances resulting from fairly routine design activity are often successfully patented.

Inventor An inventor is an individual, or sometimes a team of people, who acts as both the client and the designer for a technology. Inventors are people who identify a need for a new technology and create it. See also Entrepreneur.

Inventory analysis The second stage in an LCA. In this stage an inventory is made of all of the mass and energy going into and out of each process in the life cycle.

Isolationist A type of behavior that can create problems with teamwork. An isolationist is a team member who works competently, but does not interact well with other team members. The work delivered, while generally acceptable, may not fit very well with the rest of the project because the isolationist has not adequately communicated with the rest of the team. Isolationists are not involved in team decisions and do not want to take the time to listen to other teammates.

Iteration The use of an iterative process. One iteration is one loop through a set of processes. The last iteration through the design process results in either a solution that will be developed further or cancellation or suspension of the project. See also Iterative.

Iterative Referring to a set of processes that is repeated over and over in a loop. In the design process, iteration is used to enhance understanding of the design problem, to improve the proposed design, and to increase the quality and quantity of information generated in the process.

Job shop A company that specializes in doing contract work. For example, if some parts need to be welded but the business does not want to buy welding equipment, the parts could be sent to a job shop that has welders and welding equipment. There are job shops for all kinds of manufacturing and design activities that handle one step, or just a few steps, in the design or manufacturing process.

Label A short piece of descriptive text used on diagrams and graphs to add textual information to these visual elements. The labels help the reader interpret the visual information. Labels may be as short as one letter, number, or symbol, or as long as a phrase.

Lag A term used in project management referring to a delay of the start of a task after its predecessor. Typically, a lag is necessary due to some characteristic of the predecessor task (e.g., the predecessor task is to pour concrete and the successor is to install posts into the concrete, so the concrete must harden before mounting holes can be drilled).

Lateral thinking A term coined by Edward de Bono, a creativity guru, in 1967. The underlying motivation for lateral thinking is summed up nicely in his quote: "You cannot dig a hole in a different place by digging the same hole deeper." Thinking about a problem when you already have a solution in mind is like standing in a hole with a shovel. Lateral thinking means creating an entirely new solution approach, i.e., starting a new hole in a new place. See also Divergent thinking.

Lessons learned Analysis of past activity to determine what parts of it were successful and what parts need improvement. Lessons learned activities may be undertaken individually (e.g., analyze the results of an exam to see what you did right or wrong) or as a team (e.g., the team may run a debriefing session after a project to analyze the experience and the results). See also Reflection.

Liable (liability) Literally, this means "likely," but in the legal sense it means "responsible." Anyone paid to do professional work is responsible for the outcome. For this reason many independent consulting engineers carry liability insurance.

Licensing It is possible for an inventor to sell the rights to commercialize an intellectual property (i.e., an invention) to an existing company. This is referred to as licensing the intellectual property. For example, Fred Morrison, the inventor of the Frisbee, licensed the Frisbee to Wham-O, which is a toy manufacturing company. This allowed Wham-O to market, manufacture, and sell Frisbees during the period when the technology was covered by a patent.

Life cycle assessment (LCA) (also called life cycle analysis) A rigorous method for analyzing the production, operating, and end-of-life environmental costs for a technology in terms of mass and energy.

Life cycle diagram A diagrammatic representation of all of the processes that contribute to the life cycle of a technology. The diagram shows the mass and energy streams entering and exiting each process in the life cycle, from the extraction of raw materials through to the disposal of the technology at the end of its operating life.

Linear regression Regression analysis is the use of statistics to find a relationship between variables; linear regression is the technique of finding a linear relationship between two variables, often used in statistics to fit a line to a set of data points. See also Regression analysis and Linear relationship.

Linear relationship When two parameters (variables) have a relationship to one another that can be mathematically described by $y = ax + b$, where a and b are constants and the variables are x and y.

Lockout A type of safety system. Maintenance staff have a personal lock, for which only they have the key. The lock is used to disable controls and other components of a system before the staff member works on it. If many people are working at the same time, each one uses a personal lock in parallel with the others so that all the locks must be retrieved before the machine can be operated.

Logic A structured means for developing new, valid ideas from pre-existing facts.

Look and feel See Aesthetic.

Loss In economic analysis, if the costs of a project exceed the revenue generated, then the difference is a net loss. Essentially a loss is a negative profit where profit equals revenue minus costs.

Lower cost estimate A term used in project management to denote a realistic estimate of the minimum likely cost of a project. See also Upper cost estimate and Budgeted cost.

Magic solution A design solution that ignores the laws of nature and/or what is possible with current technology. It is not possible, for example, to move a book from one place to another at the speed of light. This is a magic solution. However, imagining this idea may suggest physically possible ideas such as moving the information in the book at the speed of light.

Main message A statement that encompasses both the topic and the purpose of a presentation.

Maintenance Taking care of a system; extending the expected lifetime of the design by replacing worn parts and used fluids, such as lubricating fluids.

Managing Guiding the work and collaboration between team members to make sure everyone is working effectively together toward shared goals.

Man-hours See Manpower.

Manpower (also called man-hours or worker-hours) The number of hours it takes a human to do a specified amount of work. The term is often used in association with estimates of how many people should be assigned to a specific task or how long a specific task will take to finish.

Manufacturing Producing copies of a product.

Market share The percentage of sales of your technology in a technology class (i.e., category). For example, if 23% of all tablets sold last year were produced by your company, then your company has a 23% market share.

Material A substance that has been designed and/or manufactured from raw materials. Almost everything you have is probably made of an engineered material with the exception of plants (wood and natural food), animals, water (oceans, rivers), and rocks.

Materials Safety Data Sheet (MSDS) A form provided by a manufacturer detailing information about a material (chemical substance). The MSDS will give information about the chemical formulation, properties, and toxicity. In many jurisdictions companies are required to keep a binder of all MSDS sheets that pertain to the chemical substances on site.

Matrix See Table.

Mean time between failures (MTBF) The average time between failures of a product or system. The MTBF can be extended through routine testing and maintenance, which can repair problems before they result in failure.

Mean time to failure (MTTF) See Mean time between failures.

Means A specific solution idea, or a means of solving a design problem, is called a means statement. Means can also refer to how a solution is implemented. For example, if the design requirements call for a system that heats food, then means could include a microwave oven, an electrical resistance heating element, or an infrared heat source.

Means analysis One variation of decomposition methods. Means analysis can be used to discover or generate additional functions, objectives, or constraints to add to a requirements list. To perform a means analysis, decompose the design problem into sub-functions or sub-structures, then generate as many potential solution ideas as possible for each of the sub-functions or sub-structures. These design solutions are "means" for solving the problem. Examine the solution ideas to identify commonalities. Functions, objectives, or constraints that all of the solutions have in common are probably integral to the design problem and should be included in the requirements.

Means statement A specific solution idea or a means of solving the design problem. Means can also refer to the how a solution is implemented. See also Means and Solution driven.

Measures of effectiveness See Objectives.

Memo A short written message used in business to convey a concept such as an action item or policy, e.g., details about an upcoming meeting or notification of issues that need to be further addressed in a project. Memos serve to document communication on important issues.

Metaphor A figure of speech that makes a comparison between two different ideas without signaling the comparison with the word "like." An example would be "This project is going downhill fast." The metaphor implies the project is heading for disaster. Metaphors tend to be heavily value-laden, because the same word can mean different things in different contexts. For example, "She is a monster pianist" means that the person is a great pianist, whereas "He is a monster" may mean the person has terrifying characteristics. Metaphors should avoided in engineering communication and should serve as a warning about the value of information when doing research.

Method of least squares A method for performing a linear regression; i.e., for finding a linear relationship between two variables.

Metrics A test procedure or performance measure. A metric is a methodology for measuring, calculating, or estimating the characteristics of a design. Metrics can be quantitative (e.g., estimating the speed of a microprocessor) or qualitative (e.g., using a focus group to get people's opinions on taste or visual appeal). In industry, the results of this type of testing are also sometimes referred to as metrics.

Micro-managing A style of managing that is highly and specifically directive. Being very controlling or directive rarely works well, particularly if you are managing highly competent and creative people. See also Hijacker.

Midpoint assessment An assessment that quantifies and characterizes the potential impact of a technology or system, but does not try to predict the precise consequences of this impact. This concept is used in Life Cycle Assessment (LCA).

Milestone A well-defined checkpoint in the execution of a project. Milestones are used in a project management plan to identify due dates or other key dates in the project schedule by which a set of tasks must be finished.

Minutes Notes taken during a meeting. Like notes taken during a lecture, these are not an exact reproduction of everything said at the meeting; instead they record major discussion points, what decisions were made, and the action items coming out of the meeting, e.g., who will do what by when.

Mitigate To reduce the impact (consequences) of an event or the risk associated with the possibility of an event.

Mode A set of states implemented for a particular purpose. For example, for an aircraft to go into landing mode the pilot will generally lower the nose of the aircraft, reduce speed, lower the landing gear, and lower the flaps on the wings. Or a pipeline may operate in high pressure mode or low pressure mode by changing the set of operating characteristics of the compressor stations along the pipeline.

Model (modeling) A model is a simulation of a real system that captures some of the key characteristics of the real system so that the engineer can explore or solve a problem and then apply the results derived from the model back to the real system. Models can be mathematical (analytic, numerical, or probabilistic) or physical (i.e., a prototype). It is also possible to model decision-making processes or other qualitative systems.

Monitoring cycle A cycle of overview during project execution consisting of monitoring the activities of the project, analysis, and necessary actions prompted by the analysis.

Monte Carlo analysis or simulation Analysis that uses random number generation and probability relationships to create simulated experimental data. "Monte Carlo" in the name refers to the famous casino in that city because the outcomes from casino games are a result of probability operating on random events.

Morphological chart (or morph chart) A graphical representation of the process of putting solution ideas for sub-systems (or sub-functions) together to create a wide variety of integrated solutions for a whole problem. This method builds on decomposition.

Multi-voting A decision-making method that is effective for reducing a very long list down to a more manageable list; used extensively by professional design teams. In this method each team member is given several votes to assign to items on the list. Items with the most votes move up to the top of the list, and items with few or no votes are discarded or moved to the bottom of the list.

Net Revenue Also called profit or net income. It is the monetary benefit flowing into a enterprise minus the costs.

Network diagram A term used in project management to denote the presentation of the tasks that make up a project by showing the connections between the tasks.

Non-disclosure agreement (NDA) An agreement that establishes confidentiality. In an intellectual property context, it is intended to prevent the use of a design in public or its offer for sale prior to a patent being applied for.

Non-governmental organization (NGO) An organization that is neither a companies nor a government agency, but that may have an interest in the design problem (i.e., may be a stakeholder). NGOs are typically not-for-profit organizations with political, social, humanitarian, or environmental missions.

Non-obvious A term used to describe an invention that is patentable. For a design to be non-obvious, it must be more than the result of the typical routine design process. In Europe, this extra something is called an inventive step.

Normal distribution A symmetric probability distribution curve centered around an average (mean) value. It is also referred to as a Gaussian distribution and informally referred to as a "bell curve" because of the shape.

Norming The third stage in the Tuckman team model. In the norming stage of team development, the team is essentially "getting its act together." It has determined its common goal; in the case of a design project, the project has been defined and planned out. While team members will still have diagreements, they are developing ways of working together effectively.

Novelty A term used with reference to intellectual property. It means that at no previous time, anywhere in the world, was a particular invention in the public domain. The specific definition of novelty depends on the laws of the particular jurisdiction. To be patentable, a technology must be novel.

Numbered headings Headings in a document that are numbered sequentially, as in this text. Section 1, for example, would have a subsection 1.1. Numbered headings impose a hierarchy on the information and identify main ideas and the supporting ideas that belong to them.

Numerical simulation Simulation carried out using a computer-generated model. A simulation is a model that mimics the behavior of an actual system that can be used to test the performance and behavior of a technology under a variety of conditions.

Object In sentence structure, the word or phrase that receives the action. In a simple sentence, the object follows the verb (or action). In the sentence "I kicked the football," the football is the object.

Objective goal The minimum target level for an objective. Objective goals are not absolute constraints but are meant to motivate the improvement of the design toward a goal set; they set an approximate threshold for acceptable levels of performance or characteristics of the design solution.

Objective tree A diagram that organizes objectives by decomposing a general objective into specific measurable objectives and placing them into a how-why tree structure. See also How-why tree.

Objective What the design solution should be. Objectives are used to judge how well an idea solves the design problem. They are the evaluation criteria. Some designers call objectives "measures of effectiveness" to emphasize this evaluative aspect of objectives.

Objectivity A perspective or viewpoint that is as removed as possible from bias or personal prejudices or feelings.

Occupational Safety and Health Administration (OSHA) The U.S. government agency responsible for workplace safety and health.

Off-shoring Contracting to a company overseas to perform one or several steps in the design or manufacturing process.

Off-the-shelf A component that is already on the market and is available from a distributer to buy or license. See also Custom off-the-shelf and Commercial off-the-shelf.

Ongoing costs Maintenance and operating costs of the design, including recurring costs such as rent and supplies, in contrast to a capital cost, which is paid once, typically at the beginning of the project.

Open question (or open-ended question) A question designed to elicit a long, more complex answer. For example, What is the company's growth strategy over the next few years? Why did your company decide to start a new product line? Under what circumstances might you decide to discontinue a product line? This is in contrast to closed questions, which are meant to elicit short, specific answers.

Open-ended problem A problem that has many possible solutions, none of which is exactly perfect. However, some possible solutions are better and some are worse in terms of meeting the problem criteria. Design problems are typically open-ended.

Operating costs The costs incurred by operating the technology after design and construction or implementation. See also Ongoing costs.

Operator The individual or group that actually works with or controls a technology: a product, device, system, plant, or process. While operators may be considered users, not all users are operators; for example, the users of clean water are a different group of people from the operators of a water treatment plant. Operators are often tasked with both running (operating) a technology and maintaining it. See also Users.

Opinion A statement based on personal feelings, views, or experience; sometimes supported and sometimes not supported by evidence.

Opportunity cost The concept that one has to give up an opportunity to do one thing in order to achieve another; for example, deciding between going to a friend's birthday party or going out to dinner with your cousin. There is an opportunity cost to each of the choices. Opportunity cost can also refer to the loss of materials in their natural state or in a state suitable for future use: This is a "cost" of using a nonrenewable natural resource now rather than saving it for the future.

Optimal solution A solution that is closest to ideal, given the circumstances (budget, timeline, and so on) and the technologies available.

Order of magnitude answer A number that is within a factor of 10 of the true answer.

Organizing The category in the Tuckman team model that includes forming, storming, and norming. In the organizing stages the team is learning to work effectively together.

Out-loud editing One team member talks out loud while editing a piece of work (e.g., a section of a document, or presentation slides). The team member who drafted the work listens carefully to what is being said, using this information to revise the work. Out-loud editing can help a team share expectations about quality of work if it is done in a respectful and constructive manner.

Outsourcing Contracting another company to perform one or several steps in the design or manufacturing process. For example, a company that specializes in mechanical design may out-source the design of an interface (e.g., a touch screen) for one of its products to another design company that specializes in this area.

Overspecifying (related to over-engineering) A more generic term for a safety factor applied to things other than material strength. An overspecified design exceeds the minimum requirements. Systems are sometimes overspecified or over-engineered to improve safety or reliability.

Overage The cost or amount of a resource in excess of what was planned or expected.

Overengineering A more generic term for a safety factor applied to things other than material strength. An overengineered design exceeds the minimum requirements. Systems are sometimes overspecified or overengineered to improve safety or reliability. See Overspecifying.

Overhead (also called indirect costs) Typically, fixed costs that must be covered by revenue but are not directly attributable to any one project. For example, suppose a company is involved in five different projects simultaneously. The cost of keeping the office space heated (or cooled) and the lights on is a cost that is not directly the result of any one of the projects, but must be covered by the total revenue flowing into the company. When companies bid on contracts, they factor this into their bid as an overhead rate, typically as a percentage added to the cost of the job.

Pairwise comparison method A simple method for ranking a list. A pairwise comparison simplifies the task by breaking a complex decision up into a set of more simple decisions; each decision compares just one item to one other on the list.

Pareto's rule A rule without formal justification that seems to approximately describe some situations. Pareto's rule is also known as the 80-20 rule because it postulates an 20/80 proportionality. For example, the last 20% of a project will take 80% of the time. Or in a fundraising campaign 20% of the donors will contribute 80% of the donations.

Participatory design A process that brings people from the target user group into the design process as members of the design team. Participatory design is becoming more common, particularly for the design of technologies for developing communities that have historically not had a voice in the design process. It is a strategy for bringing first-person perspectives into the design process.

Passive voice A sentence in which the object (rather than subject) precedes the verb (or action) and the subject is implied.

Patent See Utility patent.

Patent agent A licensed professional specializing in the patent application process. Patent agents are not attorneys, so they cannot litigate IP disputes, but they can help in the preparation of a patent.

Patent attorney Lawyers who specialize in intellectual property issues (patents, copyright, and so on).

Patent Cooperation Treaty (PCT) Because of the expense and difficulty of filing a patent claim in multiple countries, the PCT allows for a single application that is essentially a notification of intent to file internationally. It allows for more time to decide if it is worth filing internationally, and if so, which countries to file in.

Patent examiner The patent office employs examiners to confirm that the patent claims are indeed novel and non-obvious. The patent examiner will do a detailed search of the prior art to see if the technology has already been publically disclosed or previously protected.

Patentability (patentable) Meeting the criteria necessary to be protected by a patent. To be patentable an invention must be a non-obvious advance, it must be new or novel, it must be useful, and it must be in a patentable domain. These criteria are assessed to determine the patentability of an invention.

Patentable domain Depending on the country, inventions in certain domains (or areas) of interest may or may not be patentable. Business practice and existing chemical structures in living organisms are examples of domains where patents may not be possible in some countries. However, new engineering technology is typically in a patentable domain.

Payback period The time from the start of the project to the break-even point; i.e., the point of zero accumulated benefit.

Performance analysis (related to benchmarking) Testing the performance of the technology you have designed or a competing system or product. Performance analysis of competing systems that are already on the market can be used to determine requirements such functionality or objective goals. Performance analysis of a prototype can be used to determine if your design is meeting (or likely to meet) the set goals.

Performing The fourth stage in the Tuckman team model. When a team enters this stage, it has achieved the balance between focus on tasks and focus on people. The team focuses on developing efficient and effective processes and procedures for accomplishing shared goals. The team members will have a shared understanding of the problem, project, or design, and their focus is on the goals. The members also have a shared sense of responsibility and ownership of the project.

Permutation A term used in probability analysis relating to an ordering of distinct elements. For example, a row containing two red balls and one green ball has three permutations or distinct orderings, because they can be ordered with the green ball first, second, or third.

Personal protection Clothing worn to keep the user safe when direct exposure to hazards is unavoidable. Examples include steel-toed boots, a hard hat, and hearing and eye protection.

PERT (Program Evaluation and Review Technique) chart Used in project management, this is a graphic method that is used to show the dependencies of tasks in a project. It usually also shows time to completion of the tasks.

Photograph An image produced by a camera or similar recording device. A photographic image may be decorative, giving the viewer an emotional impression, or it may be made more specifically informative with the addition of labels or other information.

Picture A visual representation of an actual thing. A picture may be realistic, e.g., a photograph or a very accurate drawing. However, it may also be more representational such as the signage used to indicate "no smoking" or "curve in the road ahead."

Plant A structure and its internal equipment that is designed to produce a specified product. There are manufacturing plants that produce consumer or commercial products and power plants that produce energy products such as electricity and heat. Other common types of plants include water treatment plants that produce clean water for people and wastewater plants that clean used water before it goes back into the environment.

Plant patent Legal protection for the intellectual property produced through the creation of new strains of plants (i.e. photosynthetic organisms).

Platform A base on which a custom design is built, for example, an operating system, a standard vehicle chassis, or a standard shampoo formulation. For instance, a company may produce a wide range of variations (different scents or additives) to create a line of shampoos or a model line of cars that are all designed to use the same chassis. There may be many new pieces of technology or customized systems that are built using the same platform.

Polya's problem solving model A method for solving problems that has four steps: define, plan, solve, reflect. These steps outline effective problem solving, communicate the process, and communicate the implications of the solution.

Post-consumer waste Waste that results from the operation of a system or technology; also what is left at the end of the usable life of a system or technology.

Potential team A group with some synergistic work habits. Potential teams communicate and perform better than pseudo-teams. However, they have not achieved the level of collaboration present in high performing teams.

Prejudice Literally "prejudgment"; making decisions based on factors that are peripheral to the circumstances, such as race, religion, gender, or age.

Present worth (present value) The value of something expressed in monetary values of the current time, that is, with the effects of inflation considered.

Present worth factor (PWF) The factor used to convert future values to present values (See Present worth)

Presentation A person-to-person interaction, which can be one-to-many or team-to-many. Engineering presentations are carefully prepared, but must still seem conversational, unlike a prepared speech read or memorized for public speaking purposes.

Preventive engineering Engineering that emphasizes reduction as the first key principle for sustainability, ahead of design for reuse or recyclability. Preventive engineering is a strategy for reducing the environmental impact of a technology's life cycle.

Primary functions The most essential comprehensive expression of what a design does; i.e., the main reason it exists or is being designed. For example, the primary function of a bike lock is to secure a bike; the primary function of a can opener is to open a can; the primary function of a cup is to contain a liquid; the primary function of a cellular network is to transmit information. See also Secondary functions that enable or result from primary functions.

Prior art (also called state-of-the-art) The sum total of all information relevant to the design challenge at hand. This may take the form of commercial products or services, patents, standard industry practices and codes, published research reports and articles, and the personal knowledge of experienced engineers in the field.

Priority date The date of invention established through filing of a patent application or other type of disclosure.

Probability Measure of the chance of occurrence of an event or outcome, expressed as a percentage from 0% (no chance of the event happening or no chance of a particular outcome) to 100% (the event will happen for sure or a specific outcome is assured).

Problem statement A restatement of the client statement or design brief that describes the gap in technology that the design will fill. Key to a successful problem statement is a clear definition of the client's need. The problem statement serves as an introduction to the project requirements and is solution independent.

Process The sequence of operations needed to transform material, information, or energy from one form to another. Process design generally involves not only the design of the operational steps of information, material, or energy transformation, but also consideration of the timing and scheduling, logistics, supply chain, and quality control aspects of the process.

Process management Managing a process, such as an assembly process, to optimize objectives (such as minimizing waste, reducing storage requirements, and reducing failure rates).

Process simulation Using a model (usually a computer model) of a process to simulate what happens in reality. Using these models, the designer can see how the process will behave under different conditions and thus change the design of the process to optimize behavior.

Producing Category that encompasses the final two phases of Tuckman team development model: performing and adjourning. These stages characterize a team that has learned to work together and is primarily focused on the success of the project.

Product A physical or virtual thing that is the result of a design process. There are consumer products, such cell phones, shampoo, and software packages; commercial products, such as oil drilling equipment or supercomputers; energy products, such as electricity or natural gas; and digital products, such as network services or online financial services. There are also agricultural products such as corn, beef, or wheat, which are the result of a natural process that has been engineered to enhance the production of a desired product.

Product line A set of products for a single market to exploit differences in that market, such as a line of automobiles from one manufacturer that range from less expensive to very expensive, from small to large, and so on.

Production blocking The inhibition or reduction of ideas within a brainstorming session. An example of a cause for this reduction would be having to wait for your turn to contribute your idea.

Production waste Waste that is generated as a result of processes used to produce a technology. This could be waste generated during manufacturing or construction of the technology or system. The disposal of production waste is part of the impact of the technology on the environment and needs to be accounted for in a life cycle assessment.

Profit (also called Net Revenue or Net Income) The monetary benefit derived from a project after the deduction of the costs.

Project management A process of predicting the time, funds, and resources required to execute a project with the goal of saving time and/or money through efficiency. Project management involves planning a project so that tasks are scheduled in an efficient order and monitoring the project as it proceeds to keep it on schedule and on budget.

Project manager The person who is in charge of supervising or managing a project or large portion of a project. The manager oversees planning and progress of the project from start to finish.

Project plan What will be done in the project, the order of the tasks, the time scheduling of the tasks, and the costs and the resources involved with the tasks.

Project requirements A full definition of the design problem including the problem statement, functions, objectives, constraints, service environment, and important information that has been gathered. The documentation of the design problem goes by various names in different industries. The project requirements documentation may sometimes be called the problem definition, scope or scoping document, or specification document. See also Requirements.

Proprietary Information that belongs to a certain group, organization, or company and must be kept confidential. This may include aspects of a design that is intended to be patented. It may refer to a "trade secret," an aspect of the design that allows it to be profitable only as long as others cannot copy it.

Protocol A set of steps defining an appropriate order of actions for a given situation. There are decision-making protocols, crisis-response protocols, and protocols specified for automated systems to communicate with one another. Protocols play a role in design of systems that must communicate or interact with other systems. They are also typically part of safety systems.

Prototype A specially built one-off model of the proposed design or a subcomponent of it. It is typically built with just enough care that it realistically represents the proposed design at minimum cost. A prototype is a typically physical model, but could also be a virtual model (constructed using CAD software or some other type of simulation software).

Provisional patent application Informal precursors to a full utility patent filing. In the United States a provisional patent application is simply a dated and filed document with a full description of the IP, and it is used to establish a priority date for the invention.

Proximity sensor A sensor that detects the physical presence of a person. For example, the sensors used in elevator doors to prevent the doors from closing on a person's hand or foot.

Pseudo team A group of people who work as a set of isolated individuals rather than a real team. Pseudo teams generally perform below the level of the average member because they do not fully utilize the potential of the team.

Public domain Information that is not (or no longer) protected by any form of license, copyright, or patent. Information in the public domain can be used without payment of fees; however, in any publication or document associated with university or professional life, even if information is public domain, its source must be acknowledged. Using someone else's ideas without acknowledgment, even when they are public domain, can be seen as a form of misrepresentation (i.e., plagiarism).

Public work projects Engineering projects intended to benefit the general public in a particular location. These are generally funded by municipal, state, or provincial and/or federal governments, often in partnership with one another and/or in partnership with industry.

Pugh method A decision-making method that is a variation of a pairwise comparison that also builds on benchmarking by comparing alternatives against a standard solution. It is generally used for comparing potential design solutions against an existing technology or a standard set of goals using the objectives for the project.

Purpose The desired result of an action.

RACI matrix A strategy for dividing up work on a team so that every task gets done, but also so that people are not working in isolation. RACI stands for Responsible, Accountable, Consulted, and Informed. A RACI matrix (table) lists all of the tasks that need to be performed (usually in the first column) on a project and all of the people on the team (usually in the header row). R, A, C, and I are used in the table to identify the role of each person with respect to the tasks. Some people may have no role on a task, but every task must have at least an R and an A, and should probably have one or more Cs or Is.

Random stimulation A creativity method whereby some random word or idea is incorporated into a line of thinking to force divergent ideas. What if your solution had to include a monkey? Saltwater? A triangle? Music? This may not give you a solution, but it may open your mind to many divergent solutions.

Raw materials Materials (mass) as they exist in or on the earth, prior to any processing (ore, lumber, and water in their natural state are examples).

Real team A group of individuals that work in an interrelated manner toward a common goal. Real teams communicate actively and have developed cooperative, synergistic work habits.

Reality check The assessment of ideas or proposals taking into account the world around and such practical concerns as principles of physics, economics, existing technologies, obvious environmental, and/or social impacts. A reality check asks, "Will this idea, solution, or proposal work in the real world?" See also Feasibility analysis.

Recycling Reusing the material (mass) in a design later in a different form. It can also refer to reusing the mass produced as a by-product of producing or operating the design.

Reducing Creating a design that uses less mass and energy in production, operation, and/or disposal than other designs that exist for the same function.

Redundancy Including additional systems or load paths typically for safety and reliability purposes rather than for performing the basic function.

Reference section A list of all sources of information referred to in a document. Normally, the list appears at the end of the document in a Reference list. For each source, you must give the authors' names, the date of publication, title of the document, and all publication and/or electronic access information.

Referencing Acknowledging the source of the ideas of others that you have used in a document or other type of communication. Referencing is normally done using a standard format, many of which exist. One such standard was developed by IEEE (Institute of Electrical and Electronics Engineers) and is used in many engineering disciplines. Like most standards, it is made up of two components: in-text citations and a reference list.

Reflection The critical thinking process that looks at a situation and analyzes it in terms of its strengths and weaknesses. It is important for an engineer to reflect on the completed work: Was the result sufficient or not, and what is the most-effective next action? See also Lessons learned.

Regression analysis A term used in statistics analysis for a technique for analyzing the relationship between dependent variables. The most common type of regression is linear regression, where a "best fit" linear relationship is determined.

Regression testing During an incremental design process, testing to ensure that the new additions and changes have not adversely affected previously designed work.

Regulations A set of rules determined by a governing body with the authority of a government. Regulations may be defined by government statutes or laws, but they may be enforced by agencies authorized by government or by the professions involved. The rules are intended, like codes and standards, to ensure minimum levels of safety and performance.

Reliability The probability of failure-free operation of a system over a given period.

Renewable A resource, such as an energy source, is considered renewable if it can be replaced relatively rapidly (on the order of days, months, or years). If the resource takes more than 100 years to replace, it is considered nonrenewable (e.g., coal, natural gas, oil).

Request for Proposals (RFP) is issued by a client who is seeking solutions to a design problem or other type of professional work. The RFP will usually state the purpose of the project and explain the client's requirements. Anyone, or any company, can respond to an RFP by submitting a proposal, which is called a bid. The proposal explains how the contractor will meet the client's requirements, and the resources (cost, time, etc.) necessary to fulfill the proposed contract.

Requirements A formal description of what is necessary and desirable in a design in the form of a complete and organized documentation of the design problem. The requirements describe the functions that any possible solution must have in order to be considered a solution to the problem. Requirements also include the objectives for the design and the constraints. In addition, the requirements describe the context in which the design solution will operate, including the environment, the users and operators, and other existing factors that will influence the design process. See also Project requirements.

Research (also called information gathering) Both investigation of existing knowledge and the development of new knowledge. In the early stages of the design process most of the research involves investigating existing knowledge: finding written material, talking to people and listening, and learning what you can about the project. Later in the project you will develop new knowledge: build models and prototypes to test, develop test methodologies, and collect and interpret data.

Research & Development (R&D) Activity intended to draw upon and add to existing knowledge in order to improve or create novel products, systems, or processes.

Residual Mass or energy that is discarded as a result of a life cycle process. A residual may be regarded as waste if it requires disposal or if it is emitted into the environment. However, it can also be regarded as an opportunity if it can be turned into a marketable product.

Resources The people, equipment, consumables, and money needed to complete a task or project. In a Gantt Chart resource name, specifically, refers to the kind of labor required for a task, e.g., carpenter, electrician, or other type of specialized work.

Return on investment (ROI) The amount or percentage gained in net income because of an expenditure. If you spend $100 to buy a bicycle and sell it the next day for $150, then you have a 50% ROI.

Reuse A principle whereby material and equipment required for production (manufacturing, installation, or other use) of a design can be utilized again in another project. The design itself may be utilized repeatedly, or its components may be utilized individually for other purposes. Reuse means that there is no essential change to the components in order for them to be utilized for a different purpose. If the material requires transformation to be used again, then the term to be used is "recycle."

Revenue The monetary benefit coming into the enterprise (company or nonprofit organization) assessed before any costs are subtracted.

Reversal method A creativity method whereby the direction or sequence of things is reversed to generate new ideas. The reversal may not lead to a valid solution, but will stimulate divergent thinking.

Reverse engineering The disassembly of a technology (device, process, system) to discover how it was designed. Reverse engineering is a common practice in product and system design. See also Benchmarking.

Risk The combination of severity of consequences and probability of unintentional exposure. Specifically: Risk = (severity of consequence) × (probability of occurrence)

Routine design Using a commonly accepted method or an existing piece of technology to solve a problem. Some types of design problems occur so frequently in a particular industry that engineers have developed a routine process for developing solutions for this type of problem. See also Commercial off-the-shelf.

Row In a two-dimensional table (matrix) made up of columns of cells, a row is a set of cells along a horizontal path in the table.

Row header The first column in a table. The cells in this column announce the title for each row or give an essential piece of information that connects all of the information in the row.

Royalty A percentage of revenue. For example, Fred Morrison, the inventor of the Frisbee, reports that he licensed the Frisbee to Wham-O in exchange for 6.6% of wholesale sales, i.e., a 6.6% royalty.

Rule of thumb Approximate guideline, based on experience or practice.

Safety Factor A multiplier that is used to ensure that materials are sized to more than adequately carry a load or operate without failing. For example, elevator systems are typically sized to carry several times the rated load. See also Factor of safety.

Sample space A term used in probability analysis meaning the set of all possible outcomes from a single trial or set of trials. For example, if you are analyzing the probability that you will have a green light when you get to an intersection, then the sample space is green light, yellow light, and red light, because these represent the three possible outcomes that could occur when you arrive at the traffic signal.

Scaling Estimating how one variable in a project changes in relation to another. For example, you can estimate that if you double the area you need to paint, then you will need twice the paint. This is an example of linear scaling. In project management scaling is often used to estimate things such as the resource needs of the project or of the time and cost of execution of the project. In general scaling may be linear or nonlinear.

SCAMPER A creativity method; an acronym for a set of suggestions proposed by Robert Eberle to trigger novel design ideas: Substitute, Combine, Adapt, (Modify, Magnify, Minify), Put to other uses, Eliminate, Rearrange or Reverse.

Scenarios (also called user stories) In use cases, scenarios or user stories describe different ways in which a user may interact with a technology. Scenarios help the design team imagine the interaction so they can create a design that better fits the user's needs. A scenario may be long, but usually is only a few sentences.

Schedule variance (SV) A term used in project management that is equal to the budgeted cost of work scheduled minus budgeted cost of work performed. SV is how much remains to be spent to get the work scheduled, but not yet done, finished if behind schedule. If ahead of schedule, then SV is the amount of money spent to get ahead of schedule.

Schematic A type of diagram that typically uses visual vocabulary to represent components in a system pictorially. A common type of schematic in engineering is a circuit schematic or wiring diagram, which shows all of the components in an electrical system. In this type of schematic resistors are shown as zigzag line segments, capacitors are shown as short parallel lines, and so on.

Scope The definition of the breadth and depth of the problem to solve. Typically, the project requirements define the scope of the project, which in turn specifies what a design team will do for the client and what they will not do, i.e., the boundaries of the project. Defining a design problem is sometimes called a scoping activity, and in some industries the project requirements document is called a scoping document.

Scope creep Common project management issue in which additional client requests or new research cause the project to expand as it progresses. Scope creep can include the addition of new functionalities or features during the design process.

Scoping Deciding what to include and what to exclude in a project, life cycle analysis, or similar endeavor. A scoping document is a synonym for project requirements in some industries. The process of scoping in a life cycle analysis means choosing which processes to include in the analysis and which to leave out.

Secondary functions Functions that *enable* or *result* from the primary function. An example of a secondary function that results from a primary function is the heat that is generated by the operation of your computer. This is not the intended function of a computer, but it is a function that necessarily *results* from operating your computer. This makes it a secondary, unintended function. Another secondary function of a computer is that it enables the user to input information. This is not the primary function of a computer, but it is a necessary secondary function that *enables* the primary function.

Self-warning system A system designed to provide warning before failure. An example would be the "check oil" or "low fuel" indicators in a car.

Sensitivity analysis A method for assessing the relative sensitivity of a result to changes in an input variable in a model. When conducting a simulation or assessing a model, inputs can be varied to see what effect the variation has on the output (results). Simulations and models will typically be more sensitive to changes in some variables and less sensitive to changes in others.

Services Utilities or other services such as water, electricity, or Internet access. Services are an important consideration in the operating environment of a design and should be part of the description of the service environment. Availability of services such as high-speed data connections or transportation services in the operating environment may be important in the design process.

Service environment A description of the environment in which the technology will operate. The description of a service environment includes a description of all aspects of the environment. Typically, the environment is described using a range of conditions typical for the operating locations. The description of a service environment typically includes the physical environment, the virtual environment, and the living things the are active in the environment.

Service life A specified lifetime for a component in a design. For example, components in an airplane engine each have a specified service life. Each component is replaced at the end of its service life, rather than waiting for it to fail. Other common examples are filters, such as air filters or the oil filter in a car, which have a recommended service life after which the performance of the component will begin to degrade.

Servicemark Intellectual property protection for symbols and specific graphic images. See also Trademark.

Severity Seriousness of a consequence; typically the consequence of an accident.

Significant figures (also called significant digits) The digits in a number that are actually meaningful and valid. The result of a calculation should never be stated with more implied accuracy than the least accurate value that went into the calculation. Any additional digits are not valid and therefore are not significant digits.

Simile A figure of speech that compares one thing with another and signals the comparison with a word such as "like."

Simple sentence A sentence that has only a single action or idea.

Simulation See Numerical simulation.

Single source (also called sole source) A vendor who is the only vendor selling a particular technology. Usually a component for a design can be purchased from any of a number of sources (vendors). However, in some instances there may only be one source (sole source) for the component or service, which can effect the price and also potentially the project timeline.

Six sigma In probability analysis, sigma is the symbol used to represent standard deviation. Six sigma (i.e., ± six standard deviations) is often used as a way of describing the possible outcomes from an experiment because the six sigma range will predict 99.9997% of the trial results of an experiment that has a normal distribution of outcomes.

Sizing Performing a calculation to determine the size or capacity of the equipment or system needed.

Skeptical thinking (skepticism) A process of questioning information.

Skepticism See Skeptical thinking.

Slack A term used in project management signifying the difference between duration (estimated time of) a path and the duration of the parallel critical path. By definition the critical path has zero slack. See also Float.

Snapshot A description of a technology or design at a specific point in its use, as if you took a picture of the design in use and then answered the question, "What is going on here that affects the technology?" A snapshot is used to determine characteristics required of the design and is related to the concept of a user story (scenario), storyboard, and use case.

Social inhibition When team members do not voice all of their ideas for fear of a negative reaction from their teammates. This can effect the creativity of the team and is cited as one of the reasons for the relatively poor performance of some groups using free brainstorming.

Solution driven (the opposite of solution independent) A solution-driven statement of the design problem limits the type of solutions being considered, rather than focusing on the fundamental needs. A solution-driven statement focuses on the final design or implementation method, i.e., the means for solving the problem.

Solution independent (the opposite of solution driven) A solution-independent statement identifies the essential characteristics of the design problem and removes any unnecessary solution-driven information, i.e., does not imply a specific design solution. A solution-independent statement of a design problem leaves open the widest possible set of potential solutions while still capturing the essential features (i.e., requirements) of the project.

Source The site where something originates or the origin or place where something can be found. In design, source has three meanings: 1. An information source refers to the origin of a piece of information. Information is considered more credible if it comes from

a primary source rather than from a secondary source. A primary source is where the information was generated—in science or engineering that means the laboratory or research facility (university or industry) where the experiments or tests were conducted and data compiled. A secondary source is a commentator on the primary, someone writing about the primary material or referring to it in a report about something else.
2. Source can refer to a vendor or place to purchase an item or service such as a piece of hardware or software. See also Single source.
3. Source code refers to the computer code that is used to generate a piece of software or application.

Specification (spec) the term specifications is used in several different ways in engineering. Two meanings that are important in the design process are: 1. It can be used as a synonym for requirements. 2. More frequently it is used to refer to the detailed technical description of a system or product. You will typically find the specifications listed on the box, and in the instructions, for a product.

Spiral development A development process where successive prototypes are developed, tested, and improved iteratively until the process converges on an optimal solution.

Stakeholder interest See Interest and Stakeholders.

Stakeholders People or organizations that have a stake or interest in the technology to be created. A stakeholder interest is an aspect of an organization or person's ongoing welfare that might be affected by a design. See also Interest. The stakeholder's interest may be economic, physical, or psychological. An interest implies that there may be benefit or loss due to the design. [Note: By convention, the client, design team, and users (or operators) are not listed as stakeholders.]

Standard deviation (s.d.) The square root of the variance, thus a measure of the spread of a probability distribution. The symbol used for standard deviation is usually sigma (σ). See also Variance.

Standards (also referred to as industry standards) A set of rules determined by government or by a professional organization and intended to ensure minimum levels of safety and performance. Mandatory standards are issued by government; violations are considered offences subject to fines or prosecution. Voluntary standards are determined by groups or councils made up of professionals that may define desirable outcomes rather than minimal levels; adhering to these standards then becomes an advantage of the design; i.e., indicating a higher than required level of quality.

State Specific characteristics of the system at a particular point in time or at a particular location in the system. For example, on an aircraft the landing gear is normally in one of four states: fully up, fully down, moving down, or moving up. In a fluid flow system, such as a natural gas pipeline, the state at a particular location along the pipeline would be described by the flow rate, the temperature, and the pressure.

State of the art The sum total of all information currently available that is relevant to the design challenge at hand. Used this way, state of the art is a synonym for prior art. This may take the form of commercial products or services, patents, standard industry practices and codes, published research reports and articles, and the personal knowledge of experienced engineers in the field. State of the art is also used to indicate the most advanced technology currently in use, e.g., "our company uses a state-of-the-art inventory system."

Statement (also called credible statement) The expression of an idea. A credible statement expresses a central idea in one or two sentences, then explains the idea and gives evidence of the idea's truth and validity.

Statistics The handling of large amounts of data in a way to determine or seek to determine its characteristics through representation by fewer pieces of numeric information, for example, the statistical probability of rain today in Los Angeles. There are over a hundred years worth of data on how frequently it has rained in Los Angeles on this particular date. However, this data can be represented by one number, a percentage, that represents the probability based on the historic data.

Status report A brief report produced periodically (e.g., weekly or monthly) during the project that describes recent activities and upcoming activities of the design team. The status report can be used to update a supervisor on project activities and serve as a focus for discussion of progress and obstacles.

Steady state A system operating in steady state is not gaining or losing energy or mass over time. Components such as mass or energy may be flowing into and out of the system, but the amount flowing in is equal to the amount flowing out. As a result, the state of the system is constant (steady) over time.

Storming The second stage in the Tuckman team mode where the team is in conflict because it has not yet figured out how to work together effectively and is still developing working relationships.

Storyboard (storyboarding, also called user stories) Used in use cases and scenarios. Storyboards are sketches that illustrate the main points of a story, i.e., the interaction of a user with a technology. The storyboards help the design team imagine how a user will interact with a technology so they can design the technology to better fit the user's needs. Usually storyboards are rough sketches that convey enough information for the design team to understand the key ideas.

Strategic A term used to describe an approach or plan that is based on longer term goals. A strategy may require short-term sacrifices to achieve a broader desired outcome. Strategic moves are often contrasted with tactical moves, which solve a short-term problem but do not address the longer term or broader issue.

Strategy of least commitment This strategy states that at any point in the design process the design space should only be narrowed if necessary, which means leaving every possible design idea on the table for discussion until it is ruled out through deliberate decision making.

Structural decomposition A method that involves splitting up the presumed structure of the design into a set of sub-structures so each can be analyzed separately. Structural decomposition can be used to identify requirements (functions, objectives, and constraints), or it can be used to generate solution ideas. Related topic: Morphological chart method. See also Decomposition and Functional decomposition.

Structure An engineered product or system designed to support a load (i.e., counter a force such as gravity). Structures include built systems such as bridges or buildings. There are also virtual structures, such as the structure of a network system, which are designed to support a virtual load, e.g., a large amount of network traffic.

Structured brainstorming An idea-generation method used to develop the design requirements for a project or generate solution ideas. In structured brainstorming, team members record their own ideas silently or before the team meeting. Then they contribute the ideas to a group list. The team has the opportunity to look at the compiled list and then repeat the individual idea-generation process again. There may be several rounds of generation, and compiling ideas before the group completes the brainstorming activity. See also Brainstorming.

Style guide A set of specific instructions for the formatting and language appropriate to a certain situation, whether it is a university course, professional journal, or business.

Subjectivity Personal perspective or point of view based on one's own values, feelings, and experiences.

Sustainable (sustainable design) Refers to a design "that meets the needs of the present without compromising the ability of future generations to meet their own needs" (The World Commission on Environment and Development, *Our Common Future*. Oxford University Press, 1987).

System A set of organized components or elements that operate together as a unit. There are natural systems, such as the solar system, and engineered systems, such as a subway system or a communication network. The elements in an engineered system are designed and arranged to perform a specified set of functions.

Systematic doubt An objective process for testing ideas that may include looking for evidence or agreement from more than one independent source. Scientific testing is an example of systematic doubt.

Table Typically used in engineering communication (documents or presentations) to organize information into rows and columns to show the relationship between items. They serve the purpose of allowing concepts to be compared and contrasted. The term *table* and the term *matrix* are often used interchangeably, and there is no formal distinction between them.

Take-away message The last impression of an oral presentation or what people will remember. It is the final point made before taking questions. In a document, it is the main message that the reader should remember. This is often restated in the conclusion section and in the executive summary.

Task A term used in project management to describe the individual activities managed in a project plan. It is the smallest unit of managed work.

Task analysis An analysis of the way a user performs a task. This is usually done by observing or videoing a user performing a task. Then the engineer writes up an analysis of what the user did and how he or she did it.

Task sampling A common approach to data collection for task analysis. The user is prompted to report on what he or she is doing. This method obtains information about the frequency of tasks and the amount of time spent on different tasks.

Team A group of people who come together to work in an inter-related manner toward a common goal.

Team beliefs A set of expectations that the team members agree will govern their interactions with each other. Beliefs may include items such as "we expect everyone to do the best work possible in the time available" and "we expect everyone to contribute equally." The beliefs can be used to deal with behavior that is getting in the way of achieving the task. Negotiating beliefs is an important team forming activity. See also Team rules and Team charter.

Team charter A document that describes a set of behaviors that the team members agree will govern their interactions with each other. See also Team rules and Team beliefs.

Team collocation Having the whole team working together in the same location at the same time. There is research evidence that suggests that team collocation can improve productivity.

Team decision making How a team collectively makes decisions. Team decision making is part of the teamwork process. Team decision-making methods include consensus, voting, and let the team leader decide. See also Deciding how to decide.

Team rules A set of rules that the team members agree will govern their interactions with each other. Such rules often include items such as come prepared to meetings, don't be late, and no texting during team meetings. The rules can be used to deal with behavior that is getting in the way of achieving the task. Even if rules are not reread frequently, the act of negotiating them is an important team forming activity. See also Team beliefs and Team charter.

Technical analogy Using a technical solution to a different but related problem to inspire new ideas for solving a design problem. To use a technical analogy, ask the question: "What other technical problems have some similarity to the current problem, and how were they solved?"

Technology Anything real or virtual (e.g., a computer program) that does not occur naturally. A technology is any product, system, or process designed and made by humans or by human-made production facilities.

Test-based design A technique in which the design process starts by specifying a set of tests that the design must pass before the design choices are made.

Testing Experimenting on a technology (an actual system in operation or a prototype) to characterize its performance. In maintenance operations, testing is checking for faults to detect them before they become failures and interfere with intended operation of a system. See also Faults and Failures.

Time blocking Arranging an extended meeting time for a team that will be used for a work session. The team meets with a specific set of goals that must be accomplished before the end of the work session. The length of the work session, the goals, and the agenda for the session should be decided in advance. See also Team collocation.

Time to completion Total time necessary to complete a project (or designated section of it).

Time value of money Calculation to predict the value of a unit of currency (money) in the past or future. Inflation and interest rates change the value of money over time.

Title Short descriptive text that goes at the top or before the main information or body text in a document or presentation. If a table is used in a document or presentation, the title goes just above the table and may be one or more sentences long.

Tolerance An allowable deviation from the prescribed requirement, e.g., ± 0.1 mm.

Tone The sound qualities of a communication; for example, loud, soft, harsh, unfriendly, or comforting. Tone is naturally associated with spoken communication, but it can also appear in written communication as well, most notably, in angry emails written using all capital letters: WHAT IS GOING ON HERE, TEAM? Tone is next to body language in terms of the impact it has on communication (the words themselves having the least impact).

Topic sentence Identifies the idea to be developed in the paragraph as well as signals the mode of development (i.e., compare/contrast, sequence, cause/effect). It is usually as close as possible to the start of the paragraph, although it may not always be the actual first sentence.

Toxic A substance that causes immediate harm.

Toxicity The condition of being toxic or the degree to which a substance is toxic.

Trade secret Ideas (intellectual property) that are not disclosed publically and hence protected only because they are secret.

Trademark Intellectual property protection for symbols and specific graphic images. See also Servicemark.

Trial A term used in testing, experimental method, and probability analysis. A trial is one performance of an experiment. If you run an experiment five times, then you have performed five trials.

Triangulate To use data from multiple sources (sometimes meaning two in addition to the original source) in order to corroborate the original information.

TRIZ An acronym for the Russian phrase "Theory of Inventive Design." The TRIZ method is a creativity method developed for product design in the 1940s and still in use today. It is a set of 40 inventive principles that were identified as appearing over and over in the solution of inventive problems.

Tuckman team model Developed by Bruce Tuckman in 1965, it describes the stages in team development. The stages in the Tuckman model are forming, storming, norming, performing, and adjourning.

Turnkey solution A situation in which the contracted company handles all of the details of the design, implementation, and commissioning. When it is turned over to the client, the system is completely ready for use by the operators with minimal further implementation work required.

Under-determined problems Problems that do not have quite enough information to be solved exactly. For example, the area of a triangle where the base length is known, but the height is unknown. Under-determined problems typically have a set of possible solutions: In this example the solution set is a line representing the area as a function of the height of the triangle and all possible solutions fall on this line.

Unintended functions Functions of the design solution that were not deliberately enabled. Some untended functions are useful and can be marketed as special features of the technology. However, more often unintended functions point to possible safety hazards or failure modes and can result in legal action if the technology is improperly used.

Unit 1. Unit of production in manufacturing: A unit is a single copy of a product. For example, if a production line produces 20 washing machines, then the line has produced 20 units.

2. Unit of measure such as kg, liter, or meter. In engineering it is important to indicate units on every numerical value or indicate if a number is dimensionless (e.g., Reynold's number in fluid mechanics). See also Dimensions.

Universal design An approach that uses a set of principles to design for a broad range of users. The intent is to create design that is accessible and usable not just by the average user (if such a person actually really exists), but instead to take into account the broadest range of users possible.

Universal Markup Language (UML) A descriptive language for design used most often for software development, but now moving into other areas of design.

Unsolvable problems Problems for which there is no physically possible solution. In mathematics there are also problems that a judged to be "difficult," which means that there is theoretically a way to solve to find a unique correct solution to a problem, but it is virtually impossible to solve the math to find the solution. From an engineering perspective, this type of problem is also unsolvable, and engineers often use approximations, simplifications, or numerical methods to find the best possible solution (although not exact) to these types of problems.

Upper cost estimate A term used in project management to denote a realistic estimate of the maximum potential cost of a project. How much this is above the budgeted cost will depend on the project type and the project details. See also Lower cost estimate and Budgeted cost.

Use case (use case analysis) A description of how a person or system achieves a goal using the technology being designed. A use case describes the interaction and allows developers to predict user error and create options for dealing with it. The use case method is related to scenarios (user stories), snapshots, and storyboarding.

User-centered design (also called human-centered design) Design approach that puts the users at the center of the process and designs the technology to fit them, rather than the other way around. A user-centered approach carefully considers the user's needs at every level: physical, psychological, social, organizational, and political. These needs are the essential requirements in the design process.

User story A verbal description of the way a user interacts with a proposed technology. See also Story board, Use case, Snapshot.

Users The individuals or groups that utilize, consume, or operate a technology. See also Operators.

Utility In order for an invention to be patentable, it must have utility, which means some practical use or the potential for practical use at some time in the future.

Utility patent Legal protection for the underlying function of inventions and of new engineering designs. A utility patent will protect this type of intellectual property typically for 20 years. When people talk about "patented technology," they usually mean the technology is covered by a utility patent.

Validity The measure of the degree to which a conclusion, or claim, is supported by evidence or reasons; the quality that describes an argument that is not characterized by any form of fallacy.

Value-free language Objective language that seeks to avoid specific social, political, or cultural values.

Value-laden language Language that carries a lot of implied meanings based on social, political, or cultural values.

Variable costs The costs that depend on sales or other activities of a company. The variable costs are dependent on quantity of work, unlike the fixed costs. Example: The difference between the fixed cost of basic telephone service and the variable cost of long distance phone calls.

Variance A measure of the spread (or width) of a probability distribution. The square root of variance is called the standard deviation.

Vendor A company that sells a product or service. As a design engineer, you will frequently be purchasing off-the-shelf technology components and services from vendors. See also Source.

Venture capital Commercial ventures that pool money from a number of backers to invest in risky early-stage companies.

Virtual environment Part of the service environment. The virtual environment includes infrastructure elements, such as (but not limited to) wireless networks, satellite coverage (e.g., GPS), or cell phone signals. If your design is likely to utilize these services, then you need to describe their presence (or lack of presence) in the service environment. See also Service environment.

Visual vocabulary A pictorial representation of a concept often used in diagrams. For example, in engineering there are standard shapes that are used to represent common components: in a mechanical engineering diagram, a zigzag line is a spring; in an electrical engineering schematic, a zigzag line is a resistor. The type of diagram and shape indicate the properties or characteristics of the component.

Voting A team decision-making method. Each person involved indicates his or her choice (i.e., vote), and the choice that gets the majority support is selected. In certain situations, more than a simple majority may be required (i.e., 70% or 80%). Voting may be open (e.g., by raising hands) or by secret ballot (writing the vote on a piece of paper). When there are an even number of people voting, a method for breaking a tie must be determined.

War room A common term used in industry to describe the room where a design team is working together on a project. The team uses the room like a headquarters or operation center for the time while the project is in active development. See also Team collocation.

Warning label A sign, label, or part of packaging describing specific dangers inherent in the design. It usually has clear simply language and often a representative visual icon, symbol, or picture.

Waste Mass or energy that is unused and discarded into the environment or requires disposal.

Waterfall design process Traditional design technique where requirements are completely generated before design ideas are generated and the selected design is fully detailed before execution begins.

Wear The loss of material where two solid surfaces slide over each other. Wear can result in failure of a technology. A common example is the wear on car tire treads or on the soles of your shoes, which lose material (become thinner) over time until they fail.

Wear and tear Damage to a technology that is caused through routine use of the technology over time.

Webinar A form of presentation delivered online where the participants are able to join from remote locations.

Weighted decision matrix (also called a numerical decision matrix method) A decision-making method that uses weights (numerical values) for each objective to score alternative ideas. Important objectives are weighted higher (valued more), and less important objectives are weighted lower (valued less). This affects the decision, giving more priority to solutions that better fit the most important objectives. This method can be used in any decision-making process when trying to make a decision among a set of alternatives based on a set of objectives.

Why, why, why Repeatedly questioning "why" things are the way they are allows you to get at the root cause of a problem, or uncover hidden assumptions. The "why, why, why" method is a creativity technique that is also used to generate new, original ideas. This method is also called 5 whys, because of the typical number of "whys" you need to ask to get to the root cause of a problem.

Within-function team A team made up of members from a single discipline, for example, a team of specialized civil engineers or a team made up of only coding experts.

Wood's problem-solving model A method for successfully solving problems using seven steps: motivate, define, research, plan, execute, check, and reflect. This method helps people successfully solve problems, communicate the problem-solving process to an audience (reader), and understand and communicate the implications of the results. It is designed to help people become expert problem solvers.

Work breakdown structure (WBS) A project management technique that uses a hierarchical structure of tasks with multiple levels. Each level is a decomposition of the previous level. The top node is a project, or sub-project, and each node down the tree represents a smaller set of work until it reaches the individual task level.

Workaround A way of achieving the same result without using a particular technological solution. Engineers create workarounds if a particular technology is unavailable: not available on the market currently, not available because of circumstances (i.e., a scarce commodity), or not available because it is protected by a patent.

Index